AI Developments for Industrial Robotics and Intelligent Drones

Brij B. Gupta
Asia University, Taichung, Taiwan

Francesco Colace
University of Salerno, Italy

Vice President of Editorial	Melissa Wagner
Managing Editor of Acquisitions	Mikaela Felty
Managing Editor of Book Development	Jocelynn Hessler
Production Manager	Mike Brehm
Cover Design	Phillip Shickler

Published in the United States of America by
IGI Global Scientific Publishing
701 East Chocolate Avenue
Hershey, PA, 17033, USA
Tel: 717-533-8845
Fax: 717-533-8661
E-mail: cust@igi-global.com
Website: https://www.igi-global.com

Library of Congress Cataloging-in-Publication Data

CIP Pending
ISBN: 979-8-3693-2707-4
EISBN: 979-8-3693-2708-1

British Cataloguing in Publication Data
A Cataloguing in Publication record for this book is available from the British Library.

Dedicated to my wife and kids for their constant support during the course of this book
- Brij Gupta
Dedicated to G. who taught me to read the world around us with new eyes
- Francesco Colace

Table of Contents

Chapter 1
Brij B. Gupta, Asia University, Taiwan
Akshat Gaurav, Ronin Institute, USA
Francesco Colace, University of Salerno, Italy

Chapter 2
Akshat Gaurav, Ronin Institute, USA
Varsha Arya, Asia University, Taiwan, & Hong Kong Metropolitan University, Hong Kong

Chapter 3
Brij B. Gupta, Asia University, Taiwan
Jinsong Wu, Universidad de Chile, Chile

Chapter 4
Purwadi Agus Darwinto, Instrumentation, Control, and Optimization, Sepuluh Nopember
Institute of Technology, Indonesia
Agung Mulyo Widodo, Universitas Esa Unggul, Indonesia
Nilla Perdana Agustina, Laboratorium Instrumentasi dan Pengukuran, Indonesia
Kadek Dwi Wahyuadnyana, Sepuluh Nopember Institute of Technology, Indonesia
Mosiur Rahaman, International Center for AI and Cyber Security Research and Innovations,
Asia University, Taiwan & Computer Science and Information Engineering, Asia
University, Taiwan

Chapter 5
Mosiur Rahaman, International Center for AI and Cyber Security Research and Innovations,
Asia University, Taiwan & Computer Science and Information Engineering, Asia
University, Taiwan
Karisma Trinda Putra, Universitas Muhammadiyah Yogyakarta, Indonesia
Bambang Irawan, Universitas Esa Unggul, Indonesia
Totok Ruki Biyanto, Institut Teknologi Sepuluh Nopember, Indonesia

Detailed Table of Contents

Chapter 1
 Brij B. Gupta, Asia University, Taiwan
 Akshat Gaurav, Ronin Institute, USA
 Francesco Colace, University of Salerno, Italy

This chapter introduces the revolutionary impact of artificial intelligence (AI) on industrial robotics and drones, providing a foundation for understanding the integration of AI technologies in these fields. It covers the evolution of AI-driven systems, highlighting significant milestones and projecting future trends. The chapter aims to set the stage for the ensuing discussion by presenting the broad applications of AI in enhancing the capabilities of robots and drones, from manufacturing and logistics to surveillance and agriculture. It also addresses the societal and economic implications of widespread AI adoption in industrial robotics and drones, underscoring the potential for transformative change across various sectors.

Chapter 2
 Akshat Gaurav, Ronin Institute, USA
 Varsha Arya, Asia University, Taiwan, & Hong Kong Metropolitan University, Hong Kong

This chapter explores the application of AI technologies in the domain of robotics, focusing on how these advancements enable robots to perform complex tasks with high efficiency and autonomy. It covers critical areas such as perception, where AI allows robots to interpret sensory data; navigation and motion planning, enabling robots to move through and interact with their environments safely; and manipulation, where AI-driven robots achieve precision in handling objects. The chapter also discusses the role of AI in facilitating human-robot interaction, enhancing the ability of robots to work alongside humans in various industrial settings. Through detailed explanations and examples, the chapter demonstrates the transformative potential of AI in robotics, showcasing how it drives innovation and efficiency in industrial applications. This chapter explores the application of AI technologies in the domain of robotics, focusing on how these advancements enable robots to perform complex tasks with high efficiency and autonomy. It covers critical areas such as perception, where AI allows robots to interpret sensory data; navigation and motion planning, enabling robots to move through and interact with their environments safely; and manipulation, where AI-driven robots achieve precision in handling objects. The chapter also discusses the role of AI in facilitating human-robot interaction, enhancing the ability of robots to work alongside humans in various industrial settings. Through detailed explanations and examples, the chapter demonstrates the transformative potential of AI in robotics, showcasing how it drives innovation and efficiency in industrial applications.

Chapter 3

Brij B. Gupta, Asia University, Taiwan
Jinsong Wu, Universidad de Chile, Chile

This chapter explores the synergistic integration of the internet of things (IoT) with robotics and drones, highlighting how this convergence is revolutionizing industrial operations and capabilities. It delves into the mechanisms through which IoT devices and sensors enhance the autonomy, efficiency, and intelligence of robotic systems and drones, enabling real-time data exchange and analysis. The chapter discusses the implementation of IoT for advanced monitoring, predictive maintenance, and seamless operational control, illustrating its impact through practical examples across various sectors. It also addresses the challenges of scalability, security, and interoperability, presenting forward-looking strategies to navigate these hurdles. By emphasizing the transformative potential of IoT in augmenting the capabilities of robotics and drones, the chapter underscores the pivotal role of IoT in driving innovation and operational excellence in the digital age.

Chapter 4

Purwadi Agus Darwinto, Instrumentation, Control, and Optimization, Sepuluh Nopember Institute of Technology, Indonesia
Agung Mulyo Widodo, Universitas Esa Unggul, Indonesia
Nilla Perdana Agustina, Laboratorium Instrumentasi dan Pengukuran, Indonesia
Kadek Dwi Wahyuadnyana, Sepuluh Nopember Institute of Technology, Indonesia
Mosiur Rahaman, International Center for AI and Cyber Security Research and Innovations, Asia University, Taiwan & Computer Science and Information Engineering, Asia University, Taiwan

This article discusses the integration of artificial intelligence (AI) in drone systems, highlighting AI's ability to interpret data, learn, and adapt to achieve goals. AI's broad applications range from natural language processing to autonomous vehicles and industrial robotics. AI enables drones to autonomously perform tasks like takeoff, navigation, and landing, even in complex environments. The study also covers AI's role in real-time data analysis for various purposes, including security and agrotechnology. It emphasizes the importance of algorithm development and sensor data processing for drone operations, enhancing their performance, efficiency, and safety. The use of simulations in languages like C++, Python, MATLAB, and Simulink is mentioned, with specific reference to reducing chattering in Sliding Mode Control (SMC) for energy efficiency and extended flight times. The article concludes with the challenges of aligning drone design with market-available components and the successful application of AI in drone control systems.

Mosiur Rahaman, International Center for AI and Cyber Security Research and Innovations,
Asia University, Taiwan & Computer Science and Information Engineering, Asia
University, Taiwan
Karisma Trinda Putra, Universitas Muhammadiyah Yogyakarta, Indonesia
Bambang Irawan, Universitas Esa Unggul, Indonesia
Totok Ruki Biyanto, Institut Teknologi Sepuluh Nopember, Indonesia

This chapter presents an in-depth analysis of the transformative role of artificial intelligence (AI) in the fields of industrial robotics and drone technology, offering a comprehensive overview of the integration and evolution of these technologies. Beginning with a historical perspective, the study traces the development of robotics and drones within industrial contexts, laying the foundation for understanding the significant impact of AI. The analysis reveals how the advent of AI has revolutionized these technologies, shifting from basic mechanization to advanced, intelligent systems capable of complex tasks, and decision-making. The study explores the progression from initial AI applications in robotics to the current state-of-the-art implementations, demonstrating the profound changes in efficiency, capability, and functionality. By examining the interplay between AI, robotics, and drones, the study provides insights into the future trajectory of these technologies and their potential to redefine industrial processes.

Shaurya Katna, Chandigarh College of Engineering and Technology, Chandigarh, India
Sunil K. Singh, Chandigarh College of Engineering and Technology, Chandigarh, India
Sudhakar Kumar, Chandigarh College of Engineering and Technology, Chandigarh, India
Divyansh Manro, Chandigarh College of Engineering and Technology, Chandigarh, India
Amit Chhabra, Chandigarh College of Engineering and Technology, Chandigarh, India
Sunil Kumar Sharma, Indian Railway, India

The world is progressively moving towards the smart city concept. Drones are central to this movement. Hence, this research on communication systems for drone swarms is imperative to enhance the operational efficiency and autonomy of unmanned aerial vehicles (UAVs). Alongside, it addresses the unique challenges posed by dynamic network topologies, limited bandwidth, and ensuring seamless collaboration in diverse applications. This study examines the complex domain of communication systems for drone swarms and remote operations. Issues such as bandwidth constraints and changing network configurations are assessed with a focus on innovative technologies like AI-powered decision-making, blockchain security, and edge computing. The assessment looks at the effects of specialized signal processing methods on swarm performance. Case studies authenticate these strategies' efficacy while offering vital real-world insights. Further, this study assists those in the field by guiding them through the challenges associated with drone swarm technology.

Chapter 7

    Kwok Tai Chui, Hong Kong Metropolitan University, Hong Kong
    Varsha Arya, Asia University, Taiwan, & Hong Kong Metropolitan University, Hong Kong
    Akshat Gaurav, Ronin Institute, USA
    Shavi Bansal, Insights2Techinfo, India
    Ritika Bansal, Insights2Techinfo, India

This chapter delves into the pivotal role of machine vision and image processing in the realm of artificial intelligence for industrial robotics and intelligent drones. It begins by introducing the fundamental concepts and techniques of machine vision, including image capture, analysis, and interpretation methods that enable machines to 'see' and make sense of their surroundings. The discussion extends to advanced image processing strategies, such as edge detection, feature extraction, and object recognition, highlighting their critical applications in enabling autonomous operation of robots and drones. The chapter emphasizes the importance of 3D mapping and scene reconstruction in creating detailed environmental models for navigation and task execution. Furthermore, it explores the emerging use of augmented reality (AR) for industrial applications, offering insights into how AR enhances human-machine interaction by overlaying digital information onto the physical world.

Chapter 8

    Tushar Singh, Chandigarh College of Engineering and Technology, Chandigarh, India
    Sudhakar Kumar, Chandigarh College of Engineering and Technology, Chandigarh, India
    Sunil K. Singh, Chandigarh College of Engineering and Technology, Chandigarh, India
    Priyanshu, Chandigarh College of Engineering and Technology, Chandigarh, India
    Brij B. Gupta, Asia University, Taichung, Taiwan
    Jinsong Wu, Universidad de Chile, Chile
    Arcangelo Castiglione, University of Salerno, Fisciano, Italy

Exploring the intersection of transparency and security in autonomous systems, this chapter examines the dynamic landscape of industrial robots and intelligent drones. Advanced technologies such as machine learning, AI, robotics, and deep learning shape this intricate domain. As autonomous systems gain prominence across sectors, a focus lies on understanding decision-making frameworks. Methodologies for achieving algorithmic transparency and strengthening security protocols are outlined, emphasizing the fusion of technological innovation with ethical considerations. Real-world case studies offer practical insights and best practices. Ethical responsibilities in AI and robotics integration are emphasized, alongside a forward-looking view on emerging trends and technologies, providing a tailored roadmap for researchers, practitioners, and enthusiasts navigating the evolving realm of autonomous systems. This chapter provides a thorough analysis of transparency and security challenges and opportunities in autonomous systems, benefiting policymakers and industry stakeholders.

    Agung Mulyo Widodo, Universitas Esa Unggul, Indonesia
    Andika Wisnujati, Universitas Muhammadiyah Yogyakarta, Indonesia
    Eko Prasetyo, Universitas Muhammadiyah Yogyakarta, Indonesia
    Mosiur Rahaman, International Center for AI and Cyber Security Research and Innovations,
       Asia University, Taiwan & Computer Science and Information Engineering, Asia
       University, Taiwan

This study investigates the integration of unmanned aerial vehicles (UAVs) in air combat, focusing on their role in the battlefield management system (BMS) for effective communication and data management. Utilizing UAVs minimizes pilot casualties and enables real-time decision-making. The chapter examines resource allocation in ultra-dense networks using active-reconfigurable intelligent surfaces (A-RIS) assisted non-orthogonal multiple access (NOMA). It explores the coverage performance and ergodic capacity in a NOMA network under Nakagami-m fading channels, employing a multi-input multi-output (MIMO) system with RIS elements. The results demonstrate the superiority of RIS-assisted NOMA over conventional methods, offering enhanced coverage probabilities and ergodic capacity. The study concludes that the integration of A-RIS in UAVs significantly improves battlefield communication, highlighting its potential in military applications.

    Princy Pappachan, National Chengchi University, Taiwan
     Sreerakuvandana, Jain University, India
    Siwada Piyakanjana, Asia University, Taiwan
    Harlinda Syofyan, Esa Unggul University, Indonesia
    Gunawan Nugroho, Institut Teknologi Sepuluh Nopember, Indonesia

The introduction of robotics and AI represents a paradigm shift in the quickly developing field of special education, providing unprecedented opportunities for improving learning opportunities. The chapter highlights this by examining how these technologies are changing how special education is taught to students with special needs. The transformative role of AI and robotics is evident in how they have moved from auxiliary tools to central elements in fostering adaptive and inclusive learning environments. The chapter delves into personalized learning, highlighting AI's role in customizing educational content to individual styles and enhancing strategy efficacy through data-driven insights. The chapter also explores the role of robotic assistants in interactive teaching and therapy, enhancing physical, sensory, and cognitive skills. Additionally, by addressing ethical concerns and advocating a balanced approach to technology use, the chapter provides educators, researchers, and policymakers with a thorough and forward-looking perspective on AI and robotics in special education.

Chapter 11

This chapter provides an in-depth study at the latest developments in deep reinforcement learning (DRL) as applied to drones and unmanned aerial vehicles (UAVs), with a focus on safety standards and overcoming challenges like object detection and cybersecurity. Real-world instances across sectors demonstrate how DRL significantly improves efficiency and adaptability in autonomous systems. By contrasting various DRL techniques, the chapter highlights their effectiveness and potential to advance drone and UAV capabilities. The practical implications and advantages of DRL applications are emphasized, showcasing their transformative influence on industries. Moreover, the chapter explores future research paths to stimulate innovation and enhance self-piloting vehicle technology. In essence, it provides a thorough overview of DRL's role in drone and UAV operations, underscoring the importance of safety standards, DRL's capacity to boost efficiency and safety in autonomous systems, and its contribution to the evolving autonomous technology landscape.

Chapter 12

The integration of multi-modal sensor fusion with CRNNs in agile industrial drones addresses the need for improved object detection and SLAM capabilities. This technology enhances the drone's ability to navigate and map complex industrial environments with greater accuracy and reliability. It is crucial for tasks such as inspection, monitoring, and autonomous navigation in dynamic and challenging industrial settings. By integrating visual, LiDAR, and inertial measurement unit (IMU) sensor information, the approach enhances situational awareness, facilitating safer navigation and intelligent interaction amidst challenging conditions. The integration of LiDAR, IMU, and optical sensors provides awareness of the environment, enabling the drone to adjust to changing circumstances instantly. The suggested method meets the needs of industrial applications, which demand dependable and durable solutions. It achieves this by offering improved precision, dependability, and flexibility that eventually expands the potential of industrial drones in a variety of operational contexts.

Chapter 13

Bhupinder Singh, Sharda University, India
Christian Kaunert, Dublin City University, Ireland

The future of space exploration is defined by a close partnership between intelligent autonomous robots and humans. These robots serve as extensions in space which do tasks in dangerous or harsh environments where direct human participation is impossible. In recent years, researchers, commercial firms and space organizations have made considerable advances in the creation of autonomous and remotely controlled space robots. These robots serve an important role in space exploration, assisting people on the International Space Station (ISS) and exploring distant celestial bodies. These robots require highly developed tele-operation interfaces and HMI designs to bridge the knowledge gap between human competency and machine execution. This chapter provides a thorough analysis of the dynamic interactions between intelligent autonomous robots and humans. It also offers insights for space agencies, academics, engineers and everyone else interested in the fascinating field of space technology and exploration which lays out a vision for the future of autonomous systems and space robots.

Chapter 14

Akshat Gaurav, Ronin Institute, USA
Brij B. Gupta, Asia University, Taiwan
Varsha Arya, Asia University, Taiwan, & Hong Kong Metropolitan University, Hong Kong
Arcangelo Castiglione, University of Salerno, Fisciano, Italy

This chapter addresses the critical issues of safety, ethics, and regulation surrounding the deployment of AI in industrial robotics and intelligent drones. It highlights the importance of establishing robust safety standards to mitigate risks associated with mechanical and software failures. Ethical considerations are explored, focusing on accountability, privacy, and the socio-economic impacts of automation. The chapter also navigates the complex regulatory landscape, underscoring the need for adaptive frameworks that balance innovation with public safety. Through case studies and discussions, it calls for a multidisciplinary approach to develop ethical guidelines and effective legislation. The chapter emphasizes the collective responsibility of stakeholders to ensure the responsible advancement of AI technologies, advocating for policies that protect human welfare while fostering technological growth.

Preface

Welcome to *AI Developments for Industrial Robotics and Intelligent Drones*, a comprehensive guide to the groundbreaking advancements in artificial intelligence transforming the fields of industrial robotics and autonomous drones. As AI technology evolves, its applications within these domains continue to expand, offering unparalleled opportunities for innovation, efficiency, and automation.

This book aims to bridge the gap between theoretical research and practical implementation, providing readers with a thorough understanding of how AI technologies are being integrated into robotics and drones. It covers a wide range of topics, from the integration of IoT and AI technologies to the intricacies of machine vision, communication systems, and security enhancements. Each chapter delves into the latest developments, challenges, and future directions, offering valuable insights for researchers, practitioners, and students alike.

Authored by leading experts in the field, each chapter provides unique perspectives and expertise. We begin with an introduction to AI in robotics and drones, setting the stage for detailed discussions on specific technologies and applications. Subsequent chapters explore the integration of AI in industrial systems, communication systems for drone swarms, machine vision, and the critical aspect of security. We also delve into innovative applications, such as the role of AI in special education and the use of reinforcement learning for drone automation.

As AI continues to reshape the landscape of industrial robotics and intelligent drones, understanding the ethical, safety, and regulatory implications of these advancements is crucial. This book addresses these concerns, ensuring that the integration of AI technologies aligns with societal values and legal frameworks.

We hope this book serves as a valuable resource, inspiring further innovation and collaboration in the exciting fields of AI, robotics, and drones. The chapters covered in this book include:

- Introduction to AI in Robotics and Drones
- AI Technologies in Robotics
- Integration of IoT with Robotics and Drones
- Artificial Intelligence (AI) for Autonomous Drones
- Industrial Evolution: The Integration of AI in Robotics and Drone Systems
- Communication Systems for Drone Swarms and Remote Operations
- Machine Vision and Image Processing for Drones
- Enhancing Autonomous System Security with AI and Secure Computation Technologies
- Active-Reconfigurable Intelligent Surfaces for Unmanned Aerial Vehicles - BMS Data Transmission
- Innovative Horizons: The Role of AI and Robotics in Special Education
- Drones and Unmanned Aerial Vehicles Automation using Reinforcement Learning: Automation System of Drones and UAVs using Reinforcement Learning

- Multi-Modal Sensor Fusion with CRNNs for Robust Object Detection and Simultaneous Localization and Mapping (SLAM) in Agile Industrial Drones
- Climbing Human-Machine Interaction and Wireless Tele-Operation in Smart Autonomous Robots: Exploring the Future of Space Robotics and Autonomous Systems
- Safety, Ethics, and Regulation in Intelligent Drones

We invite you to explore the diverse and exciting topics covered in this book and join us on the journey of discovering how AI is reshaping the landscape of industrial robotics and intelligent drones.

Acknowledgment

Many people have contributed greatly to this book on AI Developments for Industrial Robotics and Intelligent Drones. We, the editors, would like to acknowledge all of them for their valuable help and generous ideas in improving the quality of this book. With our feelings of gratitude, we would like to introduce them in turn. The first mention is the authors and reviewers of each chapter of this book. Without their outstanding expertise, constructive reviews and devoted effort, this comprehensive book would become something without contents. The second mention is the IGI Global staff for their constant encouragement, continuous assistance and untiring support. Without their technical support, this book would not be completed. The third mention is the editor's family for being the source of continuous love, unconditional support and prayers not only for this work, but throughout our life. Last but far from least, we express our heartfelt thanks to the Almighty for bestowing over us the courage to face the complexities of life and complete this work.

Chapter 1
Introduction to AI in Robotics and Drones

Brij B. Gupta
Asia University, Taiwan

Akshat Gaurav
Ronin Institute, USA

Francesco Colace
https://orcid.org/0000-0003-2798-5834
University of Salerno, Italy

ABSTRACT

This chapter introduces the revolutionary impact of artificial intelligence (AI) on industrial robotics and drones, providing a foundation for understanding the integration of AI technologies in these fields. It covers the evolution of AI-driven systems, highlighting significant milestones and projecting future trends. The chapter aims to set the stage for the ensuing discussion by presenting the broad applications of AI in enhancing the capabilities of robots and drones, from manufacturing and logistics to surveillance and agriculture. It also addresses the societal and economic implications of widespread AI adoption in industrial robotics and drones, underscoring the potential for transformative change across various sectors.

INTRODUCTION

Artificial intelligence (AI) is a rapidly evolving field that has garnered significant attention in recent years. The concept of AI revolves around the development of computa- tional systems that possess the ability to interpret external data accurately, learn from such data, and utilize these learnings to accomplish specific goals and tasks through flexible adaptation (Haenlein and Kaplan 2019; Nhi, Le et al. 2022). This involves the capacity of machines to think, reason, comprehend, and understand human intellect, enabling them to perform tasks that are typically associated with human intelligence, such as recogniz- ing patterns, planning, and critical analysis based on collected data (Nabi and Xu 2021). Furthermore, AI encompasses the study and development of the- oretical methods and techniques for simulating and

DOI: 10.4018/979-8-3693-2707-4.ch001

expanding human intelligence, as well as the imitation of certain human functions and behaviors through computers or electronic devices (Cao 2017; Bisht and Vampugani 2022; Cao et al. 2021).

The integration of AI into various domains, including healthcare, education, and industry, has been a significant area of focus. In the medical field, AI has shown potential in revolutionizing patient care and administrative tasks, with the capability to aid healthcare providers in diagnosis and treatment decisions (Anwar et al. 2022; Okafor et al. 2022). Similarly, in the education sector, AI has gradually been integrated into major aspects of schooling and academic learning, leveraging breakthroughs in algorithmic machine learning to enhance educational processes (Toncic 2021; Kumb- hojkar and Menon 2022). Moreover, the tourism industry has also been impacted by AI, with scholars highlighting the transformative potential of AI in shaping the future of tourism through the rapid development of computer technology (Tuo, Ning, and Zhu 2021; Colace et al. 2022).

The historical perspective of AI is crucial in understanding its evolution. The emergence of expert systems has been identified as a major advance in the field of AI, transforming the enterprise of AI and shaping its trajectory over the years (Brock 2018; Liao et al. 2024). Additionally, the intertwined histories of AI and education have been evident since the early days of AI, signifying the deep connection between these two fields (Doroudi 2022). Furthermore, the role of AI in shaping the future of various industries, including healthcare, radiology, and mental health, has been a subject of extensive research, with a focus on the potential for AI to enhance medical practice and improve patient care (Sorantin et al. 2021) (Shazly et al. 2022; Kumari et al. 2024; Milne-Ives et al. 2022).

The definition of AI has been a topic of extensive analysis, with scholars emphasizing the importance of establishing a clear and comprehensive definition that aligns with its common usage, draws a sharp boundary, leads to fruitful research, and is as simple as possible (Wang 2019). The concept of AI has also been explored in the context of legal systems, with discussions on the challenges of developing a legal definition of AI that meets the requirements of modern technological development and can be effectively used in the process of legal regulation (Minbaleev 2022; Mishra, Kong, and Gupta 2024; Arkhipov 2022).

Artificial intelligence (AI) has significantly impacted the field of robotics and drone technology (Table 1). The integration of AI in unmanned aerial vehicles (UAVs) has enhanced their capabilities and applications. AI-enabled IoT-based drone-aided health- care services have been developed, allowing for tasks such as sample collection and medical supply delivery Wazid et al. (2020); Sharma et al. (2024). Additionally, the application of AI and IoT has increased the popularity of drones globally, indicating the widespread adoption of AI in drone technology (Yaramala et al. 2022). Furthermore, AI has been utilized in real-time autonomous drone operations, demonstrating its role in enhancing the intelligence and autonomy of drones (Kovari and Ebeid 2021; Chui et al. 2024). The use of AI algorithms in solving various problems related to drones has been a focus of the research community, highlighting the integration of intelligence at the core of UAV networks (Lahmeri, Kishk, and Alouini 2021). Moreover, AI applications in AgriTech drones have been recognized as effective tools for smart farming, providing precision evaluations and enhancing food security (Spanaki et al. 2021). In the context of robotics, AI technology has been effectively utilized in critical places such as clinics, hospitals, and logistics, contributing to the diagnosis and prevention of the spread of diseases, including the COVID-19 pandemic (Mahdi 2021).

Additionally, the fusion of blockchain and AI has been proposed to secure drone communication, highlighting the potential for advanced security and intelligent com- munication architecture in drone networks (Gupta, Kumari, and Tanwar 2020). The development of machine-learning techniques for UAV-based communications has fur- ther demonstrated the integration of AI in drone technology, emphasizing its role in enhancing communication and networking capabilities (Bithas et al. 2019). Furthermore, AI and robotics have been integrated into systems engineering education, providing job opportunities for professionals in various engineering fields (Alvarez- Dionisi, Mittra, and Balza 2019). The use of AI in robotics has also been explored in the context of teleoperation, tele-assessment, and tele-training for surgery, indicating its potential in advancing medical practices (Feizi et al. 2021). Additionally, AI has been applied in livestock management using UAVs, showcasing its role in enhancing agricultural practices and animal husbandry (Alanezi et al. 2022).

Table 1. Economic and societal implications of AI in robotics and drones

Impact Type	Description	Affect	Remarks
Job Cre- ation	Automation leads to new roles in AI main- tenance and manage- ment.	Positive	Requires re-skilling and training.
Efficiency Improve- ments	Higher throughput and accuracy in manufac- turing and logistics.	Positive	Can lead to work- force reductions in tra- ditional roles.
Safety Enhance- ments	Drones and robots per- form dangerous tasks, reducing human risk.	Positive	Reliability and mal- function concerns per- sist.
Ethical Considera- tions	AI decision-making in critical applications like healthcare.	Negative	Raises concerns about privacy, consent, and accountability.

EVOLUTION OF AI IN ROBOTICS

History of AI and Robotics

The initial applications of artificial intelligence (AI) in robotics marked a significant advancement in the capabilities and functionalities of robots. The development of AI in robotics has gone through several generations, from the Turing test and logic theory machine to expert systems and self-driving cars Chen and Luca (2021)(Table 2). The integration of AI in robotics has paved the way for robust applications, including the use of augmented reality (AR) and AI in various robotic applications, indicating a promising future for their integration (Bassyouni and Elhajj 2021). Furthermore, the initial development of intent-based deployment for robotic applications in 5G- enabled non-public networks has been presented, signifying the early stages of AI integration in robotic deployment (Qiu et al. 2023). The application of AI and robotics has also been evident in the European restaurant sector, ranging from data-driven table planning to sales forecasting and the development of diverse robots, showcasing the early adoption of AI in the hospitality industry (Bl¨ocher and Alt 2020). Additionally, the development of intelligent unmanned autonomous systems has been highlighted as one of the most important applications of AI, emphasizing the early stages of AI integration in autonomous systems (Zhang et al. 2017).

Moreover, the initial applications of AI in robotics have extended to the medical industry, with the primary applications of AI being in intelligent screening, diagnosis, risk prediction, and supplemental treatment, indicating the early utilization of AI in medical robotics (Vanak 2022). Furthermore, the perception of medical students and faculty regarding AI and robotics has emphasized the belief that most applications of AI and robotics are and would be in the field of surgery, indicating the early recognition of AI's potential in surgical robotics (Sassis et al. 2021). The convergence of robotics and AI has been underscored as a key factor in the development of self-adaptable robots, highlighting the early stages of integration between robotics and AI (Mayoral 2018). Additionally, the initial applications of AI and robotic coaches in physical re- habilitation therapy have promised improved engagement of patients through social interaction, indicating the early stages of AI integration in rehabilitation robotics (Lee et al. 2022).

Table 2. Key milestones in AI development for robotics and drones

Year	Milestone	Description	Impact on Field
1956	Inception of AI	Formal recognition of AI as an academic discipline at the Dartmouth Conference.	Established the foundation of AI research.
1980s	Emergence of expert systems	Widespread development and deployment of AI systems that emulated the decision-making abil- ities of a human expert.	Enhanced automation in various industries, setting a precedent for complex AI applications.
2000s	Advancements in machine learning	Improvement in algorithms and increase in computational power.	Facilitated the development of more advanced, autonomous robots and drones.
2010	AI integration in drones	First significant use of AI for commercial drone navigation and opera- tion.	Expanded the capabilities and applications of drones in commercial and military sectors.
2020	AI and IoT in healthcare drones	Development of AI-enabled IoT drones for tasks like sample collection and delivery during the COVID-19 pandemic.	Showcased the potential of drones in emergency healthcare services and crisis management.

Development of Robotics Using AI

The evolution of AI technology in robotics has been marked by significant advance- ments in capabilities and complexity, leading to transformative applications across various domains. Recent AI advances have propelled robot development further, en- abling robots to navigate more complex scenarios due to improved image recognition and processing techniques and facilitating sophisticated interactions with humans through increased processing capabilities of natural language Rosete et al. (2020). The fundamental conceptualization behind the designed architecture is to leverage recent technological advancements in AI, focusing mainly on the use of the FL paradigm and closely related technologies, to extend the capabilities of current techniques in robotics applications (Papadopoulos, Antona, and Stephanidis 2020).

The evolution of robots, from simple automation to superior AI, has been driven by AI's role in gaining knowledge, sensing, reasoning, and communication for robots (Geetha 2024). The combination of Robotics, IoT, and AI has resulted in robots with higher capabilities to perform more complex tasks, enhancing their functionality in emerging robotic communication systems (Alsamhi, Ma, and Ansari 2019). In indus- trial robotics technology, the improvement of productivity requires enhancing the rigid, inflexible capabilities of industrial robots, showcasing the need for increased complex- ity and adaptability in AI-enabled industrial robots (Benotsmane, Dud´as, and Kov´acs 2020).

Robotics is the next frontier in the progress of AI, as the real world in which robots operate represents an enormous, complex, continuous state space with inherent real- time requirements, highlighting the need for AI to evolve to meet the demands of complex robotic environments (Wagter 2021). Swarm-based algorithms have become increasingly popular due to their capabilities to provide solutions to complex problems,

indicating the evolution of AI in addressing complex challenges in robotics (Chamoso et al. 2015). While AI technologies are still in their developing stages, they have become increasingly popular in many areas, signifying the ongoing evolution of AI in robotics and its potential for further advancements (Hong et al. 2021).

The latest developments in AI and machine learning, and the parallel advances in robotics, have contributed to a shift in the scientific approach to modeling human intelligence, indicating the evolution of AI in shaping the capabilities of robotic sys- tems (Cangelosi and Schlesinger 2018). Human-centered AI and robotics represent a complete field of research that deals with managing the algorithmic complexity in structurally complex robotic systems, emphasizing the need for AI to address the complexities of dynamic environments in robotics (Doncieux et al. 2022). The range of action units achieved by robots is analyzed to discover their expressive capabili- ties and limitations, highlighting the evolution of AI in evaluating and controlling the capabilities of humanoid robot faces (Auflem et al. 2022).

The foreseeable evolution and development prospects of diversified and customized applications for modern robots and AI indicate the continuous expansion, develop- ment, and specialization of AI-enabled robotic systems, reflecting the ongoing evo- lution of AI in robotics (Munteanu et al. 2022). Insect-inspired AI for autonomous robots represents an innovative approach to enhancing the autonomy and adaptabil- ity of robots, showcasing the evolution of AI in shaping the capabilities of autonomous robotic systems (Croon et al. 2022). The evolution of digitalization, big data analy- sis, and AI has propelled surgical safety into a new era, indicating the transformative impact of AI on the capabilities of robotic surgical procedures (Kipnis et al. 2022).

To understand the impact of AI, it is important to draw lessons from its past successes and failures, providing a comprehensive explanation of the evolution of AI, its current status, and future directions, highlighting the ongoing evolution of AI in shaping the capabilities of robotic systems (Suarez-Ibarrola and Miernik 2020). The majority of current field robots utilize GPS-based auto-steer systems that follow predetermined paths with limited use of AI capabilities, indicating the potential for further evolution of AI in enhancing the autonomy and decision-making capabilities of field robots (Kunze et al. 2018). The advancements in AI, particularly in DL and neural networks, are likely to yield more sophisticated models capable of precise medical interventions and early disease detection, signifying the ongoing evolution of AI in enhancing the capabilities of medical robotics (Madadi 2024).

AI INTEGRATION IN DRONE TECHNOLOGY

AI Technologies in Drones

The implementation of AI technologies in drones has indeed revolutionized their ca-pabilities and applications. Drones have benefited from the integration of small and lightweight imaging devices and sensors that can detect airborne pollutants and char- acterize features of aquatic and terrestrial envi- ronmentsBogue (2023). Additionally, AI approaches for UAV navigation have enhanced the navigation models and applications of UAVs, emphasizing the role of AI in UAV navigation and control systems (Rezwan and Choi 2022). Furthermore, the potential for AI integration in medical drone ap- plications, such as percutaneous nephrolithotomy, is being explored, indicating the impact of AI on the treatment of patients in the coming years (Hameed et al. 2023).

Moreover, the application of artificial immune systems in swarm robotic systems has demonstrated the utilization of biologically inspired computation systems in robotics, showcasing the potential for bio-inspired AI technologies in drone swarm applica-tions(Daudi 2015). The convergence of AI and robotics has played a pivotal role in the development of self-adaptable robots, emphasizing the integration of AI in en- hancing the adaptability and autonomy of robotic systems, including drones (Mayoral 2018). Additionally, the implementation of AI technology in the operation of smart farm robots has aimed to manage the operation of farm robots using AI technology, indicating the potential for AI integration in agricultural drone applications (Chen and Hengjinda 2019). The development of intent-based deployment for robot applications in 5G-enabled non-public networks has highlighted the integration of AI in the deploy- ment and operation of robotic systems, including drones, in advanced communication networks (Qiu et al. 2023). Table 3 presents the comparison of application of AI in drones and robotics.

Table 3. Comparison of AI technologies used in drones vs. robotics

Feature	Drones	Robotics	Notes
Navigation	GPS, AI-enhanced visual and sensor-based navigation.	Lidar, radar, and AI-enhanced sensor fusion for spatial awareness.	Drones require morePrecise navigation in three-dimensional space.
Machine Learning Algorithms	Real-time processing for object detection and avoidance.	Complex algorithms for task automation and decision-making.	Robotics tend to use more varied and com- plex algorithms due to diverse operational en- vironments.
Sensor Integration	High-resolution cameras, thermal imaging.	Tactile sensors, force sensors, and advanced vision systems.	Sensor needs vary sig- nificantly based on ap- plication specifics.
Real-Time Processing	Essential for flight stability and obstacle avoidance.	Critical for interactive tasks and collaborative robotics.	Both fields require substantial real-time processing capabilities, but the context differs.

Advantages of AI Technologies

The integration of AI technologies has significantly enhanced the functionalities of drones across various domains. AI has played a pivotal role in advancing the capabil- ities of drones, leading to trans-formative applications in diverse fields (Figure 1). The following references provide insights into the specific AI technologies that have been implemented to enhance the functionalities of drones:

Kumar et al. (2021) discusses the opportunities of AI and machine learning in the food industry, highlighting how AI-based systems can efficiently handle food produc- tion and delivery processes, thereby enhancing operational competence. This demon- strates the potential for AI to optimize and streamline drone operations in food supply chain management and delivery.

Almalki, Alotaibi, and Angelides (2021) focuses on coupling multifunction drones with AI in the fight against the coronavirus pandemic. The AI framework is designed to optimize the elevation angle and altitude to enhance wireless connectivity between drones and ground stations, leading to improved throughput and power consumption.

Figure 1. Advantages of AI

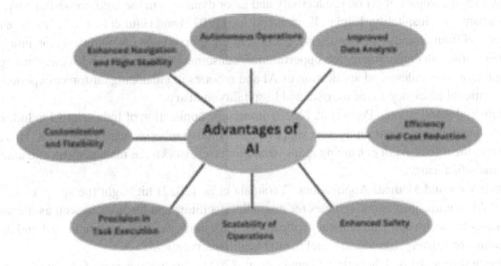

This showcases the use of AI to enhance communication and connectivity capabilities in drones, particularly in critical applications such as pandemic response.

Mozaffari et al. (2019) provides insights into the application of AI and IoT in un- manned aerial vehicles, emphasizing how AI and IoT have enhanced the popularity of drones worldwide. This highlights the role of AI in improving the overall performance and capabilities of drones, contributing to their widespread adoption.

Wang and Ye (2022) discusses the applications of AI-enhanced drones in distress pavement, pothole detection, and healthcare monitoring with service delivery. This reference underscores the potential for AI to enhance the capabilities of drones in various critical applications, including healthcare monitoring and infrastructure maintenance. Gupta, Kumari, and Tanwar (2020) explores the fusion of blockchain and artificial intelligence for secure drone networking underlying 5G communications. The integra- tion of AI and blockchain technologies aims to improve the security and intelligence of drone communication networks, showcasing the potential for AI to enhance the security and reliability of drone operations.

Kovari and Ebeid (2021) presents an FPGA-based platform for intelligent real-time autonomous drone operations. The platform utilizes reconfigurable FPGA chips to run heavy AI algorithms, demonstrating the use of AI for real-time decision-making and autonomous operations in drones.

CURRENT APPLICATIONS AND CASE STUDIES

The current broad applications of AI in robotics and drones across various sectors encompass a wide range of domains, showcasing the transformative impact of AI tech- nologies. The references provide insights into the diverse applications of AI in robotics and drones, highlighting their implications across different sectors:

- Automotive Sector: Moniz, Boavida, and Candeias (2022) discuss the potential changes in pro- ductivity and labor relations in the automotive sector due to the deployment of broad-spectrum algorithms, which can lead to considerable changes in work patterns and labor relations. This signifies the impact of AI on productivity and labor dynamics in the automotive industry.

- Tourism and Hospitality: Kumar, Kumar, and Attri (2021) and Gaur et al. (2021) explore the current and future trends of AI and service robots in the tourism and hospitality sector, emphasizing their impact on employment, job opportunities, consumers, businesses, and society at large. This highlights the widespread application of AI and robotics in enhancing customer experiences and operational efficiency in the tourism and hospitality industry.

- Industrial Applications: Peres et al. (2020) discuss the application of Industrial AI in Industry 4.0, focusing on workforce training, task support, collaborative robotics, and ergonomics. This demonstrates the role of AI in enhancing indus- trial processes, workforce management, and automation in manufacturing.

- Healthcare and Medical Applications: Yaramala et al. (2022) highlight the ap- plication of IoT and AI in unmanned aerial vehicles for upgrading facilities and features, as well as the use of AI for mental health care and clinical applications. This showcases the potential of AI and drones in healthcare delivery, monitoring, and mental health interventions.

- Agriculture and Food Security: Spanaki et al. (2021) discuss the use of AI appli- cations in AgriTech drones for smart AgriFood operations, emphasizing precision evaluations and smart farming. This illustrates the role of AI in enhancing agri- cultural practices, food security, and environmental monitoring.

- Construction and Urban Planning: Ahmad et al. Rudd (2022) and Adu-Gyamfi, Gyasi, and Darkwa (2021) focus on the application of AI and drones in envi- ronmental monitoring, urban soundscapes perception, and historicizing medical drones in Africa. This highlights the diverse applications of AI and drones in construction, environmental monitoring, and urban planning.

- Education and Learning: Ottun et al. (2022) and Knox (2020) explore the im- pact of AI and robotics on teaching, learning, and education, emphasizing the potential for AI to revolutionize higher education, educational technology, and learning environments.

FUTURE TRENDS AND PROJECTIONS

The emerging trends in AI technologies for robotics and drones encompass a wide array of applications across various sectors. The references provide insights into the diverse and transformative impact of AI technologies in robotics and drones, highlighting the following emerging trends:

- Smart Cities and Security: Rawat et al. (2022) discuss the use of AI-based drones for security concerns in smart cities, emphasizing the potential of emerging tech- nologies such as blockchain to improve the management of smart cities. This signifies the role of AI in enhancing security and surveillance capabilities in ur- ban environments.

- Public Libraries and Education: Tella and Ajani (2022) recommend the strategic planning for new and emerging technologies such as humanoid robots in public li- braries, showcasing the potential for AI and robotics to revolutionize educational technology and learning environments.

- Disaster Management and Emergency Response: Abid et al. (2021) highlight the role of AI in boosting disaster management, emphasizing the evolution of robots from basic decision-making devices to truly automated and artificially intelligent machines, showcasing the potential for AI to enhance emergency response and situation awareness.

- Environmental Monitoring and Hazardous Waste Detection: Marturano et al. (2021) discuss the numerical fluid dynamics simulation for drones' chemical de- tection and the ecosystem to maximize the "storm effects" in case of CBRNe dispersion, highlighting the potential of AI and drones in environmental moni- toring, hazardous waste detection, and disaster response.
- Industrial Revolution and Industry 4.0: Shahroom and Hussin (2018) discuss the impact of the Fourth Industrial Revolution (4IR) on education, emphasizing the transformative role of AI, big data, cloud computing, and IoT in reshaping the educational landscape and industrial practices.
- Agriculture and Food Production: Eli-Chukwu (2019) reviews the applications of AI in agriculture, emphasizing the integration of AI-based robotics technology and the interfacing of subsystems into an integrated environment, showcasing the potential for AI to revolutionize agricultural practices and productivity.
- Healthcare and Pandemic Response: Almalki, Alotaibi, and Angelides (2021) discuss the coupling of multifunction drones with AI in the fight against the coronavirus pandemic, highlighting the potential of AI and drones in healthcare delivery, pandemic response, and agricultural automation. Hiraguri et al. (2023) also discuss the shape classification technology of pollinated tomato flowers for robotic implementation, showcasing the potential of AI and robotics in agricul- tural automation.

CONCLUSION

The integration of Artificial Intelligence (AI) into robotics and drones represents a significant stride toward innovative and autonomous technologies that continue to re- define industrial capabilities and operational efficiencies across diverse sectors. As AI technologies evolve, they not only enhance the functionality of robots and drones but also present new avenues for economic growth, societal benefits, and advancements in sectors like agriculture, healthcare, and logistics. However, alongside these advancements, ethical considerations and regulatory frameworks must evolve to address the challenges posed by autonomous systems. The ongoing development and application of AI in robotics and drones signal a transformative future, where the symbiosis of human and machine intelligence can solve complex global challenges more efficiently.

ACKNOWLEDGMENT

This research work is supported by National Science and Technology Council (NSTC), Taiwan Grant No. NSTC112-2221-E-468-008-MY3.

REFERENCES

Abid, S., Sulaiman, N., Chan, S., Nazir, U., Abid, M., Han, H., Ariza-Montes, A., & Vega-Munˇoz, A. (2021). Toward an integrated disaster management approach: How artificial intelligence can boost disaster management. *Sustainability (Basel)*, 13(22), 12560. DOI: 10.3390/su132212560

Adu-Gyamfi, S., Gyasi, R., & Darkwa, B. (2021). Historicizing medical drones in africa: A focus on ghana. *History of Science and Technology*, 11(1), 103–125. DOI: 10.32703/2415-7422-2021-11-1-103-125

Alanezi, M., Shahriar, M., Hasan, M., Ahmed, S., Sha'aban, Y., & Bouchekara, H. (2022). Live- stock management with unmanned aerial vehicles: A review. *IEEE Access : Practical Innovations, Open Solutions*, 10, 45001–45028. DOI: 10.1109/ACCESS.2022.3168295

Almalki, F., Alotaibi, A., & Angelides, M. (2021). Coupling multifunction drones with ai in the fight against the coronavirus pandemic. *Computing*, 104(5), 1033–1059. DOI: 10.1007/s00607-021-01022-9

Alsamhi, S., Ma, O., & Ansari, M. (2019). Survey on artificial intelligence based tech-niques for emerging robotic communication. *Telecommunication Systems*, 72(3), 483–503. DOI: 10.1007/s11235-019-00561-z

Alvarez-Dionisi, L., Mittra, M., & Balza, R. (2019). Teaching artificial intelligence and robotics to un-dergraduate systems engineering students. *International Journal of Modern Education and Computer Science*, 11(7), 54–63. DOI: 10.5815/ijmecs.2019.07.06

Anwar, S., U. (2022). *Artificial intelligence in healthcare: an overview*. InTechOpen. .DOI: 10.5772/intechopen.102768

Arkhipov, V. (2022). Definition of artificial intelligence in the context of the russian legal system: A critical approach. *Государство и право*, 168(1), 168. DOI: 10.31857/S102694520018288-7

Auflem, M., Kohtala, S., Jung, M., & Steinert, M. (2022). Facing the facs—Using ai to evaluate and control facial action units in humanoid robot face development. *Frontiers in Robotics and AI*, 9, 887645. DOI: 10.3389/frobt.2022.887645 PMID: 35774595

Bassyouni, Z., & Elhajj, I. (2021). Augmented reality meets artificial intel- ligence in robotics: A system-atic review. *Frontiers in Robotics and AI*, 8, 724798. DOI: 10.3389/frobt.2021.724798 PMID: 34631805

Benotsmane, R., Dud'as, L., & Kov'acs, G. (2020). Survey on artificial intelligence al- gorithms used in industrial robotics. *Multidiszciplin'aris Tudom'anyok*, 10(4), 194–205. DOI: 10.35925/j.multi.2020.4.23

Bisht, J., & Vampugani, V. S. (2022). Load and cost-aware min-min workflow scheduling algorithm for heterogeneous resources in fog, cloud, and edge scenarios. *International Journal of Cloud Applications and Computing*, 12(1), 1–20. DOI: 10.4018/IJCAC.2022010105

Bithas, P., Michailidis, E., Nomikos, N., Vouyioukas, D., & Kanatas, A. G. (2019). A survey on machine-learning techniques for uav-based communications. *Sensors (Basel)*, 19(23), 5170. DOI: 10.3390/s19235170 PMID: 31779133

Blöcher, K., & Alt, R. (2020). Ai and robotics in the european restaurant sector: Assessing potentials for process innovation in a high-contact service industry. *Electronic Markets*, 31(3), 529–551. DOI: 10.1007/s12525-020-00443-2

Bogue, R. (2023). The role of robots in environmental monitoring. *The Industrial Robot*, 50(3), 369–375. DOI: 10.1108/IR-12-2022-0316

Brock, D. (2018). Learning from artificial intelligence's previous awakenings: The history of expert systems. *AI Magazine*, 39(3), 3–15. DOI: 10.1609/aimag.v39i3.2809

Cangelosi, A., & Schlesinger, M. (2018). From babies to robots: The contribution of develop- mental robotics to developmental psychology. *Child Development Perspectives*, 12(3), 183–188. DOI: 10.1111/cdep.12282

Cao, Y. (2021). Technical composition and cre- ation of interactive installation art works under the background of artificial intelligence. *Mathematical Problems in Engineering*, 1–11. .DOI: 10.1155/2021/7227416

Chamoso, P. (2015). *Swarm agent-based architecture suitable for internet of things and smartcities*. Springer. ₃.DOI: 10.1007/978-3-319-19638-1

Chen, J., & Hengjinda, P. (2019). Applying ai technology to the operation of smart farm robot. *Sensors and Materials*, 31(5), 1777. DOI: 10.18494/SAM.2019.2389

Chen, Y., & Luca, G. (2021). Technologies supporting artificial intelligence and robotics application development. *JAIT*, 1(1), 1–8. DOI: 10.37965/jait.2020.0065

Chui, K. T., Gupta, B. B., Arya, V., & Torres-Ruiz, M. (2024). Selective and Adaptive Incremental Transfer Learning with Multiple Datasets for Machine Fault Diagno- sis. *Computers, Materials & Continua*, 78(1), 1363–1379. DOI: 10.32604/cmc.2023.046762

Colace, F., Guida, C. G., Gupta, B., Lorusso, A., Marongiu, F., & Santaniello, D. (2022). A BIM-based approach for decision support system in smart buildings. In *Proceedings of Seventh International Congress on Information and Communication Technology: ICICT 2022*. Springer.

Croon, G., Dupeyroux, J., Fuller, S., & Marshall, J. (2022). Insect-inspired ai for autonomous robots. *Science Robotics*, 7(67), eabl6334. DOI: 10.1126/scirobotics.abl6334 PMID: 35704608

Daudi, J. (2015). An overview of application of artificial immune system in swarm robotic systems. *Automation Control and Intelligent Systems*, 3(2), 11. DOI: 10.11648/j.acis.20150302.11

Doncieux, S., Chatila, R., Straube, S., & Kirchner, F. (2022). Human-centered ai and robotics. *Ai Perspectives*, 4(1), 1. DOI: 10.1186/s42467-021-00014-x

Doroudi, S. (2022). The intertwined histories of artificial intelligence and educa- tion. *International Journal of Artificial Intelligence in Education*, 33(4), 885–928. DOI: 10.1007/s40593-022-00313-2

Eli-Chukwu, N. (2019). Applications of artificial intelligence in agriculture: A re- view. *Engineering Technology & Applied Science Research*, 9(4), 4377–4383. DOI: 10.48084/etasr.2756

Feizi, N. (2021). Robotics and ai for teleoperation, tele-assessment, and tele-training for surgery in the era of covid-19: existing challenges, and future vision. *Frontiers in Robotics and AI,8*. DOI: 10.3389/frobt.2021.610677

Geetha, D. (2024). The future is now: ai powers next-generation robots. *Interantional Journal of Scientific Research in Engineering and Management*. .DOI: 10.55041/IJSREM28890

Gupta, R., Kumari, A., & Tanwar, S. (2020). Fusion of blockchain and artificial intelligence for secure drone networking underlying 5g communications. *Transactions on Emerging Telecommunications Technologies*, 32(1), e4176. DOI: 10.1002/ett.4176

Haenlein, M., & Kaplan, A. (2019). A brief history of artificial intelligence: On the past, present, and future of artificial intelligence. *California Management Review*, 61(4), 5–14. DOI: 10.1177/0008125619864925

Hameed, B., Shah, M., Pietropaolo, A., Coninck, V., Naik, N., Skolarikos, A., & Somani, B. (2023). The technological future of percutaneous nephrolithotomy: A young academic urologists endourology and urolithiasis working group update. *Current Opinion in Urology*, 33(2), 90–94. DOI: 10.1097/MOU.0000000000001070 PMID: 36622261

Hiraguri, T., Kimura, T., Endo, K., Ohya, T., Takanashi, T., & Shimizu, H. (2023). Shape classification technology of pollinated tomato flowers for robotic implementation. *Scientific Reports*, 13(1), 2159. DOI: 10.1038/s41598-023-27971-z PMID: 36750598

Hong, Y., Lian, J., Li, X., Min, J., Wang, Y., Freeman, L., & Deng, X. (2021). *Statistical perspectives on reliability of artificial intelligence systems*. https://doi.org//arxiv.2111.05391.DOI: 10.48550

Kipnis, E., McLeay, F., Grimes, A., Saille, S., & Potter, S. (2022). Service robots in long-term care: A consumer-centric view. *Journal of Service Research*, 25(4), 667–685. DOI: 10.1177/10946705221110849

Knox, J. (2020). Artificial intelligence and education in china. *Learning, Media and Technology*, 45(3), 298–311. DOI: 10.1080/17439884.2020.1754236

Kovari, B. & Ebeid, E. (2021). *Mpdrone: fpga-based platform for intelligent real-time au- tonomous drone operations*. IEEE. .DOI: 10.1109/SSRR53300.2021.9597857

Kumar, I., Rawat, J., Mohd, N., & Husain, S. (2021). Opportunities of artificial intelli- gence and machine learning in the food industry. *Journal of Food Quality*, 2021, 1–10. DOI: 10.1155/2021/4535567

Kumar, S. (2021). Impact of artificial intelligence and service robots in tourism and hospitality sector: current use & future trends. *Administrative Development a Journal of Hipa Shimla,8*, 59–83. DOI: 10.53338/ADHIPA2021.V08.Si01.04

Kumbhojkar, N. R., & Menon, A. B. (2022). Integrated predictive experience management framework (IPEMF) for improving customer experience: In the era of digital transformation. [IJCAC]. *International Journal of Cloud Applications and Computing*, 12(1), 1–13. DOI: 10.4018/IJCAC.2022010107

Kunze, L., Hawes, N., Duckett, T., Hanheide, M., & Krajn'ık, T. (2018). Artificial intelligence for long-term robot autonomy: A survey. *IEEE Robotics and Automation Letters*, 3(4), 4023–4030. DOI: 10.1109/LRA.2018.2860628

Lahmeri, M., Kishk, M., & Alouini, M. (2021). Artificial intelligence for uav-enabled wire- less networks: A survey. *IEEE Open Journal of the Communications Society*, 2, 1015–1040. DOI: 10.1109/OJCOMS.2021.3075201

Lee, M., Siewiorek, D., Smailagic, A., Bernardino, A., & Badia, S. (2022). Enabling ai and robotic coaches for physical rehabilitation therapy: Iterative design and evaluation with therapists and post-stroke survivors. *International Journal of Social Robotics*, 16(1), 1–22. DOI: 10.1007/s12369-022-00883-0

Liao, M., Tang, H., Li, X., Pandi, V., & Arya, V. (2024). A lightweight network for abdominal multi-organ segmentation based on multi-scale context fusion and dual self-attention. *Information Fusion*, 108, 102401. DOI: 10.1016/j.inffus.2024.102401

Madadi, Y., Delsoz, M., Khouri, A. S., Boland, M., Grzybowski, A., & Yousefi, S. (2024). Applications of artificial intelligence-enabled robots and chatbots in oph- thalmology: Recent advances and future trends. *Current Opinion in Ophthalmology*, 35(3), 238–243. DOI: 10.1097/ICU.0000000000001035 PMID: 38277274

Mahdi, I. (2021). Evaluation of robot professor technology in teaching and business. *Information Technology in Industry, 9*, 1182–1194. .DOI: 10.17762/itii.v9i1.255

Marturano, F., Martellucci, L., Chierici, A., Malizia, A., Giovanni, D., D'Errico, F., Gaudio, P., & Ciparisse, J. (2021). Numerical fluid dynamics simulation for drones' chemical detection. *Drones (Basel)*, 5(3), 69. DOI: 10.3390/drones5030069

Mayoral, V. (2018). *Towards self-adaptable robots: from programming to training machines*. https://doi.org//arxiv.1802.04082.DOI: 10.48550

Milne-Ives, M., Selby, E., Inkster, B., Lam, C., & Meinert, E. (2022). Artificial intelligence and machine learning in mobile apps for mental health: A scoping review. *PLOS Digital Health*, 1(8), e0000079. DOI: 10.1371/journal.pdig.0000079 PMID: 36812623

Minbaleev, A. (2022). The concept of "artificial intelligence" in law. *Bulletin of Udmurt University Series Economics and Law,32*. .DOI: 10.35634/2412-9593-2022-32-6-1094-1099

Mishra, A., & Kong, K. T. C. H. (2024). Tempered Image Detection Using ELA and Convolutional Neural Networks. In *2024 IEEE International Conference on Consumer Electronics (ICCE)*, (pp. 1–3). IEEE. DOI: 10.1109/ICCE59016.2024.10444440

Moniz, A., Boavida, N., & Candeias, M. (2022). Changes in productivity and labour relations: Artificial intelligence in the automotive sector in portugal. *International Journal of Automotive Technology and Management*, 22(2), 1. DOI: 10.1504/IJATM.2022.10046022

Mozaffari, M., Saad, W., Bennis, M., Nam, Y., & Debbah, M. (2019). A tutorial on uavs for wireless networks: Applications, challenges, and open problems. *IEEE Communications Surveys and Tutorials*, 21(3), 2334–2360. DOI: 10.1109/COMST.2019.2902862

Munteanu, I. S., & Ungureanu, L. M. (2022). Analysis Of The Evolution And Prospects Of Introducing Robots And Artificial Intelligence In The Activities Of The Modern Competitive Society. *International Journal of Mechatronics and Applied Mechanics*https://api.semanticscholar.org/CorpusID:250544380

Nabi, W., & Xu, B. (2021). Applications of artificial intelligence and ma- chine learning approaches in echocardiography. *Echocardiography (Mount Kisco, N.Y.)*, 38(6), 982–992. DOI: 10.1111/echo.15048 PMID: 33982820

Nhi, N. T. U., & Le, T. M. (2022). A model of semantic-based image retrieval using C-tree and neighbor graph. [IJSWIS]. *International Journal on Semantic Web and Information Systems*, 18(1), 1–23. DOI: 10.4018/IJSWIS.295551

Okafor, N. (2022). *Business demand for a Cloud enterprise data warehouse in electronic Healthcare Computing*. IGI Global.

Ottun, A., Yin, Z., Liyanage, M., Boerger, M., Asadi, M., Hui, P., Tarkoma, S., Tcholtchev, N., Nurmi, P., & Flores, H. (2022). *Toward trustworthy and responsible autonomous drones in future smart cities*. TechRxiv. DOI: 10.36227/techrxiv.21444102

Papadopoulos, G., Antona, M., & Stephanidis, C. (2020). *Towards open and expandable cognitive ai architectures for large-scale multi-agent human-robot collaborative learning*. https://doi.org//arxiv.2012.08174.DOI: 10.48550

Peres, R., Jia, X., Lee, J., Sun, K., Colombo, A., & Barata, J. (2020). Industrial artificial intelligence in industry 4.0 - systematic review, challenges and outlook. *IEEE Access : Practical Innovations, Open Solutions*, 8, 220121–220139. DOI: 10.1109/ACCESS.2020.3042874

Qiu, R., Li, D., Ibez, A., Xu, Z., & Tarazn, R. (2023). Intent-based deployment for robot applica- tions in 5g-enabled non-public networks. *ITU Journal : ICT Discoveries*, 4(1), 209–220. DOI: 10.52953/AYMI1991

Rawat, B., Bist, A., Apriani, D., Permadi, N., & Nabila, E. (2022). Ai based drones for se- curity concerns in smart cities. *Aptisi Transactions on Management (Atm)*, 7(2), 125–130. DOI: 10.33050/atm.v7i2.1834

Rezwan, S., & Choi, W. (2022). Artificial intelligence approaches for uav nav- igation: Recent advances and future challenges. *IEEE Access : Practical Innovations, Open Solutions*, 10, 26320–26339. DOI: 10.1109/ACCESS.2022.3157626

Rosete, A. (2020). *Service robots in the hospitality industry: an exploratory literature review*. Springer. 3.DOI: 10.1007/978-3-030-38724-2

Rudd, I. (2022). Leveraging artificial intelligence and robotics to improve mental health. *IAJ*. .DOI: 10.32370/IAJ.2710

Sassis, L., Kefala-Karli, P., Sassi, M., & Zervides, C. (2021). Exploring medical students' and faculty's perception on artificial intelligence and robotics. a questionnaire survey. *Journal of Artificial Intelligence for Medical Sciences*, 2(1-2), 76–84. DOI: 10.2991/jaims.d.210617.002

Shahroom, A., & Hussin, N. (2018). Industrial revolution 4.0 and education. *International Journal of Academic Research in Business & Social Sciences*, 8(9). DOI: 10.6007/IJARBSS/v8-i9/4593

Sharma, A. (2024). Revolutionizing Healthcare Systems: Synergistic Multimodal Ensemble Learning & Knowledge Transfer for Lung Cancer Delineation & Taxonomy. In *2024 IEEE International Conference on Consumer Electronics (ICCE)*, (pp. 1–6). IEEE. DOI: 10.1109/ICCE59016.2024.10444476

Shazly, S., Trabuco, E., Ngufor, C., & Famuyide, A. (2022). Introduction to ma- chine learning in obstetrics and gynecology. *Obstetrics and Gynecology*, 139(4), 669–679. DOI: 10.1097/AOG.0000000000004706 PMID: 35272300

Sorantin, E. (2021). The augmented radiologist: artificial intelligence in the practice of radiology. *Pediatric Radiology,52*. .DOI: 10.1007/s00247-021-05177-7

Suarez-Ibarrola, R., & Miernik, A. (2020). Prospects and challenges of artificial intelligence and computer science for the future of urology. *World Journal of Urology*, 38(10), 2325–2327. DOI: 10.1007/s00345-020-03428-0 PMID: 32910230

Tella, A., & Ajani, Y. (2022). Robots and public libraries. *Library Hi Tech News*, 39(7), 15–18. DOI: 10.1108/LHTN-05-2022-0072

Toncic, J. (2021). Advancing a critical artificial intelligence theory for schooling. *Teknokultura Revista De Cultura Digital Y Movimientos Sociales,19*, 13–24. .DOI: 10.5209/tekn.71136

Tuo, Y. (2021). *How artificial intelligence will change the future of tourism industry: the practice in China*. Springer. ₇.DOI: 10.1007/978-3-030-65785-7

Vanak, J. (2022). Artificial intelligence and medicine. *Science Insights*, 41(1), 567–575. DOI: 10.15354/si.22.re068

Wagter, C. (2021). The artificial intelligence behind the winning entry to the 2019 ai robotic racing competition. https://doi.org//arxiv.2109.14985.DOI: 10.48550

Wang, P. (2019). On defining artificial intelligence. *Journal of Artificial General Intelligence*, 10(2), 1–37. DOI: 10.2478/jagi-2019-0002

Wang, Y., & Ye, T. (2022). Applications of artificial intelligence enhanced drones in distress pavement, pothole detection, and healthcare monitoring with service delivery. *Journal of Engineering*, 2022, 1–16. DOI: 10.1155/2022/7733196

Wazid, M. (2020). *Private blockchain- envisioned security framework for ai-enabled iot-based drone-aided healthcare services*. ACM. .DOI: 10.1145/3414045.3415941

Yaramala, D., Khan, D., Vasanthakumar, N., Koteswararao, K., Sridhar, D., & Ansari, M. (2022). Application of internet of things (iot) and artificial intelligence in unmanned aerial vehicles. *International Journal of Electrical and Electronics Research*, 10(2), 276–281. DOI: 10.37391/ijeer.100237

Zhang, T., Li, Q., Zhang, C., Liang, H., Li, P., Wang, T., Li, S., Zhu, Y., & Wu, C. (2017). Current trends in the development of intelligent unmanned autonomous sys- tems. *Frontiers of Information Technology & Electronic Engineering*, 18(1), 68–85. DOI: 10.1631/FITEE.1601650

Chapter 2
AI Technologies in Robotics

Akshat Gaurav
Ronin Institute, USA

Varsha Arya
Asia University, Taiwan, & Hong Kong Metropolitan University, Hong Kong

ABSTRACT

This chapter explores the application of AI technologies in the domain of robotics, focusing on how these advancements enable robots to perform complex tasks with high efficiency and autonomy. It covers critical areas such as perception, where AI allows robots to interpret sensory data; navigation and motion planning, enabling robots to move through and interact with their environments safely; and manipulation, where AI-driven robots achieve precision in handling objects. The chapter also discusses the role of AI in facilitating human-robot interaction, enhancing the ability of robots to work alongside humans in various industrial settings. Through detailed explanations and examples, the chapter demonstrates the transformative potential of AI in robotics, showcasing how it drives innovation and efficiency in industrial applications.

This chapter explores the application of AI technologies in the domain of robotics, focusing on how these advancements enable robots to perform complex tasks with high efficiency and autonomy. It covers critical areas such as perception, where AI allows robots to interpret sensory data; navigation and motion planning, enabling robots to move through and interact with their environments safely; and manipulation, where AI-driven robots achieve precision in handling objects. The chapter also discusses the role of AI in facilitating human-robot interaction, enhancing the ability of robots to work alongside humans in various industrial settings. Through detailed explanations and examples, the chapter demonstrates the transformative potential of AI in robotics, showcasing how it drives innovation and efficiency in industrial applications.

INTRODUCTION

The global artificial intelligence (AI) market was valued at USD 93.5 billion in 2021, projected to reach USD 1,011.7 billion by 2030, with a 38.1% estimated compound annual growth rate (CAGR), indicating the rapid adoption and growth of advanced AI technologies across industries, including ro-

DOI: 10.4018/979-8-3693-2707-4.ch002

botics(Sravya et al. n.d.; Chui et al. 2023). The integration of AI and robotics has become a focal point of research and devel- opment in various fields, leading to significant advancements in technology (Table 1). This convergence has paved the way for innovative applications, ethical considerations, and the evolution of machine intelligence. In industrial settings, AI and blockchain technologies are being explored in the realm of industrial robotics, indicating the po- tential for enhanced automation and data management (Samantaray 2023). The use of AI and robotics in healthcare has raised ethical implications, prompting the need for a comprehensive ethical framework to ensure responsible deployment in patient care (Elendu 2023; Xie et al. 2023; **?**).

The impact of AI and robotics has also been evident in the development of con- versational AI and knowledge graphs for social robot interaction, exemplified by the CityTalk robot dialogue system (Wilcock and Jokinen 2022; Casillo et al. 2024). Ad- ditionally, the application of AI in long-term robot autonomy has been identified as an enabler for enhancing the capabilities of robotic systems, presenting future chal- lenges and opportunities in this domain (Kunze et al. 2018; Gupta et al. 2023). The integration of AI and robotics has extended to the field of education, offering diverse job opportunities for professionals in various engineering disciplines (Alvarez-Dionisi,

Table 1. Timeline of key events and projections in the robotics industry

Year	Event
2015	**Electronics Sector:** 18% increase in industrial robot installations.
2018	**Global Investment:** A world record for investment in industrial robots, worth about $16.5 billion, with around 422,000 robots installed.
2019	•**China:** Leads the industrial robotics market, holding 48% of the top- ten market share. •**Global Installation:** Over 1.4 million new industrial robots to be installed in factories worldwide.
2025	•**Global Market Size:** Projected to exceed $61.4 billion. •**UK Market Size:** Identified by the UK Government as a key tech- nology for growth, predicting a market size of £13 billion. •**Industrial Robotics:** Global robotics market estimated to be around $45 billion.
2026	**Industrial Robotics Market:** Forecasted value of $52.85 billion.

Mittra, and Balza 2019). This indicates the broad impact of AI and robotics on differ- ent sectors, creating new avenues for employment and specialization. Furthermore, the development of autonomous systems, such as socially assistive robots for individuals with specific needs, emphasizes the pressing need for systems that support the elderly and people with disabilities (Wilson, Tickle-Degnen, and Scheutz 2020). Ethical and regulatory frameworks for the development of standards in AI and autonomous robotic surgery have been a subject of discussion, highlighting the need to address legal, ethi- cal, and regulatory aspects in the deployment of AI and robotics in surgical procedures (O'Sullivan et al. 2019).

Historical Context and Evolution of AI in Robotics

The integration of various technologies such as mechanical engineering, control sys- tems, electronics, and software in robotics has spanned roughly half a century, sig- nificantly influencing the automation of manufacturing industries and enhancing pro- ductivity(Hart 2022; Gupta et al. 2024). The roots of artificial intelligence (AI) can be traced back to the thoughts of the ancient Greek philosopher Aristotle, and have since been developed by notable figures like Thomas Bayes, George Boole, and Charles Bab- bage(Jaakkola et al. 2019). A pivotal moment in the birth of AI was marked by Alan Turing's work in solving the Enigma code during World War II. The Dartmouth conference in 1956 also played a crucial role in AI's emergence, heralding a new era of scientific exploration in this field(Jaakkola et al. 2019; Behera et al. 2023). AI's journey through various epochs, including the periods of good old fashioned AI, com- putational intelligence, two AI winters, and significant strides in machine learning and deep learning, highlight its evolutionary path(Jaakkola et al. 2019). From the first golden years of AI between 1956 to 1974, through the first AI winter, and into another period of rebirth from 1980 to 1987 focusing on expert systems, AI's impact has been profound. Notable advancements during this time include the creation of the super-computer Deep Blue in 1997 and the application of cognitive computing in Jeopardy! and deep reinforcement learning in the game Go(Gasparetto and Scalera 2019).

Influence of Historical Context on the Evolution of AI in Robotics

In recent times, the rapid advancement of information and communication technol- ogy (ICT), coupled with the emergence of the Internet of Things (IoT) and AI, is expected to enhance the intelligence and autonomy of robots. This evolution is poised to transform a variety of sectors including healthcare, construction, logistics, and agri- culture(Hart 2022; Kumar et al. 2022). The development of AI and robotics has been categorized into three generations—from the Turing test and logic theory machine to expert systems and autonomous vehicles—and these technologies have found applica- tions across industry, business, research, and education(Kose and Sakata 2017; Kiran, Pasupuleti, and Eswari 2022). The evolution of industrial robotics is intricately tied to advancements in industrial communication systems, which have become a fundamental technology enabling the modern capabilities of industrial robots(Pecora 2014).

Major Technological Advancements Shaping the Evolution of AI in Robotics

The third wave of significant investments in AI has brought about notable successes, marking a progression from logical reasoning with symbols and rule-based expert sys- tems to a revival in neural networks, spurred by new training algorithms(Callaos et al. 2023). This period has also seen the rise of model-centered approaches in AI, where intelligence is derived from reasoning within models of the world. This shift signifies a gradual erosion of the traditional modeling abstractions that underpin AI problem- solving techniques used by robots(Pecora 2014; Khanam, Tanweer, and Khalid 2022). AI's influence now permeates various sectors such as healthcare, manufacturing, trans- portation, logistics, security, retail, agri-food, and construction. In these fields, AI is in- creasingly employed for tasks such as scheduling, routing, and forecasting future move- ments in transportation systems(Gasparetto and Scalera 2019; Darapureddy, Kurni, and Saritha 2021; Hu et al. 2022).

FOUNDATIONS OF AI IN ROBOTICS

Core AI Technologies

The integration of machine learning, deep learning, reinforcement learning, computer vision, natural language processing (NLP), and speech recognition has significantly advanced the capabilities of robotics and artificial intelligence (AI) systems (Table 2). These technologies have been instrumental in enabling robots to perceive and in- teract with their environment, understand and process natural language, and make autonomous decisions. Machine learning, including deep learning and reinforcement learning, has been pivotal in enhancing the cognitive abilities of robots. Deep rein- forcement learning (DRL) has been proposed as a means to achieve end-to-end learn- ing from perception to robot manipulation, allowing robots to learn complex tasks in the real world (Jiang et al. 2021). This has laid the foundation for the applica- tion of deep reinforcement learning in machine vision to guide robots in recognizing and grasping objects, thereby advancing the capabilities of robotic systems in object

Table 2. Comparison of core AI technologies in robotics

Technology Type	Key Characteristics	Typical Applications in Robotics	Key Advantages
Machine Learning	Algorithms that learn from and make predic- tions on data	Pattern recognition, anomaly detection	Versatility, adaptabil- ity
Deep Learning	Layered neural net- works analyzing large amounts of data	Image and speech recognition, navigation	High accuracy in com- plex environments
Reinforcement Learning	Learning through trial and error using reward feedback	Autonomous decision- making, optimization	Ability to develop strategy and adapt
Computer Vi- sion	Enables computers to interpret and process visual data	Object detection, envi- ronment navigation	Enhances perception capabilities
Natural Lan- guage Process- ing (NLP)	Processing and analyz- ing human language data	Human-robot interac- tion	Enables understanding and generation of hu- man language
Speech Recog- nition	Converts spoken lan- guage into text	Voice commands, in- teractive communica- tion	Facilitates natural user interface

detection and manipulation tasks (Bai et al. 2020). Moreover, the integration of rein- forcement learning with machine learning techniques such as deep Q-networks has been widely discussed, demonstrating the potential of reinforcement learning in enhancing the learning capabilities of robotic systems (Kono et al. 2019). Reinforcement learning has also shown promise in enabling multi-robot collaboration, indicating its poten- tial to revolutionize the coordination and cooperation of robotic systems in various applications (Fan et al. 2022).

In the domain of computer vision, the surge of deep learning has significantly im- pacted robotics, particularly in the development of perception systems. The recent developments in machine learning, particularly deep learning approaches, have led to the evolution of robotic perception systems, enabling robots to perform new applica- tions and tasks (Premebida, Ambru¸s, and M´arton 2019). Additionally, the utilization of deep learning in robotics has been explored in the context of autonomous mobile robot navigation, where deep reinforcement learning approaches have been proposed for indoor navigation tasks (Lee and Yusuf 2022).

Furthermore, the application of reinforcement learning in robotic manipulation skill acquisition has been a subject of research, highlighting the potential of reinforcement learning in enabling robots to acquire and improve manipulation skills autonomously (Liu et al. 2020). Additionally, reinforcement learning has been identified as a valuable approach for robot control and navigation in complex envi- ronments, showcasing its potential in enhancing the autonomy and adaptability of robotic systems (Hu 2024). In the realm of natural language processing (NLP) and speech recognition, the inte- gration of machine learning and deep learning has paved the way for advancements in human-robot interaction. The development of intelligent robotic perception systems has been driven by deep learning approaches, enabling robots to understand and re- spond to natural language commands, thereby enhancing their ability to interact with humans in various settings (Alatabani et al. 2022).

Figure 1. Type of robotic sensors

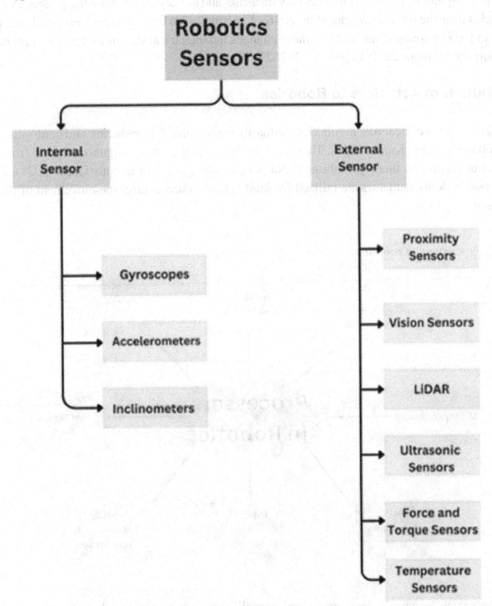

Key Functions of Sensors in Robotics

Sensors in robotics are pivotal for gathering critical data about both the environment and the robots themselves(Ramachandran 2016). They measure a variety of physical quantities such as position, orientation, velocity, distance, and temperature, which are essential for the robot's interaction with its surroundings(Kumar 2014; Zhang et al. 2022). Sensors can be broadly categorized into internal and external types, depending on the parameters they measure, and are key in detecting the presence or absence of obstacles near the robot(Ramachandran 2016). Functioning as input devices, sensors detect physical changes in the environment and convert these changes into electrical signals that are then processed by microcontrollers(Gupta 2021; Kogelis et al. 2022) (Figure 1).

Contribution of Actuators to Robotics

Actuators play a critical role in robotics, primarily responsible for producing and controlling motion within robotic systems(Kumar 2014). They work by converting electrical signals received from sensors into mechanical forces, thereby enabling robots to execute a range of actions(Gupta 2021). In multijoint robotic systems, actuators are utilized for joint actuation and control the movement of the robot's operational

Figure 2. Processors in robotics

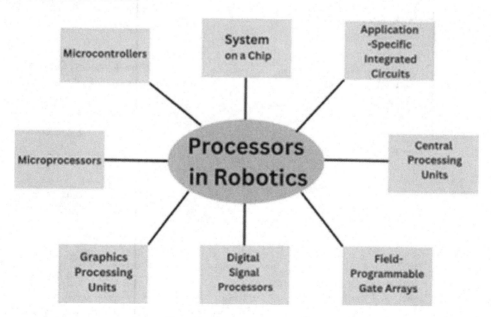

elements such as wheels, legs, arms, and propellers found in service robots(Kogelis et al. 2022) (Kumar 2014; Husak 2021). Various types of actuators, including pneumatic, hydraulic, mechanical, and electric, are employed in robotics to facilitate different forms of motion and control(Husak 2021; Prasad et al. 2024).

Significance of Processors in Robotics

Processors such as Central Processing Units (CPUs), Graphics Processing Units (GPUs), and Field Programmable Gate Arrays (FPGAs) are fundamental to the op- eration of modern robotic systems(Podlubne and Gohringer 2023)(Figure 2). They process the data collected from sensors, make decisions based on this information, and subsequently control the actuators to execute real-time actions(Gupta 2021). These processors enable robots to adapt to changes in their environments and carry out complex tasks based on sensor inputs(Zhang et al. 2022). Furthermore, in robotics, processors are crucial in the design of complex systems and the integration of paral- lelizable and energy-efficient devices, enhancing the capabilities of cutting-edge robotic platforms(Podlubne and Gohringer 2023).

Integration of Sensors, Actuators, and Processors in Robotics

The integration of sensors, actuators, and processors in robotics creates highly au- tonomous systems capable of performing intricate tasks. Sensors collect data about the environment and the robot, which processors then analyze to make informed de- cisions(Ramachandran 2016; Gupta 2021). This processed information is crucial for controlling the actuators, which in turn allow the robot to interact dynamically with its surroundings and perform tasks as required(Zhang et al. 2022). Driven by processors, actuators produce the necessary mechanical actions for the robot to execute human- like and sensible movements, based on the insights provided by sensors(Gupta 2021). This synergistic integration results in robotic systems that can operate autonomously, adapting to real-time environmental changes and executing tasks based on continuous data updates(McGinn et al. 2015; Kogelis et al. 2022).

CHALLENGES AND ETHICAL CONSIDERATIONS

Technical Challenges

Limitations in AI and machine learning models are a critical area of study that has garnered significant attention due to the potential implications for decision-making, interpretability, and reliability. Several key limitations have been identified, shedding light on the challenges and considerations associated with the deployment of AI and machine learning models.

One of the fundamental limitations in AI and machine learning models is the issue of interpretability and explainability. The black-box nature of certain machine learning models has been a subject of con- cern, as it hinders the ability to understand the rationale behind the model's predictions and decisions Carvalho, Pereira, and Cardoso (2019). This lack of interpretability poses barriers to adoption and raises ethical and regulatory concerns, particularly in high-stakes decision-making scenarios (Chanda and Banerjee 2022). The field of Explainable AI (XAI) has emerged to address this limitation by focusing on research into machine learning interpretability and aiming to shift towards a more transparent AI (Kamath and Liu 2021).

Another significant limitation lies in the potential for errors and failures in AI sys- tems, particularly those operating on machine learning and deep learning technologies. Omission and commission errors underlying AI failures have been investigated, high- lighting the challenges associated with the complexity of machine-learnt algorithms and neural networks (Chanda and Banerjee 2022). The canonical dictum that "all models are wrong, but some are useful" encapsulates the logical limitations of statis- tics and machine learning, emphasizing the inherent imperfections in modeling and prediction (Pasquinelli and Joler 2020). These limitations underscore the need for robust error detection and mitigation strategies in AI and machine learning systems. The reproducibility of machine learning models in healthcare and clinical applica- tions has also been identified as a limitation, particularly as high-capacity machine learning models begin to demonstrate early successes in clinical settings (Beam, Man- rai, and Ghassemi 2020). Ensuring the reproducibility and reliability of machine learn- ing models is crucial for their widespread adoption and integration into healthcare decision-making processes.

Furthermore, the limitations in deep learning models without explainability have been recognized as a barrier to their adoption, particularly in high-stakes decision- making scenarios (Babko-Malaya, Planer, and Li 2023). The lack of explainability in deep learning models hinders their acceptance and trustworthiness, necessitating the development of strategies to enhance their interpretability and transparency.The limi- tations of AI and machine learning models also extend to their application in surgical decision-making and individualized risk assessment. While machine learning has sig- nificantly impacted surgical decision-making, its applications continue to expand, and the development and assessment of machine learning models for individualized risk assessment present challenges in ensuring their accuracy and reliability (Hassan et al. 2022a).

In addition, the limitations of AI and machine learning models are intertwined with the challenges of integrating these tools into an economic framework, being clear about the link between predictions and decisions, and specifying the scope of payoff functions (Kleinberg et al. 2017). These considerations highlight the complexity of leveraging machine learning predictions in decision-making processes and the need for a comprehensive understanding of their implications.

Moreover, the limitations of machine learning models in healthcare have prompted the development of AI and machine learning techniques to facilitate shared decision- making and improve patient out- comes by accurately predicting patient-reported out- comes following surgery (Hassan et al. 2022b). While these techniques hold promise, addressing the limitations associated with their interpretability and reliability is cru- cial for their successful integration into clinical practice.

Ethical and Social Implications

The ethical and social implications of robotics have become a subject of significant dis- course, re- flecting the multifaceted impact of robots on society, individuals, and various domains. The integration of robots into human environments has raised complex eth- ical considerations, necessitating a comprehensive understanding of the implications associated with their deployment.

The societal impact of robotics, along with the ethical and moral implications, has been a focal point of discussion, emphasizing the need for a thorough examination of the broader implications of integrating robots into human environments Cross, Hort- ensius, and Wykowska (2019). This includes considerations related to the potential transformation of social dynamics, labor markets, and human-robot interaction, ne- cessitating a comprehensive understanding of the ethical and societal implications of robotics.

The use of artificial intelligence and robotics has been associated with ethically relevant implications for individuals, groups, and society, reflecting the need for a nu- anced approach to address the ethical considerations associated with the deployment of these technologies (Sætra and Danaher 2022). This encompasses a wide array of ethical considerations, including privacy, autonomy, and the impact on societal norms and values. The ethical implications of robot interactions extend beyond the physical level, necessitating the development of new regulations and guidelines to address the ethical considerations associated with human-robot interactions (Shukla et al. 2017). This underscores the need for a comprehensive ethical framework to guide the devel- opment and deployment of robots in various settings.

The development of customizable humanlike robots has prompted discussions on the ethical limits and legal implications, emphasizing the need to address the ethical con- siderations associated with the creation and deployment of sentient beings (Macken- zie 2018). This highlights the ethical complexities associated with the development and utilization of humanlike robots, necessitating a comprehensive ethical framework to guide their creation and deployment.The use of AI robots has triggered societal changes, yielding ethical implications that necessitate careful consideration and assess- ment to ensure the responsible deployment of these technologies (T´oth et al. 2022). This underscores the need for a comprehensive understanding of the societal and ethi- cal implications associated with the integration of AI robots into various domains.The ethical and legal implications of sex robots have been explored, reflecting the need to address the ethical considerations associated with the development and deployment of these technologies (Rodriguez-Amat and Duller 2019). This includes considerations related to the ethical responsibilities of creators and the potential impact on societal norms and values.

The development of an engineering-based design methodology for embedding ethics in autonomous robots has been proposed, emphasizing the need to align the capabil- ity of robots to recognize ethical implications with their devolved level of autonomy (Robertson et al. 2019). This underscores the im- portance of integrating ethical con- siderations into the design and development of autonomous robots. The ethical impli- cations of designing virtuous sex robots have been discussed, reflecting the need to address the ethical considerations associated with the development and deployment of these technologies (Peeters and Haselager 2019). This includes considerations related to user autonomy and responsibility, necessitating a comprehensive understanding of the ethical implications associated with the development of sex robots.The implemen- tation and use of robots and AI have presented myriad legal and ethical implications, reflecting the need for a comprehensive understanding of the ethical considerations as- sociated with the deployment of these technologies (Cain, Thomas, and Alonso 2019). This includes considerations related to privacy, autonomy, and the impact on societal norms and values.

The value of social robots in services has been explored from a social cognition per- spective, reflecting the need to address the ethical considerations associated with the deployment of social robots in various service contexts (C´ai´c, Mahr, and Oderkerken-Schr¨oder 2019). This includes considerations related to the impact on social inter- actions, user autonomy, and responsibility.The privacy implications of social robots have been examined, reflecting the need to address the ethical considerations associ- ated with the deployment of these technologies (Lutz, Sch¨ottler, and Hoffmann 2019). This includes considerations related to autonomy, privacy by design, and agency, ne- cessitating a comprehensive understanding of the ethical implications associated with the deployment of social robots.The use of social robots for clinical practice has im- portant implications for social support and engagement, reflecting the need to address the ethical considerations associated with the deployment of these technologies (Zhang et al. 2021). This includes considerations related to the impact on social interaction, user autonomy, and responsibility.In conclusion, the ethical and social implications of robotics encompass a wide array of considerations, including privacy, autonomy, societal impact, and the transformation of social dynamics. Addressing these implica- tions is essential for ensuring the responsible deployment of robots and AI technologies across various domains.

FUTURE DIRECTIONS AND INNOVATIONS

The future roles of robots in society are poised to have a profound impact on various aspects of human life, ranging from labor and healthcare to social interaction and eth- ical considerations. As robots continue to evolve and integrate advanced technologies, their potential roles in society are expected to expand, presenting both opportunities and challenges. The potential roles of robots in future society encompass a wide ar- ray of applications, including co-worker, law enforcement personnel, war soldiers, and caregivers. The impact of these roles on society and the diverging behavior humans might exhibit is an area that requires thorough investigation Karnouskos (2021). As robots take on diverse roles, the ethical and societal implications of their integration into various domains will necessitate careful consideration and assessment.

The development of liquid metal transformable machines and nature-inspired soft robots is shaping the role models for future robots, with the rapid development of science and technology turning these ideas into reality (Wang et al. 2021). These ad- vancements are expected to influence the design and capabilities of future robots, impacting their potential roles in society. Debates about the future of society should consider the potential of cyborgs and robots replacing human beings, reflecting the need to address the ethical considerations associated with the potential displacement of human roles by robots (Fox 2018). This raises important questions about the so- cietal impact of robots and the ethical implications of their integration into various domains. Industrial robots, household robots, medical robots, assistive robots, socia- ble/entertainment robots, and war robots all play important roles in human life and raise crucial ethical problems for society (Tzafestas 2018). The diverse roles of robots in society underscore the need for a comprehensive understanding of the ethical im- plications associated with their deployment and integration into various domains.

The many potential applications and roles of robots, in combination with steady advances in their uptake within society, are expected to cause various unprecedented issues, increasing the demand for new policy measures (Pagter 2021). As robots take on more prominent roles in society, the need for new governance and regulatory frame- works to address the ethical and societal implications of their integration becomes increasingly important. Recent technological developments have led to an increased interest in the discourse on human-robot interactions, as robots equipped with ar- tificial intelligence are expected to take a more prominent role in society (Gonzalez- Jimenez 2018). The evolving roles of robots in society underscore the need for a deeper understanding of their impact and the ethical consid- erations associated with their in- tegration into various domains. The future roles of robots in society are expected to have a significant impact on various aspects of human life, including work, healthcare, and social interaction. The potential displacement of human roles by robots raises important ethical and societal considerations that necessitate careful examination and assessment.

CONCLUSION

This chapter has elucidated the transformative role of AI technologies in robotics, em- phasizing their pivotal role in enhancing robot efficiency, autonomy, and interaction capabilities across various domains. From enabling sophisticated perception and in- tricate motion planning to facilitating precise manipulations and robust human-robot collaborations, AI stands as a cornerstone in advancing robotic systems. The integra- tion of AI not only revolutionizes industrial applications but also propels the ongoing innovation within the field, pointing towards a future where robots are integral and versatile participants in both everyday tasks and complex operations. This evolution underscores the critical importance of continuing advancements in AI to meet growing demands and emerging challenges in robotics

REFERENCES

Alatabani, L. (2022). *Robotics architectures based machine learning and deep learning approaches.* CrossRef. .DOI: 10.1049/icp.2022.2274

Alvarez-Dionisi, L., Mittra, M., & Balza, R. (2019). Teaching artificial intelligence and robotics to undergraduate systems engineering students. *International Journal of Modern Education and Computer Science*, 11(7), 54–63. DOI: 10.5815/ijmecs.2019.07.06

Babko-Malaya, O. (2023). *Assigning semantic meaning to machine derived competency controlling topics.* SPIE. .DOI: 10.1117/12.2663821

Bai, Q., Li, S., Yang, J., Song, Q., Li, Z., & Zhang, X. (2020). Object detection recognition and robot grasping based on machine learning: A survey. *IEEE Access : Practical Innovations, Open Solutions*, 8, 181855–181879. DOI: 10.1109/ACCESS.2020.3028740

Beam, A. (2020). Challenges to the reproducibility of machine learning models in health care. *JAMA, 323*. . DOI: 10.1001/jama.2019.20866

Behera, T. (2023). The NITRDrone dataset to address the challenges for road extraction from aerial images. *Journal of Signal Process-ing Systems, 95*(2).

Cai´c, M. (2019). Value of social robots in services: social cognition perspective. *Journal of Services Marketing, 33*. .DOI: 10.1108/JSM-02-2018-0080

Cain, L., Thomas, J., & Alonso, M.Jr. (2019). From sci-fi to sci-fact: The state of robotics and ai in the hospitality industry. *Journal of Hospitality and Tourism Technology*, 10(4), 624–650. DOI: 10.1108/JHTT-07-2018-0066

Callaos, N., Cowin, J., Erkollar, A., & Oberer, B. (2023). Cybernetic Rela- tionships between Technological Innovations, Ethics, and the Law. Touro Scholar.

Carvalho, D., Pereira, E., & Cardoso, J. (2019). Machine learning interpretability: A survey on methods and metrics. *Electronics (Basel)*, 8(8), 832. DOI: 10.3390/electronics8080832

Casillo, M., & Colace, F. (2024). Securing Digital Ecosystems: Harnessing the Power of Intelligent Machines in a Secure and Sustainable Environment. In *Handbook of Research on AI and ML for Intelligent Machines and Systems*, (50–74). IGI Global.

Chanda, S., & Banerjee, D. (2022). Omission and commission errors underlying ai failures. *AI & Society*. DOI: 10.1007/s00146-022-01585-x PMID: 36415822

Chui, K. T., Gupta, B. B., Liu, J., Arya, V., Nedjah, N., Almomani, A., & Chaurasia, P. (2023). A survey of internet of things and cyber-physical systems: Standards, algorithms, applications, security, challenges, and future directions. *Information (Basel)*, 14(7), 388. DOI: 10.3390/info14070388

Cross, E., Hortensius, R., & Wykowska, A. (2019). From social brains to social robots: applying neurocognitive insights to human–robot interaction. *Philosophical Transactions of the Royal Society B Biological Sciences, 374*. ..DOI: 10.1098/rstb.2018.0024

Darapureddy, N. (2021). A comprehensive study on artificial intelligence and robotics for machine intelligence. In *Methodologies and Applica-tions of Computational Statistics for Machine Intelligence*, (203–222). IGI Global.

Elendu, C., Amaechi, D. C., Elendu, T. C., Jingwa, K. A., Okoye, O. K., John Okah, M., Ladele, J. A., Farah, A. H., & Alimi, H. A. (2023). Ethical implications of ai and robotics in healthcare: A review. *Medicine*, 102(50), e36671. DOI: 10.1097/MD.0000000000036671 PMID: 38115340

Fan, Z., Yang, H., Liu, F., Liu, L., & Han, Y. (2022). Reinforcement learning method for target hunting control of multi-robot systems with obstacles. *International Journal of Intelligent Systems*, 37(12), 11275–11298. DOI: 10.1002/int.23042

Fox, S. (2018). Cyborgs, robots and society: Implications for the future of soci- ety from human enhancement with in-the-body technologies. *Technologies*, 6(2), 50. DOI: 10.3390/technologies6020050

Gasparetto, A., & Scalera, L. (2019). From the unimate to the delta robot: The early decades of industrial robotics. *History of Mechanism and Machine Science*, 37, 284–295. DOI: 10.1007/978-3-030-03538-9_23

Gonzalez-Jimenez, H. (2018). Taking the fiction out of science fiction: (self-aware) robots and what they mean for society, retailers and marketers. *Futures*, 98, 49–56. DOI: 10.1016/j.futures.2018.01.004

Gupta, B. B., Gaurav, A., Chui, K. T., & Arya, V. (2023). Optimized Edge- cCCN Based Model for the Detection of DDoS Attack in IoT Environment. In *International Conference on Edge Computing*, (pp. 14–23). Springer.

Gupta, B. B., Gaurav, A., Chui, K. T., Arya, V., & Choi, C. (2024). Au- toencoders Based Optimized Deep Learning Model for the Detection of Cyber Attack in IoT Environment. In *2024 IEEE International Conference on Consumer Electronics (ICCE)*, (pp. 1–6). IEEE.

Gupta, S. (2021). *Internet of Things (IOT)*. Hands-On Sensing, Actuating, and Output Modules in Robotics.

Hart, P. E. (2022). An Artificial Intelligence Odyssey: From the Research Lab to the Real World. *IEEE Annals of the History of Computing*, 44(1), 57–72. DOI: 10.1109/MAHC.2021.3077417

Hassan, A., Biaggi, A., Asaad, M., Andejani, D., Li, J., Selber, J., & Butler, C. (2022a). Development and assessment of machine learning models for individualized risk as- sessment of mastectomy skin flap necrosis. *Annals of Surgery Open: Perspectives of Surgical History, Education, and Clinical Approaches*, 278, e123–e130. DOI: 10.1097/SLA.0000000000005386 PMID: 35129476

Hassan, A., Biaggi-Ondina, A., Rajesh, A., Asaad, M., Nelson, J., Coert, J., Mehrara, B., & Butler, C. (2022b). Predicting patient-reported outcomes following surgery using machine learning. *The American Surgeon*, 89(1), 31–35. DOI: 10.1177/00031348221109478 PMID: 35722685

Hu, B., Gaurav, A., Choi, C., & Almomani, A. (2022). Evaluation and com- parative analysis of semantic web-based strategies for enhancing educational system devel- opment. [IJSWIS]. *International Journal on Semantic Web and Information Systems*, 18(1), 1–14. DOI: 10.4018/IJSWIS.302895

Hu, H. (2024). *Position control of mobile robot based on deep reinforcement learning*. SPIE. .DOI: 10.1117/12.3024732

Husak, E. (2021). *Actuators in Service Robots*.

Jaakkola, H. (2019). *Artificial intelligence yesterday, today and tomorrow*. IEEE.

Jiang, H. (2021). Learning for a robot: deep reinforcement learning, imitation learning, transfer learning. *Sensors,21*. .DOI: 10.3390/s21041278

Karnouskos, S. (2021). Symbiosis with artificial intelligence via the prism of law, robots, and society. *Artificial Intelligence and Law*, 30(1), 93–115. DOI: 10.1007/s10506-021-09289-1

Khanam, S., Tanweer, S., & Khalid, S. S. (2022). Future of internet of things: Enhancing cloud-based iot using artificial intelligence. [IJCAC]. *International Journal of Cloud Applications and Computing*, 12(1), 1–23. DOI: 10.4018/IJCAC.297094

Kiran, M. A., Pasupuleti, S. K., & Eswari, R. (2022). Efficient pairing-free identity-based signcryption scheme for cloud-assisted iot. [IJCAC]. *International Journal of Cloud Applications and Computing*, 12(1), 1–15. DOI: 10.4018/IJCAC.305216

Kleinberg, J., Lakkaraju, H., Leskovec, J., Ludwig, J., & Mullainathan, S. (2017). Human decisions and machine predictions*. *The Quarterly Journal of Economics*. DOI: 10.1093/qje/qjx032 PMID: 29755141

Kogelis, M., Fuge, Z. J., Herron, C. W., Kalita, B., & Leonessa, A. (2022). *Design Of Low-Level Hardware For A Multi-Layered Control Architecture* (Vol. 5). ASME. DOI: 10.1115/IMECE2022-94614

Kono, H., Katayama, R., Takakuwa, Y., Wen, W., & Suzuki, T. (2019). Activation and spreading sequence for spreading activation policy selection method in transfer reinforce- ment learning. *International Journal of Advanced Computer Science and Applications*, 10(12). DOI: 10.14569/IJACSA.2019.0101202

Kose, T., & Sakata, I. (2017). *Identifying technology advancements and their linkages in the field of robotics research*. IEEE. DOI: 10.23919/PICMET.2017.8125283

Kumar, N. (2014). *High-temperature motors*. Research Gate.

Kumar, R., Singh, S. K., Lobiyal, D. K., Chui, K. T., Santaniello, D., & Rafsanjani, M. K. (2022). A novel decentralized group key management scheme for cloud-based vehicular IoT networks. [IJCAC]. *International Journal of Cloud Applications and Computing*, 12(1), 1–34. DOI: 10.4018/IJCAC.311037

Kunze, L., Hawes, N., Duckett, T., Hanheide, M., & Krajn'ık, T. (2018). Artificial intelligence for long-term robot autonomy: A survey. *IEEE Robotics and Automation Letters*, 3(4), 4023–4030. DOI: 10.1109/LRA.2018.2860628

Lee, M., & Yusuf, S. (2022). Mobile robot navigation using deep reinforcement learning. *Processes (Basel, Switzerland)*, 10(12), 2748. DOI: 10.3390/pr10122748

Liu, D., Wang, Z., Lu, B., Cong, M., Yu, H., & Zou, Q. (2020). A reinforcement learning- based framework for robot manipulation skill acquisition. *IEEE Access : Practical Innovations, Open Solutions*, 8, 108429–108437. DOI: 10.1109/ACCESS.2020.3001130

Lutz, C., Sch¨ottler, M., & Hoffmann, C. (2019). The privacy implications of social robots: Scoping review and expert interviews. *Mobile Media & Communication*, 7(3), 412–434. DOI: 10.1177/2050157919843961

Mackenzie, R. (2018). Sexbots: Customizing them to suit us versus an ethi- cal duty to created sentient beings to minimize suffering. *Robotics (Basel, Switzerland)*, 7(4), 70. DOI: 10.3390/robotics7040070

McGinn, C. & Kelly, K. (2015). *Towards an embodied system-level architecture for mobile robots.* Research Gate.

O'Sullivan, S., Nevejans, N., Allen, C., Blyth, A., L'eonard, S., Pagallo, U., Holzinger, K., Holzinger, A., Sajid, M., & Ashrafian, H. (2019). Legal, regulatory, and ethical frame- works for development of standards in artificial intelligence (ai) and autonomous robotic surgery. *International Journal of Medical Robotics and Computer Assisted Surgery*, 15(1), e1968. DOI: 10.1002/rcs.1968 PMID: 30397993

Pagter, J. (2021). Speculating about robot moral standing: On the constitution of social robots as objects of governance. *Frontiers in Robotics and AI*, 8, 769349. DOI: 10.3389/frobt.2021.769349 PMID: 34926591

Pasquinelli, M., & Joler, V. (2020). The nooscope manifested: Ai as instrument of knowledge extractivism. *AI & Society*, 36(4), 1263–1280. DOI: 10.1007/s00146-020-01097-6 PMID: 33250587

Pecora, F. (2014). "Is Model-Based Robot Programming a Mirage? A Brief Sur- vey of AI Reasoning in Robotics." *KI -. Kunstliche Intelligenz*, 28(4), 255–261. DOI: 10.1007/s13218-014-0325-0

Peeters, A., & Haselager, P. (2019). Designing virtuous sex robots. *International Journal of Social Robotics*, 13(1), 55–66. DOI: 10.1007/s12369-019-00592-1

Podlubne, A., & Gohringer, D. (2023). A Survey on Adaptive Computing in Robotics: Modelling, Methods and Applications. *IEEE Access : Practical Innovations, Open Solutions*, 11, 53830–53849. DOI: 10.1109/ACCESS.2023.3281190

Prasad, B., Kanojia, R., Mishra, P., Singh, P., & Rathi, V. (2024). *Ad- vancement of actuators in today's world.* AcSIR.

Premebida, C. (2019). *Intelligent robotic perception systems.* InTech Open. .DOI: 10.5772/intechopen.79742

Ramachandran, A. (2016). *Robot's Sensors and Instrumentation.*

Robertson, L., Alici, G., Mun~oz, A., & Michael, K. (2019). Engineering-based design method- ology for embedding ethics in autonomous robots. *Proceedings of the IEEE*, 107(3), 582–599. DOI: 10.1109/JPROC.2018.2889678

Rodriguez-Amat, J. & Duller, N. (2019). *Responsibility and resistance.* Springer. .DOI: 10.1007/978-3-658-26212-9

Sætra, H., & Danaher, J. (2022). To each technology its own ethics: The problem of ethical proliferation. *Philosophy & Technology*, 35(4), 93. DOI: 10.1007/s13347-022-00591-7

Samantaray, R. (2023). *AI and blockchain fundamentals.* IGI Global. .DOI: 10.4018/979-8-3693-0659-8.ch001

Shukla, J. (2017). *Effectiveness of so-cially assistive robotics during cognitive stimulation interventions: impact on caregivers.* IEEE. .DOI: 10.1109/ROMAN.2017.8172281

Sravya, P. (n.d.). Influence of Artificial Intelligence in Robotics. In *Artificial Intelligence and Knowledge Processing*. CRC Press.

T'oth, Z., Caruana, R., Gruber, T., & Loebbecke, C. (2022). The dawn of the ai robots: Towards a new framework of ai robot accountability. *Journal of Business Ethics*, 178(4), 895–916. DOI: 10.1007/s10551-022-05050-z

Tzafestas, S. (2018). Roboethics: Fundamental concepts and future prospects. *Information (Basel)*, 9(6), 148. DOI: 10.3390/info9060148

Wang, H., Chen, S., Yuan, B., Liu, J., & Sun, X. (2021). Liquid metal transformable machines. *Accounts of Materials Research*, 2(12), 1227–1238. DOI: 10.1021/accountsmr.1c00182

Wilcock, G. (2022). *Conversational ai and knowledge graphs for social robot interaction*. IEEE. .DOI: 10.1109/HRI53351.2022.9889583

Wilson, J., Tickle-Degnen, L., & Scheutz, M. (2020). Challenges in designing a fully au- tonomous socially assistive robot for people with parkinson's disease. *ACM Transactions on Human-Robot Interaction*, 9(3), 1–31. DOI: 10.1145/3379179

Xie, Y., Li, P., Nedjah, N., Gupta, B. B., Taniar, D., & Zhang, J. (2023). Pri- vacy protection framework for face recognition in edge-based Internet of Things. *Cluster Computing*, 26(5), 3017–3035. DOI: 10.1007/s10586-022-03808-8

Zhang, D., Shen, J., Li, S., Gao, K., & Gu, R. (2021). I, robot: Depression plays differ- ent roles in human–human and human–robot interactions. *Translational Psychiatry*, 11(1), 438. DOI: 10.1038/s41398-021-01567-5 PMID: 34420040

Zhang, Y., Al-Fuqaha, A., Humar, I., & Wan, J. (2022). Special Issue on Advanced Sensors and Sensing Technologies in Robotics. *IEEE Sensors Journal*, 22(18), 17334. DOI: 10.1109/JSEN.2022.3198783 PMID: 36346095

Chapter 3
Integration of IoT With Robotics and Drones

Brij B. Gupta
Asia University, Taiwan

Jinsong Wu
https://orcid.org/0000-0003-4720-5946
Universidad de Chile, Chile

ABSTRACT

This chapter explores the synergistic integration of the internet of things (IoT) with robotics and drones, highlighting how this convergence is revolutionizing industrial operations and capabilities. It delves into the mechanisms through which IoT devices and sensors enhance the autonomy, efficiency, and intelligence of robotic systems and drones, enabling real-time data exchange and analysis. The chapter discusses the implementation of IoT for advanced monitoring, predictive maintenance, and seamless operational control, illustrating its impact through practical examples across various sectors. It also addresses the challenges of scalability, security, and interoperability, presenting forward-looking strategies to navigate these hurdles. By emphasizing the transformative potential of IoT in augmenting the capabilities of robotics and drones, the chapter underscores the pivotal role of IoT in driving innovation and operational excellence in the digital age.

INTRODUCTION

The Internet of Things (IoT) is a transformative technology that integrates various do- mains such as sensor technology, embedded systems, computing, and communication, enabling the networking of devices to capture, process, and act on data autonomously (Swamy and Raju 2020; Tolcha et al. 2021; Chui et al. 2023; ?). IoT aims to embed communication capabilities within a highly distributed, ubiquitous, and dense network of heterogeneous devices, transforming real-world objects into smarter devices (Aman et al. 2020; Farooq et al. 2022). The ultimate goal of IoT is to generate intelligence from data, and it has been instrumental in various domains such as healthcare, envi- ronmental monitoring, and industrial control systems (Liu et al. 2019; Alsaedi et al. 2020; Xie et al. 2023). IoT has been widely

DOI: 10.4018/979-8-3693-2707-4.ch003

applied in smart cities, home environments, agriculture, industry, intelligent buildings, and more (Barriga et al. 2022; Casillo et al. 2024).

On the other hand, robotics is a field that involves the design, construction, op- eration, and use of robots to perform tasks in various industries and applications. It encompasses a wide range of technologies, including artificial intelligence, machine learning, computer vision, and mechanical engineering. Robotics has been reshaping various fields such as transportation, healthcare, manufacturing, household, and agri- culture (Guo et al. 2019). The integration of edge computing into low Earth orbit satellite networks has also been explored to support robotics applications that require large computing resources (Li et al. 2021; Gupta et al. 2023).

Drones, also known as unmanned aerial vehicles (UAVs), have gained significant attention due to their wide-ranging applications in surveillance, agriculture, disaster management, and product delivery. The use of IoT technology fused with 5G and arti- ficial intelligence has enabled the development of drones for applications in smart cities and smart factories (Jeon, Park, and Jeong 2020). Additionally, the implementation of a LoRa mesh library has facilitated low-power and long-range communications for drones in IoT applications (Sole et al. 2022; Hu et al. 2022).

The integration of IoT, robotics, and drones has significantly advanced automation and data exchange in various sectors, revolutionizing industries and enabling new capa- bilities. The convergence of these technologies has led to transformative changes in sec- tors such as healthcare, transportation, manufacturing, and agriculture.In the health- care sector, the integration of automation and data exchange through IoT, robotics, and drones has led to the development of closed-loop control systems for anesthe- sia, enabling anesthesiologists to provide increased attention to alarming situations and improve precision medicine Ghita et al. (2020); Gupta et al. (2024). Additionally, the use of blockchain-based distributed networks has facilitated patient information sharing, addressing privacy concerns and enhancing data sharing operational controls (Barbaria et al. 2022).

In the transportation sector, the integration of IoT, robotics, and drones has enabled trusted orchestration for smart decision-making in Internet of Vehicles (IoV), allowing vehicles to make quick decisions in real-time and enhancing communication between vehicles (Rathee et al. 2020; Khanam, Tanweer, and Khalid 2022). Furthermore, the development of vehicle automation and cooperative perception for connected and au- tomated vehicles has been facilitated, emphasizing the importance of an adequate configuration of congestion control and interdependence in communication (Domeyer, Lee, and Toyoda 2020; Thandavarayan, Sepulcre, and Gozalvez 2020; Komorkiewicz et al. 2023; Behera et al. 2023).

In the industrial sector, the integration of IoT, robotics, and drones has signifi- cantly impacted data exchange and automation in Industry 4.0. The interoperability of enterprise and control applications has been achieved through the Industry 4.0 As- set Administration Shell, enabling the conversion between different data formats and ensuring interoperable data exchange (Ye et al. 2022; Kumar et al. 2022). Moreover, data management in Industry 4.0 has been enhanced, addressing open challenges and pro- viding insights into the application of data management in networked industrial environments (Raptis, Passarella, and Conti 2019).

In the agriculture sector, the integration of IoT, robotics, and drones has led to the development of automated distribution grids and the application of new tech- nologies, emphasizing the fundamental role of communication network infrastructure for the correct functioning of automated electricity distribution grids (Chaves et al. 2022; Kiran, Pasupuleti, and Eswari 2022). Additionally, the development of intelli- gent robotic process automation has been instrumental in smart home systems, leveraging IoT-enabled automation for efficient data flow control and network load balancing (Vajgel et al. 2021; Jabbar et al. 2019; Weichlein et al. 2023).

Table 1. Overview of IoT, robotics, and drones in integration

Feature	IoT	Robotics		Drones	Integrated pact	Im-
Primary Function	Data capture and process- ing from networked devices	Design and op- eration of auto- mated machines		Unmanned aerial vehicle operations	Enhanced tomation real-time exchange	au- and data
Key Technolo- gies	Sensors, embedded systems, commu- nication networks	AI, machine learning, com- puter vision		GPS, real-time control systems	Combining technologies for complex operations	
Main Ap- plications	Smart cities, healthcare, industrial control	Manufacturing, healthcare, agriculture		Surveillance, agriculture, disaster man- agement	Cross-sectoral innovations and efficiency improvements	
Benefits	Increased efficiency, enhanced monitoring	Improved accu- racy, reduced human labor		Access to dif- ficult terrains, rapid response	Synergistic enhancements in operational capabilities	
Challenges	Security, privacy concerns	High complex tenance	costs, main-	Regulatory restrictions, safety concerns	Integration complexities, managing di- verse data streams	
Future Potential	Smart en- vironment interac- tions, ubiquitous connectiv- ity	Autonomous systems, ad- vanced human- robot collabora- tion		Expanded ap- plication in commercial and emergency contexts	Seamless and intelligent sys- tem interactions across sectors	

FUNDAMENTALS OF IOT, ROBOTICS, AND DRONES INTEGRATION

Conceptual Framework

Architectural Frameworks and Challenges

The architecture of IoT systems is pivotal in realizing the seamless integration of diverse devices and platforms. According to T Margaret Mary, A Sangamithra, and G Ramanathan (2021), IoT architecture is an ecosystem comprising multiple layers, each responsible for distinct functions ranging from data acquisition to application-specific processing.

This layered architecture, while facilitating modularity and scalability, introduces complex challenges, particularly in terms of interoperability and standardization.

Rashmi Kushwah et al. (2020) extend this discussion by identifying key architectural elements of IoT and their inherent challenges.

The study emphasizes the need for robust frameworks that can address issues related to data privacy, security, and the efficient management of resources across the IoT spectrum. Security Design Patterns

As IoT devices proliferate, securing these devices and their data becomes paramount. Bogdan-Cosmin Chifor et al. (2021) explore IoT cloud security design patterns, highlighting the critical role of security in ensuring the trustworthiness and reliability of IoT systems

The research presents design patterns that aim to mitigate security risks, thereby protecting data integrity and user privacy in cloud-based IoT environments.

Technological Convergence and IoT Evolution

The convergence of technologies such as artificial intelligence (AI), machine learning (ML), and blockchain with IoT architecture is steering the evolution of IoT towards more intelligent and autonomous systems. Brenda Solange Gomez Bustamante (2023) provides a concise review of how this amalgamation shapes the IoT architectures, enabling advanced capabilities like predictive analytics and secure, decentralized data management .

Survey on IoT Architectures

A survey by Fatima Zahra Fagroud (2022) offers a comprehensive overview of IoT architectures, focusing on the various layers and their respective challenges. This work contributes to a better understanding of the architectural complexities and provides a foundation for future research aimed at optimizing the design and functionality of IoT systems.

ENABLING TECHNOLOGIES AND ARCHITECTURES

Communication Protocols and Standards

Communication protocols for drones are crucial for ensuring secure and efficient data exchange between drones, ground stations, and other devices (Table 2). Several re- search works have focused on developing and analyzing communication protocols for drones, addressing various aspects such as security, authentication, throughput, and standardization. One approach proposed in the literature is the use of location-aware mutual authentication mechanisms for secure inter-drone and drone-to-ground commu- nication (Nair and Thampi 2023). This method leverages physical unclonable functions and Chebyshev chaotic maps to establish secure communication channels. Addition-ally, the use of lightweight hardware-based solutions has been suggested to secure communication between ground control stations (GCS) and drones (Koubˆaa et al. 2019). Furthermore, the implementation of the Advanced Encryption Standard (AES) protocol has been proposed to enhance the security of communication between GCS and drones (Allouch et al. 2019).

In terms of data communication protocols, there is a focus on enabling commu- nication between drones and the Internet of Things (IoT) technologies (Aldeen and Abdulhadi 2021). This highlights the importance of developing protocols that facili- tate seamless integration between drones and IoT devic- es, emphasizing the need for efficient and reliable data exchange. Moreover, the need for standardized communica- tion protocols for reporting methods when using drones for wildlife research has been emphasized (Barnas et al. 2020). This highlights the diverse applications of drones and the necessity for standardized protocols to ensure effective communication and dissemination of drone-related methods across various domains. Security is a critical aspect of drone communication protocols, especially in sensitive applications such as military zones. Research has emphasized the imperative focus on security standards for drone communication as their applications increase (Hassija 2021). Additionally, the development of proxy signature-based authentication for drones' secure device-to- device (D2D) communication compatible with 5G D2D ProSe standard mechanisms has been proposed (Abdel-Malek et al. 2022). This demonstrates the ongoing efforts to align drone communication protocols with the latest advancements in wireless commu- nication technologies. Furthermore, the potential of 5G technology to impact drones' communications has been highlighted, indicating the need to adapt communication protocols to leverage the capabilities of advanced wireless technologies (Lagkas et al. 2018).

Additionally, the use of cellular-connected UAVs and the challenges and promis- ing technologies associated with their communication have been discussed, emphasiz- ing the evolving landscape of drone communication protocols (Zeng, Lyu, and Zhang 2019).

Data Processing and Analytics

The role of AI and machine learning in data analysis for drones is pivotal in en- hancing various aspects of drone operations, including communication, security, and application-specific functionalities. Several studies have highlighted the significance of AI and machine learning in optimizing drone data analysis and enabling advanced capabilities.Machine learning has been proposed as a novel approach for data anal- ysis and modeling in drone communications, aiming to reduce the effort and cost of model establishment while capturing more detailed information about drone commu- nications Shan et al. (2019). This emphasizes the potential of machine learning in streamlining data analysis processes for drone communication systems.Furthermore, the fusion of blockchain and artificial intelligence has been suggested to secure drone communication, presenting a comprehensive survey of secure and intel- ligent drone communication architecture underlying 5G communication networks (Gupta, Kumari, and Tanwar 2020). This underscores the potential of AI in ensuring the security and intelligence of drone communication systems through innovative technological integra-

Table 2. Comparison of communication protocols for drones

Protocol	Security Features	Authentication Method	onThroughput	IoT Compatibility	Standardizatio	nApplications	References
Location-Aware Mutual Authentication	Uses physical unclonable functions and Chebyshev chaotic maps for secure channels	Mutual authentication	Moderate	Low	Emerging	Inter-drone and drone-to-ground communication	(Nair and Thampi 2023)
Lightweight Hardware-Based Solutions	Hardware-based security	Device-specific authentication	High	Moderate	Developed for specific devices	Ground control station to drone communication	(Koubˆaa et al. 2019)
Advanced Encryp-tion Stan-dard (AES)	High-level encryption	Standard cryptographic authentication	High	High	Widely standard-ized	Secure communication GCS and drones	(Allouch et al. 2019)
IoT Com-muni-cation Integra-tion	Moderate security tailored to IoT	Various IoT au-thentication mechanisms	Moderate to High	High	In develop-ment	Drone to IoT devices	(Aldeen and Abdul-hadi 2021)
Standard Proto-cols for Wildlife Research	Standard security protocols	Standardized authentication	Low to Moderate	Low	Highly standard-ized	Wildlife research and reporting	(Barnas et al. 2020)
Proxy Signature-Based D2D	Proxy signatures compatible with 5G	5G D2D ProSe standard mecha-nisms	Very High	High	Emerging for 5G	Military zones, sensitive applica-tions	(Abdel-Malek et al. 2022)
5G Tech-nology Impact	Enhanced security features with 5G	5G network authentication	Extremely High	Very High	Developing for next-gen net-works	High-demand applications	(Lagkas et al. 2018)
Cellular-Connected UAVs	Cellular network security	Cellular network authentication	High	High	Adapting to cellular standards	Broad range including urban envi-ronments	(Zeng, Lyu, and Zhang 2019)

tions.In the context of specific applications, AI-enabled drones have been utilized for distress pavement and pothole detection, healthcare monitoring, and livestock man-agement, leveraging AI, machine learning, and deep learning tools for data collection, analysis, and real-time decision-making processes (Wang and Ye 2022), (Alanezi et al. 2022). These applications demonstrate the diverse domains where AI and machine learning contribute to enhancing drone functionalities and data analysis capabilities. Moreover, the potential of AI and machine learning in precision agriculture has been emphasized, highlighting machine learning as the driving force behind cutting-edge technology in this domain (Sharma et al. 2021). This underscores the transformative impact of machine learning in optimizing agricultural drone operations and data anal-ysis for improved productivity and resource management.In the realm of security, the use of machine learning algorithms for drone authentication via acoustic fingerprint has been explored, achieving high accuracy in detecting drones in noisy environments (Diao et al. 2022). This showcases the potential of machine learning in enhancing security measures for drone operations through advanced authentication techniques. Additionally, the application of AI and machine learning in neuroscience-inspired ar- tificial intelligence has demonstrated the transformative impact of machine learning techniques in the analysis of neuroimaging datasets, promising expedited connectomic analysis and other advanced techniques (Hassabis et al. 2017). This highlights the in- terdisciplinary nature of AI and machine learning, extending their impact to diverse fields, including neuroscience and cognitive science.

Security and Privacy

Securing data exchange in drone networks presents several challenges as represented in Figure 1. These challenges necessitate the development and implementation of ro- bust strategies to ensure secure data exchange in drone networks. Several research works have addressed these challenges and proposed strategies to mitigate the associ- ated risks.One of the critical challenges in secure data exchange for drones is mutual authentication. Nair and Thampi (2023) discuss the development of a location-aware physical unclonable function and Chebyshev map-based mutual authentication mecha- nism for secure data exchange in drone networks. This approach focuses on establishing secure communication channels through mutual authentication protocols, addressing the challenge of verifying the identity of communi- cating entities. Furthermore, the generation of heavy computation and communication load for drones and other IoT devices poses a significant challenge to drone security (Majeed, Abdullah, and Mush- taq 2021). To address this, propose an IoT-based cybersecurity approach using the Naïve Bayes algorithm, aiming to optimize the computational and communication overhead while ensuring robust security mea- sures for drone networks. Handover oper- ations between controllers and the security of the exchanged data are also identified as challenges for implementing drones in various applications (Filho et al. 2023).

Strategies to address this challenge involve the development of secure handover architectures validated in a software-in-the-loop environment, emphasizing the need for seamless and secure data exchange during handover processes.To ensure secure and reliable communications in the Internet of Drone (IoD) environment, propose the Lightweight Authenticated Key Exchange (AKE) protocol (Tanveer et al. 2020). This strategy focuses on establishing secure and authenticated communication channels, addressing the challenge of reliable data exchange in drone networks.Security-related attacks, such as DoS attacks, Man-in-the-middle attacks, and De-Authentication at-

Figure 1. Drone security and privacy issues

tacks, pose significant threats to drone communication (Hassija 2021). Strategies to mitigate these challenges involve the development of comprehensive security proto- cols and mechanisms to detect and prevent such attacks, ensuring the integrity and availability of data exchange in drone networks.Privacy concerns and lack of trust col- laboration paradigm between drone controllers also present challenges to secure data exchange in drone networks (Liao et al. 2021).

Blockchain-enabled secure monitoring and collaboration frameworks have been pro- posed to address these challenges, emphasizing the importance of trust and privacy preservation in drone communication. Moreover, the use of machine learning and AI has been explored to diversify and improve the performance of RF sensors for drone de- tection mechanisms, addressing the evolving challenges encountered in the war against drone terrorism (Alsifiany 2023). This highlights the potential of advanced technologies in enhancing security measures for drone networks.

APPLICATIONS

Industrial Automation

Smart manufacturing and logistics are increasingly integrating drone technology to en- hance efficiency, automation, and data-driven decision-making processes. Drones play a pivotal role in revolutionizing manufacturing and logistics operations, offering ca- pabilities such as inventory management, predictive maintenance, intelligent sensors, and aerial delivery systems. The concept of self-similar architectures, as discussed by (Sprock 2018), presents an intriguing approach for designing smart manufacturing and logistics systems. This concept, rooted in fractal manufacturing systems, offers a unique perspective on the integration of drones within manufacturing and logis- tics environments, emphasizing the potential for innovative and scalable system ar- chitectures.Furthermore, the utilization of simulations for cyber-physical production systems, as highlighted by (Kim et al. 2020), underscores the significance of leverag- ing advanced technologies, including drones, to simulate and optimize manufacturing and logistics processes. This approach aligns with the trend towards smart manufac- turing, where simulations play a crucial role in enhancing operational efficiency and decision-making.The integration of drones in smart logistics management systems, as

proposed by (Chong et al. 2018), emphasizes the need for efficient and agile logistics operations in smart factories. Drones are positioned as key enablers for improving lo- gistics efficiency, accelerating manufacturing processes, and ensuring product quality, thereby contributing to the overall advancement of smart manufacturing and logis- tics.Moreover, the application of machine vision technology for intelligent logistics, as discussed by (He, Wang, and Liu 2022), highlights the role of drones in smart warehousing and distribution systems.

Machine vision, coupled with drone technology, offers opportunities for optimizing warehouse management, achieving lean logistics, and enhancing distribution processes within smart factories.The semantic interface model for integrating drones in a cyber- physical factory, as presented by .S.A et al. (2023), underscores the increasing pop- ularity of drones as intelligent logistics tools in manufacturing and other industries. This highlights the growing significance of drones in facilitating seamless inte- gration within cyber-physical systems, contributing to the advancement of smart manufactur- ing and logistics.Furthermore, the study by Ali et al. (2023) delves into the utilization of drones in smart ware- house management, emphasizing their socio-economic bene- fits and diverse applications in inventory management, intra-logistics, inspections, and surveillance. This underscores the multifaceted role of drones in enhancing operational efficiency and automation within smart manufacturing and logistics environments.The review by Pech, Vrchota, and Bednˇaˇr (2021) on predictive maintenance and intelligent sensors in smart factories aligns with the trend towards leveraging drones for proactive maintenance and sensor-based data collection. Drones equipped with intelligent sensors offer opportunities for real-time monitoring, predictive maintenance, and data-driven insights, contributing to the evolution of smart manufacturing and logistics.

4.2. Agriculture Drones have emerged as valuable tools in modern agriculture, offering a wide range of applications that contribute to improved efficiency, productivity, and sustainabil- ity. The integration of drone technology in agriculture has revolutionized traditional farming practices, enabling farmers to make data-driven decisions, optimize resource management, and enhance crop health. Several studies have highlighted the diverse applications of drones in agriculture, emphasizing their potential to transform the in- dustry.Drones are utilized in agriculture for various purposes, including mapping soil properties, assessing crop health, and monitoring livestock Otto et al. (2018). These ap- plications enable farmers to gain valuable insights into the condition of their fields and livestock, facilitating informed decision-making and proactive management practices. Additionally, drones play a crucial role in precision farming by determining the density of weed populations for site-specific herbicide treatments and optimizing the logistics of agricultural resources (Dutta et al. 2023). The ability of drones to provide real- time, high-resolution data contributes to the implementation of precision agriculture techniques, leading to improved crop yields and resource utilization.Furthermore, the use of drones for spraying pesticides and crop dusting has been extensively explored, offering a more efficient and targeted approach to pest and disease management in advanced agriculture (Borikar, Gharat, and Deshmukh 2022). Drones equipped with pesticide spraying systems can cover large areas with precision, reducing chemical us- age and minimizing environmental impact. Additionally, the application of drones in pest control and insect sampling has been identified as a promising area for further research and development in agriculture (Ryu, Clements, and Neufeld 2022).

Drones have the potential to revolutionize pest management practices by providing real-time monitoring and targeted interventions. The adoption of drones in German agriculture has been studied, highlighting the economic and environmental benefits that drones offer to farmers (Michels et al. 2021). The multi-functionality of drones provides opportunities for diverse applications in agricultural production, ranging from crop monitoring to environmental impact assessment. Moreover, the use of unmanned aerial vehicles in agriculture has been recognized as a means to contribute to the effi- cient management of agricultural farms, emphasizing the role of drones in optimizing farming operations (Berner and Chojnacki 2017).

In precision agriculture, drones equipped with advanced imaging technologies, such as multispec-tral and thermal cameras, have been employed for crop monitoring and analysis (Daponte et al. 2019). These imaging capabilities enable farmers to gather detailed information about crop health, water stress, and nutrient deficiencies, facil- itating targeted interventions and yield optimization. Additionally, the development of highly maneuverable drones for agricultural applications has been noted, signifying the progress made in leveraging drones for monitoring, aerial mapping, and precision agriculture (Kangunde, Jamisola, and Theophilus 2021).

References:

4.3. Environmental Monitoring and Conservation Wildlife tracking and habitat monitoring using drones have become increasingly preva- lent in ecological research and conservation efforts. Drones offer a versatile and non- intrusive means of observing wildlife, collecting data, and monitoring habi-tats, pro- viding valuable insights for wildlife management and environmental protection. The diverse applications of drones in wildlife tracking and habitat monitoring have been extensively explored in the literature, showcasing the potential of drone technology in advancing ecological research and conserva-tion practices. Studies such as Stark et al. (2017) and Rahman, Sitorus, and Condro (2021) have demon-strated the effectiveness of drones in informing policy change and conducting multi-species research in various ecosystems, including riparian habitats and montane forest ecosystems. These applica- tions highlight the versatility of drones in facilitating wildlife monitoring across diverse ecological settings, contributing to a comprehensive understanding of wildlife popu- lations and their habitats. The use of drones for census and population monitoring of wildlife, particularly in hard-to-access habitats, has been emphasized in research by Blight, Bertram, and Kroc (2019) and (Chabot and Bird 2015). Drones offer a non-intrusive and efficient approach to wildlife monitoring, enabling researchers to gather data on wildlife populations and habitat conditions in challeng- ing environments, where traditional survey methods may be impractical. Furthermore, the integration of drone-mounted thermal infrared cameras for monitoring animals, as discussed by Burke et al. (2019) and (Brunton, Leon, and Burnett 2020), has provided valuable insights into the behavior and distribution of wildlife. Advances in drone technolo-gy, particularly the use of thermal imaging, have enhanced the precision and effectiveness of wildlife monitoring, offering new opportunities for conservation and ecological research. The potential impact of drones on wildlife behavior and their ef- fectiveness in wildlife monitoring have been subjects of investigation in studies such as (Egan et al. 2020), (Hodgson et al. 2018), and (Lyons et al. 2018). These stud- ies have contributed to understanding the implications of drone use on wildlife and have provided insights into optimizing observing strategies for monitoring animals using drone-mounted cameras and sensors. Ethical considerations and guidelines for approaching wildlife with drones have also been addressed in the literature, as ev- idenced by studies such as Resnik and Elliott (2018) and (Vas et al. 2015). These studies underscore the importance of ethical and responsible use of drones in wildlife research, emphasizing the need to minimize disturbance and ensure the well-being of wildlife during drone-based monitoring activities.

4.4. Urban Planning and Infrastructure Management Drones have emerged as pivotal components in the development and implementation of smart city applications, offering a wide array of capabilities that contribute to the ef- ficiency, safety, and sustainability of urban environments. The integration of drones in smart cities has been extensively explored in the literature, showcasing their potential to revolutionize various aspects of urban management and infrastructure.The study by Mirzaeinia and Hassanalian (2019) highlights the role of drones in energy manage- ment and efficiency enhancement in smart cities. The authors discuss the application of drones for traffic monitoring, policing, package delivery, ambulance services, pol- lution control, and firefighting and rescue operations, emphasizing their multifaceted contributions to urban management and public safety.Alsamhi et al. (2019a) explore the connectivity issues related to drones for smart cities and discuss how drones en- hance smart city applications, such as tracking, surveillance, object detection, data collection, path planning, navigation, and collision prevention. This study underscores the diverse roles of drones in enhancing the smartness and operational efficiency of urban environments.The book by Dai and Nguyen (2021) provides a com- prehensive overview of security vulnerabilities and countermeasures for data communication in smart cities, emphasizing the importance of addressing security concerns related to drone applications. This highlights the critical role of drones in urban security and the need for robust cybersecurity measures to safeguard drone-enabled smart city op- erations.Ragab et al. (2022) discuss the use of unmanned aerial vehicles (UAVs), or drones, as smart city components that provide surveillance and smart mobility for citizens, underscoring their potential in enhancing urban monitoring and mobility in- frastructure. The study by Baig, Syed, and Mohammad (2022) emphasizes the increasing adop- tion of drones to serve smart cities through their ability to render quick and adaptive services, highlighting their potential to revolutionize urban service delivery and emer- gency response systems.Furthermore, Rawat et al. (2022) discuss the application of AI-based drones for addressing security concerns in smart cities, underscoring the potential of advanced technologies to enhance urban security and surveillance capa- bilities.The review by Alsamhi et al. (2019b) explores the greening of the Internet of Things (IoT) for smarter and more sustainable cities, emphasizing the potential of drones and IoT technologies to contribute to en- vironmental monitoring and sustain- ability initiatives in urban environments.The study by Menouar et al. (2017) delves into the applications and challenges of UAV-enabled intelligent transportation systems for smart cities, highlighting the transformative impact of drones on urban mobility and transportation infrastructure.

CHALLENGES AND SOLUTIONS

Technical Challenges

Drones, also known as unmanned aerial vehicles (UAVs), have gained significant atten- tion across various domains due to their diverse applications. However, the widespread adoption of drones is ac- companied by several technical challenges that need to be ad- dressed to ensure their effective and safe utilization. The following references provide insights into the technical challenges associated with drone technology:Filho et al. (2019) highlight major challenges for the use of drones in precision agriculture, in- cluding the costs of drones and associated sensors and materials, limited flight time and payload, and continuously changing regulations. These challenges impact the widespread adoption of drones in precision agriculture and emphasize the need for cost-effective and regulatory-compliant solutions.Umar (2020) discusses the techni- cal challenges associated with the use of drones for safety improvement, rating them as the second most important barrier. This underscores the significance of addressing technical hurdles to enhance the safety aspects of drone operations in various domains. Al-Ghaithi, Hamid, and Slimi (2021) and Aldeen and Abdulhadi (2021) mention sig- nificant technical challenges, including limited battery life, short coverable distance, limited payload capacity, physical collision, and privacy concerns. These challenges are critical considerations for the effective deployment of drones in various applications, such as delivery services, environmental monitoring, and Internet of Things (IoT) inte- gration.The study by Mozaffari et al. (2019) explores key technical challenges such as three-dimensional deployment, performance analysis, channel modeling, and energy efficiency in the context of unmanned aerial vehicles (UAVs) for wireless networks. These challenges are crucial for optimizing the performance and reliability of UAV-enabled communication systems.Furthermore, Coluccia, Parisi, and Fascista (2020) discuss the main challenges related to the problem of drone identification, including detection, possible verification, and classification. Addressing these challenges is es- sential for the accurate and reliable identification of drones in various scenarios, such as security and surveillance applications.In addition, technical challenges related to drone battery life are highlighted by (Ali 2023), which is fundamental for ensuring the sustained and efficient operation of drones in diverse environments and applications.

Regulatory and Ethical Considerations

The integration of drones in various domains raises significant ethical considerations that need to be carefully addressed to ensure responsible and beneficial use. The fol- lowing references provide insights into the ethical considerations associated with drone technology:Aldeen and Abdulhadi (2021) discuss the ethical challenges of drone imple- mentation in smart cities, emphasizing safety, privacy, and legal issues. This highlights the importance of addressing ethical concerns to ensure the responsible deployment of drones in urban environments.Sindiramutty (2024) scrutinizes the ethical intricacies of autonomous decision-making and the dual-use nature of drones in cybersecurity, shedding light on the ethical complexities associated with drone technology.Comtet and Johannessen (2022) emphasize the essential consideration of ethical questions in the integration of drones into health care systems, reflect- ing the significance of ethical awareness in the deployment of drones for health care applications.Wang, Christen, and Hunt (2021) and Wang et al. (2022) address the ethical challenges associated with human- itarian drone activities, providing frameworks and assessments to support value sensitivity and ethical decision-making in the humanitarian use of drones. The study by West and Bowman (2016) delves into the ethical ramifications of drone surveillance, emphasizing the need to consider ethical implications that are often over- looked in decision-making criteria.Molina et al. (2018) discuss the ethical implications of civil indoor drones, highlighting the importance of security and privacy by design for the ethical use of drones.The ethical considerations associated with drone war- fare are explored by Williams (Mitrea and Lecturer 2020), emphasizing the need for ethical assessment to understand the impact of drone tech- nology on warfare and hu- man–robot interaction.Strawbridge (2022) analyzes continuing ethical and sustainable issues related to drones, reviewing existing literature focused on ethical challenges and their recommended solutions.The study by Novitzky, Kokkeler, and Verbeek (2018) focuses on the dual-use of drones, aiming to better inform the responsible research and innovation (RRI) and value-sensitive design (VSD) of drones and drone technologies.

FUTURE TRENDS AND DIRECTIONS

The potential applications of drones continue to expand, offering opportunities for innovation and advancement across various domains. The following references provide insights into potential new domains for the application of drones:Bulusu, Aryafar, and Feng (2021) discuss the potential for networked drones to transform various application domains, particularly in indoor and forest environments. The lack of accurate maps and autonomous navigation abilities in the absence of GPS, reliable wireless commu- nications, and visually inferring environments are highlighted as challenges that need to be addressed for the adoption of drones in these environments.Jasim, Kasim, and Mahmoud (2022) empha- size the potential applications of drones in various domains, including military, entertainment, services, environmental monitoring, and security. This highlights the versatility of drones and their potential to revolutionize diverse in- dustries and services.Johal et al. (2022) envision the application of "social drones" in education, suggesting the potential for drones to serve as educational tools in academic set- tings.Maghazei (2022) and and Netland Maghazei and Netland (2019) explore the potential applications of drones in the manufacturing industry, examining the benefits, challenges, and research opportunities associated with the use of drones in manufac- turing processes.Gupta, Jain, and Vaszkun (2016) discuss the enormous potential of unmanned aerial vehicles (UAVs) in public and civil domains, highlighting the wide- ranging applications of drones in various sectors.Siju, Shafiyia, and Maaouia (2022) conduct a systematic literature review to study the potential utilization of drones in the construction industry, aiming to understand the benefits and impacts of drones as a new trend in construction projects.Poikonen and Campbell (2020) discuss future di- rections in drone routing research, exploring opportunities for better modeling of drone capabilities, constraints on drone performance, and alternative delivery modes, as well as methodological advances and new applications.Chandhar and Larsson (2019) focus on the potential of massive MIMO for connectivity with drones, highlighting the need for scalable technology to meet future demands on connectivity ranges and support for fast-moving drones.Alhussan et al. (2022) aim to develop a high abstract model for the drone forensic domain, addressing the diverse and complex nature of drone field standards, operating systems, and infrastructure-based networks.Agrawal et al. (2020) co-design an emergency response system, emphasizing the potential for human- drone partnerships in achieving situation awareness in multi-stakeholder, multi-UAV emergency response applications.

CONCLUSION

The integration of IoT with robotics and drones represents a transformative leap for- ward in technology, enabling enhanced connectivity, improved data collection, and more autonomous operations. By harnessing the power of the IoT, robots and drones can communicate more efficiently, share data in real-time, and execute complex tasks with greater precision and reliability. This synergy not only enhances the capabilities of individual machines but also fosters a more interconnected and intelligent techno- logical ecosystem. As this integration deepens, it will pave the way for innovations across industries, including agriculture, healthcare, and logistics, ultimately leading to smarter, more efficient systems that could reshape our approach to challenges and opportunities in the digital age.

ACKNOWLEDGMENT

This research work is supported by National Science and Technology Council (NSTC), Taiwan Grant No. NSTC112-2221-E-468-008-MY3.

REFERENCES

Abdel-Malek, M., Akkaya, K., Bhuyan, A., & Ibrahim, A. (2022). A proxy signature-based swarm drone authentication with leader selection in 5g networks. *IEEE Access : Practical Innovations, Open Solutions*, 10, 57485–57498. DOI: 10.1109/ACCESS.2022.3178121

Agrawal, A. (2020). *The next generation of human-drone partnerships: co-designing an emergency response system*. ACM. .DOI: 10.1145/3313831.3376825

Al-Ghaithi, R., Hamid, A., & Slimi, Z. (2021). Drone delivery efficiency, challenges, and poten- tial in oman during covid-19. *Journal of University of Shanghai for Science and Technology*, 23(7), 811–830. DOI: 10.51201/JUSST/21/07214

Alanezi, M., Shahriar, M., Hasan, M., Ahmed, S., Sha'aban, Y., & Bouchekara, H. (2022). Live- stock management with unmanned aerial vehicles: A review. *IEEE Access : Practical Innovations, Open Solutions*, 10, 45001–45028. DOI: 10.1109/ACCESS.2022.3168295

Aldeen, Y., & Abdulhadi, H. (2021). Data communication for drone-enabled internet of things. *Indonesian Journal of Electrical Engineering and Computer Science*, 22, 1216. DOI: 10.11591/ijeecs.v22.i2.pp1216-1222

Alhussan, A., Al-Dhaqm, A., Yafooz, W., Razak, S., Emara, A., & Khafaga, D. (2022). Towards development of a high abstract model for drone forensic domain. *Electronics (Basel)*, 11(8), 1168. DOI: 10.3390/electronics11081168

Ali, M., Jamaludin, J., Ahmedy, I., & Awalin, L. J. (2023). Energy performance review of battery-powered drones for search and rescue (sar) operations. *IOP Conference Series. Earth and Environmental Science*, 1261(1), 012021. DOI: 10.1088/1755-1315/1261/1/012021

Ali, S., Khan, S., Fatma, N., Ozel, C., & Hussain, A. (2023). Utilisation of drones in achiev-ing various applications in smart warehouse management. *Benchmarking*, 31(3), 920–954. DOI: 10.1108/BIJ-01-2023-0039

Allouch, A. (2019). *Mavsec: securing the mavlink protocol for ardupilot/px4 unmanned aerial systems*. IEEE. .DOI: 10.1109/IWCMC.2019.8766667

Alsaedi, A., Moustafa, N., Tari, Z., Mahmood, A., & Anwar, A. (2020). Ton iot telemetry dataset: A new generation dataset of iot and iiot for data-driven intrusion detection systems. *IEEE Access : Practical Innovations, Open Solutions*, 8, 165130–165150. DOI: 10.1109/ACCESS.2020.3022862

Alsamhi, S., Ma, O., Ansari, M., & Almalki, F. (2019a). Survey on collaborative smart drones and internet of things for improving smartness of smart cities. *IEEE Access : Practical Innovations, Open Solutions*, 7, 128125–128152. DOI: 10.1109/ACCESS.2019.2934998

Alsamhi, S., Ma, O., Ansari, M., & Meng, Q. (2019b). Greening internet of things for greener and smarter cities: A survey and future prospects. *Telecommunication Systems*, 72(4), 609–632. DOI: 10.1007/s11235-019-00597-1

Alsifiany, F. (2023). *Use of ai to diversify and improve the performance of rf sensors drone detection mechanism*. IEEE. .DOI: 10.5121/csit.2023.130504

Aman, A., Yadegaridehkordi, E., Attarbashi, Z., Hassan, R., & Park, Y. (2020). A survey on trend and classification of internet of things reviews. *IEEE Access : Practical Innovations, Open Solutions*, 8, 111763–111782. DOI: 10.1109/ACCESS.2020.3002932

Baig, Z., Syed, N., & Mohammad, N. (2022). Securing the smart city airspace: Drone cyber attack detection through machine learning. *Future Internet*, 14(7), 205. DOI: 10.3390/fi14070205

Barbaria, S., Mont, M., Ghadafi, E., Mahjoubi Machraoui, H., & Rahmouni, H. B. (2022). Leveraging patient information sharing using blockchain-based distributed networks. *IEEE Access : Practical Innovations, Open Solutions*, 10, 106334–106351. DOI: 10.1109/ACCESS.2022.3206046

Barnas, A., Chabot, D., Hodgson, A., Johnston, D., Bird, D., & Ellis-Felege, S. (2020). A stan- dardized protocol for reporting methods when using drones for wildlife research. *Journal of Unmanned Vehicle Systems*, 8(2), 89–98. DOI: 10.1139/juvs-2019-0011

Barriga, J., Clemente, P., Hern'andez, J., & P'erez-Toledano, M. (2022). Simulateiot-fiware: Domain specific language to design, code generation and execute iot simulation environments on fiware. *IEEE Access : Practical Innovations, Open Solutions*, 10, 7800–7822. DOI: 10.1109/ACCESS.2022.3142894

Behera, T. K., Bakshi, S., Sa, P. K., Nappi, M., Castiglione, A., Vijayakumar, P., & Gupta, B. B. (2023). The NITRDrone dataset to address the challenges for road extraction from aerial images. *Journal of Signal Processing Systems for Signal, Image, and Video Technology*, 95(2), 197–209. DOI: 10.1007/s11265-022-01777-0

Blight, L., Bertram, D., & Kroc, E. (2019). Evaluating uav-based techniques to census an urban-nesting gull population on canada's pacific coast. *Journal of Unmanned Vehicle Systems*, 7(4), 312–324. DOI: 10.1139/juvs-2019-0005

Borikar, G., Gharat, C., & Deshmukh, S. (2022). Application of drone systems for spraying pesticides in advanced agriculture: A review. *IOP Conference Series. Materials Science and Engineering*, 1259(1), 012015. DOI: 10.1088/1757-899X/1259/1/012015

Brunton, E., Leon, J., & Burnett, S. (2020). Evaluating the efficacy and optimal deployment of thermal infrared and true-colour imaging when using drones for monitoring kangaroos. *Drones (Basel)*, 4(2), 20. DOI: 10.3390/drones4020020

Bulusu, N. (2021). *Towards adaptive, self-configuring networked unmanned aerial vehicles*. ACM. .DOI: 10.1145/3469259.3470488

Burke, C., Rashman, M., Wich, S., Symons, A., Theron, C., & Longmore, S. (2019). Optimizing observing strategies for monitoring animals using drone-mounted thermal infrared cameras. *International Journal of Remote Sensing*, 40(2), 439–467. DOI: 10.1080/01431161.2018.1558372

Casillo, M., & Colace, F. (2024). Securing Digital Ecosystems: Harnessing the Power of Intelligent Machines in a Secure and Sustainable Environment. In *Handbook of Research on AI and ML for Intelligent Machines and Systems*. IGI Global.

Chabot, D., & Bird, D. (2015). Wildlife research and management methods in the 21st century: Where do unmanned aircraft fit in? *Journal of Unmanned Vehicle Systems*, 3(4), 137–155. DOI: 10.1139/juvs-2015-0021

Chandhar, P., & Larsson, E. (2019). Massive mimo for connectivity with drones: Case studies and future directions. *IEEE Access : Practical Innovations, Open Solutions*, 7, 94676–94691. DOI: 10.1109/ACCESS.2019.2928764

Chaves, T., Martins, M., Martins, K., & Macedo, A. (2022). Development of an automated distribution grid with the application of new technologies. *IEEE Access : Practical Innovations, Open Solutions*, 10, 9431–9445. DOI: 10.1109/ACCESS.2022.3142683

Chong, Z., Low, C., Mohammad, U., Rahman, R., & Shaari, M. (2018). Conception of logistics management system for smart factory. *IACSIT International Journal of Engineering and Technology*, 7(4.27), 126. DOI: 10.14419/ijet.v7i4.27.22499

Chui, K. T., Gupta, B. B., Liu, J., Arya, V., Nedjah, N., Almomani, A., & Chaurasia, P. (2023). A survey of internet of things and cyber-physical systems: Standards, algorithms, applications, security, challenges, and future directions. *Information (Basel)*, 14(7), 388. DOI: 10.3390/info14070388

Coluccia, A. (2020). Detection and classification of multirotor drones in radar sensor networks: a review. *Sensors,20*, 4172. .DOI: 10.3390/s20154172

Dai, N. (2021). *Drone application in smart cities: the general overview of security vulnerabilities and countermeasures for data communication*. Springer. DOI: 10.1007/978-3-030-63339-4

Daponte, P., Vito, L., Glielmo, L., Iannelli, L., Liuzza, D., Picariello, F., & Silano, G. (2019). A review on the use of drones for precision agriculture. *IOP Conference Series. Earth and Environmental Science*, 275(1), 012022. DOI: 10.1088/1755-1315/275/1/012022

Diao, Y. (2022). *Drone authentication via acoustic fingerprint*. ACM. .DOI: 10.1145/3564625.3564653

Domeyer, J., Lee, J., & Toyoda, H. (2020). Vehicle automation–other road user com- munication and coordination: Theory and mechanisms. *IEEE Access : Practical Innovations, Open Solutions*, 8, 19860–19872. DOI: 10.1109/ACCESS.2020.2969233

Dutta, S. (2023). *Perspective chapter: digital inclusion of the farming sector using drone technology*. InTech Open. .DOI: 10.5772/intechopen.108740

Egan, C., Blackwell, B., Fern'andez-Juricic, E., & Klug, P. (2020). Testing a key assumption of using drones as frightening devices: Do birds perceive drones as risky? *The Condor*, 122(3), duaa014. DOI: 10.1093/condor/duaa014

Farooq, M., Sohail, O., Abid, A., & Rasheed, S. (2022). A survey on the role of iot in agri- culture for the implementation of smart livestock environment. *IEEE Access : Practical Innovations, Open Solutions*, 10, 9483–9505. DOI: 10.1109/ACCESS.2022.3142848

Filho, E., Gomes, F., Monteiro, S., Severino, R., Penna, S., Koubaa, A., & Tovar, E. (2023). A drone secure handover architecture validated in a software in the loop environ- ment. *Journal of Physics: Conference Series*, 2526(1), 012083. DOI: 10.1088/1742-6596/2526/1/012083

Filho, F., Heldens, W., Kong, Z., & Lange, E. (2019). Drones: Innovative technology for use in precision pest management. *Journal of Economic Entomology*, 113(1), 1–25. DOI: 10.1093/jee/toz268 PMID: 31811713

Ghita, M., Neckebroek, M., Muresan, C., & Copot, D. (2020). Closed-loop control of anesthe- sia: Survey on actual trends, challenges and perspectives. *IEEE Access : Practical Innovations, Open Solutions*, 8, 206264–206279. DOI: 10.1109/ACCESS.2020.3037725

Guo, X., Zeng, T., Wang, Y., & Jie, Z. (2019). Fuzzy topsis approaches for assess- ing the intelligence level of iot-based tourist attractions. *IEEE Access : Practical Innovations, Open Solutions*, 7, 1195–1207. DOI: 10.1109/ACCESS.2018.2881339

Gupta, B. B., Gaurav, A., Chui, K. T., & Arya, V. (2023). Optimized Edge- cCCN Based Model for the Detection of DDoS Attack in IoT Environment. In *International Conference on Edge Computing*. Springer.

Gupta, B. B., Gaurav, A., Chui, K. T., Arya, V., & Choi, C. (2024). Au- toencoders Based Optimized Deep Learning Model for the Detection of Cyber Attack in IoT Environment. In *2024 IEEE International Conference on Consumer Electronics (ICCE)*. IEEE.

Gupta, L., Jain, R., & Vaszkun, G. (2016). Survey of important issues in uav communication networks. *IEEE Communications Surveys and Tutorials*, 18(2), 1123–1152. DOI: 10.1109/COMST.2015.2495297

Gupta, R., Kumari, A., & Tanwar, S. (2020). Fusion of blockchain and artificial intelligence for secure drone networking underlying 5g communications. *Transactions on Emerging Telecommunications Tech- nologies*, 32(1), e4176. DOI: 10.1002/ett.4176

Hassabis, D., Kumaran, D., Summerfield, C., & Botvinick, M. (2017). Neuroscience-inspired artificial intelligence. *Neuron*, 95(2), 245–258. DOI: 10.1016/j.neuron.2017.06.011 PMID: 28728020

Hassija, V. (2021). *Fast, reliable, and secure drone communication: a comprehensive survey*. https://doi .org//arxiv.2105.01347.DOI: 10.48550

He, S., Wang, Y., & Liu, H. (2022). Image information recognition and classification of ware- housed goods in intelligent logistics based on machine vision technology. *TS. Traitement du Signal*, 39(4), 1275–1282. DOI: 10.18280/ts.390420

Hodgson, J., Mott, R., Baylis, S., Pham, T., Wotherspoon, S., Kilpatrick, A., Segaran, R., Reid, I., Ter- auds, A., & Koh, L. (2018). Drones count wildlife more accurately and precisely than humans. *Methods in Ecology and Evolution*, 9(5), 1160–1167. DOI: 10.1111/2041-210X.12974

Hu, B., Gaurav, A., Choi, C., & Almomani, A. (2022). Evaluation and com- parative analysis of semantic web-based strategies for enhancing educational system devel- opment. [IJSWIS]. *International Journal on Semantic Web and Information Systems*, 18(1), 1–14. DOI: 10.4018/IJSWIS.302895

Jeon, J., Park, J., & Jeong, Y. (2020). Dynamic analysis for iot malware de- tection with convolution neural network model. *IEEE Access : Practical Innovations, Open Solutions*, 8, 96899–96911. DOI: 10.1109/ACCESS.2020.2995887

Johal, W., Gatos, D., Yanta̧c, A., & Obaid, M. (2022). Envisioning social drones in education. *Frontiers in Robotics and AI*, 9, 666736. DOI: 10.3389/frobt.2022.666736 PMID: 36093212

Kangunde, V., Jamisola, R.Jr, & Theophilus, E. (2021). A review on drones con- trolled in real-time. *International Journal of Dynamics and Control*, 9(4), 1832–1846. DOI: 10.1007/s40435-020-00737-5 PMID: 33425650

Khanam, S., Tanweer, S., & Khalid, S. S. (2022). Future of internet of things: Enhancing cloud-based iot using artificial intelligence. [IJCAC]. *International Journal of Cloud Applications and Computing*, 12(1), 1–23. DOI: 10.4018/IJCAC.297094

Kim, B., Nam, S., Jin, Y., & Seo, K. (2020). Simulation framework for cyber-physical production system: Applying concept of lvc interoperation. *Complexity*, 2020, 1–11. DOI: 10.1155/2020/4321873

Kiran, M. A., Pasupuleti, S. K., & Eswari, R. (2022). Efficient pairing-free identity-based signcryption scheme for cloud-assisted iot. [IJCAC]. *International Journal of Cloud Applications and Computing*, 12(1), 1–15. DOI: 10.4018/IJCAC.305216

Komorkiewicz, M., Chin, A., Skruch, P., & Szelest, M. (2023). Intelligent data handling in current and next-generation automated vehicle development—A review. *IEEE Access : Practical Innovations, Open Solutions*, 11, 32061–32072. DOI: 10.1109/ACCESS.2023.3258623

Koub^aa, A., Allouch, A., Alajlan, M., Javed, Y., Belghith, A., & Khalgui, M. (2019). Mi- cro air ve- hicle link (mavlink) in a nutshell: A survey. *IEEE Access : Practical Innovations, Open Solutions*, 7, 87658–87680. DOI: 10.1109/ACCESS.2019.2924410

Kumar, R., Singh, S. K., Lobiyal, D. K., Chui, K. T., Santaniello, D., & Rafsanjani, M. K. (2022). A novel decentralized group key management scheme for cloud-based vehicular IoT networks. [IJCAC]. *International Journal of Cloud Applications and Computing*, 12(1), 1–34. DOI: 10.4018/IJCAC.311037

Lagkas, T., Argyriou, V., Bibi, S., & Sarigiannidis, P. (2018). Uav iot framework views and challenges: Towards protecting drones as "things". *Sensors (Basel)*, 18(11), 4015. DOI: 10.3390/s18114015 PMID: 30453646

Li, C., Zhang, Y., Xie, R., Hao, X., & Huang, T. (2021). Integrating edge computing into low earth orbit satellite networks: Architecture and prototype. *IEEE Access : Practical Innovations, Open Solutions*, 9, 39126–39137. DOI: 10.1109/ACCESS.2021.3064397

Liao, S., Wu, J., Li, J., Bashir, A., & Yang, W. (2021). Securing collaborative environment monitoring in smart cities using blockchain enabled software-defined internet of drones. *Ieee Internet of Things Magazine*, 4(1), 12–18. DOI: 10.1109/IOTM.0011.2000045

Liu, Y., Hassan, K., Karlsson, M., Pang, Z., & Gong, S. (2019). A data-centric in- ternet of things framework based on azure cloud. *IEEE Access : Practical Innovations, Open Solutions*, 7, 53839–53858. DOI: 10.1109/ACCESS.2019.2913224

Lyons, M., Brandis, K., Callaghan, C., McCann, J., Mills, C., Ryall, S., & Kingsford, R. (2018). Bird interactions with drones, from individuals to large colonies. *Australian Field Ornithology*, 35, 51–56. DOI: 10.20938/afo35051056

Maghazei, O. (2022). *Drones in manufacturing: opportunities and challenges*. IEOM Society. .DOI: 10.46254/EU05.20220073

Maghazei, O., & Netland, T. (2019). Drones in manufacturing: Exploring opportunities for research and practice. *Journal of Manufacturing Technology Management*, 31(6), 1237–1259. DOI: 10.1108/JMTM-03-2019-0099

Majeed, R., Abdullah, N., & Mushtaq, M. (2021). Iot-based cyber-security of drones using the naïve bayes algorithm. *International Journal of Advanced Computer Science and Applications*, 12(7). DOI: 10.14569/IJACSA.2021.0120748

Menouar, H., Gu¨ven¸c, I. ˙., Akkaya, K., Uluagac, A., Kadri, A., & Tuncer, A. (2017). Uav-enabled intelligent transportation systems for the smart city: Applications and challenges. *IEEE Communications Magazine*, 55(3), 22–28. DOI: 10.1109/MCOM.2017.1600238CM

Michels, M., Hobe, C., Ahlefeld, P., & Mußhoff, O. (2021). The adoption of drones in german agriculture: A structural equation model. *Precision Agriculture*, 22(6), 1728–1748. DOI: 10.1007/s11119-021-09809-8

Mirzaeinia, A., & Hassanalian, M. (2019). Minimum-cost drone–nest matching through the kuhn–munkres algorithm in smart cities: Energy management and efficiency enhancement. *Aerospace (Basel, Switzerland)*, 6(11), 125. DOI: 10.3390/aerospace6110125

Mitrea, G., & Lecturer, S. (2020). Drones - ethical and legal issues in civil and military research as a future opportunity. *JESS*, 4(1), 83–98. DOI: 10.18662/jess/4.1/30

Molina, M., Campos, V., Montagud, M., & Molina, B. (2018). Ethics for civil indoor drones: A qualitative analysis. *International Journal of Micro Air Vehicles*, 10(4), 340–351. DOI: 10.1177/1756829318794004

Mozaffari, M., Saad, W., Bennis, M., Nam, Y., & Debbah, M. (2019). A tutorial on uavs for wireless networks: Applications, challenges, and open problems. *IEEE Communications Surveys and Tutorials*, 21(3), 2334–2360. DOI: 10.1109/COMST.2019.2902862

Nair, A., & Thampi, S. (2023). A location-aware physical unclonable function and chebyshev map-based mutual authentication mechanism for internet of surveillance drones. *Concurrency and Computation*, 35(19), e7564. DOI: 10.1002/cpe.7564

Novitzky, P., Kokkeler, B., & Verbeek, P. (2018). The dual-use of drones. *Tijdschrift Voor Veiligheid*, 17(1-2), 79–95. DOI: 10.5553/TvV/187279482018017102007

Otto, A., Agatz, N., Campbell, J., Golden, B., & Pesch, E. (2018). Optimization approaches for civil applications of unmanned aerial vehicles (uavs) or aerial drones: A survey. *Networks*, 72(4), 411–458. DOI: 10.1002/net.21818

Pech, M., Vrchota, J., & Bednář, J. (2021). Predictive maintenance and intelligent sensors in smart factory: Review. *Sensors (Basel)*, 21(4), 1470. DOI: 10.3390/s21041470 PMID: 33672479

Poikonen, S., & Campbell, J. (2020). Future directions in drone routing research. *Networks*, 77(1), 116–126. DOI: 10.1002/net.21982

Ragab, M. (2022). A drones optimal path planning based on swarm intelligence algorithms. *Computers, Materials & Continua*, 72, 365–380. DOI: 10.32604/cmc.2022.024932

Rahman, D., Sitorus, A., & Condro, A. (2021). From coastal to montane forest ecosystems, using drones for multi-species research in the tropics. *Drones (Basel)*, 6(1), 6. DOI: 10.3390/drones6010006

Raptis, T., Passarella, A., & Conti, M. (2019). Data management in indus-try 4.0: State of the art and open challenges. *IEEE Access : Practical Innovations, Open Solutions*, 7, 97052–97093. DOI: 10.1109/ACCESS.2019.2929296

Rathee, G., Garg, S., Kaddoum, G., Choi, B., & Hossain, M. (2020). Trusted orchestra- tion for smart decision-making in internet of vehicles. *IEEE Access : Practical Innovations, Open Solutions*, 8, 157427–157436. DOI: 10.1109/ACCESS.2020.3019795

Rawat, B., Bist, A., Apriani, D., Permadi, N., & Nabila, E. (2022). Ai based drones for se- curity concerns in smart cities. *Aptisi Transactions on Management (Atm)*, 7(2), 125–130. DOI: 10.33050/atm.v7i2.1834

Resnik, D., & Elliott, K. (2018). Using drones to study human beings: Ethical and regulatory issues. *Science and Engineering Ethics*, 25(3), 707–718. DOI: 10.1007/s11948-018-0032-6 PMID: 29488061

Ryu, J., Clements, J., & Neufeld, J. (2022). Low-cost live insect scouting drone: Idrone bee. *Journal of Insect Science*, 22(4), 5. DOI: 10.1093/jisesa/ieac036 PMID: 35793373

Shan, L., Kagawa, T., Ono, F., Li, H., & Kojima, F. (2019). Machine learning-based field data analy-sis and modeling for drone communications. *IEEE Access : Practical Innovations, Open Solutions*, 7, 79127–79135. DOI: 10.1109/ACCESS.2019.2922544

Sharma, A., Jain, A., Gupta, P., & Chowdary, V. (2021). Machine learning applica- tions for precision agriculture: A comprehensive review. *IEEE Access : Practical Innovations, Open Solutions*, 9, 4843–4873. DOI: 10.1109/ACCESS.2020.3048415

Siju, N. (2022). Applications of drone technology in construction projects: a systematic literature re-view. *International Journal of Research -Granthaalayah, 10*, 1–14. .DOI: 10.29121/granthaalayah.v10.i10.2022.4810

Sindiramutty, S. (2024). *Eyes in the sky*. IGI Global. .DOI: 10.4018/979-8-3693-0774-8.ch017

Sole, J., Centelles, R., Freitag, F., & Meseguer, R. (2022). Implementation of a lora mesh library. *IEEE Access : Practical Innovations, Open Solutions*, 10, 113158–113171. DOI: 10.1109/ACCESS.2022.3217215

Sprock, T. (2018). *Self-similar architectures for smart manufacturing and logistics systems.*

Stark, D., Vaughan, I., Evans, L., Kler, H., & Goossens, B. (2017). Combining drones and satellite tracking as an effective tool for informing policy change in riparian habitats: A proboscis monkey case study. *Remote Sensing in Ecology and Conservation*, 4(1), 44–52. DOI: 10.1002/rse2.51

Strawbridge, D. (2022). Civil drone ethics and sustainability. *Proceedings of the Wellington Faculty of Engineering Ethics and Sustainability Symposium*. Victoria University. DOI: 10.26686/wfeess.vi.7660

Swamy, S., & Raju, K. (2020). An empirical study on system level aspects of internet of things (iot). *IEEE Access : Practical Innovations, Open Solutions*, 8, 188082–188134. DOI: 10.1109/ACCESS.2020.3029847

Tanveer, M., Zahid, A., Ahmad, M., Baz, A., & Alhakami, H. (2020). Lake-iod: Lightweight authenticated key exchange protocol for the internet of drone environment. *IEEE Access : Practical Innovations, Open Solutions*, 8, 155645–155659. DOI: 10.1109/ACCESS.2020.3019367

Thandavarayan, G., Sepulcre, M., & Gozalvez, J. (2020). Cooperative perception for connected and automated vehicles: Evaluation and impact of congestion control. *IEEE Access : Practical Innovations, Open Solutions*, 8, 197665–197683. DOI: 10.1109/ACCESS.2020.3035119

Tolcha, Y., Montanaro, T., Conzon, D., Schwering, G., Maselyne, J., & Kim, D. (2021). Towards interoperability of entity-based and event-based iot platforms: The case of ngsi and epcis standards. *IEEE Access : Practical Innovations, Open Solutions*, 9, 49868–49880. DOI: 10.1109/ACCESS.2021.3069194

Umar, T. (2020). Applications of drones for safety inspection in the gulf cooperation coun- cil construction. *Engineering, Construction, and Architectural Management*, 28(9), 2337–2360. DOI: 10.1108/ECAM-05-2020-0369

Vajgel, B. (2021). Development of intelligent robotic process automation: a utility case study in brazil. *IEEE Access*, 9. DOI: 10.1109/ACCESS.2021.3075693

Wang, N., Christen, M., & Hunt, M. (2021). Ethical considerations associated with "hu- manitarian drones": A scoping literature review. *Science and Engineering Ethics*, 27(4), 51. DOI: 10.1007/s11948-021-00327-4 PMID: 34342721

Wang, N., Christen, M., Hunt, M., & Biller-Andorno, N. (2022). Supporting value sensitivity in the humanitarian use of drones through an ethics assessment framework. *International Review of the Red Cross*, 104(919), 1397–1428. DOI: 10.1017/S1816383121000989

Wang, Y., & Ye, T. (2022). Applications of artificial intelligence enhanced drones in distress pavement, pothole detection, and healthcare monitoring with service delivery. *Journal of Engineering*, 2022, 1–16. DOI: 10.1155/2022/7733196

Weichlein, T., Zhang, S., Li, P., & Zhang, X. (2023). Data flow control for network load balancing in ieee time sensitive networks for automation. *IEEE Access : Practical Innovations, Open Solutions*, 11, 14044–14060. DOI: 10.1109/ACCESS.2023.3243286

West, J., & Bowman, J. (2016). The domestic use of drones: An ethical analysis of surveillance issues. *Public Administration Review*, 76(4), 649–659. DOI: 10.1111/puar.12506

Xie, Y., Li, P., Nedjah, N., Gupta, B. B., Taniar, D., & Zhang, J. (2023). Pri- vacy protection framework for face recognition in edge-based Internet of Things. *Cluster Computing*, 26(5), 3017–3035. DOI: 10.1007/s10586-022-03808-8

Ye, X., Song, W., Hong, S., Kim, Y., & Yoo, N. (2022). Toward data interoperability of enter- prise and control applications via the industry 4.0 asset administration shell. *IEEE Access : Practical Innovations, Open Solutions*, 10, 35795–35803. DOI: 10.1109/ACCESS.2022.3163738

Zeng, Y., Lyu, J., & Zhang, R. (2019). Cellular-connected uav: Potential, chal- lenges, and promising technologies. *IEEE Wireless Communications*, 26(1), 120–127. DOI: 10.1109/MWC.2018.1800023

Chapter 4
Artificial Intelligence (AI) for Autonomous Drones

Purwadi Agus Darwinto
Instrumentation, Control, and Optimization, Sepuluh Nopember Institute of Technology, Indonesia

Agung Mulyo Widodo
Universitas Esa Unggul, Indonesia

Nilla Perdana Agustina
Laboratorium Instrumentasi dan Pengukuran, Indonesia

Kadek Dwi Wahyuadnyana
https://orcid.org/0000-0002-7280-672X
Sepuluh Nopember Institute of Technology, Indonesia

Mosiur Rahaman
https://orcid.org/0000-0003-0521-2080
International Center for AI and Cyber Security Research and Innovations, Asia University, Taiwan & Computer Science and Information Engineering, Asia University, Taiwan

ABSTRACT

This article discusses the integration of artificial intelligence (AI) in drone systems, highlighting AI's ability to interpret data, learn, and adapt to achieve goals. AI's broad applications range from natural language processing to autonomous vehicles and industrial robotics. AI enables drones to autonomously perform tasks like takeoff, navigation, and landing, even in complex environments. The study also covers AI's role in real-time data analysis for various purposes, including security and agrotechnology. It emphasizes the importance of algorithm development and sensor data processing for drone operations, enhancing their performance, efficiency, and safety. The use of simulations in languages like C++, Python, MATLAB, and Simulink is mentioned, with specific reference to reducing chattering in Sliding Mode Control (SMC) for energy efficiency and extended flight times. The article concludes with the challenges of aligning drone design with market-available components and the successful application of AI in drone control systems.

DOI: 10.4018/979-8-3693-2707-4.ch004

INTRODUCTION

Over the past decade, the growth and application of Unmanned Aerial Vehicle technology has been widely used in various fields. As in the civilian and military sectors it is used for air patrols, SDA exploration, as well as for tracking victims of natural disasters, area mapping, natural disaster monitoring and can also be used to detect humans and help in search and rescue scenarios. (Lu et al., 2022)(Lappas et al., 2022). One of the UAVs that's being researched and developed today is a quadcopter (Marshall et al., 2021). With many advantages over other unmanned aerial vehicles (UAVs), namely the capacity to take off and land vertically, the high level of agility and maneuverability, the small size, and the cheap price, quadcopter UAVs have become increasingly popular in recent decades. (Xuan-Mung et al., 2022). The implementation of an autonomous system on a quadcopter model aircraft will improve the effectiveness and efficiency of surveillance as the flight route of the aircraft has been predetermined. Besides, it can minimize the risk of life casualties when carrying out andgerous missions. (Beck et al., 2016).

The following examples are some of the research that has been done in order to develop quadcopter technology. The first, the research carried out by (Dogru & Marques, 2020) that is to design a quadcopter for military needs, capable of detecting and chasing an opponent's or enemy's quadcopter automatically using two-dimensional radar. (radar 2D). Then, the next example, a quadcopter has also been used in agriculture to perform monitoring on agricultural land that is large enough, which by farmers is not possible to carry out monitoring manually, as is done by (Manuylova & Bulychev, 2023) and (Polo et al., 2015). Behind the widespread use of such quadcopters, there is an important thing that needs to be taken into account to improve the performance of the quadcopter when flying in the air, namely the design of the control system used. External interference is one of the factors that cannot be ignored in the quadcopter's dynamic system. These external disturbances consist of a variety of types, namely, wind interferences from the environment, interferencies of objects or objects that can prevent quadcopters from flying, and so on. The complexity of the tasks faced by these UAVs makes Artificial Intelligence (AI) systems crucial to controlling quadcopters. Quadcopter must be able to process information from a variety of sensors quickly and accurately for navigation in dynamic and often challenging airspace. In this case, the presence of an AI system is crucial. AI systems allow quadcopter to navigate complex environments with high efficiency and accuracy by studying patterns, processing data, and making decisions in real time. Additionally, adaptability which is crucial in dealing with a variety of changing environmental conditions is provided by AI systems. With the ability to learn from experience and change navigation strategies to adapt to changing environments, AI-controlled quadcopters can operate effectively in a range of conditions without constant human intervention.

Basic Quadcopter Concepts

Before further discussing the control system applied to quadcopter UAVs, it is important that we also know the basis of the physics and mathematics that underpins these quadcopters to fly and apply to meet their needs.

UAV Type Quadcopter

The Quadrotor uses four brushless motors with propellers as drives. Quadrotors can have a plus (+) or cross (x) flight configuration as shown in Figure 1. Quadrotor curves that have a rotating direction are placed opposite each other. This is intended to avoid unwanted roll and pitch rotation movements if the quadcopter is interrupted while hovering.

Figure 1. (a) Cross Configuration (b) Plus Configuration (Mangsatabam, 2018)

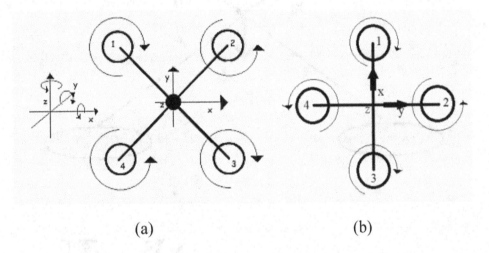

(a)   (b)

Quadcopter movements set-point, maneuver, rotate, take-off, landing, etc. are done by adjusting the rotation speed of each rotor with the DC electric motor. The rotation direction of each of the rotors is opposite to each other, so it can be said that the purpose of the controlling system designed is basically to adjust the speed of rotation of every rotor consistently so that the quadcopter can fly according to the given command. Each rotor rotates in the opposite direction of the clock, and the counter-clockwise or opposite (Idrissi et al., 2022). Before designing a suitable control system, it is necessary to first understand the mathematical model of the plant dynamic system to be controlled, in this case it is a mathematic model of quadcopter dynamic systems. In quadcopter dynamics in general, there are two reference coordinate systems being considered, namely the body-fixed frame reference system and the earth- fixed frame (Figure 2). Both of these reference co-ordinates will subsequently be reconfigured with the rotation matrix operator (R), which can change the shape from a body- Fixed Frame reference to an earth-fixing frame, so is the opposite.

Coordinate System

It takes a way to get the equation of the motion of the quadcopter relative to the ground, then the equations of the movement of the quadricopter will be directly related to the change in the coordinate system.

Figure 2. Quadcopter Reference Frame

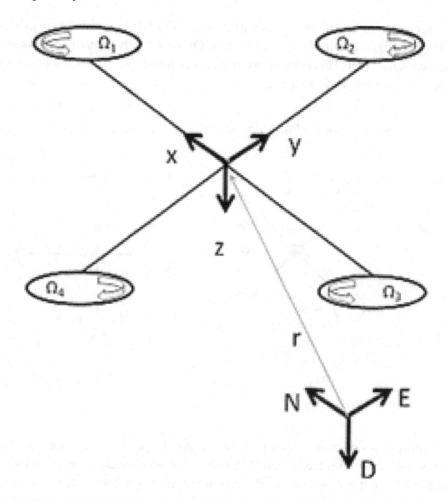

Figure 2 shows an earth reference frame with axes N, E and D and a body frame with the axes x, y, and z. A earth frame is an inertial frame mounted at a certain place on the surface of the earth as its name implies. Using the N notation that indicates the north direction, E shows the east direction, D shows the bottom direction. (Downwards). On the other hand, the body frame is in the middle of the quadrotor body, with the x axis pointing to the sidewalk 1, the y axis points to the sidelines 2 and the z axis is pointing towards the ground. (ground). The distance between the earth frame and the body frame describes the absolute position of the center of the quadrotor's mass.

$$r = [x\, y\, z]^T \tag{2.1}$$

xyz is the translation movement of a quadcopter that produces a linear position (r) equation (2.1). The angular position of Θ is determined by the Equation (2.2) of the orientation of the body frame towards the earth frame resulting in a motion of *roll* (ϕ), *pitch* (θ), *yaw* (ψ).

$$\Theta = [\phi\,\theta\,\psi]^T \tag{2.2}$$

Both of the above coordinates form the basis of the quadopter dynamics system, the absolute Earth frame, which requires the conversion of the motion of quadcopter based on the body frame to the earth reference frame.

Based Motion Quadcopter

Quadcopter consists of four rotors, in Figure 1, each rotor consists from beams mounted on the DC motor. The PWM signal is used to change the propulsion of the motor so that the angle of the quadcopter changes. Quadcopter is a 6 DOF dynamic system that represents 6 variables of motion equations $(x, y, z, \phi, \theta, \psi)$ based on Figure 3, with 4 input variables namely the speed of each motor on the quadcopter (Ω) that will control the quadricopter at the position (x, y, z) according to the Earth coordinates, and the angle (ϕ, θ, ψ) representing the attitude roll, pitch, and yaw positions on the quadrcopter.

Figure 3. System Underactuated Quadcopter

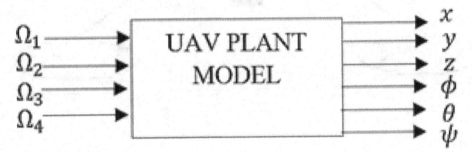

This makes the quadcopter an underactuated system. Underactuated systems are systems that have a smaller number of control inputs compared to the degrees of freedom of the system, x, y and z represent the distance of the central quadrotor along the axes x, y and z of the Earth's skeleton, and ϕ, θ and ψ are the three Euler angles that represent the orientation of the quadrotor (Agustina & Darwito, 2023). This research uses cross configuration as a reference base to form mathematical equations on quadcopter systems. Here is a form of scheme that can represent between Earth coordinates or called earth coordinates and body coordinates. The motion of these two coordinates is the basis for the formation of the equation of the motion system on the quadcopter, i.e. the translation movement system, which can be represented by the axes x, y, z and the rotation motion system, or the motion of the quadcopter on the roll, pitch and yaw movements.

Figure 4. Sudut Euler dari Quadcopter

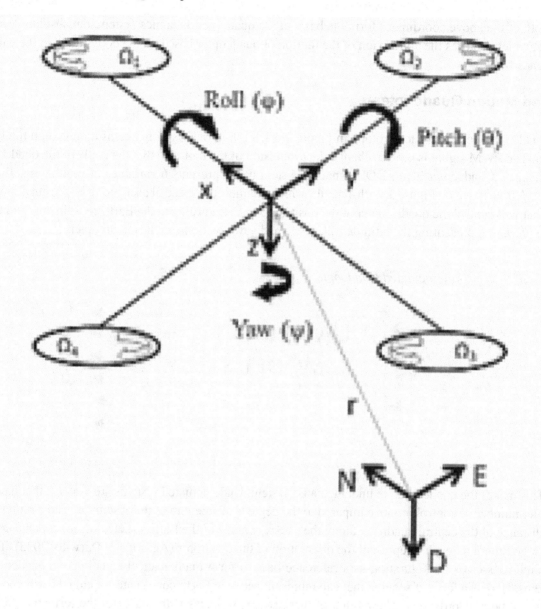

The motion of the quadcopter in Figure 4 that causes a change in the ϕ angle is called the roll which is the angle to the x axis, the change of the angle θ is called a pitch which is an angle the y axis and the change in ψ are called yaw which is the angle to the z axis. Roll and pitch movements are categorized as the attitude of the quadcopter, whereas the yaw angle is categorised as the heading of the quadricopter due to the motion of yaw relative to the earth's inertia frame. The vertical distance from the earth, the z co-ordinate, is called the altitude, while the coordinate points x and y are classified as the positions of the quadcopter.

Figure 5. (a) Altitude Movement; (b) Roll Movement; (c) Pitch Movement; (d) Yaw Movement

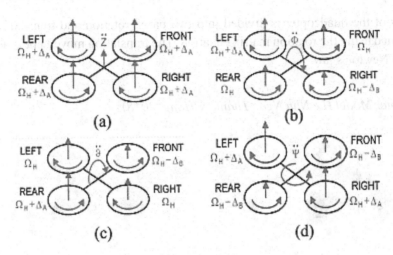

Altitude movement Figure 5(a) is an ascending and descending motion, the quadrotor must increase or decrease the rotation speed of each motor by the same amount. Increasing the speed will cause a quadrotor to fly up, and decreasing a speed will result in a quadrator to descend.

Roll movement Figure 5(b) is a right-to-left slope. To perform such movements, it is necessary to change the speed of the motor on one of the motors pairs. The change of speed is done on the left and right motor pairs (motor 2 and motor 4). The change is that one motor pair member decreases its speed and the other motor pairing member is raised with the same speed difference, while the other motors are left at a fixed speed. With a setup like this, the quadrotor will move rotating from the direction of a lower-speed motor pair, toward a higher speed motor pair.

Pitch motion Figure 5(c) is a tilt forward-back motion. To do this, you need to change the speed of the front and rear motorcycle pairs. (motor 1 and motor 3). The change is that one member of the motorcycle pair is reduced speed and the other member is raised at the same speed difference, while the other motorcycles are left at a fixed speed. With a setup like this, the quadrotor will move rotating from the direction of the lower-speed motor pair, towards the higher speed motor pair.

Yaw motion Figure 5(d) is a rotating movement with the core staying in the same position. This movement is done by lowering the speed of one pair of motors and increasing the velocity of another pair. Then, the quadrotor will move rotating in the direction of the rotation of the motor pair which is slower than the other pair.

Quadcopter Mathematical Model

The mathematical model of the quadcopter system is obtained by knowing the kinematical and dynamic models of the quadricopter.

Kinematical Model

The kinematics of the quadcopter is divided into two parts, rotation and translation. The rotation movement is obtained using the rotation matrix equation. The translation movement is derived from the equation related to Newton's law.

Figure 6. Quadcopter Model (Le Nhu Ngoc Thanh & Hong, 2018)

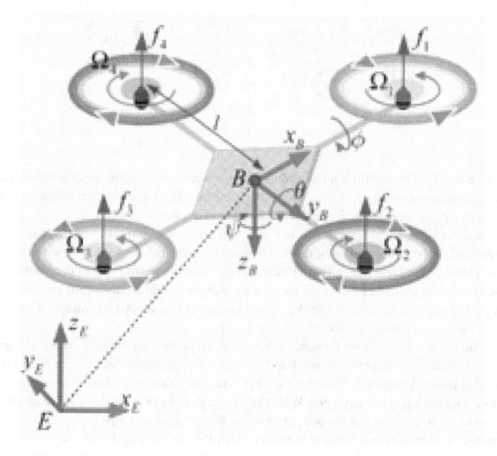

Based on this modeling based on Figure 5, two coordinate frameworks will be used, namely the Earth's skeleton with the axis x_E, y_E, z_E and the body's framework with the x_B, y_B, z_B axis represented in Fig. 6.

Rotary Movement Kinematics

The rotational motion of a quadcopter as a rigid object in space can be parameterized using an Euler angle, which is a mathematical representation of three consecutive rotations against different axis possibilities..

Figure 7. Rotate the Quadcopter against the X-axis

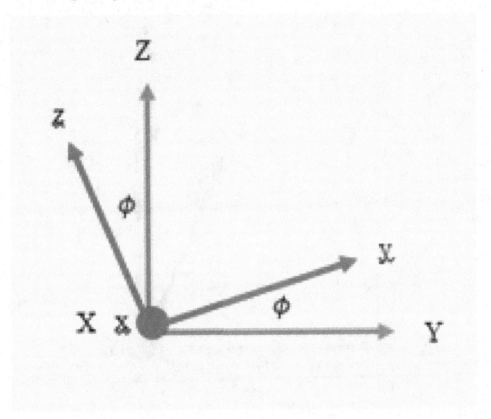

The rotation movement of the quadcopter against the x axis can be represented in Figure 7. The angle of the roll is formed by each red-coloured body frame rotating towards the axis x earth frame coordinates.

$$X = 1\,x + 0\,y + 0\,z$$

$$Y = 0\,x + cos\,\phi\,y + \;- sin\,\phi\,z$$

$$Z = 0\,x + sin\,\phi\,y + \;cos\,\phi\,z \qquad (3.1)$$

Equations (3.1) can be made into matrix models according to equations (3.2).
- $R(x, \phi)$, rotation around the x axis obtained the rotation matrix as follows

$$R\,(x,\,\phi) = \begin{bmatrix} 1 & 0 & 0 \\ 0 & cos\phi & -sin\,\phi \\ 0 & sin\phi & cos\phi \end{bmatrix} \qquad (3.2)$$

Next is a rotational motion scheme centered on the y axis and against the pitch angle.

Figure 8. Rotate the Quadcopter against the Y axis

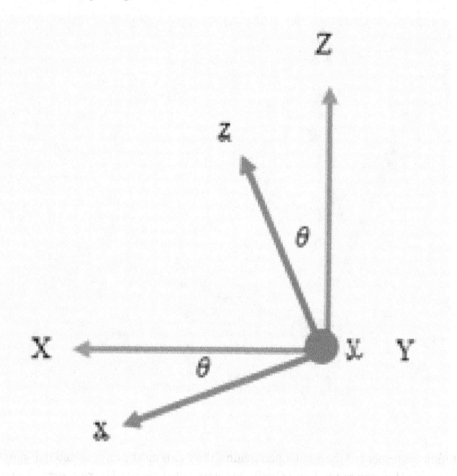

The rotation movement of the quadcopter against the y axis can be represented in Figure 8. The pitch angle is formed by each red-coloured body frame rotating towards the earth frame coordinates of the y axis.

$$X = \cos\theta\, x + 0\, y + \sin\theta\, z$$

$$Y = 0\, x + 1\, y + 0\, z$$

$$Z = -\sin\theta\, x + 0\, y + \cos\theta\, z \qquad (3.3)$$

The equation (3.3) can make a matrix model according to the equation. (3.4)
- $R(y,\theta)$, rotation around the y axis obtained rotation matrix equation (3.4)

$$R(y,\theta) = \begin{bmatrix} \cos\theta & 0 & \sin\theta \\ 0 & 1 & 0 \\ -\sin\theta & 0 & \cos\theta \end{bmatrix} \qquad (3.4)$$

Here's a rotational movement scheme centered at the z axis and against the yaw angle.

Figure 9. Rotate the Quadcopter against the Z axis

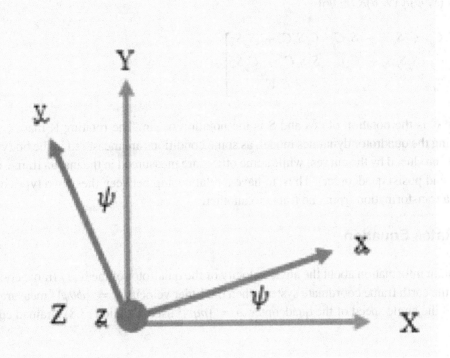

The rotation movement of the quadcopter against the z axis can be represented in Figure 9. The yaw angle is formed from each red-coloured body frame coordinate rotating against the earth frame coordinates of the z-axis.

$$X = \cos\psi\, x + -\sin\psi\, y + 0\, z$$

$$Y = \sin\psi\, x + \cos\psi\, y + 0\, z$$

$$Z = 0\, x + 0\, y + 1\, z \tag{3.5}$$

The equation (3.5) can make a matrix model according to the equation. (3.6)
- $R(z,\psi)$, rotation around the z axis obtained rotation matrix equation (3.6)

$$R(z, \psi) = \begin{bmatrix} \cos\psi & -\sin\psi & 0 \\ \sin\psi & \cos\psi & 0 \\ 0 & 0 & 1 \end{bmatrix} \tag{3.6}$$

To obtain the matrix rotation of the rotation and translation relations of the body frame coordinates to the inertial frame co-ordinates, then subtitles Equations (3.2), (3.4) and (3.6). Matrix rotations are invers of the combination of rotation matrix equations of each coordinate system as in the Equation (3.7):

$$R = [R(x, \phi)R(y, \theta)R(z, \psi)]$$

$$R = \begin{bmatrix} C_\psi C_\theta & C_\psi S_\theta S_\phi - S_\psi C_\phi & C_\psi S_\theta C_\phi + S_\psi S_\phi \\ S_\psi C_\theta & S_\psi S_\theta S_\phi + C_\psi C_\phi & S_\psi S_\theta C_\phi - C_\psi S_\phi \\ -S_\theta & C_\theta S_\phi & C_\theta C_\phi \end{bmatrix} \tag{3.7}$$

where C is the notation of cos and S is the notation of sin. The rotating R matrix will be used in formulating the quadrotor dynamics model, as some conditions are measured in the body frame (e.g. the push style produced by the curves) while some others are measureed in the inertia frame. (misalnya gaya gravitasi and posisi quadcopter). Thus, to have a relationship between these two types of conditions, it requires a transformation from one frame to another.

Euler Rates Equation

To obtain information about the angle velocity of the quadrotor of the body frame coordinates modified into the earth frame coordinate system, then the Euler velocity $\dot{\eta} = [\dot{\phi}\dot{\theta}\dot{\psi}]^T$ measured in the inertia frame and the angle speed of the quadcopter $\omega = [pqr]^T$ used equation (3.8) attained equation (2.11).

$$\omega = E\dot{\eta} \tag{3.8}$$

$$\begin{bmatrix} p \\ q \\ r \end{bmatrix} = E \begin{bmatrix} \dot{\phi} \\ \dot{\theta} \\ \dot{\psi} \end{bmatrix} \quad \begin{bmatrix} \dot{\phi} \\ \dot{\theta} \\ \dot{\psi} \end{bmatrix} = E^{-1} \begin{bmatrix} p \\ q \\ r \end{bmatrix} \tag{3.9}$$

$$E = \begin{bmatrix} 1 & 0 & -sin\theta \\ 0 & cos\phi & sin\phi cos\theta \\ 0 & -sin\phi & cos\phi cos\theta \end{bmatrix} \tag{3.10}$$

Decreasing the matrix equation of angular transformation speed from the body frame to the inertia frame requires E^{-1}.

$$E^{-1} = \begin{bmatrix} 1 & S_\phi T_\theta & C_\phi T_\theta \\ 0 & C_\phi & -S_\phi \\ 0 & \dfrac{S_\phi}{C_\theta} & \dfrac{C_\phi}{C_\theta} \end{bmatrix} \tag{3.11}$$

The linear and angular positions of the quadcopter based on the earth frame reference coordinates are implemented in the vector $[xyz\phi\theta\psi]^T$, and the linear position as well as the angular position of the quadricopter based upon the body frame coordinates is implemented within the vectors $[uvwpqr]^T$. To connect the two reference co-ordinates are performed matrix multiplication operations based on equations (3.12) and (3.13):

$$v = R \cdot v_B \tag{3.12}$$

$$\dot{\eta} = E^{-1} \cdot \omega \tag{3.13}$$

where,

$$v = [\dot{x} \quad \dot{y} \quad \dot{z}]^T \tag{3.14}$$

$$\dot{\eta} = [\phi \quad \dot{\theta} \quad \dot{\psi}]^T \tag{3.15}$$

$$v_B = [u \quad v \quad w]^T \tag{3.16}$$

$$\omega = [p \quad q \quad r]^T \tag{3.17}$$

The result of the linear position and angle of the equations (3.12) and (3.13) produces the kinematical model equations at $,y,z,\phi,\theta,\psi$, for a more complete deduction of the formula found in appendix A.2 which results in the equation (3.18):

$$\dot{x} = (C_\psi C_\theta)u + (C_\psi S_\theta S_\phi - S_\psi C_\phi)v + (C_\psi S_\theta C_\phi + S_\psi S_\phi)w$$

$$\dot{y} = (S_\psi C_\theta)u + (S_\psi S_\theta S_\phi + C_\psi C_\phi)v + (S_\psi S_\theta C_\phi - C_\psi S_\phi)w$$

$$\dot{z} = (-S_\theta)u + (C_\theta S_\phi)v + (C_\theta C_\phi)w$$

$$\dot{\phi} = p + (S_\phi T_\theta)q + (C_\phi T_\theta)r$$

$$\dot{\theta} = (C_\phi)q + (-S_\phi)r$$

$$\dot{\psi} = \left(\frac{S_\phi}{C_\theta}\right)q + \left(\frac{C_\phi}{C_\theta}\right)r \tag{3.18}$$

Model Dinamika

The most common mathematical model used to model quadcopter dynamics is the Euler-Newton model, which is based on Newton's and Euler's laws. This model takes into account the speed, acceleration, momentum, aerodynamic style, and torque applied to quadcopters. The rotational motion equation is deduced in the body frame using Newton Euler's method with the following general formalis:

$$m(\omega x v_B + \dot{v}_B) = f_B \tag{4.1}$$

with $f_B = [f_x \quad f_y \quad f_z]^T$. Then, the total torque working on the quadcopter is given the equation as follows:

$$I \cdot \dot{\omega} + \omega \, x \, (I \cdot \omega) = \tau_M \tag{4.2}$$

where, $\tau_M = [\tau_{Mx} \quad \tau_{My} \quad \tau_{Mz}]^T$ is the total torque working on the quadcopter and I is the moment of inertia expressed in the form of a diagonal matrix

$$I = \begin{bmatrix} I_x & 0 & 0 \\ 0 & I_y & 0 \\ 0 & 0 & I_z \end{bmatrix} \tag{4.3}$$

The rotational motion equation resulting from the body frame of the quadcopter of the equation (4.1) produces the total f_B of the following style:

$$f_x = m(qw - rv + \dot{u})$$

$$f_y = m(ru - pw + \dot{v})$$

$$f_z = m(pv - qu + \dot{w}) \tag{4.4}$$

The total torque that works on the quadcopter using the Equation (3.15) produces the equation of torque at each position.

$$\tau_{Mx} = \dot{p}I_x - qrI_y + qrI_z$$

$$\tau_{My} = \dot{q}I_y + prI_x - prI_z$$

$$\tau_{Mz} = \dot{r}I_z - pqI_x + pqI_y \tag{4.5}$$

Some styles that work on the body frame are highly influential in the motion of the quadcopter, such as the external style that is present on (f_B) and the moment of external styles or the τ_M torse.

$$f_B = mg R^T \cdot \hat{e}_x - f_t \hat{e}_3 + f_w \tag{4.6}$$

$$\tau_M = \tau_B - g_a + \tau_w \tag{4.7}$$

The gyroscopic moment g_a is produced from a combination of rotation by the four rotors and rotation of the skeleton. Based on research (Mellinger et al., 2011) found when the experiments were of very small value, so the g_a value is ignored. From the style equations that influence the motion of the quadcopter obtained the dynamics model of the quadricopter based on the body frame reference coordinates resulting in Equation (4.8) to Equation (4.14):

$$f_x = -mg S_\theta + f_{wx}$$

$$m(qw - rv + \dot{u}) = -mg S_\theta + f_{wx} \tag{4.9}$$

$$f_y = -mg\,C_\theta S_\phi + f_{wy}$$

$$m(ru - pw + \dot{v}) = -mg\,C_\theta S_\phi + f_{wy} \tag{4.10}$$

$$f_z = -mg\,C_\theta C_\phi + f_{wz} - f_t$$

$$m(pv - qu + \dot{w}) = -mg\,C_\theta C_\phi + f_{wz} - f_t \tag{4.11}$$

$$\tau_{Mx} = \tau_x + \tau_{wx}$$

$$\dot{p}I_x - qrI_y + qrI_z = \tau_x + \tau_{wx} \tag{4.12}$$

$$\tau_{My} = \tau_y + \tau_{wy}$$

$$\dot{q}I_y + prI_x - prI_z = \tau_y + \tau_{wy} \tag{4.13}$$

$$\tau_{Mz} = \tau_z + \tau_{wz}$$

$$\dot{r}I_z - pqI_x + pqI_y = \tau_z + \tau_{wz} \tag{4.14}$$

Dinamika Aktuator

Identifying the styles and moments generated by the beams, the τ_B moment generated from each actuator is obtained using the right hand rule. Figure 6 shows the style and moment that works on the quadrotor. Each rotor generates a push force over F_i d and produces a M_i moment in the opposite direction to the rotor's corresponding direction of rotation.

$$F_i = b\Omega_i^2 \tag{4.15}$$

$$M_i = d\Omega_i^2 \tag{4.16}$$

Where b is the aerodynamic constant working on the quadcopter, d is the momentum constant and Ω_i is the angle speed of the rotor on each actuator. So with equation (4.4) and (4.5) got momentum on each axis of x,y,z on the quadcopter.

$$\tau_x = bl\left(\Omega_4^2 - \Omega_2^2\right)$$

$$\tau_y = bl\left(\Omega_3^2 - \Omega_1^2\right)$$

$$\tau_z = -\left(d\Omega_1^2\right) + \left(d\Omega_2^2\right) - \left(d\Omega_3^2\right) + \left(d\Omega_4^2\right) \tag{4.17}$$

For the moment against the z body frame axis, the rotor push style does not generate the moment. So the torque that works on the quadcopter body is adapted to the matrix shape as follows:

$$\tau_B = \begin{bmatrix} bl(\Omega_4^2 - \Omega_2^2) \\ bl(\Omega_3^2 - \Omega_1^2) \\ d(\Omega_2^2 + \Omega_4^2 - \Omega_1^2 - \Omega_3^2) \end{bmatrix} \tag{4.18}$$

For the moment against the z body frame axis, the rotor push style does not generate the moment $x, y, z, \phi, \theta, \psi$.

$$U_1 = f_t = b(\Omega_1^2 + \Omega_2^2 + \Omega_3^2 + \Omega_4^2) \tag{4.19}$$

$$U_2 = \tau_x = b(\Omega_4^2 - \Omega_2^2) \tag{4.20}$$

$$U_3 = \tau_y = b(\Omega_3^2 - \Omega_1^2) \tag{4.21}$$

$$U_4 = \tau_z = d(\Omega_2^2 + \Omega_4^2 - \Omega_1^2 - \Omega_3^2) \tag{4.22}$$

U_1 is the upward momentum derived from four rotors responsible for the elevation of the quadrator and its velocity of change $(z, \dot{z})$. U_2 is the momentum difference between rotors 2 and 4 responsible for rolling and changing velocity $(\phi, \dot{\phi})$ of the roll. U_3 is the distinction of momentum between rotor 1 and rotor 3 which produces the pitch rotation and its change rate $(\theta, \dot{\theta})$. U_4 is the torque difference between two rotors which are rotating in the direction of the clock and the other 2 rotors that are rotating in the opposite direction of a clock needle resulting in yaw rotations and the rate of their change $(\psi, \dot{\psi})$.

State Space

Forms the state space model of the quadcopter dynamic system on the body fixed frame reference coordinates and the earth fixed frames reference co-ordinates can be expressed in the state vector. (Araar & Aouf, 2014), as shown by equation 4.18.

$$X = \begin{bmatrix} x & y & z & u & v & w & p & q & r & \phi & \theta & \psi \end{bmatrix}^T \tag{4.23}$$

Next, implemented into the freedom degrees of the quadcopter as follows:

$$\dot{X} = \begin{bmatrix} \dot{x} & \dot{y} & \dot{z} & \dot{u} & \dot{v} & \dot{w} & \dot{p} & \dot{q} & \dot{r} & \dot{\phi} & \dot{\theta} & \dot{\psi} \end{bmatrix}^T \tag{4.24}$$

Each of the components contained in the equation 4.18 can be obtained from the equations 3.18 and 4.9 The state vector determines the position of the quadrotor in space and the velocity of its linear angle. Translation motion equations are based on Newton's second law and derived within the Earth's inertia framework.

$$m\ddot{r} = \begin{bmatrix} 0 & 0 & mg \end{bmatrix}^T + RF_B \tag{4.25}$$

When the quadrotor is in a horizontal orientation (i.e. not rotating or throwing), the only non-gravity style that works on it is the push style produced by the rotation of the rotor.

$$F_B = \begin{bmatrix} 0 \\ 0 \\ -b\left(\Omega_1^2 + \Omega_2^2 + \Omega_3^2 + \Omega_4^2\right) \end{bmatrix} \tag{4.26}$$

Obtained the state-space form of the dynamic model, several state vectors on equations (4.18) and (4.19) can be assumed to be $[\dot{\phi} \quad \dot{\theta} \quad \dot{\psi}]^T = [p \quad q \quad r]^T$, then derivative to time becoming $[\ddot{\phi} \quad \ddot{\theta} \quad \ddot{\psi}]^T = [\dot{p} \quad \dot{q} \quad \dot{r}]^T$ on a quadcopter system in a linear position $\ddot{x}, \ddot{y}, \ddot{z}, \ddot{\phi}, \ddot{\theta}, \ddot{\psi}$ as follows:

$$\ddot{x} = \frac{f_t}{m}\left(C_\psi S_\theta C_\phi + S_\psi S_\phi\right) \tag{4.27}$$

$$\ddot{y} = \frac{f_t}{m}\left(S_\psi S_\theta C_\phi - C_\psi S_\phi\right) \tag{4.28}$$

$$\ddot{z} = g - \frac{f_t}{m}\left(C_\theta C_\phi\right) \tag{4.29}$$

$$\ddot{\phi} = \frac{l}{I_x}\tau_x + \frac{I_y}{I_x}\dot{\theta}\dot{\psi} - \frac{I_z}{I_x}\dot{\theta}\dot{\psi} \tag{4.30}$$

$$\ddot{\theta} = \frac{l}{I_y}\tau_y + \frac{I_z}{I_y}\dot{\phi}\dot{\psi} - \frac{I_x}{I_y}\dot{\phi}\dot{\psi} \tag{4.31}$$

$$\ddot{\psi} = \frac{l}{I_z}\tau_z + \frac{I_x}{I_z}\dot{\phi}\dot{\theta} - \frac{I_y}{I_z}\dot{\phi}\dot{\theta} \tag{4.32}$$

with input $U_1 = f_t, U_2 = \tau_x, U_3 = \tau_y, U_4 = \tau_z$

Thus, the complete expression of the mathematical model of the quadcopter dynamic system in the form of a matrix is as follows:

$$\dot{x} = f(x) + \sum_{i=1}^{4} b_1(x)u_i \tag{4.33}$$

Based on Compression (4.33) the dynamics model of the quadcopter system equations (4.27) to Equations (4.32) can be written in the form of a matrix on the equation. (4.33).

with:

$$x = \begin{bmatrix} x & y & z & \phi & \theta & \psi & \dot{x} & \dot{y} & \dot{z} & p & q & r \end{bmatrix}^T$$

$$\dot{x} = \begin{bmatrix} \dot{x} & \dot{y} & \dot{z} & \dot{\phi} & \dot{\theta} & \dot{\psi} & \ddot{x} & \ddot{y} & \ddot{z} & \dot{p} & \dot{q} & \dot{r} \end{bmatrix}^T$$

$$
\begin{bmatrix} \dot{x} \\ \dot{y} \\ \dot{z} \\ \dot{\phi} \\ \dot{\theta} \\ \dot{\psi} \\ \ddot{x} \\ \ddot{y} \\ \ddot{z} \\ \dot{p} \\ \dot{q} \\ \dot{r} \end{bmatrix} =
\begin{bmatrix} \dot{x} \\ \dot{y} \\ \dot{z} \\ p + (S_\phi T_\theta)q + (C_\phi T_\theta)r \\ (C_\phi)q + (-S_\phi)r \\ \left(\frac{S_\phi}{C_\theta}\right)q + \left(\frac{C_\phi}{C_\theta}\right)r \\ 0 \\ 0 \\ g \\ \left(\frac{I_y - I_z}{I_x}\right)qr \\ \left(\frac{I_z - I_x}{I_y}\right)pr \\ \left(\frac{I_x - I_y}{I_z}\right)pq \end{bmatrix} +
\begin{bmatrix} 0 \\ 0 \\ 0 \\ 0 \\ 0 \\ 0 \\ \frac{1}{m}(C_\psi S_\theta C_\phi + S_\psi S_\phi) \\ \frac{1}{m}(S_\psi S_\theta C_\phi - C_\psi S_\phi) \\ -\frac{1}{m}(C_\theta C_\phi) \\ 0 \\ 0 \\ 0 \end{bmatrix} U_1 +
\begin{bmatrix} 0 \\ 0 \\ 0 \\ 0 \\ 0 \\ 0 \\ 0 \\ 0 \\ 0 \\ \frac{l}{I_x} \\ 0 \\ 0 \end{bmatrix} U_2 +
\begin{bmatrix} 0 \\ 0 \\ 0 \\ 0 \\ 0 \\ 0 \\ 0 \\ 0 \\ 0 \\ 0 \\ \frac{l}{I_y} \\ 0 \end{bmatrix} U_3 +
\begin{bmatrix} 0 \\ 0 \\ 0 \\ 0 \\ 0 \\ 0 \\ 0 \\ 0 \\ 0 \\ 0 \\ 0 \\ \frac{l}{I_z} \end{bmatrix} U_4 \qquad (4.34)
$$

Equation (4.34) is a matrix equation that represents from equation (4.27) to equation (4.32) which is a nonlinear equation of a quadcopter.

Simulation and Experiment

This section will explain how artificial intelligence (AI) is applied to autonomous drones, by providing examples of simulations and experiments. The type of drone that will be considered is a quadcopter, or also known as a quadrotor. Quadcopter is one of the variants of a multicopter drone that uses four propeller systems for its flight and main control. Although there are various other types of multicopters, such as tricopters (with three propeller system), hexacopter (with six propeller System), octorotor (with eight propeller Systems), and so on, the discussion on this section will focus on quadcopter. The Quadcopter has a symmetrical configuration in which the propellers are placed symmetrically on the four sides of the drone. In principle, research in the field of drones or other fields of robotics takes no time. This process involves several stages, ranging from conceptual design and development, system programming, simulation testing, prototype testing, experimentation, to performance evaluation. The field of robotics is a discipline that covers several branches of science such as computer science, electrical engineering, mechanics, and even artificial intelligence as will be discussed in this section. Often, research in robotics involves complex experiments and requires repeated iterations to improve robotic performance. Therefore, it requires a high degree of dedication and patience to substantial progress in the development of robotic technology.

As an early concept in starting research in robotics, an understanding of the mathematical modeling concept of the dynamic system of a robot to be developed is absolutely crucial. It involves mathematical formulations that accurately describe how robots interact with their environment and how their internal systems react to external input. In this modeling, factors such as style, momentum, acceleration, and inertia are taken into consideration to construct differential equations that represent robotic dynamics. Through careful mathematical modelling, engineers can understand the complex behavior of robots in a variety of possible conditions and environments. In addition, mathematical modeling allows developers to design effective controllers to regulate robot movements and actions as needed. By understanding how control input affects the output of a robot system, engineers can optimize robot performance in achieving specific goals, such as navigation, object manipulation, or social interaction. Furthermore, mathematical modeling provides an opportunity for developers to undertake in-depth simulations before building a physical prototype. Using sophisticated simulation software, they can test and analyze a variety of scenarios and strategies without having to spend time and money building the actual prototypes. This allows significant savings of time, cost, and resources during the development phase.

Control Algorithm

Before we start the simulation, we need to know first that we need a control algorithm that we want to test. Based on the characteristics of the system to be controlled, there are two types of control algorithms, which are linear and nonlinear. Linear control is based on the assumption that a controlled system can be described by a linear mathematical model, such as linear differential equations. Linear controlling methods, like proportional, integral, and derivative controls (PIDs), are generally stable and easily analyzed, but less effective for highly nonlinear systems. On the other hand, nonlineary controls take into account the nonliner properties of controlled systems, which can be explained by nonlinears. Although more complex in analysis and design, non-linear controls are able to deal with highly non linear systems better and can optimize the performance of systems around them (Lin et al., 2023). specific operating point. Therefore, the main difference between the two is in the approach to the mathematical properties of a controlled system: linear control assumes linearity, while nonlinear control takes into account the nonliner nature of such a system.

Since a quadcopter is a complex, and of course nonlinear system, the non-linear control algorithm is used in simulations to be carried out later. One of the nonlinear control algorithms we use is the sliding mode control (SMC). The SMC is a non-linear method of control that aims to force a dynamic system to "launch" or "shift" into the desired operating mode quickly and stably. In SMC, the controller is constructed in such a way that it moves the system towards a specified sliding surface, where the behavior of the system becomes stable. This slide surface acts as the "desirable operational mode". When the system moves to the slid surface, the controllers are designed to maintain the system above the surface, thus ensuring that the system remains in the desiated operating Mode despite any interference or uncertainty in the system. This method is known for its stability against interference and uncertainty, as well as its ability to be applied to complex and nonlinear dynamic systems. We have formulated the mathematical form of the SMC control algorithm, as given by the equation (5.1)-(5.6)

$$U_2 = \frac{1}{b_1}[-\varepsilon_2 sign(S_\phi) - k_1 S_\phi - a_1 x_4 x_6 - a_2 x_2^2 - a_3 \overline{\Omega} x_4 + \ddot{\phi}_d + w_2(\phi_d - x_2)] \qquad (5.1)$$

$$U_3 = \frac{1}{b_2}[-\varepsilon_3 sign(S_\theta) - k_2 S_\theta - a_4 x_2 x_6 - a_3 x_4^2 - a_6 \overline{\Omega} x_2 + \ddot{\theta}_d + w_3(\theta_d - x_4)] \qquad (5.2)$$

$$U_4 = \frac{1}{b_3}[-\varepsilon_4 sign(S_\varphi) - k_3 S_\varphi - a_7 x_2 x_4 - a_8 x_6^2 + \ddot{\varphi}_d + w_4(\varphi_d - x_6)] \qquad (5.3)$$

$$U_1 = \frac{m}{\cos(\phi)\cos(\theta)}[-\varepsilon_1 sign(S_z) - k_6 S_z - a_{11} x_{12} + \ddot{z}_d + w_1(\dot{z}_d - x_{12}) + g] \qquad (5.4)$$

with,

$$U_x = \frac{m}{U_1}[-sign(S_x) - k_4 S_x - a_9 x_8 + \ddot{x}_d + (\dot{x}_d - x_8)] \qquad (5.5)$$

$$U_y = \frac{m}{U_1}[-sign(S_y) - k_5 S_y - a_{10} x_{10} + \ddot{y}_d + (\dot{y}_d - y_{10})] \qquad (5.6)$$

Although it has been described earlier that the main advantage of the SMC algorithm is its ability to remain stable despite external disturbances, known as robustness, but inherently SMC Algorithms have a weakness called chattering. Chattering is an event in which the control output moves with fast around the desired sliding surface, often within narrow range. This phenomenon is caused by high oscillations in the control response, which can occur when the system repeatedly "runs" or "shifts" around the sliding surface because the controller forces the system to stay above the surface. Chattering can generate mechanical pressure and affect overall system performance. When examined more deeply, it turns out that this chattering effect is most due to the signum factor (*sign*) that is found in the Equation (5.1)-(5.6). This sign factor cannot be eliminated, except to adjust its value with the setting of the weight value. For example, in the equation (5.1)-(5.6), each sign factor is given a weight that starts from ε_1, ε_2, ε_3, to ε_4. Each of these weights subsequently adjusts the signum value to the internal conditions of the quadcopter to reduce the chattering effect, but remain robust, or resist external interference.

A specific algorithm is required that can adjust the weight values adaptively to reduce the chattering effect on the quadcopter system, by adjusting those values automatically based on the internal conditions of the system. To meet these needs, an artificial neural network (ANN) is selected as an algorithm that will reset these weight values. ANN, inspired by the structure and function of human biological nerve tissue, uses structures consisting of artificial neurons connected in layers. ANN has the ability to learn complex patterns from data, extract features, and perform various tasks such as classification, regression, and pattern recognition. ANN's ability to deal with complex and nonlinear problems makes it an important component in various AI applications such as image processing, voice recognition, and system control. After that, there was a study that proposed a method to automatically adjust the weight-boots values according to the internal conditions of the quadcopter system using ANN-based backpropagation neural network (BPNN), described in the study by (Bouadi et al., 2007) and (Razmi & Afshinfar, 2019). The new equation to replace the previous weight values presented in the equation (5.7)-(5.10). BPNN is a type of ANN that uses backpropagation algorithms to train models. This algorithm works by calculating the gradient of the error function against the network weights, and then using the gradien to update the weights through the iterative process. Thus, BPNN is an important part of the ANN that allows the model to learn from the data by correcting its weights via training.

$$\varepsilon_2(k+1) = \varepsilon_2(k) + \eta_2 e_2 \frac{1}{b_1} \tanh s_2$$

$$w_2(k+1) = w_2(k) + \eta_2 e_2^2 \frac{1}{b_1} \varepsilon_2 \frac{4e^{-2s_2}}{(1+e^{-2s_2})^2}$$

(13)   (5.7)

$$\varepsilon_3(k+1) = \varepsilon_3(k) + \eta_3 e_3 \frac{1}{b_2} \tanh s_3$$

$$w_3(k+1) = w_3(k) + \eta_3 e_3^2 \frac{1}{b_2} \varepsilon_3 \frac{4e^{-2s_3}}{(1+e^{-2s_3})^2}$$

(14)   (5.8)

$$\varepsilon_4(k+1) = \varepsilon_4(k) + \eta_4 e_4 \frac{1}{b_3} \tanh s_4$$

$$w_4(k+1) = w_4(k) + \eta_4 e_4^2 \frac{1}{b_3} \varepsilon_4 \frac{4e^{-2s_4}}{(1+e^{-2s_4})^2}$$

(15)   (5.9)

$$\varepsilon_1(k+1) = \varepsilon_1(k) + \eta_1 e_1 \frac{m}{\cos\phi \cos\theta} \tanh s_1$$

$$w_1(k+1) = w_1(k) + \eta_1 e_1^2 \frac{m}{\cos\phi \cos\theta} \varepsilon_1 \frac{4e^{-2s_1}}{(1+e^{-2s_1})^2}$$

(16)   (5.10)

This equation (5.7)-(5.10) will be replaced by the equation (5.1)-(5.6). So, we got an algorithm to minimize the chattering effect on the SMC control algority, so the performance of the quadcopter system we're going to review is expected to be more stable and have a high degree of accuracy in tracking tracks.

Simulation

Having completed all the mathematical aspects involved, we are now preparing to do the simulation. The primary purpose of this simulation is to evaluate how the SMC control algorithm performs when paired with the BPNN in an effort to reduce the chattering effect that occurs. The simulation will run using three different types of tracks: a circle, a spiral, and an eight-digit shape. (lemniscate). In addition to trajectory variations, we will also consider the impact of additional variables, such as wind disturbances. Through this simulation, we want to see if there is a difference in the algorithm performance when wind distortion is applied compared to non-disturbance conditions. The software used in this simulation is purely using MATLAB by https://www.mathworks.com/. In MATLab software, there is a differential problem solver known as ode45. ode45 is one of the functions available in MATLabb that is useful for finding numerical solutions to ordinary differentials or ordinary difference equations (ODE). The method used by ode45 are the Runge-Kutta fourth and fifth orders (RK45) to solve such ODE. ode45 generally is used to resolve ODEs that are not too sensitive to rapid changes in solutions. By automaticrdally splitting time into small control steps, ode45 ensures the accuracy of the resulting solution. This makes ode45 the perfect choice for many applications where the solution changes are not so sharp. After translating all mathematical equations that have been described, including dynamic system equations, control algorithms, as well as external interference models, into the MATLAB programming language, the program is then executed. The simulation results are as follows.

Figure 10. Three-dimensional observation of quadcopter motion on a circular trajectory

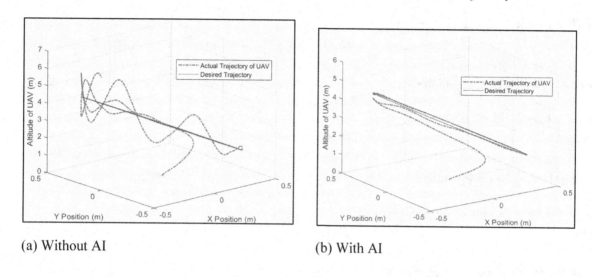

(a) Without AI   (b) With AI

Figure 11. Three-dimensional observation of the motion of the quadcopter on the spiral track

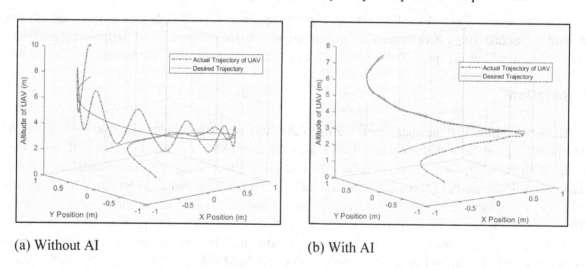

(a) Without AI   (b) With AI

Figure 12. Three-dimensional observation of the motion of the quadcopter on the tracks of lemniscate

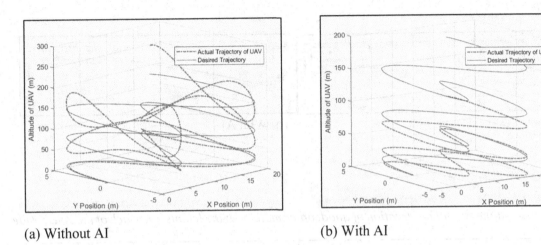

(a) Without AI   (b) With AI

From the illustration in Figure 10-12, the blue line represents the reference route to be followed by the quadcopter, while the red line reflects the actual route that the quadricopter has travelled. Visual analysis shows a considerable difference between outputs from SMC algorithms without AI and those using AI. For example, look at Figure 11, which consists of two sub-images, namely Figure 10(a) and Figue 5.1 (b). In Figure 10(a), the performance of quadcopter with SMC Algorithm without AI, and in the presence of interference, indicates persistent instability and excessive vibration, which is potentially hazardous in real-life situations. However, in Figure 10(b), a significant change is seen when AI is applied. Quadcopter is capable of flying steadily and following a reference route with high accuracy. Similar things happen in Figure 11 and Figure 23, though with different kinds of paths. From here, it can be seen that the contribution of AI has a considerable impact in dealing with the case faced, i.e. minimizing the chattering effect on the SMC algorithm.

For further reassurance, we also performed evaluations in two dimensions, focusing on the Z axis, as shown in Figure 13-15. For example, in Figure 13, there are two sub-images, Figure 14(a) and Figure 14(b). Figure 5(a) shows the performance of the quadcopter against the reference track without using AI, in which case it is seen that the quadricopter has a rather striking instability. However, when the AI is applied, as shown by Figure 14(b), the performance is improved. The same applies to simulations on other types of tracks, illustrated in Figure 14 and 15.

Figure 13. Two-dimensional observation of quadcopter motion on the circular trajectory of the Z-axis side

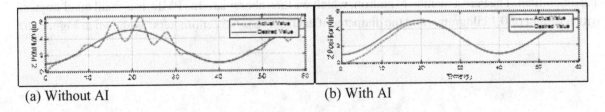

(a) Without AI   (b) With AI

Figure 14. Two-dimensional observation of the motion of the quadcopter on the spiral trajectory of the Z-axis side

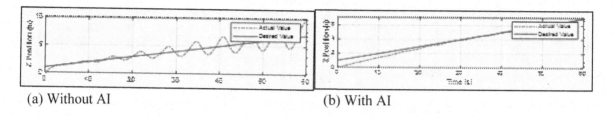

(a) Without AI   (b) With AI

Figure 15. Two-dimensional observation of quadcopter motion on the lemniscate track of the Z-axis side

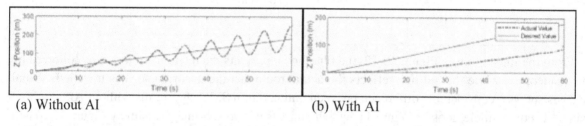

(a) Without AI   (b) With AI

Source: (Darwito & Wahyuadnyana, 2022)

Thus, it can be concluded that the application of AI to the SMC control algorithm proved effective, even in various types of tracks and in the face of external disturbances such as winds from the surrounding environment. However, it is important to note that these conclusions are based only on the simulation stage. Therefore, to test the reliability of our findings from this simulation, we conduct validation through experiments in real environments, which we will explain next..

Experiment

Since the primary objective is the development of control algorithms, so in this experiment we did not design quadcopters from scratch, but rather using quadcops that are already available on the market, and of course open-source. Open-sourced is a concept in which the design and specifications of a device are publicly available and are accessible, modified, and freely distributed by anyone. One of the open source quadcopter available in the market is the Parrot Mambo Minidrone, produced by https://www .parrot.com/en. The Parrot mambo Minidron, which we then abbreviated as PMD, is a small quad-copter, with a weight of 0.1 kilograms, and the diameter of each curve is 6.5 centimeters, as shown in Figure 16.

Figure 16. The physical appearance of Parrot Mambo Minidrone

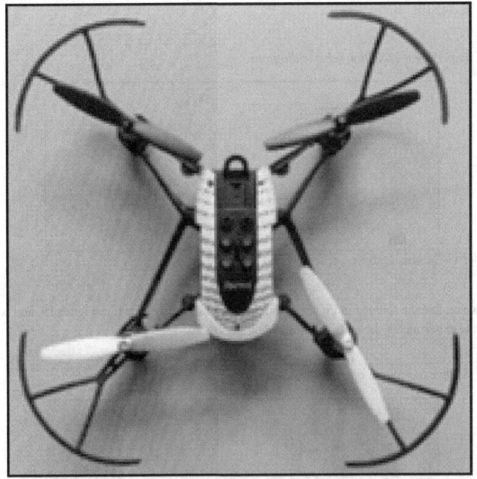

(Kadek Dwi Wahyuadnyana, 2022)

This PMD quadcopter has been equipped with a wireless communication system that supports protocols, enabling interaction with ground stations via a Bluetooth connection. The microcontroller included in the quadcopter is responsible for performing the various calculations required for the control algorithm process, including the processing of data from sensors, cameras, and the implementation of specified control algoritms.

The next question is how to insert a program that has been created into the hardware. We use the features already in MATLAB and Simulink to embed programs into the hardware. However, this discussion will be a subject of its own because the complexity of the method cannot be briefly explained in accorandce with the limits of space available in the writing of this chapter of the book. In essence, once the program is successfully embedded into the PMD quadcopter hardware, PMD will automatically take-off, float at the altitude specified in the program, and land without operator intervention. (autopilot).

To carry out the experiment, we conducted an indoor test, more precisely in the Instrumentation, Control, and Optimization (ICO) laboratory, which is located in the Department of Physical Engineering, the Institute of Technology Sepuluh November. (ITS). We also included a blower, or a wind fan, as an external source of interference for the PMD quadcopter (see Figure 17), with the aim of evaluating its performance when exposed to wind disturbance, so that we can test the stability of the algorithms that have been developed.

Figure 17. Experiment execution scheme diagram

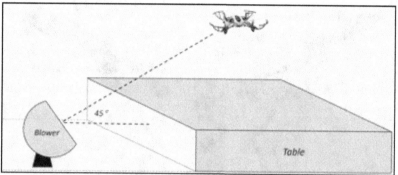

(Kadek Dwi Wahyuadnyana, 2022)

Next, we realize the scheme in Figure 17 on the actual scale, as shown in Figure 18, and we get the results of the experiments shown by Figure 19 and Figure 20.

Figure 18. Setup experiments on actual scale

(Kadek Dwi Wahyuadnyana, 2022)

Figure 19. Experimental results in a state without external interference

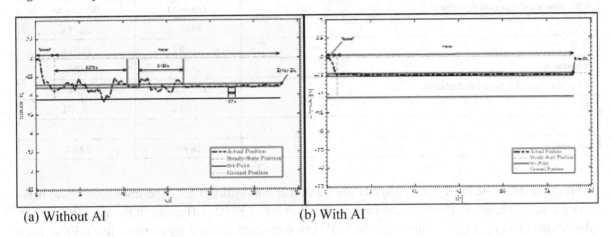

(a) Without AI   (b) With AI

Figure 20. The result of the experiment in a given condition of external interference

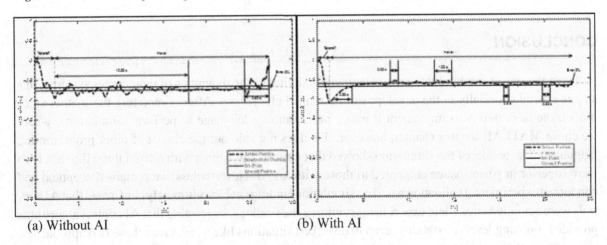

(a) Without AI   (b) With AI

Source: (Kadek Dwi Wahyuadnyana, 2022)

From Figure 19 and Figure 20, it can be seen that the black dividing line shows the actual motion of the PMD quadcopter when running with the developed control algorithm. It was observed that, both without and with external interference from the blower, control algorithms using AI showed better performance stability than those without AI. Next, we performed numerical analysis by re-checking flight data while running and conducting more in-depth analysis. The findings of the analysis are presented in Table 1.

Table 1. Numerical data from the experiment carried out

In the absence of external disturbance	Tanpa AI	Dengan AI
HO	25.89 s	26.78 s
RT	11.68 s	26.78 s
Robustness	**45.1%**	**100%**
In the presence of external disturbance	**Tanpa AI**	**Dengan AI**
HO	25.03 s	26.72 s
RT	6.75 s	20.01 s
Robustness	**26.9%**	**74.8%**

From the data listed in Table 1, you can see the numerical results of the experiments that have been carried out. HO refers to the hovering time or duration when PMD is floating in the air, while RT is robust time, that is to say, the time when the PMD can withstand external interference in air. From both data, the level of stability for each control algorithm can be calculated, whether using AI or not. The results showed that without external interference, the stability rate of the control algoritm without AI was only 45.1 percent, whereas the controlling algoritm with AI reached 100 percent. However, when external disturbances such as the wind of the blower were applied, there was a decrease in stability..

CONCLUSION

From the outset, we have done mathematical modeling of the dynamics of quadcopter systems, this step is crucial especially in the development of control algorithms. After completing the mathematics model, the next step is to implement it into a programming language to perform simulations, where we chose MATLAB for this chapter, however, this does not rule out the choice of other programming languages. The results of the simulation showed that control algorithms with artificial intelligence (AI) were superior in performance compared to those without AI. Nevertheless, we remained sceptical and conducted experimental validation based on simulation findings, which ultimately confirmed that AI controls were actually better. Field experiments reinforce the findings, suggesting that AI control algorithms provide promising levels of stability, even in disturbed situations like wind. From these two approaches, we conclude that control algorithms with artificial intelligence are promising for use in the future..

REFERENCES

Agustina, N. P., & Darwito, P. A. (2023). Autonomous Quadcopter Trajectory Tracking and Stabilization Using Control System Based on Sliding Mode Control and Kalman Filter. *2023 International Seminar on Intelligent Technology and Its Applications: Leveraging Intelligent Systems to Achieve Sustainable Development Goals, ISITIA 2023 - Proceeding*, (pp. 489–493). IEEE. DOI: 10.1109/ISITIA59021.2023.10221176

Araar, O., & Aouf, N. (2014). Full linear control of a quadrotor UAV, LQ vs H∞. *2014 UKACC International Conference on Control, CONTROL 2014*, (pp. 133–138). IEEE. DOI: 10.1109/CONTROL.2014.6915128

Beck, H., Lesueur, J., Charland-Arcand, G., Akhrif, O., Gagne, S., Gagnon, F., & Couillard, D. (2016). Autonomous takeoff and landing of a quadcopter. *2016 International Conference on Unmanned Aircraft Systems, ICUAS 2016*, (pp. 475–484). IEEE. DOI: 10.1109/ICUAS.2016.7502614

Bouadi, H., Bouchoucha, M., & Tadjine, M. (2007). Sliding mode control based on backstepping approach for an UAV type-quadrotor. *International Journal of Mechanical, Aerospace, Industrial, Mechatronic and Manufacturing Engineering, 1*(2), 39–44. http://www.waset.org/publications/11524

Darwito, P. A., & Wahyuadnyana, K. D. (2022). Performance Examinations of Quadrotor with Sliding Mode Control-Neural Network on Various Trajectory and Conditions. *Mathematical Modelling of Engineering Problems, 9*(3), 707–714. DOI: 10.18280/mmep.090317

Dogru, S., & Marques, L. (2020). Pursuing Drones with Drones Using Millimeter Wave Radar. *IEEE Robotics and Automation Letters, 5*(3), 4156–4163. DOI: 10.1109/LRA.2020.2990605

Idrissi, M., Salami, M., & Annaz, F. (2022). A Review of Quadrotor Unmanned Aerial Vehicles: Applications, Architectural Design and Control Algorithms. *Journal of Intelligent & Robotic Systems, 104*(2), 22. DOI: 10.1007/s10846-021-01527-7

Kadek Dwi Wahyuadnyana, P. A. D. (2022)... *Parallel Control System PD-SMCNN for Robust Autonomous Mini-Quadcopter.*, (July), c1–c1. DOI: 10.1109/ISITIA56226.2022.9855344

Lappas, V., Shin, H. S., Tsourdos, A., Lindgren, D., Bertrand, S., Marzat, J., Piet-Lahanier, H., Daramouskas, Y., & Kostopoulos, V. (2022). Autonomous Unmanned Heterogeneous Vehicles for Persistent Monitoring. *Drones (Basel), 6*(4), 1–27. DOI: 10.3390/drones6040094

Le Nhu Ngoc Thanh, H., & Hong, S. K. (2018). Quadcopter robust adaptive second order sliding mode control based on PID sliding surface. *IEEE Access : Practical Innovations, Open Solutions, 6*, 66850–66860. DOI: 10.1109/ACCESS.2018.2877795

Lin, C.-Y., Rahaman, M., Moslehpour, M., Chattopadhyay, S., & Arya, V. (2023). Web Semantic-Based MOOP Algorithm for Facilitating Allocation Problems in the Supply Chain Domain. *International Journal on Semantic Web and Information Systems, 19*(1), 1–23. DOI: 10.4018/IJSWIS.330250

Lu, S., Tsakalis, K., & Chen, Y. (2022). Development and Application of a Novel High-order Fully Actuated System Approach: Part I. 3-DOF Quadrotor Control. *IEEE Control Systems Letters, 7*, 1177–1182. DOI: 10.1109/LCSYS.2022.3232305

Mangsatabam, R. (2018). *Control Development for Autonomous Landing of Quadcopter on Moving Platform.*

Manuylova, N. B., & Bulychev, S. N. (2023). Space monitoring in precision agriculture. *IOP Conference Series. Earth and Environmental Science*, 1154(1), 012043. DOI: 10.1088/1755-1315/1154/1/012043

Marshall, J. A., Sun, W., & L'Afflitto, A. (2021). A survey of guidance, navigation, and control systems for autonomous multi-rotor small unmanned aerial systems. *Annual Reviews in Control*, 52(July), 390–427. DOI: 10.1016/j.arcontrol.2021.10.013

Mellinger, D., Lindsey, Q., Shomin, M., & Kumar, V. (2011). Design, modeling, estimation and control for aerial grasping and manipulation. *IEEE International Conference on Intelligent Robots and Systems*, (pp. 2668–2673). IEEE. DOI: 10.1109/IROS.2011.6094871

Polo, J., Hornero, G., Duijneveld, C., García, A., & Casas, O. (2015). Design of a low-cost Wireless Sensor Network with UAV mobile node for agricultural applications. *Computers and Electronics in Agriculture*, 119, 19–32. DOI: 10.1016/j.compag.2015.09.024

Razmi, H., & Afshinfar, S. (2019). Neural network-based adaptive sliding mode control design for position and attitude control of a quadrotor UAV. *Aerospace Science and Technology*, 91, 12–27. DOI: 10.1016/j.ast.2019.04.055

Xuan-Mung, N., Nguyen, N. P., Nguyen, T., Pham, D. B., Vu, M. T., Thanh, H. L. N. N., & Hong, S. K. (2022). Quadcopter Precision Landing on Moving Targets via Disturbance Observer-Based Controller and Autonomous Landing Planner. *IEEE Access : Practical Innovations, Open Solutions*, 10(July), 83580–83590. DOI: 10.1109/ACCESS.2022.3197181

Chapter 5
Industrial Evolution:
The Integration of AI in Robotics and Drone Systems

Mosiur Rahaman

https://orcid.org/0000-0003-0521-2080

International Center for AI and Cyber Security Research and Innovations, Asia University, Taiwan &
Computer Science and Information Engineering, Asia University, Taiwan

Karisma Trinda Putra

Universitas Muhammadiyah Yogyakarta, Indonesia

Bambang Irawan

https://orcid.org/0000-0002-9829-7548

Universitas Esa Unggul, Indonesia

Totok Ruki Biyanto

https://orcid.org/0000-0002-0023-4822

Institut Teknologi Sepuluh Nopember, Indonesia

ABSTRACT

This chapter presents an in-depth analysis of the transformative role of artificial intelligence (AI) in the fields of industrial robotics and drone technology, offering a comprehensive overview of the integration and evolution of these technologies. Beginning with a historical perspective, the study traces the development of robotics and drones within industrial contexts, laying the foundation for understanding the significant impact of AI. The analysis reveals how the advent of AI has revolutionized these technologies, shifting from basic mechanization to advanced, intelligent systems capable of complex tasks, and decision-making. The study explores the progression from initial AI applications in robotics to the current state-of-the-art implementations, demonstrating the profound changes in efficiency, capability, and functionality. By examining the interplay between AI, robotics, and drones, the study provides insights into the future trajectory of these technologies and their potential to redefine industrial processes.

DOI: 10.4018/979-8-3693-2707-4.ch005

INTRODUCTION

The integration of Artificial Intelligence (AI) into industrial robotics and drone technology marks a significant milestone in the evolution of industrial automation. This chapter offers a comprehensive exploration of this integration, underscoring the transformative impact of AI on these technologies (Banaeian Far & Imani Rad, 2024).

The journey of industrial robotics began in the mid-20th century, primarily focusing on automating repetitive and hazardous tasks. Early industrial robots were simplistic, limited to predefined tasks with minimal adaptability (Kabir et al., 2023). However, the advent of computer technology and software development brought sophistication to these machines, enabling more complex operations and broader applications in industries like automotive, manufacturing, and electronics (Javaid, Haleem, Singh, & Suman, 2022).

Similarly, the evolution of drones, or unmanned aerial vehicles (UAVs), has been remarkable. Initially developed for military applications, drones have found extensive use in various commercial and industrial sectors (Telli et al., 2023). Early industrial drones were used for simple tasks like aerial photography and basic surveillance. However, advancements in technology have expanded their capabilities to include tasks like detailed inspections, agricultural monitoring, and logistics support (Hafeez et al., 2023).

The introduction of AI has been a game-changer for both robotics and drones. AI's initial impact was seen in the enhanced autonomy and decision-making abilities of these machines. In robotics, AI enabled the development of machines capable of learning from their environment, adapting to new tasks, and making complex decisions (Emaminejad & Akhavian, 2022). For drones, AI integration meant advanced navigation systems, improved data processing capabilities, and the ability to execute more complex missions autonomously (Hodge et al., 2021).

This chapter provides an in-depth analysis of the transformative role of artificial intelligence (AI) in the fields of industrial robotics and drone technology. It begins by tracing the historical development of robotics and drones in industrial settings, setting the stage for understanding the profound impact of AI integration.

In the first section, the chapter delves into the foundational aspects of industrial robotics, charting its evolution from simple mechanized tools to sophisticated AI-integrated systems. It discusses the core technologies that underpin industrial robotics, such as automation, control systems, and sensory feedback mechanisms. The early adaptations of AI in robotics are explored, highlighting how these integrations have revolutionized manufacturing processes and efficiency.

The second section focuses on the advancements in AI-driven industrial robotics. It details key AI technologies, including machine learning and computer vision, that have been pivotal in advancing robotics. Through various case studies, the chapter illustrates successful AI implementations across different industrial sectors. It also addresses the challenges encountered in this integration, such as safety concerns, reliability issues, and ethical considerations.

The emergence of intelligent drones in industry forms the third section. This part traces the evolution of drone technology from basic automated machines to advanced AI-enabled devices capable of complex decision-making. The chapter examines diverse applications of AI-integrated drones in fields like agriculture, surveillance, and logistics, demonstrating their versatility and efficiency.

A critical aspect of the chapter is the exploration of the synergy between AI, robotics, and drones in the fourth section. It discusses how interconnectivity and data sharing between these systems enhance overall operational efficiency. The concept of collaborative robotics and drone operations is introduced, along with futuristic visions like autonomous swarms and AI collaboration networks.

The impact of AI-driven robotics and drones on the workforce is a significant focus in the fifth section. It analyzes the shift in job roles and skill requirements brought about by these technologies, highlighting the growing need for human-machine collaboration. The chapter underscores the importance of education and training to prepare the future workforce for these changes.

In the sixth and final section, the chapter forecasts future trends and predictions in AI, robotics, and drone technology. It speculates on the potential impact of emerging technologies, such as quantum computing and advanced AI algorithms, and discusses the role of AI in promoting sustainable and green industrial practices. The chapter concludes with a vision for the next decade, contemplating the direction in which robotics and drone technologies are headed.

Purpose and Scope of the Chapter

The purpose of this chapter is to provide an in-depth analysis of how AI has revolutionized industrial robotics and drone technology. It aims to trace the historical development of these technologies and examine the role of AI in enhancing their capabilities. The chapter will explore various aspects of AI integration, including technological advancements, application areas, challenges faced, and the future potential of these technologies. Additionally, it will discuss the broader implications of AI-driven robotics and drones on the industrial sector, workforce, and society at large. In summary, this chapter aims to offer a thorough understanding of the transformative impact of AI on industrial robotics and drones, highlighting the technological advancements, challenges, and future prospects of this exciting and rapidly evolving field.

THE FOUNDATIONS OF AI IN INDUSTRIAL ROBOTICS

The transformative role of Artificial Intelligence (AI) in the evolution of robotics and drone systems is a chronicle of technological advancements and the evidence of industrial automation and intelligence. From the rudimentary mechanization of systems performing simple repetitive tasks, the integration of AI has transitioned these systems into advanced, intelligent systems capable of complex tasks and decision-making.

Historical Perspective of Robotics in Industry

The foundational era of industrial robotics from the 1950s to the 1970s set the stage for advanced automated manufacturing and industrial processes. The concept of robotics as 'work that is done with robots' emerged during this time. The first electronic autonomous robots were created in the 1940s, programmed to think and respond to light stimuli. However, it was the first fully autonomous factory and commercial robot that appeared in the mid-20th century (1954-1961) known as the 'Unimate' that bore witness to the meaning of 'work that is done with robots', and set the stage for the revolution of industrial robotics (Trevelyan et al., 2008).

The Unimate, invented by George Devol, was a hydraulic manipulative arm designed to perform hazardous and repetitive tasks in a hot die-casting environment, such as handling molten metal, welding, and other repetitive tasks. The success of the Unimate marked the dawn of robotic automation in manufacturing despite the lack of intelligence or adaptability. These systems operated only through pre-programmed routines and could only be used for structured, predictable tasks since they were incapable of responding to changes in the environment (Gasparetto & Scalera, 2019).

Following this, in the late mid-20th century, robots became more sophisticated with the advent of microprocessors and computer technology. The tiny yet powerful microprocessor chips revolutionized how machines could be controlled and programmed. It also enabled more precise control and programming, facilitating robots' ability to perform a wider range of tasks. The systems in this microelectronics revolution could execute more complex assembly operations, precision tasks in electronics manufacturing, and even paint jobs in the automotive industry through Programmable Logic Controllers (PLC). These robotic systems could operate at a faster pace, with greater accuracy, and for extended periods without fatigue. The era also saw advancements in sensor technology, the integration of which allowed robots to receive and process information from their environment, leading to early stages of automated decision-making based on real-time data. Though the performance of these systems was still limited to a structured and predictable environment, it paved the way for the next generation of intelligent, interconnected robotic systems.

The last few years of the 20th century saw the introduction of complex interfaces, vision and voice systems and the integration of advanced sensors like cameras, force sensors, and laser sensors in robotics. Also, these robots saw a shift towards more sophisticated control systems, with the widespread adoption of servo controls, microprocessors and PLC, enabling sophisticated and complex task execution. The advancements in synchronized control systems enabled the coordinated control of multiple robots.

The beginning of the 21st century saw the emergence of industrial robots featuring high-level intelligent capabilities with advanced computational abilities, engaging in logical reasoning, deep learning, and implementing complex strategies. This era also marked the rise of collaborative robots, or cobots, designed to work alongside humans, enhancing safety and productivity while adapting to dynamic work environments (Pluchino et al., 2023). Advancements in sensor technology and the Internet of Things (IoT) also have facilitated seamless interaction with surroundings, making robots more intuitive and responsive. Table 1 presents a chronological view of the evolution of industrial robotics, highlighting key robots and their significance.

Table 1. A chronological view of the evolution of industrial robotics

Time Period	Inventor/Company	Name of Robot	Significance	Purpose of Robot
1950-1967: The First Generation of Industrial Robots				
Late 1940s	William Grey Walter	Elmer and Elsie	Early examples of autonomous robots capable of phototaxis	Studying neurological processes, exhibiting autonomous behavior
1954	George Devol	Programmable Article Transfer	Base for the development of Unimate, the first true industrial robot	Precursor to automated industrial tasks

continued on following page

Table 1. Continued

Time Period	Inventor/Company	Name of Robot	Significance	Purpose of Robot
1961	George Devol & Joseph Engelberger	Unimate	First industrial robot installed in General Motors for automation	Automation in manufacturing, particularly in automotive
1962	AMF Corporation	Versatran	Early cylindrical robot, popular in automotive factories	Material handling and automation in manufacturing
1969	Kawasaki Heavy Industries	Kawasaki-Unimate 2000	First industrial robot built in Japan, expanding robot use in Asia	General industrial automation, expanding use in Asian markets
1968- 1977: The Second Generation of Industrial Robots				
1973	Victor Scheinman	Stanford Arm	First prototype of an electrically actuated robot with advanced control	Research and development in robotics
1973	KUKA	Famulus	One of the early commercially available industrial robots	Various industrial applications, including assembly and handling
1974	Cincinnati Milacron	T3 (The Tomorrow Tool)	First minicomputer-controlled industrial robot	Automotive and general manufacturing automation
1974	ASEA (now ABB)	IRB-6	Known for complex tasks like machining and welding	Complex industrial tasks like machining and arc-welding
1974	Hitachi	HI-T-HAND Expert	Precision robot with force feedback control and flexible wrist	Precision assembly and material handling
1978- 1999: The Third Generation of Industrial Robots				
1978	Hiroshi Makino	SCARA	A novel robot design suitable for assembly tasks	Assembly of small objects, particularly in electronics
1981	Takeo Kanade and Haruhiko Asada	CMU Direct Drive Arm	Featured direct drive actuation for higher accuracy	Research and development, precision tasks
1984	Adept Technology	AdeptOne	First commercially available direct-driven SCARA robot	Electronic assembly and other precision tasks
1980s	Honda Motor Company	ASIMO	One of the most advanced humanoid robots, known for its bipedal motion	Research and public interaction, showcasing advanced robotics technology
1985	Yaskawa Electric Corporation	Motoman-L10	One of the first robots with a large payload capacity	Heavy industrial tasks, such as automotive assembly
1992	Reymond Clavel	Delta Robot	A parallel robot for high-speed pick-and-place operations	High-speed pick-and-place operations in packaging and manufacturing
1994	Motoman	MRC System	First system for synchronized control of multiple robots	Coordinated industrial tasks involving multiple robots
1997	NASA / JPL	Sojourner	First Mars rover, demonstrating robotic exploration in space	Exploration of Mars, collecting data for scientific research
2000 – Present: The Fourth Generation Robots			Characterized by intelligent capabilities like AI, deep learning	Diverse industrial tasks, collaboration with humans, adaptable automation

The robots mentioned above also vary in terms of being autonomous or semi-autonomous in addition to the type of robots. Autonomous robots are designed to operate independently, making decisions and performing tasks without ongoing human intervention. These robots can function over extended periods, perform maintenance, and follow directions. On the other hand, semi-autonomous robots require some level of human interaction or supervision while performing the tasks, particularly for complex decision-making, problem-solving, or when navigating uncertain environments. Though real-time inference is available, autonomous robots have the potential and sufficient computational power to decide on the best course of action to accomplish set goals.

Core Technologies Behind Industrial Robotics

Industrial robotics integrates several core technologies to function effectively. The three generations of robots discussed above can also be seen in terms of technological innovation. For instance, the first generation of robots typically consists of imitation industrial robots, the body and the teaching box where the control signal is programmed. The second generation saw robots with off-line programming features that would work based on a simulation environment. The third-generation robot went further by integrating sensors and machine vision to provide vital information about the robot's environment and its own state. Based on this account, the following can be seen to be the key technologies involved:

Mechanical Engineering and Kinematics: Mechanical engineering contributes to the design, analysis, manufacturing, and maintenance of robots, such as developing arms, joints, and end effectors like grippers or tools, considering aspects such as material science and structural analysis. Kinematics is crucial for understanding and designing the robot's movement as it allows engineers and designers to predict and plan the movements of a robot's parts. Particularly in the context of first and second-generation industrial robots dealing with robotic arms, kinematics was used to ascertain the positions, velocities, and movement patterns of robotic arms utilizing concepts like forward and inverse kinematics for control and precision in tasks. Each component must be meticulously engineered to perform specific tasks and work well harmoniously as part of a larger system.

Control Systems and Servomechanisms: Control systems and servomechanisms govern the movements and operations of robots, ensuring precision, safety and reliability in various industrial tasks. Control systems like open-loop systems operate without feedback, meaning the control action is independent of the output, while closed-loop systems continuously monitor and adjust their operation based on feedback from sensors. These advanced control systems ensure accuracy and adaptability in various tasks. Based on the feedback, servomechanisms use error-sensing feedback to correct the performance of a mechanism. These are often pivotal for precise control of position, velocity, and acceleration. Second-generation robots were further enhanced by microcontrollers, compact integrated circuits designed to govern specific operations, and Programmable Logic Controllers, which are more robust, versatile, and capable of handling complex tasks with high reliability.

Electrical and Electronic Engineering: Electrical and electronic engineering covers everything from basic circuitry to complex control systems, ensuring that robots function efficiently, reliably, and safely. Figure 1 illustrates the different aspects of industrial robotics under the scope of electrical and electronic engineering.

Figure 1. Mindmap illustrating the different key components of industrial robotics under electrical and electronic engineering

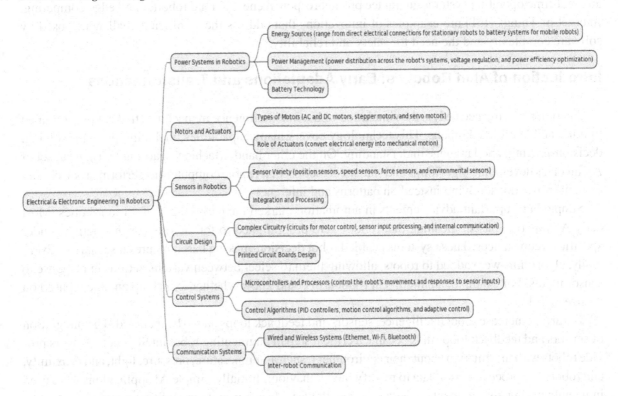

Sensors and Machine Vision: Third-generation robotics incorporated sensors and machine vision systems to enhance the different capabilities of robots, making them more efficient, flexible, and intelligent. Different types of sensors include vision systems (involving cameras and image processing systems), tactile sensors (which involve detecting pressure, texture, and temperature), force and torque sensors, proximity sensors (which involve detecting the presence or absence of objects), and environmental sensors that detect conditions like temperature, humidity, or chemical composition which are essential for process control in various industrial applications.

Pneumatics and Hydraulics: Pneumatics and hydraulics have been integral core technologies behind the early generation of industrial robotics. Pneumatics utilizing compressed air are used for tasks that require speed and repetition, while hydraulics based on pressurized fluids excel in applications demanding substantial force and precision, such as heavy manufacturing. Both systems include components like compressors, valves, and cylinders that are tailored to the required system's operational needs.

Computer Science and Software Engineering: Computer science and software engineering are key fundamental technologies in industrial robotics, covering everything from high-level algorithms for task planning and execution to low-level firmware for hardware control. Path planning, obstacle avoidance, and task-specific algorithms are essential elements created through sophisticated software engineering. Frameworks like ROS and programming languages like Python, C++, and Java are essential to this development.

AI and machine learning are becoming increasingly important, improving robot flexibility and decision-making. Furthermore, software architecture in robotics is frequently modular for scalability, and real-time operating systems guarantee prompt responsiveness. Cloud robotics and edge computing, Internet of Things (IoT) are examples of innovations that address the significant challenges posed by software complexity and the need for safety and reliability.

Introduction of AI in Robotics: Early Adaptations and Transformations

Artificial Intelligence (AI) is the simulation of human intelligence in machines that are programmed to think and learn like humans. This technology covers many capabilities, including problem-solving, decision-making, and language understanding. On the other hand, Machine Learning (ML), a subset of AI, involves developing algorithms and statistical models that enable computers to perform tasks without explicit instructions, relying instead on patterns and inference.

Simple but important advancements in automation marked the early stages of AI in robotics. Most early AI was rule-based, meaning that robots were programmed to follow rules or instructions under specific circumstances. These systems enabled robot decision-making based on preset scenarios. Eventually, algorithms were added to robots, allowing them to select between various actions in response to sensor inputs. For example, an intelligent robot arm could know whether to pick up an object based on its size or colour.

To further increase robot intelligence, sensors and feedback loops had to be included. The integration of sensors and feedback loops thus became a critical step in augmenting robot intelligence. Sensors provide robots with information about their environment, such as temperature, pressure, light, and proximity, and robots can process sensor data to modify their behaviour. Initially, simple AI applications were used in manufacturing environments, where robots with basic decision-making abilities completed jobs like welding or assembly line labour.

With time and technological advancements, robots could enhance their operations based on prior experiences or environmental changes thanks to early machine learning techniques like pattern recognition, which allowed robots to distinguish between objects or conditions, and adaptive learning, which taught robots to adapt to new tasks and environments through limited learning capabilities. Furthermore, early robotics programming languages like VAL (Victor's Assembly Language) allowed for more sophisticated control than conventional programming languages. Better programming environments and user interfaces were also added to robotics software upgrades, which made it simpler for operators to give commands and for robots to carry out increasingly complicated tasks.

This synergistic relationship between AI and ML enhances efficiency and unlocks new possibilities in robotics as it helps them to handle complex, dynamic environments, broadening the horizon of technological advancements and applications. Figure 2 showcases how this interaction works between AI/ML researchers, AI, ML, robotics systems, and data collection.

Figure 2. Sequence diagram illustrating the integration of AI and ML in robotics

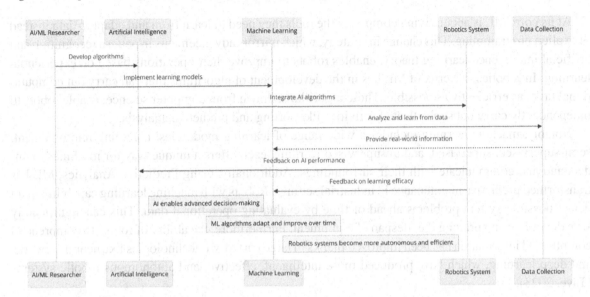

This is also evident in business process automation, where AI and ML elevate the capabilities of Robotic Process Automation (RPA) by introducing intelligent decision-making and problem-solving skills. ML, in particular, empowers RPA solutions to learn from data patterns and user interactions, leading to continuous improvement in task execution. This fusion of AI and ML with RPA helps to create more sophisticated and intelligent automation solutions that help to reshape the future of work by streamlining operations and reducing errors.

ADVANCEMENTS IN AI-DRIVEN INDUSTRIAL ROBOTICS

The application of AI in industrial robotics represents a significant transition from simple automation to giving robots the capacity to learn, adapt, and make decisions. The technological advancement in AI has enabled robots to execute complex tasks with previously unheard-of accuracy and efficiency and has played a significant role in this evolution. Robots are thus now able to go beyond the conventional limitations of preprogrammed routines and can now assess their surroundings, draw lessons from the past, and decide instantly. ML further amplifies the capabilities of these industrial robots as robots can improve their performance over time without explicit reprogramming through exposure to vast datasets and iterative learning processes.

AI Technologies Revolutionizing Robotics

In industrial robots, AI drives a transformation beyond automation to teach robots to learn, adapt, and make judgments. Technologies such as modern computer vision and ML algorithms have enabled robots to carry out more complicated jobs with high accuracy and efficiency.

Machine Learning

At its core, ML is about giving computers the tools they need to learn from and adapt to data instead of explicit programming. This change in strategy, which mirrors advancements in pattern recognition and artificial intelligence learning theory, enables robots to improve their operations based on experiential learning. In robotics, the crux of ML lies in the development of algorithms that can carry out computational tasks as efficiently as possible. These algorithms, drawn from computer science, enable robots to independently carry out sophisticated activities like sorting and predictive analysis.

Notably, machine learning includes a wide range of learning modalities, including reinforcement, semi-supervised, supervised, and unsupervised learning, each offers a unique way for machines to understand and communicate with their surroundings. Additionally, using Predictive Analytics, ML has transformed predictive maintenance in industrial settings. Robots with machine learning capabilities can detect possible system problems ahead of time by evaluating operational data. This can significantly save downtime and prolong the lifespan of equipment. Furthermore, the ability of robots to comprehend complicated information has been improved thanks in large part to ML technologies like neural networks and deep learning, which have produced more intelligent, effective, and autonomous robotic systems (Taye, 2023).

Computer Vision

Basic image processing was the first step in developing computer vision in robotics when robots were trained to recognize particular patterns or colors. These abilities grew as AI developed, enabling robots to comprehend and interact with their environment in addition to seeing. The visual skills of robots have significantly increased with the introduction of high-resolution cameras and smart sensors. These technologies, combined with sophisticated algorithms, allow robots to analyze intricate scenes instantly. Convolutional Neural Networks (CNNs), in particular, have been instrumental in improving computer vision through deep learning. Thanks to this technology, robots can also already evaluate visual data with human-like depth and complexity (Chai et al., 2021).

With advancements in computer vision, robots' capacity to discern and categorize objects according to their visual attributes has also developed, leading to better object recognition. In this process, methods such as texture analysis, edge identification, and pattern recognition are essential. Advanced computer vision systems allow robots to sense depth and do more than recognize objects in two dimensions, which is critical for activities like robotic grasping and spatial navigation.

Other vital technologies revolutionizing AI-powered robotics are Natural Language Processing (NLP), Speech Recognition and Synthesis, Affective Computing, Augmented Reality (AR), IoT, and Conversational AI.

Case Studies and Examples: Successful Implementations in Various Industries

The fourth-generation robots witnessed remarkable advancements in AI-driven robotics, marking significant milestones each year. These robots, which range from personal assistants to space explorers, have demonstrated their technological capability and created new opportunities across various industries.

Case Study: AI-Powered Farm Equipment From John Deere

John Deere's entry into this field in the latter part of the 2010s illustrates how artificial intelligence (AI) and machine learning (ML) have revolutionized agricultural operations, especially in precision farming. Their main goal was to create precision farming-capable tractors, harvesters, and other farm equipment. These devices have AI algorithms, GPS technology, and sophisticated sensors. They collect and process field data so the machinery can make informed decisions. This entails modifying planting schemes, making the most use of fertilizer, and fine-tuning irrigation systems per the particular requirements of each farm area.

Farmers utilizing John Deere equipment have seen increased crop yields while using AI to gain data-driven insights. Precision farming techniques lower waste and improve crop development by ensuring that seeds, water, and fertilizers are used as efficiently as possible. These AI-powered devices' increased efficiency has also resulted in a significant reduction in operating expenses. Above all, the influence on sustainability is arguably one of the most significant advantages since agriculture methods have a minor environmental impact when precision farming is used. It helps to create a more ecologically friendly and sustainable type of agriculture by using available resources better and lowering the need for pesticides and fertilizers. Also, this technology helps address issues brought on by population expansion and climate change.

Case Study: da Vinci Surgical System

The da Vinci Surgical System, introduced in the early 2000s, marked a significant milestone in robotic-assisted surgery. Intuitive Surgical created it to improve surgical control and precision above and beyond what is possible for human hands. The device gives surgeons unmatched precision by utilizing robotics and artificial intelligence. It consists of a console where the surgeon sits to operate, robotic arms that carry out the surgery, and a high-definition 3D vision system. The surgeon controls the arms remotely, and the AI algorithms offer real-time guidance, ensuring precise and steady movements. Machine learning algorithms, trained on data from thousands of surgeries, assist in real-time decision-making. The system can thus analyze patterns from past surgeries, providing insights that help surgeons make more informed decisions during operations (Reddy et al., n.d.).

The da Vinci system has wholly changed surgery, especially for highly precise procedures. Because of its less invasive technique, patients recover more quickly and experience fewer complications and smaller incisions. Additionally, the system has proven particularly effective in complex procedures, including those for cancer and heart disease, because of its accuracy and control. The precision with which it can operate in confined places has enhanced the results of these complex procedures. In addition to actual surgeries, the da Vinci system is utilized by surgeons for training and simulation, providing a safe environment in which to practice and hone skills. Furthermore, the feedback from the ML algorithms aids in training surgeons on specific surgical techniques.

There have also been cases where AI and ML integration was unsuccessful. For example, Walmart's Use of Bossa Nova Robots for inventory management. This failure was attributed to several factors, such as navigational challenges, inconsistency in data collection, customer and employee acceptance, and cost-effectiveness. This case study offers essential insights into the practical difficulties of integrating robotics in real-world contexts and the necessity of a comprehensive strategy considering a range of stakeholders and quickly changing market circumstances.

Table 2 lists other critical examples of fourth-generation robots that have successfully integrated the application of AI and ML.

Table 2. Key examples of AI and ML- Powered robots

Time Period	Inventor/ Company	Name of Robot	Significance	Purpose of Robot
2004	iRobot	Roomba	One of the first successful domestic robots	Vacuum cleaning in a home environment
2008	Boston Dynamics	BigDog	Advanced quadruped robot for rough-terrain mobility	Research in locomotion and balance in robotics
2012	Google (Alphabet Inc.)	Google Self-Driving Car (Waymo)	One of the pioneering projects in autonomous driving	Research and development in autonomous vehicle technology
2015	Softbank Robotics	Pepper	Humanoid robot designed for interacting with people	Customer service and personal interaction in various settings
2018	Boston Dynamics	SpotMini	Advanced mobile robot with manipulation capabilities	Inspection, data collection, and carrying out tasks in complex environments
2020	SpaceX	Dragon 2's Autonomous Docking System	Demonstrates advanced autonomous operation in space	Autonomous docking with the International Space Station (ISS)

Artificial intelligence (AI) and robotics have changed our everyday lives and pushed the boundaries of what is conceivable, from the ordinary to the spectacular. While space travel was formerly the purview of highly skilled astronauts, autonomous technologies like the Dragon 2 are redefining space exploration, and the once-futuristic idea of robotic vacuum cleaners independently roaming our homes is now a reality. This range of creativity demonstrates the adaptability of robotics powered by AI and its potential to transform a wide range of human endeavors completely. Furthermore, new opportunities arise from the confluence of robots with other cutting-edge technologies like big data, AR, and IoT. But these developments also bring obligations and difficulties. Privacy, ethics, and security concerns are becoming increasingly important as AI and robotics integrate into our daily lives. To ensure that these technologies are developed for the benefit of humanity, their societal effects must be carefully considered throughout their development.

THE EMERGENCE OF INTELLIGENT DRONES IN INDUSTRY

Unmanned Aerial Vehicles (UAVs), sometimes referred to as drones, have evolved from military instruments to multipurpose assets in a range of commercial, recreational, scientific, and civil businesses. This evolution is a testament to the rapid advancements in technology, AI integration, and the expanding range of uses for drones.

The Evolution of Drone Technology

Drones were initially developed mainly for military use during World War I. The American Kettering Bug was created in 1918, while the British engineer Archibald Low created the Aerial Target in 1915. While they weren't widely employed in battle, these early aircraft demonstrated the military potential of unmanned flight systems.

Drone technology advanced to new heights after World War II. Unmanned Aerial Vehicle (UAV) development reached a turning point in the 1950s and 60s, particularly with regard to reconnaissance applications. The evolution continued with unmanned combat aerial vehicles (UCAVs) in the 1960s, though their significant deployment in combat operations only began in the 1990s (Kumar, 2020).

Drone technology has evolved over the past few decades to serve a broad range of commercial and civilian purposes, in addition to military ones. Drones are becoming more widely available, compact, lighter, and less expensive. These days, they are essential to many professions, including search and rescue, cinematography, infrastructure inspection, agricultural crop inspection, and aerial footage. Notably, drone technology has also sparked interest in delivery services, with companies like Amazon and Google exploring drones for package delivery. Figure 3 depicts the evolution of drone technology.

Figure 3. Timeline illustrating the evolution of drone technology

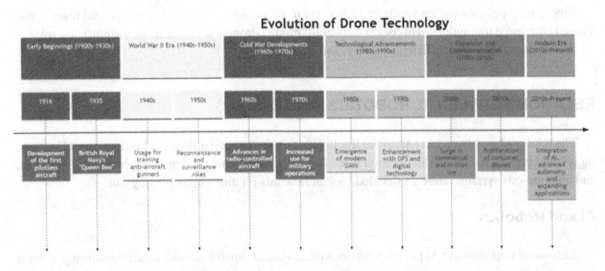

The advent of intelligent drones represents a significant turning point in artificial intelligence and robotics history. Drones have transformed from primitive military weapons to advanced tools for agriculture, environmental conservation, and surveillance, offering creative answers to challenging problems in various industries.

Integration of AI in Drones: From Basic Automation to Advanced Decision-Making

Autonomous Capabilities: Drones can now operate autonomously, having been upgraded from being remotely controlled through the integration of AI and machine learning. These intelligent systems are now capable of autonomous navigation, object detection, and complex decision-making.

Machine Learning and Real-Time Data Processing: Advanced machine learning algorithms allow drones to process enormous volumes of data for improved functioning while learning from various activities and settings. This feature is essential for real-time surveillance and threat detection in the security and agricultural sectors, respectively, as well as crop monitoring.

Advanced Computing and Onboard Technology: Drones have become more efficient by incorporating complex processing tasks, such as Qualcomm's Snapdragon Flight. Drones are now lighter, faster, and more energy-efficient thanks to these technologies, which have decreased design complexity and cost.

Application Areas: Agriculture, Surveillance, Logistics, and Beyond

Versatile Applications: Drones are used widely in policing, firefighting, surveillance, military, and environmental monitoring, demonstrating their adaptability. They support precision farming in agriculture and are transforming package delivery in logistics.

Emerging Applications: Prospects include roles in air refueling operations, search and rescue missions, and remote tour guides. This development highlights drones' potential to play significant roles in various sectors, enhancing efficiency, safety, and accessibility.

RELATION BETWEEN AI, ROBOTICS, AND DRONES

The relationship between AI (Artificial Intelligence), robotics, and drones is a fascinating and increasingly interconnected one. This integration is leading to the development of more sophisticated, efficient, and autonomous systems. Here's a detailed look at how this relationship is playing out.

AI and Robotics

Enhanced Capabilities: AI provides robots with advanced capabilities like machine learning, natural language processing, and computer vision. This enhances their ability to understand and interact with their environment, make decisions, and learn from experiences (Soori et al., 2023).

Improved Efficiency and Precision: In manufacturing and logistics, AI-driven robots can optimize production lines, handle materials with greater precision, and adapt to new tasks quickly (Licardo et al., 2024).

Autonomous Navigation: In sectors like warehousing and healthcare, robots use AI for navigation, avoiding obstacles, and performing tasks like delivering goods or assisting in surgeries (Deo & Anjankar, n.d.).

AI and Drones

Advanced Flight Control: AI algorithms enable drones to fly autonomously, make decisions in real-time, and adapt to changing environmental conditions (Zhu et al., 2024).

Data Analysis and Processing: Drones equipped with cameras and sensors, combined with AI, can process vast amounts of data for applications like agricultural monitoring, infrastructure inspection, and surveillance (Shakhatreh et al., 2019).

Object Detection and Tracking: AI allows drones to recognize and track objects and people, which is crucial in search and rescue operations, wildlife monitoring, and security applications (Buchelt et al., 2024).

Robotics and Drones

Expanded Accessibility and Reach: Drones extend the reach of robotic systems, allowing them to access hard-to-reach areas like high-altitude locations, remote terrain, or dangerous environments (Mohd Daud et al., 2022).

Collaborative Operations: In scenarios like disaster response, drones can work alongside ground robots to provide aerial views, map terrains, and assist in locating victims or hazards (Chen et al., 2020).

Integrated AI, Robotics, and Drones

Complex Task Execution: The combination allows for the execution of complex tasks. For example, drones can survey an area and send data to AI systems, which analyze this data and instruct ground robots to perform specific actions based on the analysis (Iovino et al., 2022).

Real-Time Data Sharing and Decision Making: In smart city applications, drones can monitor traffic and crowd situations, relaying this information to AI systems that process the data and enable robots on the ground to respond accordingly (Alsamhi et al., 2019).

Enhanced Communication Networks: Using AI, drones and robots can form ad-hoc networks for communication in areas lacking infrastructure, like in post-disaster scenarios or remote explorations (Hadiwardoyo et al., 2018).

Precision Agriculture: Drones collect data on crop health, which is processed by AI to guide agricultural robots in tasks like planting, weeding, and harvesting (Rejeb et al., 2022).

Environmental Monitoring and Conservation: AI-driven analysis of data collected by drones can inform robots to act in environmental monitoring and conservation efforts, such as detecting pollution or illegal logging activities (Popescu et al., 2024).

Safety and Security: In security applications, drones can provide aerial surveillance data that AI processes to alert and guide security robots on the ground (Mohsan et al., 2022).

Figure 4. Relation between AI, robotics, and drones

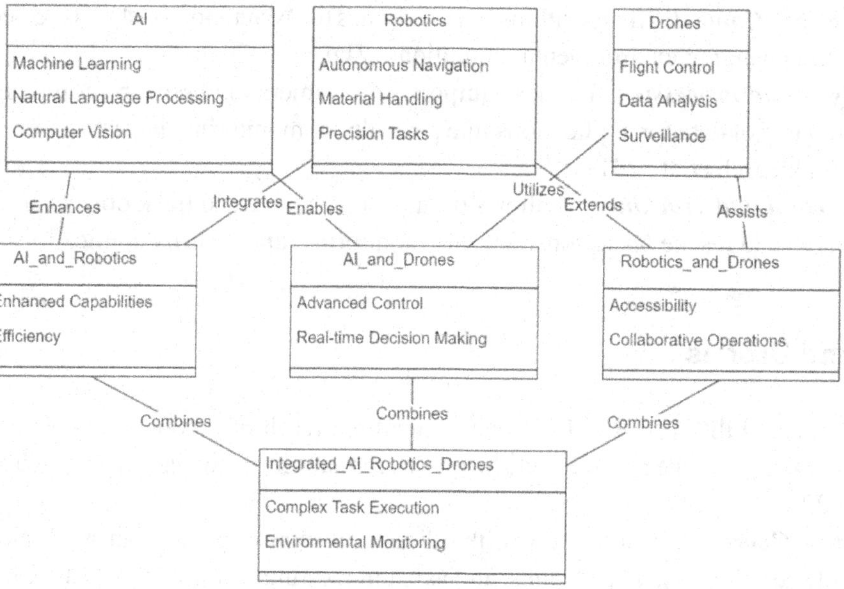

Variables Definition:

AI: Quantitative measure of Artificial Intelligence capability.
R: Quantitative measure of Robotics capability.
D: Quantitative measure of Drones' capability.
S: Overall Synergy or effectiveness of the combined system.

Nonlinear Interaction Equations

1. **AI and Robotics Interaction**:

$$S_{AI,R} = \alpha_1 . AI^\beta . R^\gamma$$

Here, $S_{AI,R}$ represents the synergy between AI and Robotics.
α_1, β, γ are coefficients/parameters to be determined.
This equation suggests that the relationship is not a simple product of AI and Robotics capabilities but is affected more complexly.

2. **AI and Drones Interaction**:

$$S_{AI,D} = \alpha_2 . AI^\delta . D^\epsilon$$

represents the relationship between AI and Drones.
are unique coefficients/parameters for this interaction.

3. **Robotics and Drones Interaction**:

   denotes the relationship between Robotics and Drones.
   are coefficients/parameters specific to this relationship.

4. **Integrated AI, Robotics, and Drones Synergy**:

   is the overall relationship of AI, Robotics, and Drones combined.
   are parameters that adjust the overall relationship level.
   This equation suggests that the total relationship is not a simple sum but is influenced more complexly by the interactions.

Considerations

The parameters would be determined based on empirical data, reflecting how these technologies interact in different scenarios.

These equations represent a model that can be adjusted or expanded based on specific applications, technological advances, or desired outcomes.

Interconnectivity and Data Sharing for Enhanced Efficiency

AI, robotics, and drones are redefining many professions through their networking and data sharing. In addition to increasing productivity, this combination creates new avenues for tackling challenging issues. As these technologies develop, we should anticipate even more creative uses and enhanced functionalities. Table 3 summarizes the ways in which artificial intelligence (AI), robotics, and drones combine to improve efficiency and effectiveness in a range of applications.

Table 3. Interconnectivity and data sharing for enhanced efficiency roles

Role and Aspects	Elaboration
AI's Role	AI serves as the brain of the operation, processing data, making decisions, and learning over time. It analyzes data from drones and robots, identifies patterns, and makes predictive analyses.
Robotics Integration	Robots, equipped with sensors and actuators, perform physical tasks based on AI's analysis and commands. They collect environmental data, feeding it back to the AI system.
Drones Adding a New Dimension	Drones extend operational reach, especially in inaccessible areas, collecting aerial data. They provide real-time surveillance and environmental data, crucial for AI applications.
Enhanced Efficiency through Interconnectivity and Data Sharing	Real-time data exchange between AI, robots, and drones allows for quick responses to dynamic situations and a comprehensive understanding of the environment. AI analyzes data to predict future outcomes and plan, accordingly, leading to adaptive learning and improved efficiency.
Application Areas	Applications include disaster management with real-time data for coordinated response, precision farming with drones and robots, infrastructure inspection and maintenance, and comprehensive security and surveillance.

Within integrated systems that combine robotics, AI, and drones, interconnectivity and data sharing greatly improve productivity. Seamless data flow is a key component that guarantees constant and ongoing information exchange between various components and is at the center of this integration. Robots and AI systems, for instance, receive real-time aerial data from drones and process and analyze it instantly. Combining different data sources—from sensors on robots to cameras on drones—into a single information repository is made easier by this synergy. With the help of this method, AI algorithms can operate on a whole dataset and are given a thorough operational picture. Additionally, this coherent operational dynamic is maintained by real-time communication between these systems, which guarantees that modifications in one segment are swiftly communicated to others.

Decision-making is significantly enhanced by this integrated framework. Robust decision-making and job execution are enhanced by AI, which has access to vast amounts of data. AI, for example, assesses crop health data from drones and soil conditions data collected by robots to make informed farming decisions in precision agriculture. Artificial Intelligence (AI) is not limited to assessments made in real-time; it can also be used to estimate maintenance needs in infrastructure management by using past and current data. Allocating resources and coordinating robots and drones to optimize delivery routes and logistics are just two examples of how effective data sharing enhances operations. Enhanced collaborative functioning is another benefit it provides. In search and rescue missions, for example, robots and drones combine their ground and aerial capabilities to perform effectively.

But overcoming major obstacles is necessary to fully realize the potential of interconnectivity and data exchange. Ensuring compatibility between systems and developing common protocols for data transmission are essential for a smooth integration. Because the information communicated across networks is sensitive, data security and privacy are critical. Systems need to be flexible and expandable to fit different operational scenarios and settings. Building resilient communication networks that can handle massive amounts of data and endure disturbances is essential. Furthermore, it is crucial to develop AI so that it can manage and evaluate the diversity and complexity of data. To summarize, the integration of AI, robotics, and drones is made more efficient by interconnection and data exchange. However, there are certain hurdles that must be overcome in order to fully utilize drones for automation and decision-making.

Collaborative Robotics and Drone Operations

Control and Communication Systems

Advanced communication and control systems are key components of drone operations and collaborative robotics. Real-time information and command interchange between drones and robots is made possible by wireless networks such as Wi-Fi, Bluetooth, or satellite communications. To coordinate their actions across different operational ranges and situations, this network is essential. An AI-powered centralized control system, which interprets data from both drones and robots and makes important choices and orders actions, is at the center of this configuration. Based on the aggregate data and overall operational objectives, this system guarantees a coordinated operation by providing customized directives to each unit.

Data Sharing and Processing

Data integration and processing are essential components of this partnership. Robots and drones are equipped with a variety of sensors that collect various environmental data, such as cameras, LiDAR, GPS, and infrared sensors. The units can react quickly to dynamic changes in their environment because AI algorithms share and analyze this sensor data in real-time. A ground robot, for example, may receive information about an impediment detected by a drone and use it to modify its course. The flexibility and effectiveness of the collaborative system depend on this degree of real-time data processing and sharing.

Task Execution and Collaboration

The assignment of divided roles and cooperative task performance are clear examples of how this synergy is expressed practically. Based on the relative strengths of drones and robots, particular duties are assigned to each; ground robots are better at physical operations, while drones are better at aerial monitoring. To ensure that every unit makes the best possible contribution to the collective effort, these roles are coordinated by a central system. When drones and robots work together to do ground-level rescue chores and aerial reconnaissance, for example, activities are often completed more efficiently. This is seen in search and rescue operations. It is not without difficulties, nevertheless, to achieve this smooth cooperation. Other important factors to take into account include energy management, interoperability, and redundancy and safety assurance.

Future Possibilities: Autonomous Swarms, AI Collaboration Networks

Significant advances in AI and robotics may be seen in both autonomous swarming and AI collaborative networks. They claim to not only boost productivity and effectiveness across the board but also create fresh avenues for creativity. The technological, moral, and security issues that arise with these sophisticated technologies must be addressed as we go toward the future. The potential is enormous, but so is the obligation to create and apply new technologies in a way that is secure and advantageous to society.

The ideas behind AI Collaboration Networks and Autonomous Swarms are summarized in the following table:

Table 4. AI collaboration networks and autonomous swarms

Category	Aspect	Details
Autonomous Swarms	Definition and Concept	Groups of robots or drones operating collectively under minimal human oversight, mimicking natural systems like flocks of birds.
Autonomous Swarms	Advanced Coordination	Each unit in the swarm makes autonomous decisions based on collective goals, enabling adaptable group dynamics.
Autonomous Swarms	Applications	Applications include environmental monitoring, military operations, search and rescue, agriculture, and logistics.
Autonomous Swarms	Challenges	Challenges involve developing robust communication networks, reliable decision-making algorithms, and addressing safety and ethics.

continued on following page

Table 4. Continued

Category	Aspect	Details
AI Collaboration Networks	Concept	Multiple AI systems working together across different platforms (robots, drones, IoT) for collaborative problem-solving.
AI Collaboration Networks	Enhanced Problem Solving	Pooling capabilities and data from various AI entities to tackle complex problems more efficiently than single AI systems.
AI Collaboration Networks	Future of Work	Could lead to sophisticated automation in workplaces, managing different aspects of business or manufacturing processes.
AI Collaboration Networks	Societal Impact	Potential to transform urban infrastructures, transportation, and healthcare systems, impacting society significantly.
General Outlook	Looking Ahead	These advancements promise increased efficiency and new innovation opportunities, but come with responsibilities in technology, ethics, and safety.

An overview of the key features uses cases, and difficulties of AI collaboration networks and autonomous swarms is given in this table, along with an analysis of the possible effects on different industries and the general public.

Establishing models that capture the essential characteristics and capacities of these structures is necessary for considering potential futures for AI collaboration networks and autonomous swarms in a more statistically valid way (Schranz et al., 2021). In terms of mathematics, let us conceive this:

1. **Modelling Autonomous Swarms**:

Let's define an autonomous swarm as a collection of agents, each with a set of behaviors governed by local rules. The performance of the swarm can be modeled as a function of the number of agents, their interaction quality, and the environmental complexity E.

where represents the contribution of each agent, considering its interaction quality and environmental factors. The performance is not linear but is influenced by the interaction quality and the environment. As increases, the function may show super-linear characteristics up to a certain point, beyond which additional agents contribute negatively due to factors like overcrowding or resource limitations.

2. **AI Collaboration Networks**:

An AI collaboration network can be represented as a graph where is a set of nodes (AI agents) and is a set of edges representing the communication links. The effectiveness of the network can be a function of the graph's topology, the data processing capability of each node, and the network's adaptability.

where represents the overall effectiveness function. The topology influences how information is shared and processed within the network. The adaptability reflects the network's ability to reconfigure itself in response to changing conditions.

3. **Proving Effectiveness**:

To prove the effectiveness of these models, we can set hypotheses and validate them through simulations or real-world experiments.

Hypothesis 1 (Autonomous Swarms): There exists a critical number of agents such that is maximized. For .

Hypothesis 2 (AI Collaboration Networks): The network effectiveness is maximized for a certain topology , processing capability , and adaptability . That is, are maximized at .

The behaviour, interactions, and overall performance of these systems:

Autonomous Swarms

1. *Agent Interactions*: The basic principle in autonomous swarms is that the collective behavior emerges from simple local rules followed by each agent. These rules are often based on proximity to other agents or environmental factors.

   Example Equation: Consider an agent in a swarm. Its position at time might be determined by its position at time and the positions of its neighbors. A simple rule could be:

   Here, represents the neighbours of agent , and is a velocity factor. This rule moves each agent towards the average position of its neighbors, a common flocking behavior.

2. **Swarm Performance**: The overall performance of the swarm is not just the sum of individual contributions but is enhanced by the interactions among agents.

   Super-Linear Scaling: This can be represented by an equation where the collective performance depends nonlinearly on the number of agents

   Where and are constants, and indicates super-linear scaling.

AI Collaboration Networks

1. *Network Efficiency*: AI networks are often modelled as graphs, where nodes represent agents and edges represent communication links. Network efficiency can depend on the topology and the processing capabilities of each agent.
   - *Example Equation*: If represents the network graph, the efficiency of the network could be modelled as:

   Here, and are the data processing capability and adaptability of agent , is the communication cost between agents is a function representing the processing effectiveness, and is a function representing the communication efficiency.

2. *Adaptability and Learning*: AI collaboration networks often adapt and learn over time, which can be modeled using learning algorithms.

   Learning Algorithm: If represents the learning model of agent its performance at time could be modelled as:

Where is a learning rate, and represents the gradient of a loss function that measures the difference between the desired and actual output, considering the agent and its neighbors.

These equations are examined in theoretical proofs to determine the ideal circumstances for the system's behaviour. This could involve demonstrating that a certain local rule produces a stable, effective universal behaviour for autonomous swarms. It could entail demonstrating that a certain network topology optimizes efficiency or adaptability for AI networks.

IMPACT OF AI-DRIVEN ROBOTICS AND DRONES ON THE WORKFORCE

Drones and robotics powered by AI have a deep and wide-ranging impact on the workforce, drastically altering employment positions, skill requirements, and the nature of work itself. As we examine this transition in more detail, we find several areas of adaptation and change.

Changes in Job Roles and Skills Requirements

Automating work and changing employment roles and skill sets are two outcomes of the integration of AI, robotics, and drones in numerous industries. There is a movement in the workforce toward more sophisticated, analytical, and creative professions because of computers taking over traditional, repetitive activities.

Higher-Level Tasks: Instead of doing manual tasks, workers are increasingly expected to supervise and manage AI-driven systems. In doing so, data-driven strategic decision-making, output analysis, and system performance monitoring are all included (Zirar et al., 2023).

New Roles Emergence: The integration of robotics and artificial intelligence is leading to the emergence of new work roles. Data analysts, drone operators, robotics programmers and maintenance personnel, and AI trainers who instruct AI systems in pattern recognition are a few examples (Joksimovic et al., 2023).

Evolution of Skills: The current workforce's skill set is changing. Programming, systems thinking, data literacy, and AI literacy are among the talents that are in greater demand. Additionally, soft skills like adaptability, critical thinking, and problem-solving are more important than ever (Poláková et al., 2023).

Opportunities for Human-Machine Collaboration

Instead of simply replacing human labour, the introduction of AI and robotics into the workplace opens novel opportunities for joint development.

Human Capabilities: Robots and artificial intelligence (AI) have the potential to improve the abilities of humans. To obtain meaningful insights, a human operator can evaluate the collected data while a drone performs hazardous survey tasks (Javaid et al., 2021).

Environments for Collaborative Work: Humans and machines collaborate on numerous projects together. To increase productivity and safety, collaborative robots, or cobots, are made to interact with human workers in a safe and productive way (Javaid, Haleem, Singh, Rab, et al., 2022).

Help with Decision-Making: AI-powered systems have the speed and capacity to handle large volumes of data, which enables people to make better, more precise decisions. Industries like healthcare, banking, and engineering can especially benefit from this(Javaid, Haleem, Pratap Singh, Suman, et al., 2022).

Educational and Training Needs for the Future Workforce

An enormous focus on training and education is necessary to fully utilize the promise of robotics, AI, and drones.

Redefining Education Curriculum: The job of incorporating robotics and artificial intelligence (AI) into curricula falls on educational institutions. Along with imparting technical information, this also teaches students how to communicate and collaborate with intelligent systems.

Upskilling and continual Learning: As technology develops quickly, so does the demand for up-skilling and continual learning. In order to remain competitive in the work market, lifelong learning becomes essential.

Public-Private Partnerships in Education: Governments, academic institutions, and private businesses working together to design training programs that meet the ever-evolving needs of the labour market are essential. These partnerships may offer practical experiences, internships, and apprenticeships.

Importance on Multidisciplinary capacities: To be competitive, the workforce requires interdisciplinary abilities that combine technical expertise with domain-specific knowledge and soft skills. Navigating the AI-driven workplace requires this adaptability.

Figure 5. The graph diagram illustrating the statistical aspects of changes in job roles, opportunities for human-machine collaboration, and educational and training needs

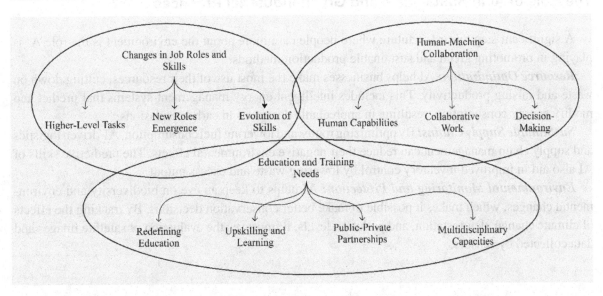

FUTURE TRENDS AND PREDICTIONS

Several trends and projections for the future are important, particularly in the areas of technology, society, and the environment. These tendencies represent the possibility of revolutionary breakthroughs and changes in society in addition to being merely extensions of existing technologies.

Emerging Technologies and Their Potential Impact

Emerging technologies offer an exciting new horizon with potentially revolutionary impacts. Quantum computing and sophisticated AI algorithms are worth exploring.

Quantum computing: This cutting-edge technology has the potential to completely transform industries including materials research, medicines, and encryption because of its extraordinary speed at which it can execute complicated computations. The huge data sets that quantum computing can handle could lead to improvements in machine learning efficacy and efficiency, as well as an exponential development of AI capabilities. Its influence may be observed in the creation of new materials, considerable improvements in medical research, and more precise weather forecasts (Gill et al., 2024).

Advanced AI Algorithms: More complex and powerful algorithms are becoming more common as AI keeps developing. More accurate forecasts and decision-making are made possible by these sophisticated systems' ability to analyse and understand data with a level of detail previously unachievable. The industries that depend heavily on accuracy and dependability, like finance, healthcare, and autonomous cars, will be greatly impacted by this change. Improved healthcare diagnostic tools enhanced financial models, and safer, more effective autonomous transportation options are some of the ramifications.

The Role of AI in Sustainable and Green Industrial Practices

A significant step towards a future where people care more about the environment is the role AI is playing in promoting green and sustainable production methods.

Resource Optimization: AI helps businesses make the most use of their resources, cutting down on waste and raising productivity. This includes intelligent energy management systems that predict and modify energy consumption, resulting in appreciable decreases in carbon emissions.

Sustainable Supply Chains: By optimizing routes and lowering fuel consumption, AI-driven logistics and supply chain management can reduce their negative environmental effects. The predictive skills of AI also aid in improved inventory control by lowering waste and excess output.

Environmental Monitoring and Protection: AI helps to keep an eye on biodiversity and environmental changes, which makes it possible to make better conservation decisions. By tracking the effects of climate change, deforestation, and pollution levels, it improves the evaluation of satellite images and data collected by sensors.

Vision for the Next Decade: Where Robotics and Drone Technologies Are Headed

Robotics and drone technologies have never had more potential than they do in the next 10 years, and new developments will likely redefine their uses.

Enhanced Intelligence and Autonomy: Drones and robotics will become more intelligent and autonomous as a result of advanced AI that will allow them to carry out difficult jobs and make judgments with little assistance from humans. Their use in fields including unmanned research missions, emergency response, and intricate surgical operations will grow as a result of this growth (Soori et al., 2023).

Integration with Daily Life: These technologies will become more easily incorporated into daily life as they get easier to use and more accessible. This includes personal care and household chore-assisting domestic robots as well as drones used for urban mobility and delivery services.

Collaboration and Swarms: The idea of swarms of robots and drones, in which several units cooperate to complete tasks, is going to gain popularity. This might revolutionize activities like search and rescue, large-scale construction projects, and agricultural management.

Regulation and Ethical Development: As a result of these technical advancements, frameworks for regulations and ethical issues will change. At the center of policy conversations will be concerns about privacy, safety, and the socioeconomic effects of automation.

Figure 6. The use-case diagram illustrating the technological perspective with specific technology actors and technology-oriented use cases

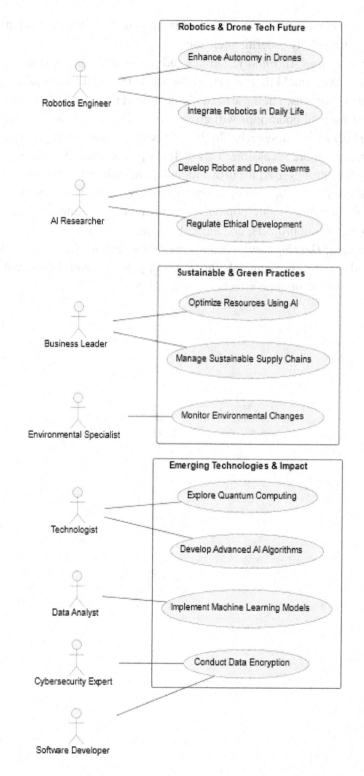

CONCLUSION

This chapter provides a thorough examination of the significant changes that artificial intelligence (AI) has brought about in the fields of industrial robots and drones. It follows the development of mechanization from the early stages to the present, when AI-powered robotics and drones are able to perform complicated functions and make decisions on their own. The experience demonstrates the transition from basic, task-specific machinery to intelligent systems that can adapt, learn, and make decisions on their own. The chapter examines the increased operational efficiency and wider application potential of these integrated technologies, highlighting the beneficial relationship between AI, robots, and drones. It demonstrates how AI is transforming autonomy and decision-making powers, enabling robotics and drones to perform more tasks in fields like logistics, surveillance, and precision agriculture.

The chapter forecasts an agenda for AI development that will result in increasingly intelligent and self-governing systems. These advancements could have a big impact on the transportation, healthcare, and environmental monitoring sectors, among others. AI-driven robotics and drone integration is changing the nature of work and the skills that are required, putting a focus on the necessity of human-machine cooperation and underscoring the significance of education and training for the workforce of the future. As these technologies improve, questions about security, privacy, and ethics become more pressing, highlighting the necessity of giving these issues careful thought. The chapter offers a thorough analysis of these revolutionary technologies, laying the groundwork for upcoming discoveries and uses that have the potential to drastically alter the industrial landscape.

REFERENCES

Alsamhi, S. H., Ma, O., Ansari, M. S., & Almalki, F. A. (2019). Survey on Collaborative Smart Drones and Internet of Things for Improving Smartness of Smart Cities. *IEEE Access : Practical Innovations, Open Solutions*, 7, 128125–128152. DOI: 10.1109/ACCESS.2019.2934998

Banaeian Far, S., & Imani Rad, A. (2024). Internet of Artificial Intelligence (IoAI): The emergence of an autonomous, generative, and fully human-disconnected community. *Discover Applied Sciences*, 6(3), 91. DOI: 10.1007/s42452-024-05726-3

Buchelt, A., Adrowitzer, A., Kieseberg, P., Gollob, C., Nothdurft, A., Eresheim, S., Tschiatschek, S., Stampfer, K., & Holzinger, A. (2024). Exploring artificial intelligence for applications of drones in forest ecology and management. *Forest Ecology and Management*, 551, 121530. DOI: 10.1016/j.foreco.2023.121530

Chai, J., Zeng, H., Li, A., & Ngai, E. W. T. (2021). Deep learning in computer vision: A critical review of emerging techniques and application scenarios. *Machine Learning with Applications*, 6, 100134. DOI: 10.1016/j.mlwa.2021.100134

Chen, J., Li, S., Liu, D., & Li, X. (2020). AiRobSim: Simulating a Multisensor Aerial Robot for Urban Search and Rescue Operation and Training. *Sensors (Basel)*, 20(18), 5223. DOI: 10.3390/s20185223 PMID: 32933186

Deo, N., & Anjankar, A. (2023, May 23). (n.d.). Artificial Intelligence With Robotics in Healthcare: A Narrative Review of Its Viability in India. *Cureus*, 15(5), e39416. DOI: 10.7759/cureus.39416 PMID: 37362504

Emaminejad, N., & Akhavian, R. (2022). Trustworthy AI and robotics: Implications for the AEC industry. *Automation in Construction*, 139, 104298. DOI: 10.1016/j.autcon.2022.104298

Gasparetto, A., & Scalera, L. (2019). From the Unimate to the Delta Robot: The Early Decades of Industrial Robotics. In Zhang, B., & Ceccarelli, M. (Eds.), *Explorations in the History and Heritage of Machines and Mechanisms* (pp. 284–295). Springer International Publishing. DOI: 10.1007/978-3-030-03538-9_23

Gill, S. S., Wu, H., Patros, P., Ottaviani, C., Arora, P., Pujol, V. C., Haunschild, D., Parlikad, A. K., Cetinkaya, O., Lutfiyya, H., Stankovski, V., Li, R., Ding, Y., Qadir, J., Abraham, A., Ghosh, S. K., Song, H. H., Sakellariou, R., Rana, O., & Buyya, R. (2024). Modern computing: Vision and challenges. *Telematics and Informatics Reports*, 13, 100116. DOI: 10.1016/j.teler.2024.100116

Hadiwardoyo, S. A., Hernández-Orallo, E., Calafate, C. T., Cano, J. C., & Manzoni, P. (2018). Experimental characterization of UAV-to-car communications. *Computer Networks*, 136, 105–118. DOI: 10.1016/j.comnet.2018.03.002

Hafeez, A., Husain, M. A., Singh, S. P., Chauhan, A., Khan, M. T., Kumar, N., Chauhan, A., & Soni, S. K. (2023). Implementation of drone technology for farm monitoring & pesticide spraying: A review. *Information Processing in Agriculture*, 10(2), 192–203. DOI: 10.1016/j.inpa.2022.02.002

Hodge, V. J., Hawkins, R., & Alexander, R. (2021). Deep reinforcement learning for drone navigation using sensor data. *Neural Computing & Applications*, 33(6), 2015–2033. DOI: 10.1007/s00521-020-05097-x

Iovino, M., Scukins, E., Styrud, J., Ögren, P., & Smith, C. (2022). A survey of Behavior Trees in robotics and AI. *Robotics and Autonomous Systems*, 154, 104096. DOI: 10.1016/j.robot.2022.104096

Javaid, M., Haleem, A., Pratap Singh, R., Suman, R., & Rab, S. (2022). Significance of machine learning in healthcare: Features, pillars and applications. *International Journal of Intelligent Networks*, 3, 58–73. DOI: 10.1016/j.ijin.2022.05.002

Javaid, M., Haleem, A., Singh, R. P., Rab, S., & Suman, R. (2022). Significant applications of Cobots in the field of manufacturing. *Cognitive Robotics*, 2, 222–233. DOI: 10.1016/j.cogr.2022.10.001

Javaid, M., Haleem, A., Singh, R. P., & Suman, R. (2021). Substantial capabilities of robotics in enhancing industry 4.0 implementation. *Cognitive Robotics*, 1, 58–75. DOI: 10.1016/j.cogr.2021.06.001

Javaid, M., Haleem, A., Singh, R. P., & Suman, R. (2022). Enabling flexible manufacturing system (FMS) through the applications of industry 4.0 technologies. *Internet of Things and Cyber-Physical Systems*, 2, 49–62. DOI: 10.1016/j.iotcps.2022.05.005

Joksimovic, S., Ifenthaler, D., Marrone, R., De Laat, M., & Siemens, G. (2023). Opportunities of artificial intelligence for supporting complex problem-solving: Findings from a scoping review. *Computers and Education: Artificial Intelligence*, 4, 100138. DOI: 10.1016/j.caeai.2023.100138

Kabir, H., Tham, M.-L., & Chang, Y. C. (2023). Internet of robotic things for mobile robots: Concepts, technologies, challenges, applications, and future directions. *Digital Communications and Networks*, 9(6), 1265–1290. DOI: 10.1016/j.dcan.2023.05.006

Kumar, A. (2020). Drone Proliferation and Security Threats: A Critical Analysis. *Indian Journal of Asian Affairs*, 33(1/2), 43–62.

Licardo, J. T., Domjan, M., & Orehovački, T. (2024). Intelligent Robotics—A Systematic Review of Emerging Technologies and Trends. *Electronics (Basel)*, 13(3), 3. DOI: 10.3390/electronics13030542

Mohd Daud, S. M. S., Mohd Yusof, M. Y. P., Heo, C. C., Khoo, L. S., Chainchel Singh, M. K., Mahmood, M. S., & Nawawi, H. (2022). Applications of drone in disaster management: A scoping review. *Science & Justice*, 62(1), 30–42. DOI: 10.1016/j.scijus.2021.11.002 PMID: 35033326

Mohsan, S. A. H., Zahra, Q., Khan, M. A., Alsharif, M. H., Elhaty, I. A., & Jahid, A. (1593). ul A., Khan, M. A., Alsharif, M. H., Elhaty, I. A., & Jahid, A. (2022). Role of Drone Technology Helping in Alleviating the COVID-19 Pandemic. *Micromachines*, 13(10), 1593. DOI: 10.3390/mi13101593 PMID: 36295946

Pluchino, P., Pernice, G. F. A., Nenna, F., Mingardi, M., Bettelli, A., Bacchin, D., Spagnolli, A., Jacucci, G., Ragazzon, A., Miglioranzi, L., Pettenon, C., & Gamberini, L. (2023). Advanced workstations and collaborative robots: Exploiting eye-tracking and cardiac activity indices to unveil senior workers' mental workload in assembly tasks. *Frontiers in Robotics and AI*, 10, 1275572. DOI: 10.3389/frobt.2023.1275572 PMID: 38149058

Poláková, M., Suleimanová, J. H., Madzík, P., Copuš, L., Molnárová, I., & Polednová, J. (2023). Soft skills and their importance in the labour market under the conditions of Industry 5.0. *Heliyon*, 9(8), e18670. DOI: 10.1016/j.heliyon.2023.e18670 PMID: 37593611

Popescu, S. M., Mansoor, S., Wani, O. A., Kumar, S. S., Sharma, V., Sharma, A., Arya, V. M., Kirkham, M. B., Hou, D., Bolan, N., & Chung, Y. S. (2024). Artificial intelligence and IoT driven technologies for environmental pollution monitoring and management. *Frontiers in Environmental Science*, 12, 1336088. DOI: 10.3389/fenvs.2024.1336088

Reddy, K., Gharde, P., Tayade, H., Patil, M., Reddy, L. S., & Surya, D. (2023, December 12). (n.d.). Advancements in Robotic Surgery: A Comprehensive Overview of Current Utilizations and Upcoming Frontiers. *Cureus*, 15(12), e50415. DOI: 10.7759/cureus.50415 PMID: 38222213

Rejeb, A., Abdollahi, A., Rejeb, K., & Treiblmaier, H. (2022). Drones in agriculture: A review and bibliometric analysis. *Computers and Electronics in Agriculture*, 198, 107017. DOI: 10.1016/j.compag.2022.107017

Schranz, M., Di Caro, G. A., Schmickl, T., Elmenreich, W., Arvin, F., Şekercioğlu, A., & Sende, M. (2021). Swarm Intelligence and cyber-physical systems: Concepts, challenges and future trends. *Swarm and Evolutionary Computation*, 60, 100762. DOI: 10.1016/j.swevo.2020.100762

Shakhatreh, H., Sawalmeh, A. H., Al-Fuqaha, A., Dou, Z., Almaita, E., Khalil, I., Othman, N. S., Khreishah, A., & Guizani, M. (2019). Unmanned Aerial Vehicles (UAVs): A Survey on Civil Applications and Key Research Challenges. *IEEE Access : Practical Innovations, Open Solutions*, 7, 48572–48634. DOI: 10.1109/ACCESS.2019.2909530

Soori, M., Arezoo, B., & Dastres, R. (2023). Artificial intelligence, machine learning and deep learning in advanced robotics, a review. *Cognitive Robotics*, 3, 54–70. DOI: 10.1016/j.cogr.2023.04.001

Taye, M. M. (2023). Understanding of Machine Learning with Deep Learning: Architectures, Workflow, Applications and Future Directions. *Computers*, 12(5), 5. DOI: 10.3390/computers12050091

Telli, K., Kraa, O., Himeur, Y., Ouamane, A., Boumehraz, M., Atalla, S., & Mansoor, W. (2023). A Comprehensive Review of Recent Research Trends on Unmanned Aerial Vehicles (UAVs). *Systems*, 11(8), 8. DOI: 10.3390/systems11080400

Trevelyan, J. P., Kang, S.-C., & Hamel, W. R. (2008). Robotics in Hazardous Applications. In Siciliano, B., & Khatib, O. (Eds.), *Springer Handbook of Robotics* (pp. 1101–1126). Springer. DOI: 10.1007/978-3-540-30301-5_49

Zhu, D., Bu, Q., Zhu, Z., Zhang, Y., & Wang, Z. (2024). Advancing autonomy through lifelong learning: A survey of autonomous intelligent systems. *Frontiers in Neurorobotics*, 18, 1385778. DOI: 10.3389/fnbot.2024.1385778 PMID: 38644905

Zirar, A., Ali, S. I., & Islam, N. (2023). Worker and workplace Artificial Intelligence (AI) coexistence: Emerging themes and research agenda. *Technovation*, 124, 102747. DOI: 10.1016/j.technovation.2023.102747

Chapter 6
Communication Systems for Drone Swarms and Remote Operations

Shaurya Katna

Chandigarh College of Engineering and Technology, Chandigarh, India

Sunil K. Singh

https://orcid.org/0000-0003-4876-7190

Chandigarh College of Engineering and Technology, Chandigarh, India

Sudhakar Kumar

https://orcid.org/0000-0001-7928-4234

Chandigarh College of Engineering and Technology, Chandigarh, India

Divyansh Manro

Chandigarh College of Engineering and Technology, Chandigarh, India

Amit Chhabra

Chandigarh College of Engineering and Technology, Chandigarh, India

Sunil Kumar Sharma

Indian Railway, India

ABSTRACT

The world is progressively moving towards the smart city concept. Drones are central to this movement. Hence, this research on communication systems for drone swarms is imperative to enhance the operational efficiency and autonomy of unmanned aerial vehicles (UAVs). Alongside, it addresses the unique challenges posed by dynamic network topologies, limited bandwidth, and ensuring seamless collaboration in diverse applications. This study examines the complex domain of communication systems for drone swarms and remote operations. Issues such as bandwidth constraints and changing network configurations are assessed with a focus on innovative technologies like AI-powered decision-making, blockchain security, and edge computing. The assessment looks at the effects of specialized signal processing methods on swarm performance. Case studies authenticate these strategies' efficacy while offering vital real-world insights. Further, this study assists those in the field by guiding them through the challenges associated with drone swarm technology.

DOI: 10.4018/979-8-3693-2707-4.ch006

INTRODUCTION

Drones or Unmanned Aerial Vehicles (UAVs) are valued for their versatility, cost-effectiveness, and ease of deployment, making them beneficial for various military and civilian uses (Larsen & Johnston, 2024). Military roles involve surveillance and scouting, while civilian roles range from disaster response to delivery of goods and traffic management. However, constraints like limited field of view and battery life hinder drones from performing prolonged tasks across extensive areas.

The idea of utilizing drone swarms has been introduced to tackle these difficulties, with multiple drones working together to complete tasks more effectively, providing broader coverage, flexibility, and backup. Projects such as the Low-cost UAV Swarm Technology and efforts by the Chinese Electronics Technology Group are examples of the progress in developing drone swarms, to improve the effectiveness of drone systems through coordinated teamwork (Parnell et al., 2023).

Background

The historical evolution of drone swarm technology is a fascinating journey that reflects the progress in both technology and military strategy. It begins with the earliest conceptions of unmanned aerial vehicles (UAVs) and then moving towards the current state of the art systems that range to the autonomous and Autonomous Aerial Systems in Figure 1. Examining the historical trajectory explains the achievements and breakthroughs that have moved the technology drone swarms into visibility. This includes the conceptualization of swarm intelligence, advancements in sensor technology, and the integration of machine learning algorithms (Tang et al., 2023) (Herm et al., 2023). The complex communications between technological progress and operational demands undermines the necessity for robust communication systems tailored to the requirements of drone swarm ecosystems. As UAVs dive into diverse sectors, ranging from surveillance to agriculture, the critical role of seamless communication becomes evident, motivating an in-depth exploration into the system of communication within drone swarms (Shah et al., 2023).

Figure 1. History of drones

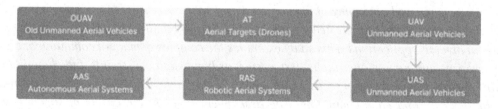

Objectives of Drone Swarm Communication

The objectives of this research chapter merge various elements to provide a systematic examination of the complex domain of communication systems for drone swarms. Firstly, it seeks to dissect the challenges that are permanently present in drone swarm communication, including but not limited to the constraints of limited bandwidth, the dynamic nature of network topologies, and potential interference issues. This involves exploration of how these challenges impact the overall efficacy of UAV operations. Secondly, the chapter aims to conduct a critical appraisal of existing communication paradigms such as mesh networks, ad-hoc communication, and multi-hop strategies (Taleb et al., 2023) (Sekhar et al., 2023) (Abedini & Al-Anbagi, 2024). This involves dissecting their applicability, strengths, and limitations within the context of the complex and dynamic network architectures characterizing drone swarm operations. Lastly, the research endeavors to navigate the forefront of technological innovation by investigating emerging solutions. Artificial intelligence, blockchain, and edge computing are scrutinized for their potential to redefine the communication landscape of drone swarms (Vats et al., 2023) (Zhang et al., 2024). Through these objectives, the chapter seeks to not only contribute to the theoretical understanding but also provide practical insights into advancing communication systems within the dynamic sphere of UAV technology.

Scope of the Chapter

The scope of this chapter is defined to encompass a comprehensive exploration of communication systems within drone swarms. This includes a detailed examination of communication protocols, network architectures, and signal processing techniques specifically tailored to the complex dynamics of drone swarm operations (Karam et al., 2022). The investigation extends beyond theoretical considerations, encompassing practical applications through the analysis of case studies and experimental findings. The chapter's focus on communication system design has various sections, Section 2 discussing Evolution of unnamed aerial vehicles, Conceptualization of swarm intelligence, Advancement in Sensor Technology, Intelligence of machine learning algorithms. In sections 3 we go through Limited bandwidth, Dynamic Network Topologies, Interference Issues, moving to Section 3 continuing with the previous information focus is given upon Mesh Networks, Ad hoc Communication, Multi hop Strategies. In Section 5 discussion is on Artificial Intelligence, Blockchain, Edge Computing. For Section 6 the chapter goes through Urban Area Surveillance and Monitoring, Communication System, Experimental Setup, Results and Insights. In conclusion, the chapter goes through practical applications and future directions of Drone Swarm Technology.

HISTORICAL DEVELOPMENT OF DRONE SWARM TECHNOLOGY

Drone swarms have become more adaptable and efficient systems for a range of applications because of these developments. Swarm intelligence concepts are applied, sensor technology is improved for improved situational awareness, and machine learning algorithms are incorporated for autonomous decision-making and adaptive behavior in drone swarm technology.

Evolution of Unmanned Aerial Vehicles (UAVs)

"Evolution of Unmanned Aerial Vehicles (UAVs)" is a general term that covers the development of UAV technology over its lifespan. UAVs were built at first for the military, including reconnaissance and surveillance. These first UAVs, although primitive with their restricted capabilities, were pioneers of unmanned aerial vehicles on account of the obvious feature of being unmanned.

The technology, through the years, evolves. It includes enhancement materials, propulsion systems, and onboard electronics, thus, developing better and more complicated UAVs. These advancements have opened up the use of drones for the civilian domain beyond the military, including agriculture, environmental monitoring, infrastructure inspection and aerial photography, among others (Kardasz & Doskocz, 2016).

The chronicle in the development of UAVs depicts a number of defining events. Researchers and engineers originally worked on the fundamental principles behind unmanned flight starting with different designs and propulsion systems that would enable controlled flight without a human pilot on board. The technological development, such as the miniaturization of the components, better batteries, and the development of the more powerful propulsion systems, heavily impacted the manufacture of lighter, more maneuverable, and longer-lasting UAVs. While UAV technology has progressed over time, it offers some very wide applications. In this way, it has been adopted for civilian use like aerial photography and videography, precision agriculture, search and rescue operations, disaster management and inspections of infrastructures. Among the most commonly witnessed functionalities in modern UAVs are autonomous features like GPS navigation, obstacle avoidance sensors and pre-programmed flight paths, which enable them to perform complex tasks without needing direct supervision by humans. This evolution has been led by a mix of technological innovations, market demand, regulatory changes, and creating a strong presence in different industries. The UAVs offer cost savings and efficient solutions for various applications making them capable of playing an important role in different industries (Coindreau et al., 2021).

Conceptualization of Swarm Intelligence

The foundation of swarm intelligence as it applies to drone swarms is the adoption of concepts derived from natural mutualistic behaviors, such as those observed in insect colonies or bird flocks. Here, the agents are individual creatures (like ants or birds) that engage in social interactions; their collective behavior is referred to as emergent behavior. A group of people's behavior may be so complicated that it is challenging to understand the norms or values that they adhere to. Swarm intelligence recognises that decentralized systems, without global knowledge and centralized control, can exhibit complex cooperative behaviors that enable them to solve issues and adjust to changing surroundings.

In the design and operation of drone swarms, the concept of distributed intelligence is considered through programming each drone to communicate and function together as a team based on local information and predefined rules. Each drone is actually operating as an independent entity that is making decisions based on what it is able to observe and from what it is receiving from neighboring drones. Drone swarms are able to divide the tasks among individual drones, coordinate the movements between drones, and adapt to the current situation in real-time using swarm intelligence mechanisms(Suh, 2018).

This method provides several benefits though compared to the conventional centralized control systems. It also improves swarms of drones' scalability and reliability, as even if a portion of the units malfunction or lose communication, others can still be operational. The swarm intelligence, further, enables drone swarms to display emergent behaviors that may be more efficient or effective than those possible for an individual drone working separately.

Conceptualizing swarm intelligence in drone swarms results in a paradigm shift in the way we approach the design and implementation of autonomous systems. In contrast to centralized control and rigid patterns of interactions, these swarms inspired by natural systems can exhibit self-organization and switch the actions considering the local context.

Advancements in Sensor Technology

The development of sensor technology has been the main driving force of the development of drone swarm technology. These innovations are technology-based, which means that new and old sensors are developed and plugged into the system thus enabling drones to collect a wide variety of data with high accuracy.

First, better sensors allow the drones in the swarm and the collective to perform more efficiently. For example, with their high resolution, it would be possible for drones to pilot camera shots required in aerial photography as well as surveillance. LiDAR (Light Detection and Ranging) sensors help drones to acquire precise distance measurements to objects within the surrounding area, thereby accomplishing tasks such as terrain mapping and 3D modeling as shown in figure 2 (Abir et al., 2023). Also, sensors for the purposes of measuring environmental variables such as temperature, humidity and air quality can be used to generate data which serve the application areas such as agricultural monitoring and pollution assessment.

Besides that, innovations in sensor technology further increase the performance capabilities and autonomy of drone swarms since they improve their situational awareness(Besada et al., 2018). Measuring and mapping the environment is possible through the deployment of drones as they can collect data from multiple sensors, detect obstacles and even potentially dangerous areas. This highly-situational awareness is essential for tasks such as navigation, clearance of obstacles, and coordination within the swarm. In this regard, drones can utilize the use of proximity sensors to stay at a safe distance from each other and not collide while engaged in a common task.

Besides this, sensors being diverse help drone swarms to be more adaptive and flexible to multiple tasks and environments. For example, drones can be fitted with advanced sensors that are made for leak detection in industrial settings or environmental conservation in ecological studies.

In general, there are technological advances in sensors that endow the drone swarms with the provision of more precise and diverse data, improving their utility for different applications. Consequently, the improved sensors result in better awareness of the situation, more precise measurements, and thereby to efficiency, effectiveness, and autonomy of drone swarm operations.

Figure 2. Working of LiDAR

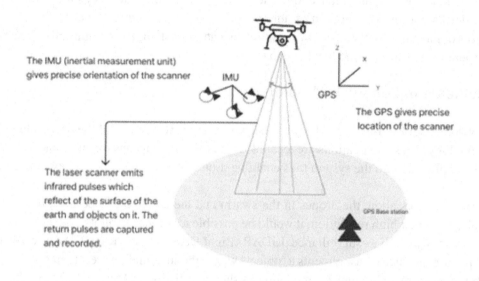

Integration of Machine Learning Algorithms

The machine learning algorithms integration into drone swarm systems is the major step in their development due to the increased capabilities and effectiveness. The machine learning algorithms allow the drones to be autonomous in the process of analyzing data and decision making as well as changing their behavior depending on the type of environment and the task (I. Singh et al., 2022) (Kumar et al., 2023b) (Kumar et al., 2022, 2023a).

The machine learning algorithms play a high-stake role in strengthening the capabilities of drone swarms in the multiple tasks. Firstly, they help in autonomous decision-making of machines by processing large volumes of data collected by drones that comprise a wide variety of data such as sensor data, imagery and environmental variables. Through data analysis, machine learning models are able to identify patterns, detect the stumbling blocks, and route optimization, task prioritization, and the swarm coordinated. Besides, machine learning promotes adaptability in drones, which enables them to change their conduct in real time according to changing scenarios or action-types. The more they learn from their past experiences, the more accurate their performance and the better equipped they are to go through a new situation. Moreover, machine learning algorithms are intended to improve upon the whole swarm's behavior by discovering data from multiple drones, better behavior and resource allocation of the ensemble of devices. Optimization is what ensures that the swarms operations, such as surveillance and search and rescue are carried out pretty well. Overall, machine learning boosts the intelligence and adaptability of the drone swarms such that they are capable of performing complex tasks and operating in a vast range of environments with greater efficiency and effectiveness, enabling them to take real-time actions autonomously, and making decisions which are always most optimal.

The implementation of machine learning algorithms into drone swarm systems makes it possible for autonomous actions, adaptive behavior and predictable performance. By applying machine learning, drone swarms can cover a broad array of tasks and scenarios being sophisticated, quicker, and more effective.

CHALLENGES IN DRONE SWARM COMMUNICATION

The efficient coordination and communication among individual drones in drone swarm communication encounters three primary challenges, namely Limited Bandwidth, Dynamic Network Topologies, and Interference Issues.

Limited Bandwidth

Limited bandwidth refers to the limited capacity of communication channels in drone swarms, making it a fundamental restriction in their communication network. This key issue in drone swarm communication infrastructure presents a crucial bottleneck, requiring a detailed examination in academic circles. The increasing amount of data shared among swarm members emphasizes the need for efficient bandwidth usage. Academic research on this challenge involves exploring new approaches like compression algorithms, prioritization methods, and adaptive communication protocols to optimize limited bandwidth resources.

There is a focused endeavor in the academic world to design new data compression algorithms so that they can optimize network bandwidth utilization without undermining the data integrity. Therefore, this interdisciplinary approach, which combines information theory, signal processing, and machine learning, means adequate data processing in drone swarms. Moreover, the bandwidth is overcome by developing intelligent prioritization mechanisms which schedule the data packet in the timely manner of the important data. On the other hand, this optimization will cause more bandwidth usage which will, however, ensure that crucial information is sent at the right time especially for a drone swarm application. In addition, adaptive communication protocols are under development to confront changing availability of bandwidth, making transmission reliable in changing communication conditions as well. Machine learning functionalities are among the pillars that lead to the solution of unpredictable conditions by fostering the autonomy of the drone swarms and promoting self-evolving communication systems. By utilizing the historical and the current data, machine learning methods enable proactive response, which is very efficient in optimizing the overall communication efficiency and in mitigating possible bottlenecks.

Dynamic Network Topologies

Dynamic Network Topologies involve the evolving spatial relationships and connectivity patterns among drones in a swarm, influenced by the mobility of drones and resulting fluctuations in proximity, line-of-sight conditions, and connectivity status. Unlike traditional networks with fixed configurations, drone swarms have fluid structures due to the continuous movement of individual drones (Liu et al., 2024). Therefore, traditional network models are not suited for the unpredictable topology changes seen in drone swarms, prompting academic research to address the complexities of maintaining strong communication connections within dynamic network topologies. Academic research focuses on the challenges of drone swarm network topologies, emphasizing the need for adaptive routing algorithms and decentralized communication structures. The ever-changing movement of drones requires self-organizing mechanisms to ensure reliable communication amid dynamic spatial relationships, impacting factors like latency and packet loss. Insights from this research are crucial for the development of resilient communication infrastructures to facilitate effective swarm navigation in evolving spatial configurations.

Interference Issues

Interference problems involve the harmful effects of competing signals and electromagnetic disturbances on communication links between drones in a swarm. The widespread use of wireless communication technologies and crowded electromagnetic spectrum make drone communication channels vulnerable. In the area of reducing the challenges of interference in drone group communication, scholars have been engaged in actively addressing the weak areas. First, the focus of frequency spectrum allocation is based on the strategic management of the frequency spectrum. To channel potential congestion to impaired communications on traditional bands, researchers examine new approaches including dynamic spectrum access and cognitive radio principles. These dynamic changes provide a spectral landscape that can enable drones to reduce interference, thus improving reliability. These strategies are also put to use to combat the negative effects of signaling interference through design-focused signal processing algorithms, advanced modulation schemes, and error correction techniques aimed at strengthening weak communications in drones. Optical and acoustic communication techniques are also emerging as promising alternatives that work in low-intensity intervention environments, providing an important learning curve to investigate the feasibility and effectiveness of drone group communication. Now as we have understood the limitation of the technologies used we will go through the existing networks, communication and strategies that are currently in use and are valuable for drone communication to work.

EXISTING COMMUNICATION PARADIGMS

The historical evolution of drone swarm technology is a fascinating journey that reflects the progress in both technology and military strategy. It begins with the earliest conceptions of unmanned aerial vehicles (UAVs) and gradually unfolds into the current state-of-the-art autonomous and collaborative systems.react with each other, ground stations and so on using different communication networks, like mesh networks, ad-hoc Communications and multi-hop communications. These paradigms are crucial for drone swarms working cooperatively and performing safely.

Mesh Networks

Mesh networking is the type of infrastructure used to allow the drones to set up decentralized connections among themselves by creating in effect a network where each drone acts both as a transmitter and a receiver. In a mesh network, one drone can communicate directly with the nearby drone or pass the message to the other drone acting as a relay station (Qu et al., 2023). This decentralizing model provides high resilience and faultlessness of the drone swarm communication.

The other major feature of mesh networks is the ability to automatically retransmit messages in case there is a direct link between the two drones broken. In addition, a drone dropping out of range or facing the interruption can be automatically rerouted through the alternative paths available via other drones used as relays. This feature of dynamic rerouting is aimed at ensuring that the swarm continues to have communication within its network even in the face of complex or dynamic conditions where conventional communication infrastructure might be non-existent or inadequate.

Mesh networks have the capacity to be applied in such scenarios as disaster response, search and rescue operations, and surveillance in the uplands area. The operation of drones in these environments may include the path to avoid obstacles or covering frequently wide areas with uneven terrain that may make the linkage between all drones and ground centers unstable. The obstacles of radio waves propagation can be overcome by drones forming a mesh network, ensuring constant communication among them, and between drone and the ground stations, paving the way for their enhancement in effectiveness and efficiency in task completion.

Figure 3. Mesh network in drone swarm communication architecture

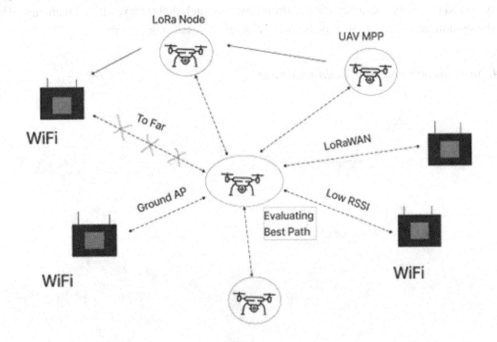

Ad-Hoc Communication

Networking drone is implemented in such a way that it does not need the existing infrastructure or the centralized coordination and it can establish the nearby drone-to-drone connectivity. Ad-hoc communication of drones does this by establishing connectivity dynamically based on the position of each one of them and the power of their signals (Sahoo et al., 2022).

This mode of operation allows drones to cooperate and communicate in real time without the need for Wi-Fi or cell phone towers in an environment where traditional communication infrastructure is not available or unreliable. Drone builds its ability to quickly respond to changing conditions by allowing an ad-hoc communication, where it is not dependent on fixed communication infrastructure but rather on collaborating effectively.

Ad-hoc communication does not require drones to follow a predefined protocol; rather, they simply interact and negotiate the best way to communicate among themselves. Having established a link, drones can then transfer data such as sensor readings, telemetry data and commands in order to synchronize their actions and to gain situational awareness.

Ad-hoc communication is good in situations of changes where drones can adapt back to back to a newly formed network which arises as they move and face new obstacles or interference. The flexibility of drone swarms enables them to stay connected and coordinate even in complicated situations, e.g. disaster response, SAR operations or advanced engineering missions.

By definition, ad-hoc communication allows drones to connect with nearby drones on the fly and share the data which can be used for exchanging information and coordinating their actions in real-time without relying on fixed infrastructure. Applying a swarm-like behavior makes it possible to improve the flexibility and adaptability characteristics of drone swarms such that they work as a team and efficiently react to the dynamism of the environment and to changing mission objectives.

Figure 4. Ad-Hoc communication in drone swarm

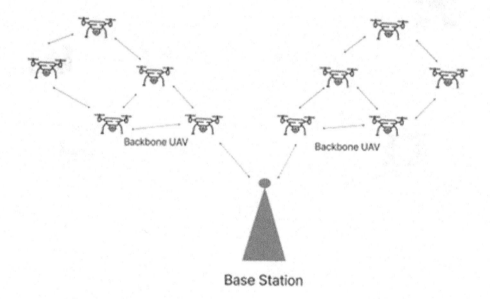

Multi-Hop Strategies

Multi-jump communication is an approach used by drones to forward messages from underlying destinations inaccessible from the same network by hopping in through different intermediate nodes within the network(Korkmaz et al., 2004). In this approach, the drones collaborate by hopping the messages forward to the intended destinations through the capabilities of the intermediate drones, so they extend the communication range beyond what the single drones can cover. Table 1 illustrates how multi-hop communication operates.

Table 1. Working multi-hop communication

Communication Strategy	Explanation
Message Relay	For a drone to pass on its message to a destination which is beyond the long range, it can use the intermediary drones. The first drone sends a message to the closest drone that is able to hear it, which then relays the message to another drone over its range. It is continued until the time when the message reaches its end destination.
Cooperative Forwarding	Drones collaborate to route and deliver messages by serving as co-forwarders of the network to the destinations. Each drone plays the role of a transceiver that flexibly optimizes the role of transmitter/receiver based on the network topology and communication requirements. Through cooperating in this way, drones will find the best route for messages ensuring removal of delays and provision of reliability.
Extended Communication Range	Multi-hop communication provides the multi-drone communications network with extra-effective communication ranges by maximally utilizing the capabilities of all drones. Through the use of multiple hops, drones can transmit messages over longer distances than would be possible to transmit straightly from one point to another by only using point-to-point communication(Dhilip Kumar et al., 2022).
Improved Coverage and Reliability	Multi-hop procedures better mine the whole of network coverage and reliability. Through using several paths and spare connections, multi-hop communication decreases the effect of obstacles, interference, or signal weakening on communication performance. Such an approach increases the availability and reliability of the network, so that the messages can get delivered even when the conditions are difficult enough.

EMERGING TECHNOLOGIES IN COMMUNICATION SYSTEMS

The recent technology is changing the ways of communication among drone swarms which is very effective, reliable, and secure. Here's how three key emerging technologies, namely Artificial Intelligence (AI), Blockchain, and Edge Computing are shaping the future of communication systems in drone swarms:

Artificial Intelligence

Artificial Intelligence (AI) is one of the key elements that contribute to the development of drone swarms to act in a unified manner in terms of performing analysis, decision-making and adaptive optimization in real-time (Chopra, 2023) (Yadav et al., 2023).

AI algorithms have the capability to analyze large quantities of data gathered by drones instantly. Through analyzing this information, patterns, trends, and irregularities can be recognised. AI algorithms can spot alterations in network traffic patterns or environmental situations that cause interference. AI can predict possible network congestion, signal degradation, or communication failures by analyzing historical data and current conditions (Kataria & Puri, 2022). It also evaluates network layout, traffic volume and environmental elements through transmission strength, frequency bands, or route paths in order to select the best communication approach. AI algorithms identify possible bottlenecks or weak signal areas and take proactive steps to resolve them, such as reallocating resources or adjusting communication parameters to maintain connectivity and coordination in the swarm. AI techniques like machine learning and deep learning allow drones to adjust communication parameters based on environmental changes and network dynamics (I. Singh et al., 2022) (Peñalvo et al., 2022) (Abulaish et al., n.d.). Drones can autonomously modify transmission power, modulation schemes, or routing strategies to adapt to signal strength variations, interference levels, or network congestion, improving communication efficiency and reliability within the drone swarm.

AI is crucial for enhancing communication in drone swarms through real-time data analysis, predictive analytics, intelligent decision-making, proactive communication management, and adaptive optimization. With the help of AI, drone swarms can improve communication efficiency, reliability, and autonomy to successfully complete tasks in various environments.

Blockchain

Blockchain technology transforms communication systems in drone swarms by offering decentralized, secure, and unchangeable protocols ("Blockchain for Data Science," 2021) (Zheng et al., n.d.).

Blockchain technology functions as a distributed ledger spread among various nodes in the network, guaranteeing no central point of failure or control and data stored cannot be changed. Which makes communication resistant to attacks or interruptions. Blockchain-powered communication protocols provide encryption features to protect data exchanged within drone swarms, safeguarding against unauthorized access and tampering. It allows verification of data integrity by preserving a clear and tamper-proof log of all drone swarm communication. It guarantees the reliability of exchanged data in the swarm, enabling drones to confirm the legitimacy and integrity of messages or orders received. These have set rules and conditions stored in blockchain transactions, allowing drones to independently discuss and uphold communication agreements like sharing resources or allocating bandwidth without needing centralized entities or human involvement. Utilizing blockchain-based communication protocols enables drones to independently oversee communication tasks, make deals, and uphold security rules in the swarm. This decreases dependence on centralized communication systems and human involvement, improving autonomy and efficiency in drone communication systems.

Blockchain technology provides decentralized, secure, and unchangeable communication protocols that improve the security, privacy, and trustworthiness of communication systems in drone swarms. Utilizing blockchain allows drones to securely share data, verify messages, and independently arrange communication agreements, guaranteeing effective and dependable communication in varied and changing settings.

Edge Computing

Edge computing revolutionizes communication systems in drone swarms by placing computational resources nearer to the drones, allowing real-time data processing and analysis at the network edge (Saini et al., 2020).

Edge computing allows drones to analyze data at the network edge, avoiding the need to send it to central servers. This make immediate decisions based on collected data. Through the use of it, computational tasks are transferred to distributed edge devices in the drone swarm, resulting in decreased communication latency and bandwidth usage. Drones send processed data to centralized servers, reducing the amount of data transferred and improving system efficiency and responsiveness. Edge computing boosts drone swarms' independence and robustness by enabling local decision-making and task execution, even in environments with limited connectivity. Drones can pool their computational resources and work together on tasks that demand significant processing power or memory. For instance, they can collectively analyze sensor data from various origins to create a detailed situational awareness map or coordinate their activities to reach shared goals. This cooperative method boosts the efficiency and effectiveness of communication systems in the drone swarm.

Edge computing moves computational resources nearer to drones, allowing real-time data processing and analysis at the network edge. By transferring computational tasks to multiple edge devices, edge computing decreases latency, saves bandwidth, boosts autonomy and resilience, and enables cooperative data processing among drone groups, ultimately enhancing overall swarm efficiency and agility in various environments.

Integration Challenges

Incorporating AI, blockchain, and edge computing into drone swarm communication systems poses numerous obstacles. These challenges are discussed as follows:

Integrating Artificial Intelligence Algorithms

AI algorithms depend on access to extensive amounts of data for their learning and decision-making processes. Ensuring that data is available in real time for communication systems in drone swarms can pose difficulties, particularly in environments that are both dynamic and resource-constrained. Computational demands are high for AI algorithms, necessitating considerable processing power and memory resources. Optimizing algorithms for efficiency and scalability is required when incorporating AI into drone swarm communication systems to ensure they can function effectively on UAV platforms. AI algorithms may also necessitate sensitive data access, leading to privacy and security worries. It is crucial to establish systems for protecting data and managing it securely in drone operations while also ensuring autonomy and privacy.

Incorporating Blockchain Technology

The decentralized aspect of Blockchain guarantees security but also brings about issues with latency and computational overhead. Agreement methods like Proof of Work or Proof of Stake need significant computational power and time to confirm transactions, possibly resulting in communication interruptions. Furthermore, the decentralized nature of blockchain requires significant storage and processing of data, contributing to computational challenges. Redesigning network architecture to incorporate blockchain into current UAV communication infrastructure calls for a major overhaul. Conventional centralized communication systems must adjust to the decentralized structure of blockchain by making changes in protocols, data handling methods, and network topology. Balancing the computational resources necessary for blockchain operations with other communication tasks is also essential for resource allocation. Developing effective resource allocation methods is essential to reduce the impact on communication performance while maintaining the security and integrity of blockchain transactions.

Obstacles in Edge Computing

Effective communication and collaboration are crucial for smooth operation in drone swarm communication systems, ensuring seamless interaction between edge devices and centralized control stations. Ensuring a secure and trustworthy connection among edge devices, with minimal latency and bandwidth usage, presents a major challenge. Resource limitations on Edge devices usually include restricted computational resources, storage capacity, and energy supply. It is crucial to optimize algorithms and data processing tasks to work within these limitations without compromising performance. Dealing with large amounts of data at the network edge and ensuring data consistency and integrity also present challenges in data management. Efficient data management techniques like caching, compression, and aggregation are necessary in edge computing solutions to minimize bandwidth usage and improve system responsiveness.

In order to overcome these obstacles, collaboration between researchers and engineers is also needed to create and implement innovative communication protocols, network designs, and data processing techniques that can successfully incorporate these technologies into drone swarm communication systems. This necessitates a multifaceted strategy that integrates knowledge in different fields, including communication networks, machine learning, cryptography, and systems engineering.

CASE STUDIES

The implementation of AI algorithms, blockchain technology, and edge computing transforms communication systems in drone swarms and remote operations. AI algorithms help drones analyze data quickly, predict network issues, and adjust communication settings autonomously. Machine learning and deep learning techniques enable drones to enhance communication abilities in various signal conditions. Blockchain technology ensures secure communication by guaranteeing data integrity and resisting unauthorized access. Edge computing allows drones to process data locally, reducing latency and reliance on central servers. Cooperative efforts in analyzing data and coordinating tasks improve efficiency and effectiveness in drone communication networks.

Urban Area Surveillance and Monitoring

The city government realizes the need for enhancements in their surveillance and monitoring capacities, to beef up public safety and boost urban management. Old devices of monitoring are often restricted and not sufficiently effective due to their immobility. Therefore, to resolve the described limitations and complete the monitoring task on the whole, the city authority employs a drone swarm communication system.

The system of drone swarm communication uses a group of unmanned aerial vehicles (UAVs) or drones which act and work together as a unit. These drones have onboard sensors, cameras and communication technology which enables them to collect data and transmit it in real- time to a central command center or control station.

The drones are used to record neighborhood areas in real-time including streets, parks, and main squares. They fly over these places using specialized equipment, taking pictures of the high-resolution videos containing thermal imaging and any other necessary data. This is, therefore, an advantage that authorities have to give them an overview of the urban area and monitor activities, traffic patterns, security threats, and so on.

The sweep of the drone swarm communication network equips the authorities of the city to upgrade its surveillance and monitoring capacities which result, in turn, in the increased public safety, better urban management and more effective actions to deal with the security threats or emergencies.

Communication System

The communication system for the drone swarm utilizes the combined properties of mesh networking, autoML and edge computing technologies to make the data layout, processing, and analysis robust and efficient (Mengi et al., 2023). Through mesh networking, each drone in the swarm becomes a node in the network pattern, thus data is transmitted to multiple nodes and relay is being made directly across all nodes and consequently, a resilient communication structure is formed which is adaptive to the dynamic urban environments. Using the edge computing neurons in each drone additionally improves the communication infrastructure by making data processing locally, reducing latency, enabling real time decision making, and making it possible for the drones to analyze the onboard sensor data, detect anomaly, and make the decision based on its sensor analysis in real time without depending on the centralized computing infrastructure. The integrated communication system equips the drone swarm with more enhanced capabilities that work together, which improve the drone's situational awareness, responsiveness,, and autonomy in a complex urban environment (Rastogi et al., 2017).

Results and Insights

There are immense benefits a drone swarm communication system offers in terms of surveillance and monitoring of urban areas. Relying on the communication system design, one can hear everything, see anything and track every movement in the urban area, in order to prevent crimes and enhance the urban management. Such immediate data transmission and processing capabilities enable quick and timely responses to incidents as the data is streamlined for analysis and decision-making in real time, developing situational awareness and crisis management. The robust communications architecture of the system, using mesh networking and edge computing technologies, assures the system's reliability even in the adverse urban areas where limited connectivity is possible. This makes it possible for data to be transmitted from drones to centralized command centers without friction, and this leads to efficient coordination and response processes. Furthermore, AI-powered surveillance helps to detect, classify, and prioritize any potential threats and enable proactive threat suppression. On the whole, practical usage of the communication system demonstrates its efficiency in maintaining security levels and raising the urban preparedness to respond to emergencies, which indicates a significant step in pushing urban communities towards higher resilience and crisis management(Kumar et al., 2024) (Setia et al., 2024) (M. Singh et al., 2023).

IMPACT ON ENVIRONMENTAL FACTORS

The ever-changing aspect of drone swarm operations implies that environmental factors may greatly affect communication systems. Different factors impact the communication systems as follows:

Types of Environmental Factors

Weather Conditions

Rainfall and Fog can cause signal attenuation, which decreases both the communication range and quality. Water droplets in the atmosphere have the ability to soak up or reflect radio waves, causing a reduction in signal strength for drone communication. Strong gusts of wind also may lead drones to stray off their intended flight paths, impacting the layout of the network and creating communication difficulties. Furthermore, turbulence caused by wind can impact the stability of drones, possibly affecting their ability to maintain steady communication connections. Extreme temperatures can impact the performance and battery duration of drones as well. Low temperatures may decrease battery performance, resulting in shorter flight durations and potentially affecting communication capabilities. Elevated temperatures can also impact electronics and may require thermal management solutions.

Physical Barriers

Structures, vegetation, and uneven ground like buildings, trees, and terrain can block the line of sight between drones, causing signal disruptions and reflections. This could lead to decreased communication distance, higher delays, and dropped packets. Signal reflections can also occur due to obstacles, resulting in multipath propagation that causes signals to travel through multiple paths before reaching their intended destination. Multipath propagation may cause interference and signal distortion that can impact the reliability of communication.

Varying Altitudes

Signal propagation can be affected by altitude variations in a group of drones, impacting line-of-sight conditions. Greater altitudes can improve communication coverage by decreasing barriers and obstacles, but they can also lead to more signal interference and weakening. Increased altitudes may enhance line-of-sight conditions among drones, potentially boosting communication reliability. Nevertheless, topographical elements or high buildings can still obstruct visibility, especially in city or heavily wooded regions.

Mitigation Strategies

Measures and technologies can adapt to reduce the effects of environmental factors on communication within drone swarms. Some of these are:

Adaptive Communication Protocols

Communication protocols that are adaptive can modify communication parameters like transmission power, modulation scheme, and routing strategy according to environmental conditions. These protocols can assist in preserving communication quality and coverage despite weather conditions, physical obstacles, and changing altitudes.

Antenna Design

Advanced antenna design, such as directional antennas, smart antennas, and multiple-input multiple-output (MIMO) systems, can enhance communication performance when environmental factors are present. These antenna designs can enhance communication distance, minimize interference, and enhance signal quality.

Signal Processing Techniques

Methods like equalization, beamforming, and diversity combining can aid in reducing the effects of environmental factors on drone swarm communication. These methods can aid in mitigating signal loss, interference, and distortion due to weather, obstacles, and altitude changes.

Machine Learning Algorithms

Utilizing machine learning algorithms to forecast and adjust to environmental factors impacting communication within drone swarms. These algorithms are capable of assessing past and current data to forecast weather conditions, obstacles, and changes in altitude, and modify communication settings as needed.

FUTURE DIRECTIONS AND EMERGING STANDARDS

As the usage of drone swarm communication systems increases, various new standards and protocols are being created to aid in compatibility and scalability among various UAV platforms and applications. Included in this list are:

MAVLink

The Micro Air Vehicle Link (MAVLink) is a protocol for communication between ground control stations (GCS) and unmanned vehicles, such as drones, that is open-source. MAVLink is now widely accepted in the drone industry as the standard because of its flexibility, reliability, and user-friendly nature.

DDS

DDS is a middleware protocol and API standard for data-centric connectivity, developed by the Object Management Group (OMG). DDS allows distributed devices, such as drones, to share data in a scalable, real-time, reliable, high-performance, and interoperable manner.

MQTT

MQTT is a lightweight messaging protocol called Message Queuing Telemetry Transport that is designed to be open, simple, and easy to implement for publish-subscribe purposes. The low bandwidth needs and capability to work with high-latency or unreliable networks make MQTT a good fit for IoT and drone swarm use cases.

5G and Beyond

The fifth-generation mobile network (5G) and beyond provides fast, minimal delay, and dependable communication, making it an appealing choice for drone swarm communication systems. 5G has the capability to back up extensive machine-type communication (mMTC) and ultra-reliable low-latency communication (URLLC), necessary for drone swarm uses.

AI and Machine Learning

AI and Machine Learning can be incorporated into drone swarm communication systems to improve routing, resource distribution, and decision-making processes. These methods can enhance the effectiveness, dependability, and flexibility of drone swarm communication systems.

Blockchain

Blockchain technology offers a secure, decentralized, and tamper-proof way to communicate and manage data in drone swarm systems. Blockchain is able to maintain the trust, privacy, and integrity of data for drones, operators, and other parties involved.

These new standards and protocols aim to tackle the issues of interoperability, scalability, and security in drone swarm communication systems. By embracing these criteria and procedures, stakeholders can encourage creativity, speed up the implementation of drone swarm technology, and discover fresh opportunities for different uses.

CONCLUSION

The potential of drone swarm communication systems in the future is enormous and has the capability to transform various industries, such as improving surveillance, facilitating effective disaster response, and streamlining agricultural practices. At the core of this revolutionary idea is the smooth incorporation of AI algorithms into the communication framework of groups of drones. By integrating artificial intelligence, these swarms acquire the capacity to analyze large quantities of real-time information, empowering them to independently make decisions and adjust flexibly to their environment.

The incorporation of artificial intelligence into communication systems of drone swarms is set to bring about a new era of adaptability and responsiveness. AI algorithms can examine environmental data, detect patterns, and anticipate changes, enabling drone swarms to enhance their operations in real-time. In a disaster response situation, AI drones working together could independently assess impacted regions, detect dangers, and organize rescue operations without human involvement. Likewise, in the realm of agriculture, drones equipped with AI could supervise the well-being of crops, identify pests or diseases, and enhance irrigation plans according to the surrounding environment.

Nevertheless, it is crucial to prioritize strong security and privacy measures as the use of drone swarms becomes more common. Future communication protocols need to have encryption, authentication, and intrusion detection features integrated to protect sensitive data and prevent unauthorized access. This is especially important in areas like surveillance or infrastructure monitoring, where the security and privacy of data are essential.

Interoperability standards are crucial in fully unlocking the potential of drone swarm technology in various applications and platforms. It is crucial to work on creating universally agreed upon protocols for communication and collaboration between drones of different manufacturers and technologies. Stakeholders can promote innovation and speed up the adoption of drone swarm technology by setting up standards for interoperability, which enables smooth integration and interaction.

Additionally, it is crucial to tackle energy efficiency issues in order to maintain the ongoing operational abilities of drone swarms. Future studies should concentrate on creating communication protocols and optimization algorithms that are energy-efficient in order to enhance the longevity of drone swarms. By utilizing improvements in low-power communication technologies and deploying effective routing strategies, stakeholders can prolong mission periods and boost system reliability.

In summary, the potential of drone swarm communication systems to transform various industries and tackle societal issues is highly promising. By adopting AI integration, enhancing security measures, setting interoperability standards, and focusing on energy efficiency, stakeholders can lay the groundwork for the responsible and ethical deployment of drone swarm technology, opening up new possibilities for innovation and advancement.

REFERENCES

Abedini, M., & Al-Anbagi, I. (2024). Enhanced Active Eavesdroppers Detection System for Multihop WSNs in Tactical IoT Applications. *IEEE Internet of Things Journal*, 11(4), 6748–6760. DOI: 10.1109/JIOT.2023.3313048

Abir, T. A., Kuantama, E., Han, R., Dawes, J., Mildren, R., & Nguyen, P. (2023). Towards Robust Lidar-based 3D Detection and Tracking of UAVs. *Proceedings of the Ninth Workshop on Micro Aerial Vehicle Networks, Systems, and Applications*, (pp. 1–7). IEEE. DOI: 10.1145/3597060.3597236

Abulaish, M., Wasi, N. A., & Sharma, S. (n.d.). The role of lifelong machine learning in bridging the gap between human and machine learning: A scientometric analysis. *WIREs Data Mining and Knowledge Discovery, n/a*(n/a), e1526. DOI: 10.1002/widm.1526

Besada, J. A., Bergesio, L., Campaña, I., Vaquero-Melchor, D., López-Araquistain, J., Bernardos, A. M., & Casar, J. R. (2018). Drone Mission Definition and Implementation for Automated Infrastructure Inspection Using Airborne Sensors. *Sensors (Basel)*, 18(4), 4. DOI: 10.3390/s18041170 PMID: 29641506

Blockchain for Data Science. (2021, October 16). *Insights2Techinfo*. https://insights2techinfo.com/blockchain-for-data-science/

Chopra, A. G. (2023). Impact of Artificial Intelligence and the Internet of Things in Modern Times and Hereafter: An Investigative Analysis. In *Advanced Computer Science Applications*. Apple Academic Press.

Coindreau, M.-A., Gallay, O., & Zufferey, N. (2021). Parcel delivery cost minimization with time window constraints using trucks and drones. *Networks*, 78(4), 400–420. DOI: 10.1002/net.22019

Dhilip Kumar, V., Kanagachidambaresan, G. R., Chyne, P., & Kandar, D. (2022). Extended Communication Range for Autonomous Vehicles using Hybrid DSRC/WiMAX Technology. *Wireless Personal Communications*, 123(3), 2301–2316. DOI: 10.1007/s11277-021-09242-0

Karam, S. N., Bilal, K., Shuja, J., Rehman, F., Yasmin, T., & Jamil, A. (2022). RETRACTED: Inspection of unmanned aerial vehicles in oil and gas industry: critical analysis of platforms, sensors, networking architecture, and path planning. *Journal of Electronic Imaging*, 32(1), 011006. DOI: 10.1117/1.JEI.32.1.011006

Kardasz, P., & Doskocz, J. (2016). Drones and Possibilities of Their Using. *Journal of Civil & Environmental Engineering*, 6(3). DOI: 10.4172/2165-784X.1000233

Kataria, A., & Puri, V. (2022). AI- and IoT-based hybrid model for air quality prediction in a smart city with network assistance. *IET Networks*, 11(6), 221–233. DOI: 10.1049/ntw2.12053

Korkmaz, G., Ekici, E., Özgüner, F., & Özgüner, Ü. (2004). Urban multi-hop broadcast protocol for inter-vehicle communication systems. *Proceedings of the 1st ACM International Workshop on Vehicular Ad Hoc Networks*, (pp. 76–85). ACM. DOI: 10.1145/1023875.1023887

Kumar, R., Singh, S. K., Lobiyal, D. K., Kumar, S., & Jawla, S. (2024). Security Metrics and Authentication-based RouTing (SMART) Protocol for Vehicular IoT Networks. *SN Computer Science*, 5(2), 236. DOI: 10.1007/s42979-023-02566-7

Kumar, S., Singh, S. K., & Aggarwal, N. (2023, September). Sustainable Data Dependency Resolution Architectural Framework to Achieve Energy Efficiency Using Speculative Parallelization. In *2023 3rd International Conference on Innovative Sustainable Computational Technologies (CISCT)* (pp. 1-6). IEEE. DOI: 10.1109/CISCT57197.2023.10351343

Kumar, S., Singh, S. K., & Aggarwal, N. (2023). Speculative parallelism on multicore chip architecture strengthen green computing concept: A survey. In *Advanced computer science applications* (pp. 3–16). Apple Academic Press. DOI: 10.1201/9781003369066-2

Kumar, S., Singh, S. K., Aggarwal, N., Gupta, B. B., Alhalabi, W., & Band, S. S. (2022). An efficient hardware supported and parallelization architecture for intelligent systems to overcome speculative overheads. *International Journal of Intelligent Systems*, 37(12), 11764–11790. DOI: 10.1002/int.23062

Larsen, G. D., & Johnston, D. W. (2024). Growth and opportunities for drone surveillance in pinniped research. *Mammal Review*, 54(1), 1–12. DOI: 10.1111/mam.12325

Liu, T., Bai, G., Tao, J., Zhang, Y.-A., & Fang, Y. (2024). A Multistate Network Approach for Resilience Analysis of UAV Swarm considering Information Exchange Capacity. *Reliability Engineering & System Safety*, 241, 109606. DOI: 10.1016/j.ress.2023.109606

Mengi, G., Singh, S. K., Kumar, S., Mahto, D., & Sharma, A. (2023). Automated Machine Learning (AutoML): The Future of Computational Intelligence. In Nedjah, N., Martínez Pérez, G., & Gupta, B. B. (Eds.), *International Conference on Cyber Security, Privacy and Networking (ICSPN 2022)* (pp. 309–317). Springer International Publishing. DOI: 10.1007/978-3-031-22018-0_28

Parnell, K. J., Fischer, J. E., Clark, J. R., Bodenmann, A., Galvez Trigo, M. J., Brito, M. P., Divband Soorati, M., Plant, K. L., & Ramchurn, S. D. (2023). Trustworthy UAV Relationships: Applying the Schema Action World Taxonomy to UAVs and UAV Swarm Operations. *International Journal of Human-Computer Interaction*, 39(20), 4042–4058. DOI: 10.1080/10447318.2022.2108961

Peñalvo, F. J. G., Maan, T., Singh, S. K., Kumar, S., Arya, V., Chui, K. T., & Singh, G. P. (2022). Sustainable Stock Market Prediction Framework Using Machine Learning Models. [IJSSCI]. *International Journal of Software Science and Computational Intelligence*, 14(1), 1–15. DOI: 10.4018/IJSSCI.313593

Qu, C., Sorbelli, F. B., Singh, R., Calyam, P., & Das, S. K. (2023). Environmentally-Aware and Energy-Efficient Multi-Drone Coordination and Networking for Disaster Response. *IEEE Transactions on Network and Service Management*, 20(2), 1093–1109. DOI: 10.1109/TNSM.2023.3243543

Rastogi, A., Singh, S., Sharma, A., & Kumar, S. (2017). *Capacity and Inclination of High Performance Computing in Next-Generation Computing*.

Sahoo, L., Panda, S. K., & Das, K. K. (2022). A Review on Integration of Vehicular Ad-Hoc Networks and Cloud Computing. [IJCAC]. *International Journal of Cloud Applications and Computing*, 12(1), 1–23. DOI: 10.4018/IJCAC.300771

Saini, T., Kumar, S., Vats, T., & Singh, M. (2020). Edge computing in cloud computing environment: opportunities and challenges. In *International Conference on Smart Systems and Advanced Computing (Syscom-2021)*. Research Gate.

Sekhar, B. V. D. S., Udayaraju, P., Kumar, N. U., Sinduri, K. B., Ramakrishna, B., Babu, B. S. S. V. R., & Srinivas, M. S. S. S. (2023). Artificial neural network-based secured communication strategy for vehicular ad hoc network. *Soft Computing*, 27(1), 297–309. DOI: 10.1007/s00500-022-07633-4

Setia, H., Chhabra, A., Singh, S. K., Kumar, S., Sharma, S., Arya, V., Gupta, B. B., & Wu, J. (2024). Securing the road ahead: Machine learning-driven DDoS attack detection in VANET cloud environments. *Cyber Security and Applications*, 2, 100037. DOI: 10.1016/j.csa.2024.100037

Shah, S. A., Lakho, G. M., Keerio, H. A., Sattar, M. N., Hussain, G., Mehdi, M., Vistro, R. B., Mahmoud, E. A., & Elansary, H. O. (2023). Application of Drone Surveillance for Advance Agriculture Monitoring by Android Application Using Convolution Neural Network. *Agronomy (Basel)*, 13(7), 7. DOI: 10.3390/agronomy13071764

Singh, I., Singh, S. K., Singh, R., & Kumar, S. (2022). Efficient Loop Unrolling Factor Prediction Algorithm using Machine Learning Models. *2022 3rd International Conference for Emerging Technology (INCET)*, (pp. 1–8). IEEE. DOI: 10.1109/INCET54531.2022.9825092

Singh, M., Singh, S. K., Kumar, S., Madan, U., & Maan, T. (2023). Sustainable Framework for Metaverse Security and Privacy: Opportunities and Challenges. In Nedjah, N., Martínez Pérez, G., & Gupta, B. B. (Eds.), *International Conference on Cyber Security, Privacy and Networking (ICSPN 2022)* (pp. 329–340). Springer International Publishing. DOI: 10.1007/978-3-031-22018-0_30

Suh, J. (2018). Drones: How They Work, Applications, and Legal Issues. *Georgetown Law Technology Review*, 3, 502.

Taleb, S. M., Meraihi, Y., Mirjalili, S., Acheli, D., Ramdane-Cherif, A., & Gabis, A. B. (2023). Mesh Router Nodes Placement for Wireless Mesh Networks Based on an Enhanced Moth–Flame Optimization Algorithm. *Mobile Networks and Applications*, 28(2), 518–541. DOI: 10.1007/s11036-022-02059-6

Tang, J., Duan, H., & Lao, S. (2023). Swarm intelligence algorithms for multiple unmanned aerial vehicles collaboration: A comprehensive review. *Artificial Intelligence Review*, 56(5), 4295–4327. DOI: 10.1007/s10462-022-10281-7

Vats, T., Singh, S. K., Kumar, S., Gupta, B. B., Gill, S. S., Arya, V., & Alhalabi, W. (2023). Explainable context-aware IoT framework using human digital twin for healthcare. *Multimedia Tools and Applications*, 83(22), 62489–62490. DOI: 10.1007/s11042-023-16922-5

Yadav, N., Singh, S. K., & Sharma, D. (2023). Forecasting Air Pollution for Environment and Good Health Using Artificial Intelligence. *2023 3rd International Conference on Innovative Sustainable Computational Technologies (CISCT)*, (pp. 1–5). IEEE. DOI: 10.1109/CISCT57197.2023.10351334

Zhang, Z., Zeng, K., & Yi, Y. (2024). Blockchain-Empowered Secure Aerial Edge Computing for AIoT Devices. *IEEE Internet of Things Journal*, 11(1), 84–94. DOI: 10.1109/JIOT.2023.3294222

Zheng, Z., Xie, S., Dai, H.-N., Chen, X., & Wang, H. (n.d.). Blockchain Challenges and Opportunities. *Survey (London, England)*.

Chapter 7
Machine Vision and Image Processing for Drones

Kwok Tai Chui
Hong Kong Metropolitan University, Hong Kong

Varsha Arya
Asia University, Taiwan, & Hong Kong Metropolitan University, Hong Kong

Akshat Gaurav
Ronin Institute, USA

Shavi Bansal
Insights2Techinfo, India

Ritika Bansal
Insights2Techinfo, India

ABSTRACT

This chapter delves into the pivotal role of machine vision and image processing in the realm of artificial intelligence for industrial robotics and intelligent drones. It begins by introducing the fundamental concepts and techniques of machine vision, including image capture, analysis, and interpretation methods that enable machines to 'see' and make sense of their surroundings. The discussion extends to advanced image processing strategies, such as edge detection, feature extraction, and object recognition, highlighting their critical applications in enabling autonomous operation of robots and drones. The chapter emphasizes the importance of 3D mapping and scene reconstruction in creating detailed environmental models for navigation and task execution. Furthermore, it explores the emerging use of augmented reality (AR) for industrial applications, offering insights into how AR enhances human-machine interaction by overlaying digital information onto the physical world.

DOI: 10.4018/979-8-3693-2707-4.ch007

INTRODUCTION

Machine vision is a rapidly advancing field with applications in various industries. It involves the use of artificial intelligence and computer vision systems to process and analyze visual information. Machine vision systems have been widely used for indus- trial quality control inspections, such as in the automotive industry, where they can monitor physical changes in products during the manufacturing process (Silva et al. 2018). Additionally, machine vision technologies have been applied in diverse areas, including agriculture for autonomous navigation of off-road vehicles and self-steering tractors (Vrochidou et al. 2022). The integration of machine learning in computer vision has also been pivotal in bringing automation and intelligence to photogram- metry and remote sensing (Qin and Gruen 2020). Furthermore, the use of machine vision in monitoring and optimizing processes, such as drying fruits and vegetables, has been investigated, demonstrating its potential in the food industry (Raponi et al. 2017). The development of machine vision systems has been facilitated by the availabil- ity of computational power and advances in machine learning algorithms, leading to high-performance artificial intelligence systems for tasks with well-annotated datasets (Hulstaert and Hulstaert 2019).

Moreover, the use of machine vision in clinical image analysis has been highlighted, particularly in the early detection of cancer and image interpretation, showcasing its potential in the medical field (Ying and Yu 2021; Chui et al. 2024). Furthermore, the plausibility of machine vision image-retrieval systems has been supported, emphasizing the significance of machine vision in image enhancement, categorization, and pattern recognition (Griffin 2005).

In the context of education, machine vision tools have been designed for educational use, validating a learning methodology based on the quick and easy design of short case studies, with potential applications in robotics and industrial engineering degrees (Sanguino and Webber 2018). Additionally, the review of computer vision education has emphasized the importance of competence with common industrial vision and ma- chine perception techniques, indicating the relevance of machine vision in educational settings (Bebis, Egbert, and Shah 2003; Nhi, Le et al. 2022; Sharma et al. 2024).

Machine vision has found extensive applications in the field of drones, revolution- izing their capabilities and functionalities. The integration of machine vision with drones has significantly enhanced their performance in various domains, including agriculture, surveillance, and autonomous navigation. In modern agriculture, the use of drones equipped with machine vision systems has become increasingly popular due to the decreasing prices and technological advancements in the field. These systems en- able drones to perform tasks such as sustainable oil palm resource assessment based on enhanced deep learning methods, allowing for efficient and accurate analysis of agricultural resources Liu, Ghazali, and Shah (2022); Bisht and Vampugani (2022). Furthermore, machine vision facilitates the detection of body movements in real-time to control drones without the need for additional instruments, showcasing its potential in gesture-controlled drone applications (Natarajan and Mete 2018; Okafor et al. 2022). Moreover, machine vision has been pivotal in the development of autonomous drone racing, where fully autonomous drones navigate through racecourses using machine vision, demonstrating the potential for advanced applications in the field of physics and computer science (Foehn et al. 2020; Kumbhojkar and Menon 2022). Additionally, the use of machine vision for target localization and measurement has been identified as a hot research topic, ensuring precise positioning of industrial robots and drones (Zheng 2023; Colace et al. 2022).

In the context of surveillance and security, machine vision has been employed for drone detection and classification using machine learning techniques, enhancing the capabilities of drones in identifying and responding to potential threats (Taha and Shoufan 2019). Furthermore, the use of machine vision for adaptive, self-configuring networked unmanned aerial vehicles has been proposed to address perception and control in challenging environments, highlighting its potential in enhancing the operational efficiency of drones (Bulusu, Aryafar, and Feng 2021). The application of machine vision in drone navigation and formation using selective target tracking- based computer vision has been investigated, showcasing its potential in enhancing the coordination and navigation capabilities of multiple drones (Upadhyay, Rawat, and Deb 2021; Mishra, Kong, and Gupta 2024; **?**). Additionally, machine vision has been utilized for autonomous human following drones, leveraging deep learning for robust drone vision systems that enable drones to autonomously follow and interact with humans (Piquero et al. 2021; Liao et al. 2024).

Furthermore, the combination of machine vision with wireless communication has opened up new possibilities for applications in vehicular/drone communications, virtual/augmented reality, gaming, and smart cities, highlighting the potential for ma- chine vision to enhance the communication and networking capabilities of drones (Alrabeiah et al. 2020). In the field of agriculture, machine vision has been instrumental in weed detection based on drone imagery, leveraging advanced machine learning algorithms for efficient and accurate weed monitoring in arable crops (Jensen et al. 2021). Additionally, the identification of weed patches in cultivated fields has been achieved through the combination of image acquisition by drones and further process- ing by machine learning techniques, demonstrating the potential for machine vision to revolutionize sustainable weed management practices (Esposito et al. 2021).

FUNDAMENTALS OF MACHINE VISION

Machine vision and image processing are integral components of modern technological advancements, with applications spanning various fields such as automation, biology, computer science, and physics. Machine vision encompasses the utilization of artificial intelligence and computer vision systems to process and analyze visual information, while image processing involves the manipulation and analysis of images using mathematical algorithms and techniques. The first task of machine vision involves the processing of images and their associated equipment, with systems incorporating devices with digital inputs/outputs (Stefanova and Komitov 2021). Current machine vision systems achieve this by processing numerous image frames or using complex algorithms (Tan and Dijken 2023). Key technologies of machine vision include image acquisition, image segmentation, and image recognition (Wang 2022). Furthermore, machine vision ap- plications are divided into four main categories: automated visual inspection, process control, parts identification, and robotic guidance and control (Herlambang, Purba, and Jaqin 2021). In the context of image processing, it involves the manipulation and analysis of images using mathematical algorithms and techniques. Sections 12.1 and

12.2 of a book provide an overview of imaging systems and digital image processing (Ben-Ari and Mondada 2017). Additionally, image processing techniques have been ap- plied in various domains, such as medical image processing, quality evaluation based on color grading, and vision measurement systems of mechanical workpieces (Wang, Wang, and Wang 2015) (Cui et al. 2020) (Guo 2022). Moreover, image processing meth- ods have been utilized for condition monitoring and fault diagnosis of machine tools, as well as for image enhancement and protection of naturalness in images (Liu et al. 2022).

The integration of machine vision and image processing has been pivotal in various applications, including the development of machine vision methods for agrorobots, dynamic machine vision with retinomorphic photomemristor-reservoir computing, and the application of machine vision in intelligent manufacturing (Stefanova and Komi- tov 2021) (Tan and Dijken 2023) (Wang 2022) (Figure 2). Furthermore, machine vision has been utilized for dimensional accuracy measurement of cylindrical spun parts, development of machine vision to increase the level of automation in the electronic component industry, and robot guidance using machine vision techniques in industrial environments (Xiao et al. 2018) (Herlambang, Purba, and Jaqin 2021) (P´erez et al. 2016). In the realm of image processing, research has been conducted on the illumination modeling for reconstructing machined surface topography, image processing in quantum computers, and generative imaging and image processing via generative en- coder (Shi et al. 2023) (Dendukuri and Luu 2018) (Chen and Yang 2019). Additionally, image processing techniques have been applied in the development of a low-cost ma- chine vision-based quality control system for a learning factory and greedy algorithms for image compression (Louw and Droomer 2019) (Bajpai, Sa, and Tewari 2017).

The synergy between machine vision and image processing has also been instrumental in various scientific and industrial domains, such as medical image processing, quality evaluation in the food industry, and real-time detection and inspection of malfunctioning pieces or processes (Wang, Wang, and Wang 2015) (Cui et al. 2020) (Zohdy et al. 2019). Moreover, the application of machine vision and image processing has extended to fields such as manufacturing engineering, computer science, and artificial intelligence, showcasing the versatility and widespread applicability of these technologies (Han and Deng 2015) (Barthelm´e and Tschumperl´e 2019) (Duong et al. 2017).

Figure 1. Components of machine vision used in drones

Figure 1. Components of machine vision used in drones

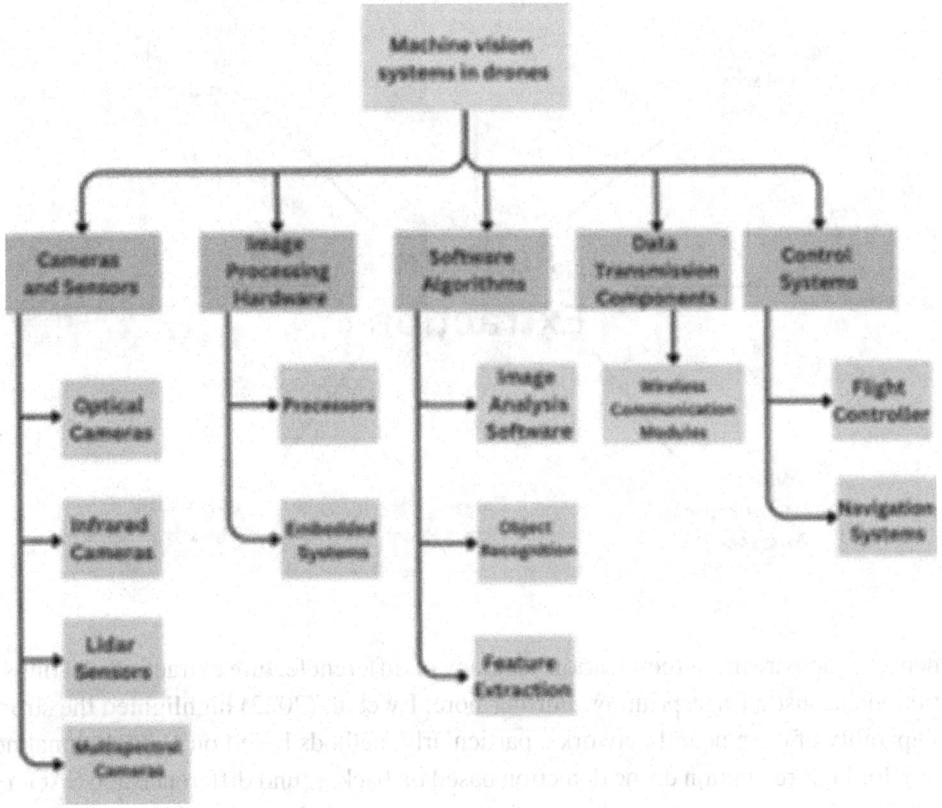

ADVANCED IMAGE PROCESSING TECHNIQUES

Feature Extraction Techinques

Feature extraction techniques are essential for the classification and discrimination of drones, enabling the identification of various drone types and their distinguishing characteristics (Figure 2). Several studies have focused on developing and evaluating feature extraction methods for drone classification, leveraging diverse technologies and signal processing approaches.

One prominent area of research involves the use of micro-Doppler radar data for feature extraction in drone classification. Studies such as those by Bernard-Cooper, Rahman, and Robertson (2022) and Rahman and Robertson (2019) have explored the use of frequency modulated continuous wave (FMCW) radar micro-Doppler data for machine learning-based classification of multiple drone types. These studies have developed feature extraction methods based on the micro-Doppler signature characteristics of in-flight targets, obtained through radar data, to discriminate between drones and other flying objects.

Additionally, feature extraction techniques have been investigated in the context of drone authentication and detection. Diao et al. (2022); Kumari et al. (2024) conducted.

Figure 2. Feature extraction techniques

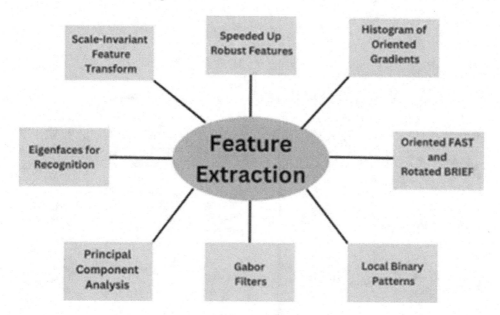

experiments to compare the authentication accuracy of different feature extraction set- tings for drone authentication via acoustic fingerprinting. Furthermore, Lv et al. (2022) highlighted the strong feature extraction capability of deep neural networks, particularly methods based on convolutional neural net- works (CNN), for high-resolution drone detection based on background difference and SAG-YOLOv5s.

In the domain of motion state classification for micro-drones, He et al. (2021) ex- plored the use of modified Mel frequency cepstral coefficient (MFCC) and hidden Markov models for feature extraction, emphasizing the impact of drone rotor characteristics on detection capabilities. Moreover, studies such as those by Bj¨orklund, Pe tersson, and Hendeby (2015) and Bennett et al. (2020) have emphasized the problem- specific nature of feature extraction, considering factors such as radar type, envi- ronment, and target characteristics. Furthermore, Clemente et al. (2021) assessed the effectiveness of Chebyshev moments-based feature extraction for drone classification, recognition, and fingerprinting, demonstrating the capability of the proposed frame- work to discriminate drones from birds and differentiate between different drone types. Similarly, Zhang and Li (2018) proposed the use of cadence velocity diagram (CVD) analysis for feature extraction, characterizing the shape, size, and variation tendency of CVD as distinguishing features for drone detection. Moreover, feature extraction techniques have been explored in the context of radar signature analysis for small UAVs or drones. Patel, Fioranelli, and Anderson (2018) employed up to 12 extracted vectors from radar signatures across multiple domains as features for classification, highlighting the significance of radar-based feature extraction in drone classification and characterization.

Object Recognition Techniques

Object recognition techniques in drones are essential for enabling drones to identify and classify objects in their operational environment. These techniques encompass a wide range of methodologies, including machine vision, deep learning, and image processing, which are crucial for enhancing the capabilities of drones in various applications. One significant area of research involves the use of machine vision and deep learning-based techniques for object recognition in drones. et al. Wang et al. (2023a) conducted a deep learning-based experiment on forest wildfire detection, highlighting the interdisciplinary nature of machine vision, which combines artificial intelligence and digital image processing methods.

Furthermore, Malche (2023) proposed an efficient method for solid waste inspection through drone-based aerial imagery and TinyML vision models, enabling the identification and classification of waste objects in the garbage discovered by the drone. Moreover, Tang et al. (2022) reviewed surface defect detection of steel products based on machine vision, emphasizing the application of machine vision techniques for detecting and classifying surface defects. Additionally, Jang et al. (2022) discussed the potential for in-sensor optoelectronic computing using electrostatically doped silicon, highlighting the role of emerging machine vision applications in data processing within the photodiode array itself.

Furthermore, Ninh et al. (2022) presented a method for the navigation of drones in GPS-denied environments based on vision processing, showcasing the significance of vision-based navigation for drones operating in challenging environments. Additionally, Citron et al. (2021) discussed the recovery of meteorites using an autonomous drone and machine learning, demonstrating the successful application of machine learning algorithms for processing images from a live video feed to achieve accurate object recognition. In the context of drone-based applications, machine vision and deep learning techniques have been instrumental in enabling drones to recognize and classify objects in their surroundings. These techniques have been applied in various domains, including wildfire detection, waste inspection, and meteorite recovery, showcasing their versatility and widespread applicability in diverse operational scenarios.

MACHINE VISION AND AUTONOMOUS DRONES

Image processing indeed plays a crucial role in supporting drone navigation and enhancing task performance through various techniques and methodologies. The utilization of image processing in drones enables them to recognize and interpret visual data, leading to improved navigation, object recognition, and overall operational efficiency. Table 1 presents the important techniques of image processing in drones.

One significant aspect of image processing in drone navigation is the use of deep learning-based techniques for visual navigation and object recognition. Kupervasser et al. (2020) demonstrated the application of deep learning for visual navigation of drones with respect to 3D ground objects, highlighting the potential of deep learning in enhancing drone navigation capabilities. Similarly, Ninh et al. (2022) presented a method for the navigation of drones in GPS-denied environments based on vision processing, showcasing the significance of vision-based navigation for drones operating in challenging environments.

Furthermore, image processing techniques have been instrumental in enabling high- resolution drone detection based on background difference and deep learning models, as demonstrated by (Lv et al. 2022). These techniques leverage the capabilities of deep neural networks for efficient object recognition and classification, contributing to enhanced drone navigation and task performance.

In addition, the application of image processing for drone navigation has been ex- tended to include reinforcement learning and imitation learning techniques. Anwar and Raychowdhury (2018) studied supervised learning for drone navigation, while Hodge, Hawkins, and Alexander (2020) explored deep reinforcement learning for drone navigation using sensor data, showcasing the diverse range of meth- odologies employed to enhance drone navigation capabilities.

Table 1. Advancements in image processing techniques for enhanced drone navigation and task performance

Study Reference	Technique Used	Application & Contribution
Kupervasser et al. (2020)	Deep Learning	Applied deep learning for visual navigation of drones with respect to 3D ground objects, enhancing drone navigation capabilities.
Ninh et al. (2022)	Vision Processing	Demonstrated navigation of drones in GPS-denied environments, highlighting vision-based navigation in challenging conditions.
Lv et al. (2022)	Background Difference and Deep Learning Models	Enabled high-resolution drone detection, leveraging deep neural networks for efficient object recognition and classification.
Anwar and Ray- chowdhury (2018)	Supervised Learning	Studied supervised learning for drone navigation, illustrating the use of established learning tech- niques in drone guidance.
Hodge, Hawkins, and Alexan- der (2020)	Deep Reinforcement Learning	Explored deep reinforcement learning for drone navigation using sensor data, enhancing autonomy in navigation.
Yulianto, Yuniar, and Prase- tyo (2022)	Deep Learning	Focused on navigation and guidance for autonomous quadcopter drones, facilitating autonomous navigation systems.
Amiri and Ramli (2022)	Fiducial Marker Detection	Developed a visual navigation system for autonomous drones, enabling drones to recognize and navigate based on visual cues.

Moreover, image processing techniques have been pivotal in enabling autonomous drone navigation and guidance, as demonstrated by Yulianto, Yuniar, and Prasetyo (2022) in their study on navigation and guidance for autonomous quadcopter drones using deep learning. These techniques have facilitated the development of autonomous navigation systems for drones, enabling them to operate effectively in various environments.

Furthermore, the use of image processing for object recognition and classification has been crucial for enhancing the performance of drones in various applications. For instance, Amiri and Ramli (2022) presented a visual navigation system for autonomous drones using fiducial marker detection, demon- strating the effectiveness of image processing techniques in enabling drones to recognize and navigate based on visual cues.

3D MAPPING AND SCENE RECONSTRUCTION

3D Mapping and Environmental Modeling

Techniques and technologies for 3D mapping and environmental modeling play a crucial role in various domains, including environmental science, computer science, and geospatial analysis. These techniques encompass a wide range of methodologies, including machine learning, chemometrics, and spatial change detection, which are essential for creating accurate 3D models of the environment and supporting environmental modeling.

One significant area of research involves the use of vision-based techniques for un- manned aerial vehicle (UAV) navigation, as highlighted by (Lu et al. 2018). These techniques enable the creation of different types of maps containing varying degrees of detail, from 3D models of the complete environment to the interconnection of environ- mental elements, contributing to the development of accurate environmental models. Furthermore, chemometrics techniques have been applied to solve environmental issues, as demonstrated by (Shafii et al. 2019). These techniques offer greater accuracy, flexibility, and efficiency, making them suitable for environmental modeling and analysis, thereby supporting the creation of detailed environmental models.

In addition, spatial change detection using normal distributions transform has been instrumental in creating accurate 3D environmental models, as discussed by (Katsura et al. 2019). These techniques utilize registration methods based on point-to-point correspondence and voxel-to-voxel correspondence, contributing to the creation of precise 3D environmental models.

Moreover, machine learning modeling has been employed for environmentally relevant chemical reactions, as highlighted by (Zhang and Zhang 2022). While most of these techniques have yet to be employed in environmental reaction modeling, they hold significant potential for creating accurate environmental models.

Furthermore, autonomous mapping and exploration with unmanned aerial vehicles (UAVs) using low-cost sensors has been pivotal in creating dense 3D maps of the environment, as demonstrated by (Ravankar et al. 2018). These techniques leverage sensor fusion methods to build accurate 3D maps of the environment, supporting environmental modeling and analysis.

Additionally, heritage smart city mapping, planning, and land administration techniques have been employed for creating accurate 3D maps, as discussed by (Suwardhi et al. 2022). These techniques, including photogrammetry and terrestrial and UAV/drone photogrammetry, contribute to the creation of detailed 3D environ- mental models.

Scene Reconstruction in Complex Environments

The accurate reconstruction of complex scenes is of paramount importance in various domains, including computer science, environmental science, and robotics. The recon- struction of complex environments requires advanced techniques and technologies to ensure precise and detailed modeling. The following references provide insights into the significance of accurate scene reconstruction in complex environments:

Mustafa et al. (2015) emphasize the importance of accurate reconstruction in con- trolled environments, where fixed and calibrated cameras are used. This highlights the need for precise reconstruction techniques in dynamic and uncontrolled environments. Kumar, Dai, and Li (2021) demonstrate the importance of dense optical flow in accurate reconstruction of dynamic scenes, underscoring the significance of detailed motion information for scene reconstruction. Chen, Lai, and Hu (2015) discuss the improvement of reconstruction results in complex scenes through the detection of points of interest (POI) and protection of local geometry, emphasizing the need for advanced techniques to enhance reconstruction accuracy. Mustafa et al. (2020) com- pare state-of-the-art approaches for accurate reconstruction in complex indoor and outdoor scenes, highlighting the importance of improved accuracy in multiview seg- mentation and dense reconstruction. Fathy et al. (2021) propose a new framework for accurate 3D model reconstruction in dynamic environments without using sensor data, addressing the challenges of reconstruction in practical applications.

Jiao et al. (2017) discuss the utilization of stereo cameras for 3D reconstruction of indoor scenes, emphasizing the need for accurate reconstruction techniques in complex environments. Borgmann, Tok, and Sikora (2015) demonstrate the achievement of high accuracy reconstructions of scene surfaces using active 3D-scanning systems based on structured light, highlighting the importance of precision in 3D reconstructions. Tianwei et al. (2020) present an approach for dynamic dense RGB-D SLAM based on optical flow, emphasizing the need for accurate and efficient reconstruction techniques in dynamic environments. Mustafa and Hilton (2019) emphasize the importance of semantic coherence in 4D scene flow estimation, segmentation, and reconstruction for complex dynamic scenes, highlighting the significance of accurate reconstruction for dynamic environments. Christie et al. (2016) discusses the development and use of 3D laser scanners for accurate large 3D scene reconstructions, underscoring the importance of advanced technologies for precise reconstructions.

AUGMENTED REALITY (AR) IN INDUSTRIAL APPLICATIONS

Augmented Reality and Machine Vision

Augmented Reality (AR) has gained significant relevance in the context of machine vi- sion, offering innovative opportunities for enhancing human perception and interaction with the environment. AR technology integrates virtual elements into the real world, providing an enriched and interactive experience. The relevance of AR to machine vision is evident in various domains, including education, healthcare, robotics, and environmental modeling. The following references provide insights into the relevance of AR to machine vision:

Wang et al. (2023b) investigate the introduction of AR technology into the bio- chemistry classroom, highlighting its potential to enhance student participation and equity in learning. This study underscores the relevance of AR in educational settings and its impact on human interaction with virtual information.

Nguyen (2021) presents a pathway for the introduction of AR technologies into the neurosurgical operating room, emphasizing the potential for AR to enhance workflow and artificial intelligence applications. This highlights the relevance of AR in healthcare and its potential to augment human capabilities in complex tasks.

Guan, Waliser, and Ralph (2023) highlight the global application of the Atmospheric River Scale, demonstrating the relevance of AR in environmental science and its impact on the perception and understanding of atmospheric phenomena. This underscores the potential of AR to enhance environmental modeling and analysis.

The integration of AR with machine vision offers new possibilities for creating im- mersive and interactive experiences, enabling the visualization of complex data and environments. AR technology has the potential to enhance human perception, inter- action, and decision-making processes, making it a valuable tool in various fields. The relevance of AR to machine vision is evident in its ability to augment human capabilities, improve learning experiences, and facilitate advanced applications in diverse domains.

Augmented Reality and Drones

Augmented Reality (AR) has emerged as a transformative technology with significant relevance to the field of drones, offering innovative opportunities for enhancing drone operations, training, and user experiences. The integration of AR with drones has the potential to revolutionize various applications, including pilot training, navigation, and data visualization. The following references provide insights into the relevance of AR to drones:

Konstantoudakis et al. (2022) present a solution based on Microsoft's HoloLens 2 headset that leverages augmented reality and gesture recognition to make drone piloting easier, more comfortable, and more intuitive. This study highlights the potential of AR to enhance the control and visualization of drones, offering a more intuitive system for single-handed gesture control and contextualized camera feed visualization.

Al-Bahri, Kishri, and Dharamshi (2021) devote their article to studying the inter- action of augmented reality applications and control methods for Unmanned Aerial Vehicles (UAVs). This work emphasizes the potential of AR to enhance the capabilities of drones and contribute to the development of smart cities, showcasing the relevance of AR in advancing drone technologies.

Zamora-Antun̆ano et al. (2022) present a methodology for the development of augmented reality applications (MeDARA) using a concrete, pictorial, and abstract approach, with a case study focusing on drone flight. This study demonstrates the potential of AR to promote knowledge, skills, and attitudes of students within the conceptual framework of educational mechatronics, highlighting the relevance of AR in drone pilot training and education.

Ribeiro et al. (2021) discuss the use of augmented reality as a web-based solution for UAV pilot training and usability testing. This study showcases the potential of AR to extend the perception of reality through the addition of a virtual layer on top of real-time images, offering new ways of representing data and enhancing the training and usability of drone pilots.

The integration of AR with drones offers new possibilities for creating immersive and interactive ex- periences, enabling the visualization of complex data and environments. AR technology has the potential to enhance the control, navigation, and training of drone pilots, making it a valuable tool in advancing drone technologies and applications.

147

CHALLENGES AND FUTURE PROSPECTS

Current challenges in machine vision and image processing for drones encompass a wide range of issues, including navigation, computational limitations, visual-inertial odometry, and autonomous racing. The following references provide insights into the current challenges in machine vision and image processing for drones:

Pfeiffer (2021) highlights the challenges in visual processing and control for human- piloted drone racing, emphasizing the need for a better understanding of human pilots' ability to select appropriate motor commands from highly dynamic visual information. Bulusu, Aryafar, and Feng (2021) discuss the unique challenges posed by drones due to payload restrictions, which severely limit on-board compute and communication, underscoring the need for innovative solutions to overcome these limitations. Foehn et al. (2021) emphasize the challenging visual environments and limited computational power of drones in autonomous drone racing, highlighting the difficulties in achieving high speeds and navigating complex visual environments. Zitar et al. (2023) address the challenging task of drone/bird detection and classification, emphasizing the com- plexities involved in distinguishing between birds and drones, which share similarities in flight altitude, velocity, and maneuverability.

Delmerico et al. (2019) discuss the challenges in visual-inertial odometry algorithms, highlighting the difficulties in accurately tracking the trajectory of drones, particularly in high-speed and dynamic racing scenarios. Madaan et al. (2020) and Moon et al. (2019) emphasize the challenging research problem of autonomous drone racing, which requires advanced capabilities in computer vision, planning, state estimation, and control to navigate complex racing environments. Alrabeiah et al. (2020) discuss the new research direction of vision-aided mmWave beam tracking, highlighting the potential for leveraging vision to enable new capabilities such as proactive hand-off and resource allocation. Jacob et al. (2019) address the challenges in effective ground- truthing of supervised machine learning for drone classification, emphasizing the difficulties in discriminating between birds and drones due to their similarities in flight characteristics.

Future trends in the field of drones encompass a wide range of innovative developments and challenges, including advancements in technology, security, ethical considerations, and applications in various domains. The following references provide insights into the future trends in the field of drones:

Han et al. (2021) discuss emerging drone trends for blockchain-based 5G net- works, highlighting open issues and future perspectives, emphasizing the potential for blockchain technology to revolutionize drone networks. Majeed et al. (2021) present an intelligent cybersecurity system for IoT-aided drones using a voting classifier, showcasing the potential for advanced cybersecurity solutions to enhance the safety and security of drone operations. Tezza and Andujar (2019) provide a comprehensive sur- vey of the state-of-the-art of human-drone interaction, presenting a discussion of cur- rent challenges and future work in the field of human-drone interaction, emphasizing the potential for enhanced human-drone collaboration. Mitrea and Lecturer (2020) address ethical and legal issues in civil and military research as a future opportunity for drone technology adoption, highlighting the importance of ethical considerations in the future development and deployment of drones.

Alsamhi et al. (2019) discuss the survey on collaborative smart drones and the Internet of Things for improving the smartness of smart cities, emphasizing the potential for drones to play a significant role in improving city life and supporting various urban activities. Ashraf et al. (2023) present an IoT-empowered smart cybersecurity framework for intrusion detection in the Internet of Drones, highlighting the potential for advanced cybersecurity solutions to protect drone networks from malicious exploitation. Ruwaimana et al. (2018) discuss the advantages of using drones over space-borne imagery in the mapping of mangrove forests, emphasizing the potential for drone technology to offer competitive advantages in the future.

CONCLUSION

This chapter has highlighted the transformative impact of machine vision and im- age processing technologies in enhancing the capabilities of drones within industrial and autonomous systems. By delving into advanced methods such as edge detection, feature extraction, and object recognition, we have explored how these technologies empower drones to perform complex tasks autonomously. The integration of augmented reality further enriches these applications, offering innovative ways to interact with and interpret environmental data. As these technologies continue to evolve, their potential to revolutionize industrial practices and enhance human-machine collaboration will undoubtedly expand, underscoring the importance of continued research and development in this dynamic field.

REFERENCES

Al-Bahri, M. (2021). Using augmented reality and drones in tandem to serve smart cities. *Artificial Intelligence & Robotics Development Journal, 147–157.* .DOI: 10.52098/airdj.202144

Alrabeiah, M., Booth, J., Hredzak, A., & Alkhateeb, A. (2020). *Viwi vision- aided mmwave beam tracking: dataset, task, and baseline solutions.* https://doi.org//arxiv.2002.02445.DOI: 10.48550

Alsamhi, S., Ma, O., Ansari, M., & Almalki, F. (2019). Survey on collaborative smart drones and internet of things for improving smartness of smart cities. *IEEE Access : Practical Innovations, Open Solutions,* 7, 128125–128152. DOI: 10.1109/ACCESS.2019.2934998

Amiri, M., & Ramli, R. (2022). Visual navigation system for autonomous drone using fiducial marker detection. *International Journal of Advanced Computer Science and Applications,* 13(9). DOI: 10.14569/IJACSA.2022.0130981

Anwar, M. (2018). *Navren-rl: learning to fly in real en- vironment via end-to-end deep reinforcement learning using monocular images.* IEEE. .DOI: 10.1109/M2VIP.2018.8600838

Ashraf, S. (2023). *Iot empowered smart cybersecurity framework for intrusion detection in internet of drones.* Research Square. .DOI: 10.21203/rs.3.rs-3047663/v1

Bajpai, P., Sa, P., & Tewari, R. (2017). Greedy algorithm for image compression in image processing. *International Journal of Computer Applications,* 166(8), 34–37. DOI: 10.5120/ijca2017914118

Barthelm´e, S. (2019). Imager: an r package for image processing based on cimg. *The Journal of Open Source Software,4.*DOI: 10.21105/joss.01012

Ben-Ari, M. (2017). *Image processing.* Springer. ₁2.DOI: 10.1007/978-3-319-62533-1

Bennett, C. (2020). *Use of symmetrical peak extraction in drone micro-doppler classification for staring radar.* IEEE. .DOI: 10.1109/RadarConf2043947.2020.9266702

Bernard-Cooper, J. (2022). *Multiple drone type classi- fication using machine learning techniques based on fmcw radar micro-doppler data.* SPIE. .DOI: 10.1117/12.2618026

Bisht, J., & Vampugani, V. S. (2022). Load and cost-aware min-min workflow scheduling algorithm for heterogeneous resources in fog, cloud, and edge scenarios. [IJCAC]. *International Journal of Cloud Applications and Computing,* 12(1), 1–20. DOI: 10.4018/IJCAC.2022010105

Bj¨orklund, S., Petersson, H., & Hendeby, G. (2015). Features for micro-doppler based activity classification. *IET Radar, Sonar & Navigation,* 9(9), 1181–1187. DOI: 10.1049/iet-rsn.2015.0084

Borgmann, T. (2015). *Image guided phase unwrapping for real-time 3d-scanning.* IEEE. .DOI: 10.1109/PCS.2015.7170058

Bulusu, N. (2021). *Towards adaptive, self-configuring networked unmanned aerial vehicles.* ACM. .DOI: 10.1145/3469259.3470488

Chen, K., Lai, Y., & Hu, S. (2015). 3d indoor scene modeling from rgb-d data: A survey. *Computational Visual Media,* 1(4), 267–278. DOI: 10.1007/s41095-015-0029-x

Chen, L., & Yang, H. (2019). Generative imaging and image processing via generative en- coder. https://doi.org//arxiv.1905.13300.DOI: 10.48550

Christie, D. (2016). *3d reconstruc- tion of dynamic vehicles using sparse 3d-laser-scanner and 2d image fusion.* IEEE. .DOI: 10.1109/IAC.2016.7905690

Chui, K. T., Gupta, B. B., Arya, V., & Torres-Ruiz, M. (2024). Selective and Adaptive Incremental Transfer Learning with Multiple Datasets for Machine Fault Diagno- sis. *Computers, Materials & Continua*, 78(1), 1363–1379. DOI: 10.32604/cmc.2023.046762

Citron, R., Jenniskens, P., Watkins, C., Sinha, S., Shah, A., Ra¨ıssi, C., Devillepoix, H., & Albers, J. (2021). Recovery of meteorites using an autonomous drone and machine learning. *Meteoritics & Planetary Science*, 56(6), 1073–1085. DOI: 10.1111/maps.13663

Clemente, C., L. Pallotta, C. Ilioudis, F. Fioranelli, G. Giunta, and A. Farina. 2021. "Chebychev moments based drone classification, recognition and fingerprinting." .DOI: 10.23919/IRS51887.2021.9466211

Colace, F., Guida, C. G., Gupta, B., Lorusso, A., Marongiu, F., & Santaniello, D. (2022). A BIM-based approach for decision support system in smart buildings. In *Proceedings of Seventh International Congress on Information and Communication Technology: ICICT 2022,* (pp. 471–481). Springer.

Cui, Y., Wang, L., Duan, L., & Suiqing, C. (2020). Quality evaluation based on color grading - relationship between chemical susbtances and commercial grades by machine version in corni fructus. *Tropical Journal of Pharmaceutical Research*, 19(7), 1495–1501. DOI: 10.4314/tjpr.v19i7.23

Delmerico, J. (2019). *Are we ready for autonomous drone racing? the uzh-fpv drone racing dataset.* IEEE. .DOI: 10.1109/ICRA.2019.8793887

Dendukuri, A., & Luu, K. (2018). *Image processing in quantum computers.* https://doi.org//arxiv.1812.11042.DOI: 10.48550

Diao, Y. (2022). *Drone authentication via acoustic fingerprint.* ACM. .DOI: 10.1145/3564625.3564653

Duong, C. (2017). *Temporal non- volume preserving approach to facial age-progression and age-invariant face recognition.* IEEE. .DOI: 10.1109/ICCV.2017.403

Esposito, M., Crimaldi, M., Cirillo, V., Sarghini, F., & Maggio, A. (2021). Drone and sensor technology for sustainable weed management: A review. *Chemical and Biological Technologies in Agriculture*, 8(1), 18. DOI: 10.1186/s40538-021-00217-8

Fathy, G., Hassan, H., Sheta, W., Omara, F., & Nabil, E. (2021). A novel no-sensors 3d model reconstruction from monocular video frames for a dynamic environment. *PeerJ. Computer Science*, 7, e529. DOI: 10.7717/peerj-cs.529 PMID: 34084931

Foehn, P., Brescianini, D., Kaufmann, E., Cieslewski, T., & Gehrig, M. (2020). *Alphapilot: autonomous drone racing.* https://doi.org//arxiv.2005.12813.DOI: 10.48550

Foehn, P., Brescianini, D., Kaufmann, E., Cieslewski, T., Gehrig, M., Muglikar, M., & Scaramuzza, D. (2021). Alphapilot: Autonomous drone racing. *Autonomous Robots*, 46(1), 307–320. DOI: 10.1007/s10514-021-10011-y PMID: 35221535

Griffin, L. (2005). Optimality of the basic colour categories for classification. *Journal of the Royal Society, Interface*, 3(6), 71–85. DOI: 10.1098/rsif.2005.0076 PMID: 16849219

Guan, B. (2023). Global application of the atmospheric river scale. *Journal of Geophysical Research Atmospheres,128*. DOI: 10.1029/2022JD037180

Han, L. (2015). *A study on flexible vibratory feeding system based on halcon machine vision software*. IEEE. .DOI: 10.2991/isrme-15.2015.3

Han, T., Ribeiro, I., Magaia, N., Preto, J., Segundo, A., Macedo, A., & Muhammad, K. (2021). Emerging drone trends for blockchain-based 5g networks: Open issues and future perspectives. *IEEE Network*, 35(1), 38–43. DOI: 10.1109/MNET.011.2000151

He, T., Dong, C., Li, Y., & Yin, H. (2021). Motion state classification for micro-drones via modified mel frequency cepstral coefficient and hidden markov mode. *Electronics Letters*, 58(4), 164–166. DOI: 10.1049/ell2.12384

Herlambang, H., Purba, H., & Jaqin, C. (2021). Development of machine vision to increase the level of automation in indonesia electronic component industry. *Journal Européen des Systèmes Automatisés*, 54(2), 253–262. DOI: 10.18280/jesa.540207

Hodge, V., Hawkins, R., & Alexander, R. (2020). Deep reinforcement learning for drone navigation using sensor data. *Neural Computing & Applications*, 33(6), 2015–2033. DOI: 10.1007/s00521-020-05097-x

Hulstaert, E., & Hulstaert, L. (2019). Artificial intelligence in dermato-oncology: A joint clinical and data science perspective. *International Journal of Dermatology*, 58(8), 989–990. DOI: 10.1111/ijd.14511 PMID: 31149729

Jacob, S. (2019). *Effective ground-truthing of supervised machine learning for drone classification*. IEEE. .DOI: 10.1109/RADAR41533.2019.171322

Jang, H., Hinton, H., Jung, W., Lee, M., Kim, C., Park, M., Lee, S., Park, S., & Ham, D. (2022). In- sensor optoelectronic computing using electrostatically doped silicon. *Nature Electronics*, 5(8), 519–525. DOI: 10.1038/s41928-022-00819-6

Jensen, S., Akhter, M., Azim, S., & Rasmussen, J. (2021). The predictive power of regression models to determine grass weed infestations in cereals based on drone imagery—Statistical and practical aspects. *Agronomy (Basel)*, 11(11), 2277. DOI: 10.3390/agronomy11112277

Jiao, J., Yuan, L., Tang, W., Deng, Z., & Wu, Q. (2017). A post-rectification approach of depth images of kinect v2 for 3d reconstruction of indoor scenes. *ISPRS International Journal of Geo-Information*, 6(11), 349. DOI: 10.3390/ijgi6110349

Katsura, U., Matsumoto, K., Kawamura, A., Ishigami, T., Okada, T., & Kurazume, R. (2019). Spatial change detection using normal distributions transform. *Robomech Journal*, 6(1), 20. DOI: 10.1186/s40648-019-0148-8

Konstantoudakis, K., Christaki, K., Tsiakmakis, D., Sainidis, D., Albanis, G., Dimou, A., & Daras, P. (2022). Drone control in ar: An intuitive system for single-handed gesture control, drone tracking, and contextualized camera feed visualization in augmented reality. *Drones (Basel)*, 6(2), 43. DOI: 10.3390/drones6020043

Kumar, S., Dai, Y., & Li, H. (2021). Superpixel soup: Monocular dense 3d reconstruction of a complex dynamic scene. *IEEE Transactions on Pattern Analysis and Machine Intelligence*, 43(5), 1705–1717. DOI: 10.1109/TPAMI.2019.2955131 PMID: 31765303

Kumari, P., Shankar, A., Behl, A., Pereira, V., Yahiaoui, D., Laker, B., Gupta, B. B., & Arya, V. (2024). Investigating the barriers towards adoption and im- plementation of open innovation in healthcare. *Technological Forecasting and Social Change*, 200, 123100. DOI: 10.1016/j.techfore.2023.123100

Kumbhojkar, N. R., & Menon, A. B. (2022). Integrated predictive experience management framework (IPEMF) for improving customer experience: In the era of digital transformation. [IJCAC]. *International Journal of Cloud Applications and Computing*, 12(1), 1–13. DOI: 10.4018/IJCAC.2022010107

Kupervasser, O., Kutomanov, H., Levi, O., Pukshansky, V., & Yavich, R. (2020). Using deep learning for visual navigation of drone with respect to 3d ground objects. *Mathematics*, 8(12), 2140. DOI: 10.3390/math8122140

Liao, M., Tang, H., Li, X., Pandi, V., & Arya, V. (2024). A lightweight network for abdominal multi-organ segmentation based on multi-scale context fusion and dual self-attention. *Information Fusion*, 108, 102401. DOI: 10.1016/j.inffus.2024.102401

Liu, X. (2022). Sustainable oil palm resource assessment based on an enhanced deep learning method. *Energies, 15*. DOI: 10.3390/en15124479

Louw, L., & Droomer, M. (2019). Development of a low cost machine vision based quality control system for a learning factory. *Procedia Manufacturing*, 31, 264–269. DOI: 10.1016/j.promfg.2019.03.042

Lu, Y., Xue, Z., Xia, G., & Zhang, L. (2018). A survey on vision-based uav navigation. *Geo-Spatial Information Science*, 21(1), 21–32. DOI: 10.1080/10095020.2017.1420509

Lv, Y., Ai, Z., Chen, M., Gong, X., Wang, Y., & Lu, Z. (2022). High-resolution drone detection based on background difference and sag-yolov5s. *Sensors (Basel)*, 22(15), 5825. DOI: 10.3390/s22155825 PMID: 35957382

Madaan, R., Gyde, N., Vemprala, S., Brown, M., Nagami, K., & Taubner, T. (2020). *Airsim drone racing lab*. Research Gate. https://doi.org//arxiv.2003.05654.DOI: 10.48550

Majeed, R., Abdullah, N., Mushtaq, M., Umer, M., & Nappi, M. (2021). Intelligent cyber-security system for iot-aided drones using voting classifier. *Electronics (Basel)*, 10(23), 2926. DOI: 10.3390/electronics10232926

Malche, T., Maheshwary, P., Tiwari, P. K., Alkhayyat, A. H., Bansal, A., & Kumar, R. (2023). Efficient solid waste inspection through drone-based aerial imagery and tinyml vision model. *Transactions on Emerging Telecommunications Technologies*, 35(4), e4878. DOI: 10.1002/ett.4878

Mishra, A., & Kong, K. T. C. H. (2024). Tempered Image Detection Using ELA and Convolutional Neural Networks. In *2024 IEEE International Conference on Consumer Electronics (ICCE)*, (pp. 1–3). IEEE. DOI: 10.1109/ICCE59016.2024.10444440

Mitrea, G., & Lecturer, S. (2020). Drones - ethical and legal issues in civil and military research as a future opportunity. *JESS*, 4(1), 83–98. DOI: 10.18662/jess/4.1/30

Moon, H. (2019). Challenges and implemented technologies used in autonomous drone racing. *Intelligent Service Robotics, 12*. DOI: 10.1007/s11370-018-00271-6

Mustafa, A. (2015). *General dynamic scene reconstruc- tion from multiple view video*. IEEE. .DOI: 10.1109/ICCV.2015.109

Mustafa, A., Volino, M., Kim, H., Guillemaut, J., & Hilton, A. (2020). Temporally coherent general dynamic scene reconstruction. *International Journal of Computer Vision*, 129(1), 123–141. DOI: 10.1007/s11263-020-01367-2

Natarajan, K. (2018). *Hand gesture controlled drones: an open source library*. IEEE. .DOI: 10.1109/ICDIS.2018.00035

Nguyen, N. (2021). *Augmented reality and human factors applications for the neurosurgical operating room*. IEEE. .DOI: 10.32920/ryerson.14643753.v1

Nhi, N. T. U., & Le, T. M. (2022). A model of semantic-based image retrieval using C-tree and neighbor graph. [IJSWIS]. *International Journal on Semantic Web and Information Systems*, 18(1), 1–23. DOI: 10.4018/IJSWIS.295551

Ninh, N. (2022). Navigation for drones in gps-denied environ- ments based on vision processing. *ACSIS, 33,* 25-27. .DOI: 10.15439/2022R46

Okafor, N. (2022). *Business demand for a Cloud enterprise data warehouse in electronic Healthcare Computing: Issues and developments.*

P'erez, L., Rodr'ıguez, '. I., Rodr'ıguez, N., Usamentiaga, R., & Garc'ıa, D. (2016). Robot guidance using machine vision techniques in industrial environments: A comparative review. *Sensors (Basel)*, 16(3), 335. DOI: 10.3390/s16030335 PMID: 26959030

Patel, J., Fioranelli, F., & Anderson, D. (2018). Review of radar classification and rcs charac- terisation techniques for small uavs or drones. *IET Radar, Sonar & Navigation*, 12(9), 911–919. DOI: 10.1049/iet-rsn.2018.0020

Pfeiffer, C. (2021). Human-piloted drone racing: visual processing and control. https://doi.org//arxiv .2103.04672.DOI: 10.48550

Piquero, J., Sybingco, E., Chua, A., Say, M., Crespo, C., Rivera, R., & Roque, M. (2021). A novel implementation of an autonomous human following drone using lo- cal context. *International Journal of Automation and Smart Technology*, 11(1), 2147–2147. DOI: 10.5875/ausmt.v11i1.2147

Qin, R., & Gruen, A. (2020). The role of machine intelligence in photogrammetric 3d mod- eling – an overview and perspectives. *International Journal of Digital Earth*, 14(1), 15–31. DOI: 10.1080/17538947.2020.1805037

Rahman, S. (2019). *Millimeter-wave radar micro-doppler fea- ture extraction of consumer drones and birds for target discrimination.* ACM. .DOI: 10.1117/12.2518846

Raponi, F., Moscetti, R., Monarca, D., Colantoni, A., & Massantini, R. (2017). Monitoring and optimization of the process of drying fruits and vegetables using computer vision: A review. *Sustainability (Basel)*, 9(11), 2009. DOI: 10.3390/su9112009

Ravankar, A. (2018). Autonomous map- ping and exploration with unmanned aerial vehicles using low cost sensors. IEEE. .DOI: 10.3390/ecsa-5-05753

Ribeiro, R., Ramos, J., Safadinho, D., Reis, A., Rabad̃ao, C., Barroso, J., & Pereira, A. (2021). Web ar solution for uav pilot training and usability testing. *Sensors (Basel)*, 21(4), 1456. DOI: 10.3390/s21041456 PMID: 33669733

Ruwaimana, M. (2018). The advantages of using drones over space-borne imagery in the mapping of mangrove forests. *PLoS One*, 13, e0200288. DOI: 10.1371/journal.pone.0200288 PMID: 30020959

Sanguino, T., & Webber, P. (2018). Making image and vision effortless: Learning method- ology through the quick and easy design of short case studies. *Computer Applications in Engineering Education*, 26(6), 2102–2115. DOI: 10.1002/cae.22003

Shafii, N., Saudi, A., Chyang, P., Abu, I., Kamarudin, M., & Saudi, H. (2019). Application of chemometrics techniques to solve environmental issues in malaysia. *Heliyon*, 5(10), e02534. DOI: 10.1016/j.heliyon.2019.e02534 PMID: 31667387

Sharma, A. (2024). Revolutionizing Healthcare Systems: Synergistic Multimodal Ensemble Learning & Knowledge Transfer for Lung Cancer Delineation & Taxonomy. In *2024 IEEE International Conference on Consumer Electronics (ICCE)*, (pp. 1–6). IEEE. DOI: 10.1109/ICCE59016.2024.10444476

Shi, W., Zheng, J., Sheng, Q., Wang, Q., Wang, L., & Li, Q. (2023). Illumination modelling for reconstructing the machined surface topography. *International Journal of Advanced Manufacturing Technology*, 125(11-12), 4975–4987. DOI: 10.1007/s00170-023-10925-0

Silva, R. (2018). *Machine vision systems for industrial quality control inspections.* Springer. ₅8.DOI: 10.1007/978-3-030-01614-2

Stefanova, V., & Komitov, G. (2021). Overview of the machine vision methods at agrorobots. *Agrarni Nauki*, 13(30), 13–19. DOI: 10.22620/agrisci.2021.30.002

Suwardhi, D., Trisyanti, S., Virtriana, R., Syamsu, A., Jannati, S., & Halim, R. (2022). Heritage smart city mapping, planning and land administration (hestya). *ISPRS International Journal of Geo-Information*, 11(2), 107. DOI: 10.3390/ijgi11020107

Taha, B., & Shoufan, A. (2019). Machine learning-based drone detection and classification: State-of-the-art in research. *IEEE Access : Practical Innovations, Open Solutions*, 7, 138669–138682. DOI: 10.1109/ACCESS.2019.2942944

Tan, H., & Dijken, S. (2023). Dynamic machine vision with retinomorphic photomemristor- reservoir computing. *Nature Communications*, 14(1), 2169. DOI: 10.1038/s41467-023-37886-y PMID: 37061543

Tang, B., Chen, L., Sun, W., & Lin, Z. (2022). Review of surface defect detec- tion of steel products based on machine vision. *IET Image Processing*, 17(2), 303–322. DOI: 10.1049/ipr2.12647

Tezza, D., & Andujar, M. (2019). The state-of-the-art of human–drone interaction: A survey. *IEEE Access : Practical Innovations, Open Solutions*, 7, 167438–167454. DOI: 10.1109/ACCESS.2019.2953900

Tianwei, Z., Zhang, H., Li, Y., Nakamura, Y., & Zhang, L. (2020). *Flowfusion: dynamic dense rgb-d slam based on optical flow*. https://doi.org//arxiv.2003.05102.DOI: 10.48550

Upadhyay, J., Rawat, A., & Deb, D. (2021). Multiple drone navigation and for- mation using selective target tracking-based computer vision. *Electronics (Basel)*, 10(17), 2125. DOI: 10.3390/electronics10172125

Vrochidou, E., Oustadakis, D., Kefalas, A., & Papakostas, G. (2022). Computer vision in self-steering tractors. *Machines (Basel)*, 10(2), 129. DOI: 10.3390/machines10020129

Wang, C. (2022). The application of machine vision in intelligent manufacturing. *Highlights in Science Engineering and Technology*, 9, 47–50. DOI: 10.54097/hset.v9i.1714

Wang, J. (2015). *Research on medical image processing*. IEEE. .DOI: 10.2991/icmii-15.2015.119

Wang, L., Zhang, Y., Zhang, H., An, K., & Hu, K. (2023a). A deep learning-based experi- ment on forest wildfire detection in machine vision course. *IEEE Access : Practical Innovations, Open Solutions*, 11, 32671–32681. DOI: 10.1109/ACCESS.2023.3262701

Wang, S., Sung, R., Reinholz, D., & Bussey, T. (2023b). Equity analysis of an augmented reality-mediated group activity in a college biochemistry classroom. *Journal of Research in Science Teaching*, 60(9), 1942–1966. DOI: 10.1002/tea.21847

Xiao, G., Zhong, X., Weixian, S., Xia, Q., & Chen, W. (2018). Research on the dimensional accuracy measurement method of cylindrical spun parts based on machine vision. *Matec Web of Conferences,167*. IEEE. DOI: 10.1051/matecconf/201816703010

Ying, H. (2021). Big data techniques for clinical image analysis. *International Journal of Advanced Information and Communication Technology*, 213–221. .DOI: 10.46532/ijaict-202108029

Yulianto, A., Yuniar, D., & Prasetyo, Y. (2022). Navigation and guidance for autonomous quadcopter drones using deep learning on indoor corridors. *Jurnal Jartel Jurnal Jaringan Telekomunikasi*, 12(4), 258–264. DOI: 10.33795/jartel.v12i4.422

Zamora-Antun~ano, M., Luque-Vega, L., Carlos-Mancilla, M., Hern'andez-Quesada, R., V'azquez, N., Carrasco-Navarro, R., Gonz'alez-Guti'errez, C., & Aguilar-Molina, Y. (2022). Methodology for the development of augmented reality applications: Medara. drone flight case study. *Sensors (Basel)*, 22(15), 5664. DOI: 10.3390/s22155664 PMID: 35957223

Zhang, K., & Zhang, H. (2022). Machine learning modeling of environmentally rel- evant chemical reactions for organic compounds. *ACS ES&T Water*, 4(3), 773–783. DOI: 10.1021/acsestwater.2c00193

Zhang, W., & Li, G. (2018). Detection of multiple micro-drones via cadence velocity diagram analysis. *Electronics Letters*, 54(7), 441–443. DOI: 10.1049/el.2017.4317

Zheng, Y. (2023). *Research on target localization method based on binocular vision technology.* SPIE. .DOI: 10.1117/12.3007964

Zitar, R., Kassab, M., Fallah, A., & Barbaresco, F. (2023). *Bird/drone de- tection and classification using classical and deep learning methods.* Authorea. DOI: 10.22541/au.168075364.45332093/v1

Zohdy, B. (2019). *Machine vision application on science and industry.* IGI Global. .DOI: 10.4018/978-1-5225-5751-7.ch008

Chapter 8
Enhancing Autonomous System Security With AI and Secure Computation Technologies

Tushar Singh

https://orcid.org/0009-0002-5053-1443

Chandigarh College of Engineering and Technology, Chandigarh, India

Sudhakar Kumar

https://orcid.org/0000-0001-7928-4234

Chandigarh College of Engineering and Technology, Chandigarh, India

Sunil K. Singh

https://orcid.org/0000-0003-4876-7190

Chandigarh College of Engineering and Technology, Chandigarh, India

Priyanshu

Chandigarh College of Engineering and Technology, Chandigarh, India

Brij B. Gupta

Asia University, Taichung, Taiwan

Jinsong Wu

https://orcid.org/0000-0003-4720-5946

Universidad de Chile, Chile

Arcangelo Castiglione

University of Salerno, Fisciano, Italy

ABSTRACT

Exploring the intersection of transparency and security in autonomous systems, this chapter examines the dynamic landscape of industrial robots and intelligent drones. Advanced technologies such as machine learning, AI, robotics, and deep learning shape this intricate domain. As autonomous systems gain prominence across sectors, a focus lies on understanding decision-making frameworks. Methodologies for achieving algorithmic transparency and strengthening security protocols are outlined, emphasizing the fusion of technological innovation with ethical considerations. Real-world case studies offer practical insights and best practices. Ethical responsibilities in AI and robotics integration are emphasized, alongside a forward-looking view on emerging trends and technologies, providing a tailored roadmap for researchers, practitioners, and enthusiasts navigating the evolving realm of autonomous systems. This chapter provides a thorough analysis of transparency and security challenges and opportunities in autonomous systems, benefiting policymakers and industry stakeholders.

DOI: 10.4018/979-8-3693-2707-4.ch008

INTRODUCTION

In the era of autonomous systems, where industrial robots streamline manufacturing processes and intelligent drones navigate complex environments with precision, the symbiotic interplay between transparency and security emerges as a foundational aspect of technological progress. This exploration, titled "Enhancing Autonomous System Security with AI and Secure Computation Technologies" delves into the intricate landscape shaped by cutting-edge technologies, ethical considerations, and practical challenges inherent in this transformative domain(Tesei et al., 2021)(He et al., 2022).

At its core, this endeavor offers a comprehensive overview of transparency and security within the context of autonomous systems(Li et al., 2021), focusing on industrial robots and intelligent drones. These systems, driven by a convergence of machine learning, artificial intelligence (AI), robotics (Santoso & Finn, 2023), and deep learning, are poised to revolutionize industries and redefine human-technology interactions.

The integration of machine learning, AI, robotics, and deep learning forms the basis for the autonomy exhibited by these systems, enabling them to perceive, reason, and act in dynamic environments. However, as autonomous systems assume increasingly complex roles, understanding decision-making processes becomes paramount as shown in figure 1. Navigating these intricacies requires a nuanced comprehension of underlying complexities and a concerted effort to address potential challenges.

Central to this discussion is the concept of algorithmic transparency, aimed at demystifying decision-making processes, fostering trust, and ensuring accountability. Methodologies for achieving algorithmic transparency are examined, offering insights into transparent algorithm implementation in autonomous systems.

Simultaneously, fortifying security measures is imperative to safeguard autonomous systems against potential threats and vulnerabilities. Understanding diverse security threats and vulnerabilities is crucial, as is adopting strategies for robust security implementation to mitigate risks and ensure system integrity. (R. Kumar et al., 2024)

Figure 1. Uses of Autonomous Drones

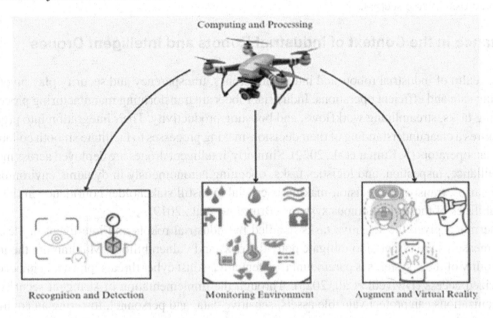

Ethical considerations play a significant role, emphasizing the ethical imperatives involved in autonomous systems' deployment and operation. Real-world case studies provide illuminating examples, showcasing transparency and security measures in action, and offering valuable insights and lessons learned from practical implementations (Michael et al., 2020).

Looking ahead, emerging trends and technologies are anticipated, envisioning a landscape where transparency and security evolve alongside the dynamic evolution of autonomous systems. Embracing transparency, fortifying security, and upholding ethical principles are crucial for charting a course towards a future where autonomous systems enhance lives while safeguarding collective well-being.

Overview of Transparency and Security in Autonomous Systems

In the realm of autonomous systems, characterized by the seamless coordination of industrial robots and intelligent drones, transparency and security are fundamental pillars. Transparency, essential for trust and accountability, signifies the clarity and comprehensibility embedded within these systems' decision-making mechanisms (Theodorou et al., 2017). It entails stakeholders understanding the reasoning behind autonomous actions, fostering confidence in their capabilities. By demystifying the algorithms driving autonomous decisions, transparency not only enhances stakeholder comprehension but also facilitates collaboration between human operators and autonomous systems (Wortham, 2020).

On the other hand, security acts as a shield against various threats and vulnerabilities jeopardizing the integrity of autonomous operations. Embracing security involves a multifaceted approach, from defending against cyber-attacks and data breaches to thwarting physical tampering and unauthorized access. Strengthening the defenses of autonomous systems safeguards sensitive information, assets, and personnel, ensuring operational continuity and resilience. The seamless integration of transparency and security forms the foundation for trustworthy and resilient autonomous systems, driving their widespread adoption across diverse sectors.

Importance in the Context of Industrial Robots and Intelligent Drones

In the realm of industrial robots and intelligent drones, transparency and security play pivotal roles in ensuring safe and efficient operations. Industrial robots are transforming manufacturing processes by automating tasks, streamlining workflows, and boosting productivity. Their integration into production lines requires a clear understanding of their decision-making processes to facilitate smooth collaboration with human operators (S. Kumar et al., 2022). Similarly, intelligent drones are deployed across industries for surveillance, inspection, and logistics tasks, operating autonomously in dynamic environments. In these scenarios, transparent decision-making is crucial to instill stakeholder confidence in the actions and capabilities of these autonomous systems (Bruzzone et al., 2019).

Furthermore, given the sensitive tasks handled by industrial robots and intelligent drones, robust security measures are essential to mitigate potential risks and vulnerabilities. Maintaining the integrity and reliability of these systems is paramount to safeguard against cyber threats, physical tampering, and unauthorized access (Dwivedi et al., 2023). Through the implementation of stringent security protocols, organizations can protect valuable assets, sensitive data, and personnel, fostering an environment conducive to the widespread adoption of autonomous technologies in industrial settings. In summary, transparency and security serve as essential components for the seamless integration and effective operation of industrial robots and intelligent drones, driving innovation and efficiency across diverse industrial sectors (Svaigen et al., 2023).

This chapter contributes to the discourse on transparency and security in autonomous systems by providing a comprehensive overview of the challenges, methodologies, and real-world implementations in this rapidly evolving field. It synthesizes existing literature, identifies key technological foundations, and delves into practical considerations for achieving algorithmic transparency and robust security measures. By elucidating the ethical implications and responsibilities incumbent upon stakeholders, this chapter aims to foster a deeper understanding of the multifaceted nature of autonomous systems and their societal impact. Additionally, the inclusion of real-world case studies offers valuable insights into the practical challenges encountered and innovative solutions devised in various domains, ranging from healthcare robotics to autonomous vehicles and delivery services.

The chapter unfolds through a structured exploration, each section sequentially addressing distinct facets of transparency and security within autonomous systems. It commences with an overarching introduction (Section 1), followed by an extensive literature review (Section 2) to establish the existing knowledge base. Subsequently, it delves into the technological underpinnings (Section 3), examines the challenges in decision-making processes (Section 4), and elucidates methodologies for achieving algorithmic transparency (Section 5). Real-world case studies (Section 5.4) and lessons learned (Section 5.5) offer practical insights, while predictions and emerging technologies (Section 6) shed light on future trends. The chapter culminates in a conclusive reflection (Section 7), synthesizing key insights and implications for future research and practice, thereby presenting a comprehensive roadmap for navigating the complex terrain of transparency and security in autonomous systems.

LITERATURE REVIEW

In the rapidly advancing realm of autonomous systems, where technologies like machine learning, AI, robotics, and deep learning converge, there's a growing interest in understanding and fortifying the core principles of transparency and security. With industrial robots revolutionizing manufacturing processes and intelligent drones navigating intricate environments, this section embarks on a detailed review of scholarly works exploring these vital areas. By examining influential studies, frameworks, and initiatives, this literature review aims to shed light on the complex discussions surrounding transparency and security within autonomous systems. Through this exploration, we seek to uncover key insights, emerging trends, and areas where research is lacking. Additionally, by comparing and contrasting these insights with the overarching themes outlined in the abstract, we hope to place our contribution within the broader technical discourse below in table 1, offering valuable perspectives on the evolving landscape of autonomous systems.

Table 1. Literature review on enhancing autonomous system security with AI and secure computation technologies

Author(s)	Year	Relevance to out paper	Main Findings
Li, B.(Li et al., 2021)	2021	Focuses on data security and encryption techniques for autonomous IoT systems	Algorithms enhance data encryption in IoT systems using AI, reducing correlation and improving security.
Janeera, D.A.(Janeera et al., 2021)	2021	Addresses challenges in governance, economy, mobility, environment, and cybersecurity in smart cities with autonomous vehicles	Analyzes challenges in smart cities and proposes solutions for autonomous vehicle deployment.
Christou, A.G.(Christou et al., 2023)	2023	Emphasizes the importance of real-time data analysis and security in smart and autonomous systems	Highlights potential of AI, IoT, and edge computing in enhancing efficiency and decision-making in various sectors
Santoso, F. & Finn, A.	2020	Evaluates the ethical, legal, and security implications of AI-driven military technology	Explores AI's impact on military autonomy, cyber vulnerabilities, legal and ethical implications.
He, H.(Siwach & Li, 2024)	2020	Highlights the importance of trust and reliability in trustworthy robots and autonomous systems	Analyzes trustworthiness properties of RAS and discusses AI's role in enhancing them.
Flammini, F. (Sifakis & Harel, 2023)	2022	Provides insights into methodologies for developing secure autonomous systems	Discusses taxonomies and methodologies for developing trustworthy autonomous systems.
Jenihhin, M.(Horeis et al., 2020)	2019	Emphasizes the need for robustness and resilience in AI-driven autonomous systems	Examines reliability challenges in cyber-physical systems and proposes solutions for enhanced autonomy.

TECHNOLOGICAL FOUNDATIONS

Technological advancements in machine learning, artificial intelligence (AI), robotics, and deep learning have revolutionized the capabilities of autonomous systems, ushering in a new era of innovation and efficiency. In this section, we delve into the foundational elements that underpin the autonomy and intelligence of modern autonomous systems. The convergence of these advanced technologies plays a pivotal role in shaping the functionality and capabilities of autonomous systems, enabling them to perceive, reason, and act autonomously in dynamic environments. By understanding the role of machine learning, AI, robotics, and deep learning in autonomous systems, we gain insights into the intricate interplay between hardware and software components that drive their operation. Furthermore, we explore the integration of these technologies in autonomous systems, highlighting the seamless fusion of hardware and software to enable autonomous decision-making and execution of tasks. Through this exploration, we aim to unravel the complexities of technological foundations in autonomous systems and appreciate their transformative potential across diverse industries and applications.

Role of Machine Learning, AI, Robotics, and Deep Learning

The integration of machine learning, artificial intelligence (AI), robotics, and deep learning technologies forms the foundation of autonomous systems, empowering them with advanced cognitive and physical capabilities. Machine learning algorithms play a crucial role in enabling autonomous systems to learn from data, identify patterns, and make predictions, thereby optimizing their decision-making processes. Supervised learning algorithms, for instance, enable autonomous systems to learn from labeled data, while unsupervised learning algorithms facilitate the discovery of hidden patterns and structures within unlabeled data. Additionally, reinforcement learning mechanisms allow autonomous agents to learn optimal behaviors through interaction with their environment, enhancing their adaptability and autonomy over time(Mengi et al., 2023). This intricate interplay of machine learning, AI, robotics, and deep learning technologies underscores the multifaceted nature of autonomous systems. The given Table 2 below summarizes the key roles of these technologies in driving the autonomy and intelligence of such systems:

Table 2. Overview of the key roles of machine learning, artificial intelligence, robotics, and deep learning technologies in autonomous systems

Technology	Role in Autonomous Systems
Machine Learning	Learn patterns, adapt to environments, optimize decision-making. Techniques: supervised, unsupervised, reinforcement learning.
Artificial Intelligence (AI)	Provides cognitive functionalities: NLP, computer vision, reinforcement learning. Enables perception, reasoning, and intelligent action.
Robotics	Hardware and software for physical interaction: sensors, actuators, motion planning.
Deep Learning	Processes complex data, recognizes patterns, makes informed decisions using neural networks. Enables image/speech recognition, autonomous decision-making.

Artificial intelligence (AI) further enhances the capabilities of autonomous systems by providing cognitive functionalities essential for perception, reasoning, and intelligent action. Natural Language Processing (NLP) enables autonomous systems to understand and generate human language, facilitating seamless communication and interaction with users. Computer vision algorithms empower autonomous systems to interpret visual data from sensors and cameras, enabling tasks such as object detection, recognition, and tracking (Chopra et al., 2022) (Sharma, Singh, Badwal, et al., 2023). Reinforcement learning techniques enable autonomous agents to learn and improve their decision-making strategies based on feedback received from their environment, enhancing their ability to navigate complex scenarios and make informed decisions.

On the hardware side, robotics technology provides the physical infrastructure necessary for autonomous systems to interact with their environment. This includes sensors for perception, actuators for manipulation, and motion planning algorithms for efficient movement. Robotics enables autonomous systems to navigate dynamic environments, manipulate objects with precision, and execute tasks with accuracy and efficiency. Moreover, advancements in deep learning techniques have revolutionized the field of autonomous systems by enabling them to process complex data, recognize intricate patterns, and extract meaningful insights. Neural networks with multiple layers facilitate advanced cognitive capabilities such as image and speech recognition, enabling autonomous systems to perceive and understand their surroundings effectively.

Integration in Autonomous Systems

Integration of various technological components is paramount in achieving the seamless functionality of autonomous systems. In the context of machine learning, AI, robotics, and deep learning, integration involves the harmonious amalgamation of hardware and software components to enable autonomous decision-making and execution of tasks.

At the core of integration lies the fusion of machine learning algorithms with the hardware infrastructure of autonomous systems. This entails the development of software frameworks capable of processing large volumes of data collected from sensors and other sources. Machine learning algorithms are then deployed to analyze this data, extract meaningful insights, and adapt the behavior of autonomous systems in response to changing environmental conditions.

Furthermore, the integration of AI techniques involves the incorporation of cognitive functionalities into the decision-making processes of autonomous systems. Natural language processing algorithms enable autonomous systems to understand and respond to human commands, facilitating seamless interaction in various applications. Computer vision algorithms, integrated with cameras and sensors, allow autonomous systems to perceive and interpret their surroundings, enabling tasks such as object detection and navigation.

Robotics technology plays a crucial role in the integration process by providing the physical infrastructure necessary for autonomous systems to interact with the environment (Aggarwal, Singh, Chopra, Kumar, et al., 2022). This includes actuators, manipulators, and locomotion mechanisms that enable autonomous systems to move, manipulate objects, and perform tasks autonomously. Additionally, software algorithms for motion planning and control are integrated with robotics hardware to enable precise and efficient movement in dynamic environments.

Deep learning techniques are seamlessly integrated into autonomous systems to enhance their cognitive capabilities. Neural networks are deployed to process complex data, recognize patterns, and make informed decisions in real-time (Gupta et al., 2023). (Aggarwal, Singh, Chopra, & Kumar, 2022) Deep learning algorithms are trained on large datasets to enable autonomous systems to perform tasks such as image recognition, speech recognition, and autonomous decision-making with high accuracy and efficiency as shown in Figure 2.

Figure 2. Integration in autonomous system

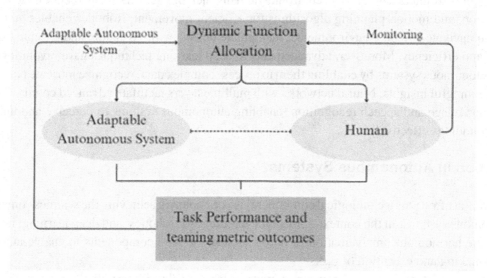

CHALLENGES IN DECISION-MAKING

The decision-making processes of autonomous systems are fraught with a myriad of challenges stemming from the complexity of their operating environments and the intricacies of their tasks. One significant challenge lies in ensuring the accuracy and reliability of decision-making algorithms amidst uncertainties and variations in real-world data. Autonomous systems must contend with noisy sensor measurements, incomplete information, and dynamic environmental conditions, posing challenges for robust decision-making. Additionally, the computational complexity of decision-making algorithms presents challenges in achieving real-time performance, especially in scenarios with high-dimensional state spaces and complex task dependencies. Furthermore, ethical considerations such as fairness, transparency, and accountability add layers of complexity to decision-making processes, requiring careful balancing of competing objectives and values. Table 3 below provides an overview of the key challenges faced by autonomous systems in their decision-making processes, along with their implications.

Table 3. Challenges in decision-making processes of autonomous systems

Challenges	Description	Implications
Uncertainties in Real-World Data	Noisy sensor measurements, incomplete information, and dynamic environmental conditions introduce uncertainties	Reduced accuracy and reliability
Computational Complexity	Computational challenges arise in scenarios with high-dimensional state spaces and complex task dependencies	Constraints on real-time decision-making and scalability
Ethical Considerations	Ethical considerations such as fairness, transparency, and accountability add complexity to decision-making	Conflicts between ethical principles and operations

Complexity of Autonomous Decision-Making Processes

The complexity inherent in autonomous decision-making processes arises from the multifaceted nature of the tasks they must perform and the dynamic environments in which they operate. Autonomous systems are tasked with processing vast amounts of sensory data in real-time, interpreting this information to make informed decisions, and executing actions with precision and efficiency. However, several factors contribute to the complexity of these processes.

Firstly, autonomous decision-making involves dealing with uncertainty and variability in data and environmental conditions. Sensors may provide noisy or incomplete information, and environmental factors such as lighting conditions, weather, and the presence of other agents can introduce unpredictability. Autonomous systems must navigate through this uncertainty to make decisions that are robust and reliable.

Secondly, the decision-making processes of autonomous systems often involve high-dimensional state spaces and complex task dependencies. This complexity arises from the need to consider multiple factors simultaneously, such as spatial and temporal relationships, object interactions, and dynamic constraints. As a result, decision-making algorithms must operate in environments with a large number of possible states and actions, requiring efficient algorithms and computational resources to handle the complexity effectively.

The interactions between perception, cognition, planning, and execution further compound the complexity of autonomous decision-making. Perception involves interpreting sensory data to understand the environment, while cognition entails higher-level reasoning and decision-making based on this information. Planning involves generating sequences of actions to achieve specific objectives, while execution involves translating these plans into physical actions. Coordinating these processes in real-time and adapting to changes in the environment adds another layer of complexity to autonomous decision-making.

Furthermore, ethical considerations such as fairness, transparency, and accountability add complexity to decision-making processes. Autonomous systems must navigate ethical dilemmas and trade-offs, making decisions that align with ethical principles and societal norms. Ensuring transparency in decision-making and accountability for actions taken by autonomous systems is essential for fostering trust and acceptance among stakeholders.

Identifying and Addressing Challenges

Identifying and addressing the myriad challenges inherent in autonomous decision-making processes is crucial for advancing the capabilities and reliability of autonomous systems. One approach involves leveraging interdisciplinary expertise to systematically identify and categorize challenges across various domains, including perception, cognition, planning, and execution. This interdisciplinary collaboration allows for a holistic understanding of the factors contributing to the complexity of decision-making processes and facilitates the development of targeted solutions.

Addressing the challenges of autonomous decision-making often involves the development and refinement of algorithms and methodologies that can effectively handle uncertainty, variability, and high-dimensional state spaces. Machine learning and AI techniques play a pivotal role in this regard, offering scalable and adaptable solutions for learning from data, optimizing decision-making policies, and reasoning under uncertainty. Deep reinforcement learning, for example, enables autonomous systems to learn optimal decision-making strategies through interaction with the environment, leveraging trial and error to improve performance over time.

Furthermore, the integration of probabilistic reasoning and uncertainty modeling techniques enables autonomous systems to explicitly represent and reason about uncertainty in decision-making processes. By incorporating probabilistic models of the environment and sensor measurements, autonomous systems can make informed decisions while accounting for uncertainties and variations in data. Bayesian inference methods, probabilistic graphical models, and Monte Carlo simulation techniques are commonly used to model uncertainty and propagate uncertainties through decision-making algorithms.

Moreover, ensuring transparency, fairness, and accountability in decision-making processes is essential for building trust and acceptance among stakeholders. Transparent decision-making algorithms provide insights into the reasoning behind autonomous systems actions, enabling users to understand and trust their behavior. Fairness-aware decision-making algorithms strive to mitigate biases and disparities in decision outcomes, ensuring equitable treatment for all stakeholders. Accountability mechanisms track and document the actions taken by autonomous systems, enabling stakeholders to hold them accountable for their behavior.

ATTAINING ALGORITHMIC TRANSPARENCY

Algorithmic transparency is paramount in ensuring accountability and trustworthiness in autonomous systems. As these systems become increasingly integrated into various aspects of our lives, understanding how decisions are made is essential for fostering trust among users and stakeholders. In this section, we explore strategies for attaining algorithmic transparency in autonomous systems. From employing interpretable machine learning models to implementing post-hoc explanation techniques, each strategy aims to shed light on the inner workings of decision-making algorithms (Peñalvo, Maan, et al., 2022). Additionally, we discuss the importance of documentation, disclosure, and collaborative development practices in promoting transparency and accountability. By embracing these strategies, autonomous systems can enhance transparency, bolster trust, and facilitate responsible deployment in real-world applications.

Methodologies and Techniques for Transparency

Ensuring transparency in the decision-making processes of autonomous systems is paramount for building trust, understanding system behavior, and addressing ethical concerns. This section explores methodologies and techniques aimed at enhancing transparency within autonomous systems, shedding light on the inner workings of algorithms and facilitating comprehension of decision outcomes. By leveraging a combination of interpretable machine learning models, comprehensive documentation, post-hoc explanation techniques, and open-source software practices, autonomous systems can achieve greater transparency and accountability.

Table 4 below outlines key strategies and techniques for enhancing transparency within autonomous systems, providing insights into methodologies aimed at promoting understandability and interpretability of decision-making algorithms.

Table 4. Strategies for transparency in autonomous systems

Strategy	Description
Interpretable Machine Learning Models	- Utilize models such as decision trees, linear regression, and rule-based systems. - Provide easily understandable results with clear explanations for decision outcomes
Documentation and Disclosure	- Offer detailed documentation of algorithmic processes including data preprocessing, feature selection, and model training. - Disclose information about the sources and characteristics of training data
Post-hoc Explanation Techniques	- Implement techniques such as feature importance analysis, local surrogate models, and counterfactual explanations. - Generate explanations for individual decisions, highlighting input features' contributions and decision-making patterns
Open-Source Software and Collaborative Development	- Embrace open-source frameworks and collaborative practices. - Enable developers to inspect, modify, and contribute to decision-making algorithms. - Foster knowledge sharing, validation, and improvement of system behavior through code reviews, documentation, and peer feedback

Implementing Transparent Algorithms in Autonomous Systems

Implementing transparent algorithms in autonomous systems necessitates meticulous attention to technical details and considerations. Here, we delineate the technical intricacies involved in integrating transparent algorithms into the decision-making processes of autonomous systems:

1. **Algorithm Selection and Customization**: Opt for algorithms known for their transparency and interpretability, such as decision trees, linear regression, and rule-based models. Customizing these algorithms to suit specific application requirements may involve fine-tuning parameters or modifying decision criteria to enhance interpretability while maintaining performance.

2. **Data Preprocessing Pipeline Transparency**: Establish a transparent data preprocessing pipeline, ensuring clarity and reproducibility in data cleaning, feature extraction, and normalization procedures. Documenting each step of the preprocessing pipeline is crucial for traceability and understanding of data transformations.

3. **Feature Selection and Engineering**: Prioritize features that are easily interpretable and relevant to the problem domain. Employ techniques like univariate or multivariate feature selection to identify salient features while minimizing complexity. Document the rationale behind feature selection decisions to provide transparency in model inputs.

4. **Transparent Model Training and Evaluation**: Adopt transparent methodologies for model training and evaluation, emphasizing reproducibility and interpretability. Document hyperparameter selection, model architecture, and optimization techniques to facilitate understanding and validation of the training process.

5. **Validation and Testing Protocols**: Employ robust validation and testing protocols to assess the performance and generalization capabilities of transparent algorithms. Utilize techniques such as k-fold cross-validation or bootstrapping to evaluate algorithm robustness and assess potential overfitting.

6. **Comprehensive Documentation**: Provide exhaustive documentation detailing all aspects of the implemented algorithm, including algorithmic assumptions, limitations, and potential biases. Document the technical specifications, implementation details, and algorithmic logic to facilitate transparency and reproducibility.

7. **Seamless Integration with Autonomous Systems**: Ensure seamless integration of transparent algorithms into the decision-making pipeline of autonomous systems. Develop interfaces that enable efficient communication between the algorithm and other system components while providing clear explanations for decision outputs.

By meticulously addressing these technical considerations, developers can effectively implement transparent algorithms in autonomous systems, bolstering accountability, trustworthiness, and interpretability in decision-making processes.

Security Measures

Within the domain of autonomous systems, characterized by intricate interconnectivity and dynamic operational environments, the fortification of security measures emerges as a critical imperative to uphold system integrity and thwart potential cyber threats (Sharma, Singh, Kumar, et al., 2023). As industrial robots and intelligent drones assume increasingly complex roles across various sectors, the implementation of robust security protocols becomes indispensable. This section delves into the technical intricacies of security fortification within autonomous systems, exploring advanced strategies and methodologies aimed at mitigating vulnerabilities and enhancing resilience against cyber intrusions. Through a meticulous examination of encryption protocols, access control mechanisms, anomaly detection algorithms, and threat intelligence integration, stakeholders can proactively safeguard autonomous systems, ensuring uninterrupted operations and safeguarding sensitive data from malicious exploitation.

Security Threats and Vulnerabilities

Now we delve into the technical intricacies of cyber risks inherent in autonomous systems, elucidating their multifaceted nature and potential implications. By dissecting the intricacies of security threats and vulnerabilities, stakeholders can equip themselves with the knowledge necessary to devise robust defense strategies tailored to the unique challenges posed by autonomous operations.

1. **Cyber Threat Landscape Analysis:** Examining the diverse array of cyber threats targeting autonomous systems, including malware, ransomware, denial-of-service (DoS) attacks, and insider threats. Analyzing the evolving tactics, techniques, and procedures (TTPs) employed by threat actors to exploit vulnerabilities and compromise system integrity.
2. **Vulnerability Assessment and Analysis:** Conducting comprehensive vulnerability assessments to identify potential weaknesses and entry points within autonomous systems. Assessing vulnerabilities stemming from software flaws, misconfigurations, insecure network protocols, and inadequate access controls.
3. **Impact Assessment and Risk Analysis:** Evaluating the potential impact of security breaches on autonomous system operations, including financial losses, reputational damage, and safety risks. Performing risk analysis to prioritize mitigation efforts based on the severity and likelihood of identified threats and vulnerabilities.

4. **Threat Intelligence Integration:** Leveraging threat intelligence feeds and security information and event management (SIEM) systems to enhance situational awareness and proactively identify emerging threats. Integrating threat intelligence into security operations to enable timely threat detection, incident response, and threat mitigation.

5. **Regulatory Compliance and Standards Adherence:** Ensuring compliance with industry-specific regulations and standards governing cybersecurity practices, such as ISO/IEC 27001, NIST Cybersecurity Framework, and GDPR. Adhering to best practices and guidelines outlined by regulatory bodies and industry consortia to enhance security posture and mitigate legal and regulatory risks.

By delving into the intricacies of security threats and vulnerabilities, stakeholders can gain valuable insights into the threat landscape and devise proactive measures to safeguard autonomous systems against potential cyber attacks.

Strategies for Robust Security Implementation

Ensuring the resilience of autonomous systems against cyber threats requires the implementation of robust security strategies tailored to mitigate risks and safeguard system integrity. This section delves into advanced methodologies and techniques employed to fortify security within autonomous systems, emphasizing proactive measures to enhance resilience and mitigate potential vulnerabilities.

1. **Secure Design Principles:** Incorporating secure design principles into the development lifecycle of autonomous systems to preemptively address security concerns. Adopting principles such as the principle of least privilege, defense-in-depth, and fail-safe defaults to minimize attack surfaces and mitigate the impact of security breaches.

2. **Threat Modeling and Risk Assessment:** Conducting comprehensive threat modeling exercises to identify potential attack vectors and prioritize security controls based on the severity and likelihood of threats. Performing risk assessments to evaluate the potential impact of security incidents on autonomous system operations and define risk mitigation strategies accordingly.

3. **Encryption and Cryptographic Protocols:** Implementing encryption mechanisms and cryptographic protocols to protect sensitive data transmitted and stored within autonomous systems(S. Kumar et al., 2021). Deploying robust encryption algorithms such as AES and RSA to ensure confidentiality and integrity of data exchanges.

4. **Access Control Mechanisms:** Enforcing stringent access control mechanisms to regulate user privileges and limit unauthorized access to critical system resources. Implementing role-based access control (RBAC) and multi-factor authentication (MFA) to authenticate users and restrict access to privileged functions.

5. **Intrusion Detection and Prevention Systems (IDPS):** Deploying intrusion detection and prevention systems (IDPS) to monitor network traffic and detect anomalous behavior indicative of cyber attacks (I. Singh et al., 2022). Utilizing machine learning and anomaly detection algorithms to identify suspicious patterns and trigger timely response mechanisms to mitigate security incidents.

6. **Continuous Monitoring and Incident Response:** Establishing continuous monitoring mechanisms to track system activity and detect security anomalies in real-time. Developing robust incident response plans and procedures to facilitate rapid containment and remediation of security incidents, minimizing their impact on autonomous system operations.

Ethical Considerations

Ethical considerations are paramount in the development, deployment, and operation of autonomous systems, particularly in the domains of artificial intelligence (AI) and robotics. This section delves into the ethical implications inherent in the advancement of AI and robotics technologies, as well as the responsibilities incumbent upon stakeholders in their deployment and operation.

Ethical Implications in AI and Robotics

Figure 3. Ethical consideration in AI

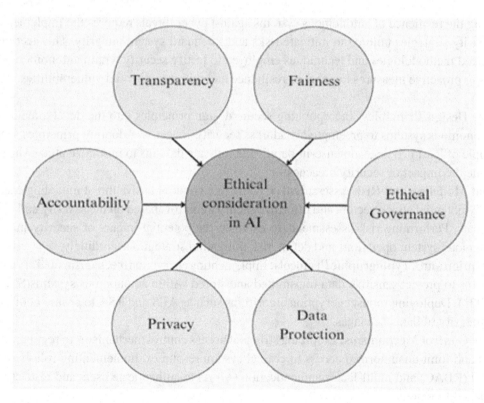

The integration of AI and robotics technologies raises profound ethical questions pertaining to the autonomy, accountability, and societal impact of autonomous systems as shown above in figure 3. Key ethical considerations include:

1. **Autonomy and Agency:** Autonomous systems possess varying degrees of autonomy, ranging from semi-autonomous to fully autonomous capabilities. Ethical frameworks must delineate the boundaries of autonomy and define mechanisms for human oversight and intervention to ensure alignment with societal values and norms.

2. **Accountability and Transparency:** Ensuring accountability in AI and robotics entails establishing mechanisms for traceability, auditability, and explainability of decision-making processes. Transparent algorithms and decision-making mechanisms foster trust and enable stakeholders to understand and scrutinize the actions of autonomous systems.

3. **Fairness and Bias Mitigation:** AI and robotics systems are susceptible to biases inherent in training data, algorithms, and decision-making processes. Ethical frameworks must address issues of fairness, equity, and bias mitigation to prevent discriminatory outcomes and promote inclusivity.

4. **Privacy and Data Protection:** Autonomous systems generate vast amounts of data, raising concerns about privacy, consent, and data protection. Ethical guidelines should safeguard individuals' privacy rights and ensure responsible data stewardship throughout the lifecycle of autonomous systems (M. Singh et al., 2023).

5. **Societal Impact and Ethical Governance:** The widespread deployment of AI and robotics technologies has far-reaching societal implications, ranging from job displacement to shifts in power dynamics. Ethical governance frameworks must consider the broader societal impacts of autonomous systems and prioritize the collective well-being of individuals and communities.

Responsibilities in Deployment and Operation

Stakeholders involved in the deployment and operation of autonomous systems bear ethical responsibilities to uphold principles of safety, transparency, accountability, and societal benefit. Key responsibilities include:

1. **Safety and Risk Mitigation:** Prioritizing the safety of individuals and communities by conducting thorough risk assessments, implementing robust safety measures, and adhering to regulatory standards and best practices.

2. **Transparency and Accountability:** Promoting transparency and accountability by providing clear documentation of system capabilities, limitations, and decision-making processes. Establishing mechanisms for monitoring, auditing, and reporting on system performance and compliance with ethical guidelines.

3. **Human-Centric Design:** Designing autonomous systems with human values and preferences in mind, ensuring user-centered interfaces, intuitive interaction modalities, and mechanisms for human oversight and intervention.

4. **Continuous Evaluation and Improvement:** Committing to continuous evaluation and improvement of autonomous systems through feedback loops, performance metrics, and stakeholder engagement. Iteratively refining algorithms, models, and operational procedures to align with evolving ethical standards and societal expectations.

5. **Ethical Leadership and Governance:** Fostering a culture of ethical leadership and governance within organizations and institutions involved in AI and robotics development. Establishing clear lines of responsibility, accountability, and ethical oversight to guide decision-making and promote ethical conduct throughout the lifecycle of autonomous systems.

Real-World Case Studies

In this section, we delve into real-world case studies that provide concrete examples of the implementation of transparency and security measures in autonomous systems. These case studies offer valuable insights into how advanced technologies such as machine learning, artificial intelligence (AI), robotics, and deep learning are applied in various domains to enhance the reliability, trustworthiness, and ethical considerations of autonomous systems.

1. **Autonomous Vehicles in Urban Environments -** One compelling case study focuses on the deployment of autonomous vehicles (AVs) in urban environments. Companies like Waymo and Tesla have pioneered the development of self-driving cars equipped with advanced sensors, AI algorithms, and machine learning models(Panja, 2021). These AVs navigate complex city streets, interact with other vehicles and pedestrians, and make real-time decisions to ensure safe and efficient transportation. Case studies in this domain highlight the challenges of integrating transparency and security measures into AV's as shown in figure 4, including ensuring robust perception capabilities, mitigating cybersecurity risks, and addressing ethical dilemmas such as decision-making in emergency situations.

Figure 4. Safety and security in autonomous vehicles

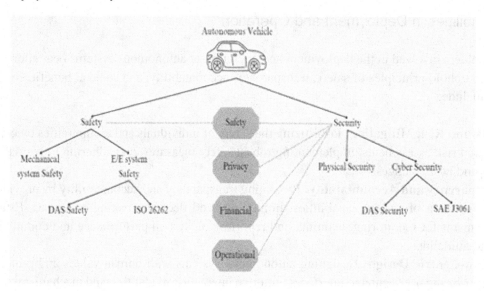

2. **Industrial Robotics in Manufacturing -** Another notable case study revolves around the use of industrial robots in manufacturing settings. Companies like Fanuc, ABB, and KUKA leverage robotic arms equipped with AI-powered vision systems to automate repetitive tasks, optimize production processes, and improve product quality. Real-world examples demonstrate how transparent decision-making frameworks and robust security protocols are implemented in industrial robotics to ensure safe human-robot collaboration, prevent unauthorized access to sensitive manufacturing data, and

maintain operational integrity [Xi]. These case studies underscore the importance of algorithmic transparency and security in industrial automation, particularly in safety-critical environments.

3. **Intelligent Drones for Surveillance and Inspection -** Intelligent drones represent another area where transparency and security measures are paramount. Organizations across industries, including agriculture, construction, and public safety, deploy drones equipped with AI algorithms and computer vision systems for surveillance and inspection tasks. Case studies in this domain showcase how transparent algorithms and secure communication protocols enable drones to perform aerial reconnaissance, detect anomalies, and transmit data securely to ground stations as shown in figure 5. Moreover, these examples shed light on the ethical considerations involved in drone operations, such as privacy concerns and adherence to regulatory frameworks.

Figure 5. Drone for security and surveillance

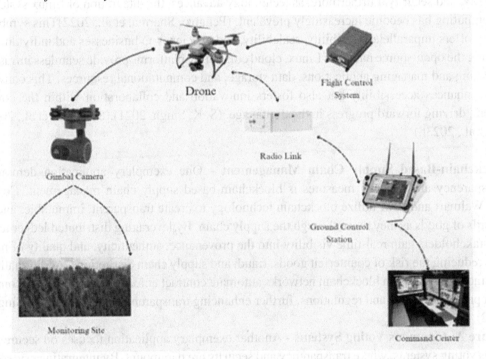

4. **Healthcare Robotics and AI-Assisted Diagnosis -** In the field of healthcare, robotics and AI play a vital role in improving patient care and medical diagnosis. Surgical robots like the da Vinci Surgical System and AI-assisted diagnostic tools leverage machine learning algorithms to assist healthcare professionals in performing minimally invasive procedures and analyzing medical imaging data. Real-world case studies highlight the challenges of ensuring algorithmic transparency and data security in healthcare robotics, particularly concerning patient privacy, data confidentiality, and regulatory compliance. These examples underscore the importance of ethical guidelines and transparent decision-making processes in healthcare AI applications.

Through these real-world case studies, we gain valuable insights into the diverse applications of transparency and security measures in autonomous systems across different industries. These examples demonstrate the practical challenges encountered, innovative solutions devised, and lessons learned from implementing transparency and security protocols in autonomous operations. Ultimately, these case studies provide valuable guidance for researchers, practitioners, and policymakers navigating the evolving landscape of autonomous systems.

Exemplary Applications Demonstrating Transparency and Security Measures

In this section, we highlight exemplary applications that serve as shining examples of transparency and security measures implemented in autonomous systems. These applications showcase innovative approaches, best practices, and successful implementations of technologies such as machine learning, artificial intelligence (AI), robotics, and deep learning, with a strong emphasis on ensuring transparency, integrity, and security. Furthermore, as technology advances, the integration of Linux systems with cloud computing has become increasingly prevalent. (Peñalvo, Sharma, et al., 2022)This symbiotic relationship offers unparalleled flexibility, scalability, and efficiency to businesses and individuals alike. Leveraging the open-source nature of Linux, cloud computing platforms provide seamless infrastructure for deploying and managing applications, data storage, and computational resources. This convergence not only enhances accessibility but also fosters innovation and collaboration within the computing ecosystem, driving forward progress in the digital age. (S. K. Singh, 2021).(S. Kumar et al., 2023a) (S. Kumar et al., 2023b)

1. **Blockchain-Based Supply Chain Management** - One exemplary application demonstrating transparency and security measures is blockchain-based supply chain management. Companies like Walmart and IBM utilize blockchain technology to create transparent, immutable, and secure records of goods as they move through the supply chain. By leveraging distributed ledger technology, stakeholders gain real-time visibility into the provenance, authenticity, and quality of products, thus reducing the risk of counterfeit goods, fraud, and supply chain disruptions. Additionally, smart contracts embedded in blockchain networks automate contract enforcement and ensure compliance with predefined rules and regulations, further enhancing transparency and security (S. Singh et al., 2022).

2. **Secure Autonomous Voting Systems** - Another exemplary application focuses on secure autonomous voting systems, where transparency and security are paramount. By integrating cryptographic techniques and distributed ledger technology, organizations can create tamper-proof and verifiable voting systems that ensure the integrity and confidentiality of electoral processes. These systems enable voters to cast their ballots securely from remote locations, while blockchain-based transparency mechanisms allow for the auditability and accountability of election results. Moreover, advanced encryption techniques protect voter privacy and prevent tampering or manipulation of voting data, ensuring trust and confidence in democratic processes.

3. **Healthcare Data Sharing Platforms** - Healthcare data sharing platforms represent another exemplary application that emphasizes transparency and security measures. Platforms like Synapse and TriNetX leverage federated learning techniques and differential privacy algorithms to enable secure and privacy-preserving collaboration among healthcare institutions, researchers, and pharmaceutical companies. By anonymizing patient data and encrypting sensitive information, these platforms facil-

itate the sharing and analysis of large-scale healthcare datasets while protecting patient privacy and confidentiality. Furthermore, transparent governance models and audit trails ensure accountability and compliance with regulatory requirements, fostering trust and collaboration in healthcare data sharing initiatives.

4. **Autonomous Financial Trading Systems** - In the realm of financial markets, autonomous trading systems exemplify the integration of transparency and security measures. High-frequency trading firms and hedge funds deploy AI-powered algorithms and machine learning models to execute trades autonomously and optimize investment strategies. These systems employ transparent decision-making frameworks and risk management protocols to ensure compliance with regulatory standards and mitigate market manipulation risks. Additionally, secure communication channels and encryption techniques safeguard sensitive trading data and prevent unauthorized access or tampering, enhancing the integrity and resilience of financial markets.

Through these exemplary applications, we witness the transformative potential of transparency and security measures in autonomous systems across diverse domains. These applications demonstrate the effectiveness of innovative technologies and robust protocols in ensuring trust, integrity, and accountability in autonomous operations. By learning from these examples and embracing best practices, stakeholders can navigate the complexities of deploying autonomous systems with confidence and resilience in an increasingly interconnected and data-driven world.

Lessons Learned From Practical Implementations

In this section, we distill key lessons learned from practical implementations of transparency and security measures in autonomous systems. Drawing from real-world experiences and case studies, we uncover valuable insights, challenges encountered, and best practices identified during the deployment and operation of autonomous systems in various domains.

1. **Importance of Interdisciplinary Collaboration** - One of the foremost lessons learned is the significance of interdisciplinary collaboration in designing and implementing transparent and secure autonomous systems. Successful projects often involve close collaboration between domain experts, data scientists, cybersecurity specialists, ethicists, and regulatory compliance professionals. By fostering interdisciplinary teamwork, organizations can leverage diverse perspectives, domain knowledge, and technical expertise to address complex challenges effectively. Moreover, interdisciplinary collaboration promotes holistic problem-solving approaches and ensures alignment with regulatory requirements and ethical principles.

2. **Continuous Monitoring and Evaluation** - Another critical lesson is the importance of continuous monitoring and evaluation throughout the lifecycle of autonomous systems. Real-world deployments often face evolving threats, changing environmental conditions, and unforeseen challenges that require adaptive responses. Implementing robust monitoring mechanisms, anomaly detection algorithms, and incident response protocols enables organizations to detect and mitigate security breaches, system failures, and performance degradation in a timely manner. Furthermore, regular audits, assessments, and post-mortem analyses facilitate continuous improvement and refinement of system architectures, algorithms, and security measures.

3. **User-Centric Design and Transparency** - User-centric design principles and transparency are essential for fostering trust, acceptance, and usability of autonomous systems among end-users and stakeholders. Practical implementations demonstrate the importance of designing intuitive user interfaces, providing clear explanations of system functionalities, and soliciting feedback from end-users throughout the development process. Transparent decision-making processes, explainable AI techniques, and visualizations of system behaviors enhance user comprehension and trust in autonomous systems' actions and recommendations. By prioritizing user needs and expectations, organizations can enhance user satisfaction, adoption rates, and overall system performance.

4. **Resilience to Adversarial Attacks** - Practical implementations underscore the need for building resilience to adversarial attacks and security threats in autonomous systems. Adversaries may exploit vulnerabilities in system architectures, manipulate input data, or launch sophisticated cyber-attacks to compromise system integrity and disrupt operations. Implementing defense-in-depth strategies, encryption techniques, and intrusion detection systems can help mitigate the impact of adversarial attacks and enhance system resilience. Additionally, fostering a security-first mindset, conducting penetration testing, and staying informed about emerging threats are crucial for proactively addressing security risks and vulnerabilities.

5. **Ethical Considerations and Responsible AI** - Finally, practical implementations highlight the importance of integrating ethical considerations and responsible AI principles into the design, development, and deployment of autonomous systems. Organizations must prioritize fairness, transparency, accountability, and privacy protection in algorithmic decision-making processes to ensure equitable outcomes and mitigate unintended consequences. Adhering to ethical guidelines, regulatory frameworks, and industry standards promotes responsible use of AI technologies and fosters public trust and confidence in autonomous systems (Sharma, Singh, Chhabra, et al., 2023).

By embracing these lessons learned from practical implementations, organizations can navigate the complexities and challenges associated with deploying transparent and secure autonomous systems effectively. Continuous learning, adaptation, and collaboration are key to realizing the transformative potential of autonomous technologies while safeguarding against risks and ensuring ethical and responsible use in diverse application domains.

TOWARDS FUTURE TRENDS

As autonomous systems continue to evolve at a rapid pace, it is imperative to anticipate future trends that will shape the trajectory of this transformative field. In this section, we delve into emerging technologies and evolving paradigms that are poised to redefine the landscape of autonomous systems in the coming years.

Predictions and Forecasts for the Evolution of Autonomous Systems

The future of autonomous systems holds immense promise, with significant advancements expected in various domains. Forecasts suggest a proliferation of autonomous technologies across industries, driven by innovations in machine learning, robotics, and sensor technologies (Chhabra et al., 2024). Key predictions for the evolution of autonomous systems include:

1. **Expansion of Autonomous Vehicles:** Autonomous vehicles are poised to revolutionize transportation systems, with forecasts indicating widespread adoption of self-driving cars, trucks, and drones. The integration of advanced sensors, AI algorithms, and communication technologies will enable autonomous vehicles to navigate complex environments safely and efficiently, transforming urban mobility and logistics.

2. **Advancements in Healthcare Robotics:** Healthcare robotics is expected to witness substantial growth, with autonomous robots playing pivotal roles in patient care, surgery, and medical logistics. Innovations in soft robotics, AI-driven diagnostics, and telemedicine technologies will enable autonomous systems to augment healthcare delivery, improve patient outcomes, and alleviate caregiver burden.

3. **Deployment of Autonomous Delivery Services:** Autonomous delivery services are poised to disrupt traditional logistics models, with drones and autonomous vehicles offering efficient and cost-effective delivery solutions. Forecasts suggest an exponential rise in autonomous last-mile delivery services, catering to e-commerce, food delivery, and supply chain logistics.

4. **Integration of Autonomous Systems in Smart Cities:** Smart cities of the future will leverage autonomous systems to optimize urban infrastructure, enhance public safety, and promote sustainability (Tran et al., 2023). Autonomous drones, robots, and sensors will facilitate data-driven decision-making, enabling cities to efficiently manage traffic, monitor environmental quality, and respond to emergencies (R. Singh et al., 2022).

5. **Emergence of Swarm Robotics:** Swarm robotics is expected to emerge as a disruptive paradigm, enabling the coordinated operation of large numbers of autonomous agents to accomplish complex tasks. Swarm robotics applications span diverse domains, including agriculture, disaster response, and environmental monitoring, offering scalable and adaptive solutions to real-world challenges.

6. **Ethical and Regulatory Considerations:** With the proliferation of autonomous systems, ethical and regulatory considerations will assume greater importance. Stakeholders must grapple with complex questions surrounding safety, accountability, privacy, and equity, necessitating the development of robust ethical frameworks and regulatory policies to govern the deployment and operation of autonomous technologies.

Emerging Technologies Shaping Transparency and Security

The future of autonomous systems is intertwined with emerging technologies that enhance transparency and security. Here's a concise overview of key advancements:

1. **Blockchain:** Enables decentralized, tamper-resistant data storage and transparent audit trails, fostering trust.

2. **Explainable AI (XAI):** Provides human-understandable explanations for system decisions, enhancing transparency.

3. **Secure Multi-Party Computation (SMPC):** Facilitates collaborative decision-making while preserving data privacy.

4. **Homomorphic Encryption:** Allows secure data processing without decrypting it, ensuring data privacy.

5. **Federated Learning:** Enables collaborative model training without sharing raw data, respecting privacy.

6. **Trusted Execution Environments (TEE):** Safeguards critical system components and data from tampering.
7. **Differential Privacy:** Protects individual privacy while enabling statistical analysis of data.
8. **Zero-Knowledge Proofs (ZKP):** Facilitates secure authentication and data integrity verification without revealing sensitive information.

CONCLUSION

The exploration of transparency and security in autonomous systems underscores their pivotal role in shaping the future of technology and society. Throughout this chapter, we have delved into the intricate landscape of industrial robots and intelligent drones, emphasizing the importance of understanding decision-making frameworks and fortifying security protocols.

Autonomous systems represent a transformative force in various industries, offering unparalleled efficiency, precision, and adaptability. However, with their increasing integration into critical domains such as manufacturing, healthcare, and transportation, ensuring transparency and security becomes paramount for their successful adoption and societal acceptance.

Key Insights

Key insights gleaned from our exploration provide a comprehensive understanding of the challenges and opportunities inherent in achieving transparency and security in autonomous systems. We have underscored the critical importance of algorithmic transparency in fostering trust and accountability. By demystifying decision-making processes, stakeholders can better comprehend the actions and behaviors of autonomous systems, leading to increased confidence in their capabilities.

Moreover, our discussion has highlighted the multifaceted nature of security threats and vulnerabilities facing autonomous systems(Panja, 2021). From cyber-attacks to physical tampering, the integrity of autonomous operations is constantly at risk. Robust security measures are indispensable for mitigating these threats and safeguarding sensitive information, assets, and personnel.

Implications for Future Research and Practice

The implications of our findings extend to future research and practice in the field of autonomous systems. As technology continues to advance, researchers and practitioners must remain vigilant in addressing the evolving challenges posed by algorithmic transparency and security.

In terms of algorithmic transparency, further research is needed to develop methodologies that provide deeper insights into decision-making processes while maintaining performance and efficiency. Explainable AI techniques, for instance, hold promise for enhancing transparency by providing human-understandable explanations for system decisions.

Similarly, in the realm of security, ongoing efforts are required to stay ahead of emerging threats and vulnerabilities. This entails the development of innovative security protocols and mechanisms capable of thwarting sophisticated cyber-attacks and ensuring the resilience of autonomous systems in dynamic operational environments.

Moreover, ethical considerations should remain central in the deployment and operation of autonomous systems. Future research endeavors should focus on establishing ethical frameworks and guidelines that promote responsible innovation and address societal concerns regarding the impact of autonomous technologies.

By addressing these implications, we can pave the way for the continued development and adoption of autonomous systems, ensuring their positive contribution to various sectors while upholding ethical standards and ensuring the well-being of society as a whole. Through collaborative efforts between academia, industry, and regulatory bodies, we can navigate the complex landscape of autonomous systems and harness their transformative potential for the benefit of humanity.

REFERENCES

Aggarwal, K., Singh, S. K., Chopra, M., & Kumar, S. (2022). Role of Social Media in the COVID-19 Pandemic: A Literature Review. In *Data Mining Approaches for Big Data and Sentiment Analysis in Social Media* (pp. 91–115). IGI Global. DOI: 10.4018/978-1-7998-8413-2.ch004

Aggarwal, K., Singh, S. K., Chopra, M., Kumar, S., & Colace, F. (2022). Deep Learning in Robotics for Strengthening Industry 4.0.: Opportunities, Challenges and Future Directions. In Nedjah, N., Abd El-Latif, A. A., Gupta, B. B., & Mourelle, L. M. (Eds.), *Robotics and AI for Cybersecurity and Critical Infrastructure in Smart Cities* (pp. 1–19). Springer International Publishing. DOI: 10.1007/978-3-030-96737-6_1

Bruzzone, A. G., Massei, M., Di Matteo, R., & Kutej, L. (2019). Introducing Intelligence and Autonomy into Industrial Robots to Address Operations into Dangerous Area. In Mazal, J. (Ed.), *Modelling and Simulation for Autonomous Systems* (pp. 433–444). Springer International Publishing. DOI: 10.1007/978-3-030-14984-0_32

Chhabra, A., Singh, S. K., Sharma, A., Kumar, S., Gupta, B. B., Arya, V., & Chui, K. T. (2024). Sustainable and intelligent time-series models for epidemic disease forecasting and analysis. *Sustainable Technology and Entrepreneurship*, 3(2), 100064. DOI: 10.1016/j.stae.2023.100064

Chopra, M., Singh, S. K., Aggarwal, K., & Gupta, A. (2022). Predicting Catastrophic Events Using Machine Learning Models for Natural Language Processing. In *Data Mining Approaches for Big Data and Sentiment Analysis in Social Media* (pp. 223–243). IGI Global. DOI: 10.4018/978-1-7998-8413-2.ch010

Christou, A. G., Stergiou, C. L., Memos, V. A., Ishibashi, Y., & Psannis, K. E. (2023). Revolutionizing Connectivity: The Power of AI, IoT, and Edge Computing for Smart and Autonomous Systems. *2023 6th World Symposium on Communication Engineering (WSCE)*, 56–60. DOI: 10.1109/WSCE59557.2023.10365771

Dwivedi, K., Govindarajan, P., Srinivasan, D., Keerthi Sanjana, A., Selvanambi, R., & Karuppiah, M. (2023). Intelligent Autonomous Drones in Industry 4.0. In Sarveshwaran, V., Chen, J. I.-Z., & Pelusi, D. (Eds.), *Artificial Intelligence and Cyber Security in Industry 4.0* (pp. 133–163). Springer Nature. DOI: 10.1007/978-981-99-2115-7_6

Gupta, A., Singh, S. K., Gupta, B. B., Chopra, M., & Gill, S. S. (2023). Evaluating the Sustainable COVID-19 Vaccination Framework of India Using Recurrent Neural Networks. *Wireless Personal Communications*, 133(1), 73–91. DOI: 10.1007/s11277-023-10751-3

He, H., Gray, J., Cangelosi, A., Meng, Q., McGinnity, T. M., & Mehnen, J. (2022). The Challenges and Opportunities of Human-Centered AI for Trustworthy Robots and Autonomous Systems. *IEEE Transactions on Cognitive and Developmental Systems*, 14(4), 1398–1412. DOI: 10.1109/TCDS.2021.3132282

Horeis, T. F., Kain, T., Müller, J.-S., Plinke, F., Heinrich, J., Wesche, M., & Decke, H. (2020). A Reliability Engineering Based Approach to Model Complex and Dynamic Autonomous Systems. *2020 International Conference on Connected and Autonomous Driving (MetroCAD)*, (pp. 76–84). IEEE. DOI: 10.1109/MetroCAD48866.2020.00020

Janeera, D. A., Gnanamalar, S. S. R., Ramya, K. C., & Kumar, A. G. A. (2021). Internet of Things and Artificial Intelligence-Enabled Secure Autonomous Vehicles for Smart Cities. In Kathiresh, M., & Neelaveni, R. (Eds.), *Automotive Embedded Systems: Key Technologies, Innovations, and Applications* (pp. 201–218). Springer International Publishing. DOI: 10.1007/978-3-030-59897-6_11

Kumar, R., Singh, S. K., Lobiyal, D. K., Kumar, S., & Jawla, S. (2024). Security Metrics and Authentication-based RouTing (SMART) Protocol for Vehicular IoT Networks. *SN Computer Science*, 5(2), 236. DOI: 10.1007/s42979-023-02566-7

Kumar, S., Karnani, G., Gaur, M. S., & Mishra, A. (2021). Cloud Security using Hybrid Cryptography Algorithms. *2021 2nd International Conference on Intelligent Engineering and Management (ICIEM)*, (pp. 599–604). IEEE. DOI: 10.1109/ICIEM51511.2021.9445377

Kumar, S., Singh, S. K., & Aggarwal, N. (2023a). Speculative Parallelism on Multicore Chip Architecture Strengthen Green Computing Concept: A Survey. In *Advanced Computer Science Applications*. Apple Academic Press. DOI: 10.1201/9781003369066-2

Kumar, S., Singh, S. K., & Aggarwal, N. (2023b). Sustainable Data Dependency Resolution Architectural Framework to Achieve Energy Efficiency Using Speculative Parallelization. *2023 3rd International Conference on Innovative Sustainable Computational Technologies (CISCT)*, (pp. 1–6). IEEE. DOI: 10.1109/CISCT57197.2023.10351343

Kumar, S., Singh, S. K., Aggarwal, N., Gupta, B. B., Alhalabi, W., & Band, S. S. (2022). An efficient hardware supported and parallelization architecture for intelligent systems to overcome speculative overheads. *International Journal of Intelligent Systems*, 37(12), 11764–11790. DOI: 10.1002/int.23062

Li, B., Feng, Y., Xiong, Z., Yang, W., & Liu, G. (2021). Research on AI security enhanced encryption algorithm of autonomous IoT systems. *Information Sciences*, 575, 379–398. DOI: 10.1016/j.ins.2021.06.016

Mengi, G., Singh, S. K., Kumar, S., Mahto, D., & Sharma, A. (2023). Automated Machine Learning (AutoML): The Future of Computational Intelligence. In Nedjah, N., Martínez Pérez, G., & Gupta, B. B. (Eds.), *International Conference on Cyber Security, Privacy and Networking (ICSPN 2022)* (pp. 309–317). Springer International Publishing. DOI: 10.1007/978-3-031-22018-0_28

Michael, K., Abbas, R., Roussos, G., Scornavacca, E., & Fosso-Wamba, S. (2020). Ethics in AI and Autonomous System Applications Design. *IEEE Transactions on Technology and Society*, 1(3), 114–127. DOI: 10.1109/TTS.2020.3019595

Panja, S. K. S. (2021). Human Factors of Vehicle Automation. In *Autonomous Driving and Advanced Driver-Assistance Systems (ADAS)*. CRC Press.

Peñalvo, F. J. G., Maan, T., Singh, S. K., Kumar, S., Arya, V., Chui, K. T., & Singh, G. P. (2022). Sustainable Stock Market Prediction Framework Using Machine Learning Models. [IJSSCI]. *International Journal of Software Science and Computational Intelligence*, 14(1), 1–15. DOI: 10.4018/IJSSCI.313593

Peñalvo, F. J. G., Sharma, A., Chhabra, A., Singh, S. K., Kumar, S., Arya, V., & Gaurav, A. (2022). Mobile Cloud Computing and Sustainable Development: Opportunities, Challenges, and Future Directions. [IJCAC]. *International Journal of Cloud Applications and Computing*, 12(1), 1–20. DOI: 10.4018/IJCAC.312583

Santoso, F., & Finn, A. (2023). An In-Depth Examination of Artificial Intelligence-Enhanced Cybersecurity in Robotics, Autonomous Systems, and Critical Infrastructures. *IEEE Transactions on Services Computing*, 1–18. DOI: 10.1109/TSC.2023.3331083

Sharma, A., Singh, S. K., Badwal, E., Kumar, S., Gupta, B. B., Arya, V., Chui, K. T., & Santaniello, D. (2023). Fuzzy Based Clustering of Consumers' Big Data in Industrial Applications. *2023 IEEE International Conference on Consumer Electronics (ICCE)*, (pp. 01–03). IEEE. DOI: 10.1109/ ICCE56470.2023.10043451

Sharma, A., Singh, S. K., Chhabra, A., Kumar, S., Arya, V., & Moslehpour, M. (2023). A Novel Deep Federated Learning-Based Model to Enhance Privacy in Critical Infrastructure Systems. [IJSSCI]. *International Journal of Software Science and Computational Intelligence*, 15(1), 1–23. DOI: 10.4018/ IJSSCI.334711

Sharma, A., Singh, S. K., Kumar, S., Chhabra, A., & Gupta, S. (2023). Security of Android Banking Mobile Apps: Challenges and Opportunities. In Nedjah, N., Martínez Pérez, G., & Gupta, B. B. (Eds.), *International Conference on Cyber Security, Privacy and Networking (ICSPN 2022)* (pp. 406–416). Springer International Publishing. DOI: 10.1007/978-3-031-22018-0_39

Sifakis, J., & Harel, D. (2023). Trustworthy Autonomous System Development. *ACM Transactions on Embedded Computing Systems, 22*(3). DOI: 10.1145/3545178

Singh, I., & Singh, S. Kr., Kumar, S., & Aggarwal, K. (2022). Dropout-VGG Based Convolutional Neural Network for Traffic Sign Categorization. In M. Saraswat, H. Sharma, K. Balachandran, J. H. Kim, & J. C. Bansal (Eds.), *Congress on Intelligent Systems* (pp. 247–261). Springer Nature. DOI: 10.1007/978-981-16-9416-5_18

Singh, M., Singh, S. K., Kumar, S., Madan, U., & Maan, T. (2023). Sustainable Framework for Metaverse Security and Privacy: Opportunities and Challenges. In Nedjah, N., Martínez Pérez, G., & Gupta, B. B. (Eds.), *International Conference on Cyber Security, Privacy and Networking (ICSPN 2022)* (pp. 329–340). Springer International Publishing. DOI: 10.1007/978-3-031-22018-0_30

Singh, R., Singh, S. K., Kumar, S., & Gill, S. S. (2022). SDN-Aided Edge Computing-Enabled AI for IoT and Smart Cities. In *SDN-Supported Edge-Cloud Interplay for Next Generation Internet of Things.* Chapman and Hall/CRC. DOI: 10.1201/9781003213871-3

Singh, S., Sharma, S., Singla, D., & Gill, S. S. (2022). *Evolving Requirements and Application of SDN and IoT in the Context of Industry 4.0.* Blockchain and Artificial Intelligence.

Singh, S. K. (2021). *Linux Yourself: Concept and Programming.* Chapman and Hall/CRC. DOI: 10.1201/9780429446047

Siwach, G., & Li, C. (2024). Unveiling the Potential of Natural Language Processing in Collaborative Robots (Cobots): A Comprehensive Survey. *2024 IEEE International Conference on Consumer Electronics (ICCE)*, (pp. 1–6). IEEE. DOI: 10.1109/ICCE59016.2024.10444393

Svaigen, A. R., Boukerche, A., Ruiz, L. B., & Loureiro, A. A. F. (2023). Security in the Industrial Internet of Drones. *IEEE Internet of Things Magazine*, 6(3), 110–116. DOI: 10.1109/IOTM.001.2200260

Tesei, A., Luise, M., Pagano, P., & Ferreira, J. (2021). Secure Multi-access Edge Computing Assisted Maneuver Control for Autonomous Vehicles. *2021 IEEE 93rd Vehicular Technology Conference (VTC2021-Spring)*, (pp. 1–6). IEEE. DOI: 10.1109/VTC2021-Spring51267.2021.9449087

Theodorou, A., Wortham, R. H., & Bryson, J. J. (2017). Designing and implementing transparency for real time inspection of autonomous robots. *Connection Science*, 29(3), 230–241. DOI: 10.1080/09540091.2017.1310182

Tran, C. N. N., Tat, T. T. H., Tam, V. W. Y., & Tran, D. H. (2023). Factors affecting intelligent transport systems towards a smart city: A critical review. *International Journal of Construction Management*, 23(12), 1982–1998. DOI: 10.1080/15623599.2022.2029680

Wortham, R. H. (2020). *Transparency for Robots and Autonomous Systems: Fundamentals, technologies and applications*. Institution of Engineering and Technology. DOI: 10.1049/PBCE130E

Chapter 9
Active–Reconfigurable Intelligent Surfaces for Unmanned Aerial Vehicles:
BMS Data Transmission

Agung Mulyo Widodo
Universitas Esa Unggul, Indonesia

Andika Wisnujati
https://orcid.org/0000-0001-6261-7936
Universitas Muhammadiyah Yogyakarta, Indonesia

Eko Prasetyo
Universitas Muhammadiyah Yogyakarta, Indonesia

Mosiur Rahaman
https://orcid.org/0000-0003-0521-2080
International Center for AI and Cyber Security Research and Innovations, Asia University, Taiwan & Computer Science and Information Engineering, Asia University, Taiwan

ABSTRACT

This study investigates the integration of unmanned aerial vehicles (UAVs) in air combat, focusing on their role in the battlefield management system (BMS) for effective communication and data management. Utilizing UAVs minimizes pilot casualties and enables real-time decision-making. The chapter examines resource allocation in ultra-dense networks using active-reconfigurable intelligent surfaces (A-RIS) assisted non-orthogonal multiple access (NOMA). It explores the coverage performance and ergodic capacity in a NOMA network under Nakagami-m fading channels, employing a multi-input multi-output (MIMO) system with RIS elements. The results demonstrate the superiority of RIS-assisted NOMA over conventional methods, offering enhanced coverage probabilities and ergodic capacity. The study concludes that the integration of A-RIS in UAVs significantly improves battlefield communication, highlighting its potential in military applications.

DOI: 10.4018/979-8-3693-2707-4.ch009

INTRODUCTION

In military or defences literature, the phrase "Battle-Air Management System" (BAMS) first appeared in mid-January 2022. On the other hand, the phrase describes a military air operations management system.

Battle Air Management System (BAMS)

The successful execution of military operations, especially those involving air forces, depends on the efficient management of airspace and air assets (Albarado et al., 2022; Chen, 2019). In order to accomplish these objectives, a "Battle Air Management System" may consist of a collection of technologies, practices, and guidelines. This could consist of:

- *Air Traffic Control (ATC) Systems*: By controlling aircraft passage inside designated airspace, these systems guarantee safe separation and effective routing.
- *Air Defence Systems*: To safeguard friendly forces and property, these systems identify, track, and, if required, destroy enemy aircraft or missiles.
- *Command and Control (C2) System*: With this system, commanders can watch and manage aviation operations in real time, which helps them make wise judgments and adapt to changing circumstances.
- *Communications Networks*: Information sharing and air operations coordination across various units and platforms depend on dependable and secure communications technologies.
- *Surveillance and reconnaissance*: Information on enemy forces and the operating environment is gathered using a variety of sensors, including drones and radar.
- *Data Fusion and Analysis*: Systems that combine information from several sources and give operators and commanders useful intelligence.
- *Tools for Mission Planning and Debriefing*: Software programs that help in mission planning, performance analysis, and recording lessons learned for use in subsequent operations.

Air Traffic Control (ATC) Systems

The part of a Battle Air Management System (BAMS) that controls the airspace and coordinates military aircraft movement during combat operations is known as the Air Traffic Control (ATC) system (Fernandes & Ullah, 2022; USAE, 2020). This system is essential for maintaining the effective and safe use of airspace, averting aircraft collisions, and enabling the carrying out of military operations. Typically, the ATC system in a BAMS has a number of functions, including:

- Surveillance: Monitoring the airspace with radar, satellites, or other sensors to identify and follow aircraft movements in real time.
- Communication: Establishing channels of communication to transfer directives, information, and alerts between pilots, air traffic controllers, and other pertinent staff members.
- Assisting aircraft with navigation: This includes providing direction and support, assigning altitudes, and helping with take-off and landing.
- Conflict resolution: Resolving problems between aircraft, such as preventing mid-air crashes or guaranteeing safe separation distances.

- Combining Command and Control (C2) Systems Integration: coordinating with more extensive command and control systems to guarantee that air operations are in line with the overarching military plans and objectives.

Depending on the particular operating context and threat environment, the ATC system's urgency inside a BAMS may change. The ATC system is in fact critical in many military circumstances, particularly during active combat operations. Operational efficiency, personnel safety, and mission success all depend on timely and efficient airspace control (Nichols et al., 2020). Incidents involving friendly fire, operational disruptions, or even catastrophic accidents can arise from improper air traffic control. To support the overall effectiveness of military air operations, it is imperative that a BAMS guarantee the robustness, reliability. Battle Air Management Systems is a specific or evolving concept in military or defense systems that refers to the technology and strategies for managing and coordinating air assets, such as aircraft, drones, and other airborne platforms (Kiohara et al., 2021; Nichols et al., 2019). These systems may involve communications, surveillance, and control mechanisms to ensure the effective use of air power in combat situations.

Air Defense Systems

Military devices known as Air Defense Systems (ADS) are created to identify, track, intercept and destroy air threats such as enemy aircraft, missiles and unmanned aerial vehicles. These systems are critical for defending against air threats to population centers, military sites, airspace, and critical infrastructure (Bronk, Reynolds, & Watling, 2022; Tuncer & Cirpan, 2022).

Common air defense system components include the following:

- *Surveillance and Detection*: Radar systems, electro-optical sensors, and other surveillance technologies are used in defended airspace to identify and track incoming threats.
- *Command and Control*: A centralized command and control center makes choices in attacking enemy targets, assesses threat information, and plans the actions of air defense units.
- *Weapon systems*: Surface-to-air missiles, anti-aircraft artillery, and other kinetic and non-kinetic weapons that can attack and eliminate air threats are examples of these systems.
- *Electronic warfare (EW) systems*: To improve the efficiency of defense operations, EW systems can be included into air defense networks to interfere with or trick adversary sensors and communications.

Air defense systems must be implemented immediately, particularly during times of conflict or high tension (Şandru, 2016). Here are a few explanations for this:

- *Protection of Assets*: In order to prevent aerial attacks on military property, infrastructure, and civilian populations, air defense systems are essential. A weak air defense would allow opponents to cause a great deal of damage and casualties.
- *Deterrence*: Strong air defense capabilities act as a deterrence to prospective aggressors, preventing them from launching airborne assaults or invading national airspace.
- *Response to Threats*: To lessen the effects and stop hostilities from escalating further, quick detection, interception, and neutralization are necessary in the case of an actual aerial threat.

- *Strategic Importance*: Retaining operational supremacy and accomplishing military goals frequently depend on airspace control. The enemy cannot move as freely while friendly forces are able to operate without interference thanks to the air defense systems.
- *Dynamic threat Environment*: The danger environment is ever-changing due to the widespread use of sophisticated aircraft, missiles, and unmanned aerial systems (UAS). Air defense systems need to be updated and adjusted frequently to properly handle new threats.

In general, the importance of air defense systems may be attributed to their vital function in safeguarding the country's security, maintaining operational preparedness, and maintaining airspace integrity against a variety of dynamic threats (Gao, Xiao, Qu, & Wang, 2022).

Command and Control (C2) System

The term "command and control" (C2) describes how a duly appointed commander exercises power and direction over specified forces in order to complete a mission. The tools, procedures, and frameworks that enable military commanders to organize, coordinate, and manage their forces in real-time during various operations are collectively referred to as a command-and-control system, or C2 system. Both the tactical and strategic levels of command are included in this (Eisenberg, Alderson, Kitsak, Ganin, & Linkov, 2018; X. Xu, Yang, & Tang, 2015).

Important parts of a command-and-control setup consist of:

- Situation Awareness: Information gathering and dissemination to give commanders a thorough grasp of the operational environment, including the whereabouts and conditions of both friendly and enemy forces, is known as situational awareness.
- Decision Support: Assisting commanders in making well-informed decisions by providing them with the information, intelligence, and analysis they need.
- Communication and Coordination: Establishing dependable communication networks would make it easier for deployed forces and various echelons of command to share directives, reports, and other vital information.
- Planning and Execution: Participating in the creation of operational plans and supervising their implementation to guarantee that resources are used efficiently to meet mission goals.
- Resource Management: Encouraging cooperation across various military branches, allies, and other pertinent organizations participating in a mission is known as coordination.

A command-and-control system is highly necessary, particularly for military operations and emergency scenarios (Park, Jeon, Sohn, & Kim, 2022). These are some of the factors making C2 urgent:

- Making Decisions in Real Time: Making decisions quickly is often necessary in military operations in order to adapt to shifting circumstances and new threats. The coherent operation of various elements within a force is ensured by a functional C2 system.
- Adaptability: Unexpected developments can happen in a dynamic operational environment. Commanders can quickly modify their plans and tactics in response to changing circumstances when they have a responsive C2 system in place.

- Reducing Friction: In military operations, friction is the term for difficulties and unknowns. Strong C2 systems ensure a cohesive effort, reduce misconceptions, and facilitate clear communication, all of which assist to reduce friction.
- Mission Success and Safety: Good command and control is essential to both the military's mission success and the personnel's safety. Making and carrying out decisions quickly is essential to accomplishing mission goals and lowering risks.

As the center for organizing, carrying out, and overseeing operations in every area of military operations, the command-and-control system is, in short, vital. In dynamic and competitive contexts, securing missions, safeguarding forces, and preserving strategic advantage depend on a flexible and responsive C2 system (Bastidas-Puga, Andrade, Galaviz, & Covarrubias, 2019; Park et al., 2022).

Communications Networks

Communications networks are essential to a Battle Air Management System (BAMS) because they allow different air operations elements to coordinate and share information (Ali, Jadoon, Changazi, & Qasim, 2020; Blasch & Bélanger, 2016). These networks provide communication between:

- Airborne platforms refer to any aerial assets that are used in combat, such as drones, manned aircraft, and unmanned aerial vehicles (UAVs).
- Ground Control Centers: Airspace surveillance, air traffic coordination, and air operations direction fall within the purview of command-and-control centers.
- Supporting Assets: These include air defense units, radar stations, surveillance systems, and other ground-based infrastructure that support air operations and provide situational awareness.
- Joint forces: Communication networks can also help allied troops taking part in joint operations and various military branches (such as the air force, army, and navy) cooperate and communicate with one another.

In a BAMS, important communications network characteristics include:

- Reliability: To guarantee continuous connectivity, communication lines need to be dependable and durable, able to function in difficult settings and unfavourable circumstances.
- Security: Maintaining the security and integrity of data transferred across the network is crucial to preventing interception, tampering, or cyberattacks, especially considering the sensitive nature of military communications.
- Bandwidth: Sufficient bandwidth is necessary to enable the transfer of substantial amounts of data, including as voice, video, and sensor data, which are needed for situational awareness and decision-making in real time.
- Interoperability: To facilitate smooth coordination and collaboration throughout the battlespace, communication networks must offer interoperability between various platforms, systems, and units.
- Scalability: Retaining efficacy and responsiveness requires the ability to scale communication networks to meet different demands, from normal operations to high-intensity conflict scenarios.

Regarding the importance of communications networks in a BAMS, it is true that these networks are necessary to guarantee the security and success of air operations (La Porte, 2019). Here are a few explanations for this:

- Authority and Management: In order to issue commands, distribute information, and coordinate air assets in real time, commanders must effectively communicate, especially in dynamic and fast-paced combat scenarios.
- Situational Awareness: Pilots and ground controllers are able to react swiftly to new threats and shifting battlefield conditions because of the timely information exchanged over communication networks.
- Mission Execution: Communication networks improve mission effectiveness and operational success by making it easier to carry out mission tasks like target acquisition, reconnaissance, surveillance, and engagement of hostile targets.
- Force Protection: Quick communication reduces the possibility of aircraft losses and casualties as a result of enemy action by allowing commanders to alert and steer aircraft away from possible hazards.

The vital role that communications networks play in facilitating efficient command and control, situational awareness, and mission execution in air operations makes them crucial components of a battle air management system. In dynamic and disputed environments, maintaining operational supremacy and accomplishing mission success need reliable, secure, and interoperable communication (Seizovic, Thorpe, & Goh, 2022).

Surveillance and Reconnaissance

Within a Battle Air Management System (BAMS), surveillance and reconnaissance (S&R) are critical functions that furnish vital data for threat assessment, situational awareness, and decision-making (Smagh, 2020). An outline of these ideas and their importance is provided below:

- *Surveillance*: In order to obtain information regarding the actions of friendly and hostile troops, surveillance entails the methodical observation of airspace, topography, and other pertinent areas. Radar, electro-optical sensors, signals intelligence, and unmanned aerial vehicles (UAVs) are some of the ways this can be accomplished. Commanders can detect possible dangers and weaknesses, as well as the behavior, movement, and goals of the opposition, with the use of surveillance data. Depending on the operational context, surveillance's urgency varies, but it is generally regarded as essential, particularly in dynamic and disputed contexts where adversaries may try to hide their operations or take advantage of weaknesses.
- *Reconnaissance*: Gathering precise and in-depth information about the enemy's forces, the terrain, the infrastructure, and other elements of the operating environment is the main goal of reconnaissance. Target selection, mission planning, and risk assessment all depend on this knowledge. Man-in-the-middle (UAV) aircraft, ground-based sensors, manned aircraft, and special operations soldiers engaged in direct observation are examples of reconnaissance assets. Similar to surveillance, reconnaissance is frequently critical, especially in the early phases of an operation when

prompt and precise information is required to guide choices and determine the best course of action.

A battle air management system's surveillance and reconnaissance capabilities are crucial because they give commanders the knowledge they need to comprehend the battlefield, evaluate threats, and organize and carry out military actions (Paucar et al., 2018). While the operational context, the type of mission, and the degree of threat all influence how urgent S&R operations should be, they are typically seen as essential to preserving situational awareness and accomplishing mission success in challenging and dynamic circumstances.

Data Fusion and Analysis

Commanders can obtain a more comprehensive picture of the situation by combining data from various sensors and platforms, which includes the whereabouts of friendly and opposing forces, topographical characteristics, and possible threats. Analysing the fused data entails looking for trends, patterns, and correlations that might help guide decisions (S. Huang et al., 2022). This could entail figuring out any openings or vulnerabilities as well as the capabilities and intentions of the opposition forces and the success of friendly actions (Azcarate, Ríos-Reina, Amigo, & Goicoechea, 2021). Predictive modelling and simulation can also be used in analysis to estimate future developments and evaluate the possible results of certain actions. A BAMS's need for data fusion and analysis depends on a number of variables, including:

In a Battle Air Management System (BAMS), data fusion and analysis entail combining and evaluating data from several sources to produce a thorough and precise picture of the operating environment. This procedure aggregates information from several platforms, sensors, and intelligence sources to give commanders useful insights for mission execution and decision-making. An outline of data fusion, analysis, and its urgency is provided below:

- Fusion of Data: The technique of merging data from several sources to create a cohesive and consistent image of the battlefield is known as data fusion. Data from surveillance planes, radar systems, unmanned aerial vehicles (UAVs), signals intelligence (SIGINT), human intelligence (HUMINT), and other sources may be included in this. By fusing data from multiple sensors and platforms, commanders can gain a more complete understanding of the situation, including the location of friendly and hostile forces, terrain features, and potential threats.
- Analysis: Analysing the fused data entails looking for trends, patterns, and correlations that might help guide decisions. This could entail figuring out any openings or vulnerabilities as well as the capabilities and intentions of the opposition forces and the success of friendly actions.

Predictive modelling and simulation can also be used in analysis to estimate future developments and evaluate the possible results of certain actions.

The urgency of data fusion and analysis in a BAMS depends on several factors:

- Timeliness: To make well-informed judgments in hectic and dynamic operational contexts, commanders need prompt access to accurate and trustworthy information. When enemies are changing tactics or positions quickly, or when operations are moving at a fast pace, the need for data fusion and analysis becomes even more critical.

- Situational Awareness: To be aware of changing threats and modify operations plans appropriately, situational awareness requires data fusion and analysis. When working in contested or hostile areas, where there is a significant chance of surprise attacks or unanticipated events, the urgency of these activities is increased.
- Mission Effectiveness: By combining and analysing data effectively, commanders may prioritize targets, allocate resources optimally, and coordinate the operations of many units to accomplish mission goals.

Data fusion and analysis are essential parts of a battle air management system that give commanders the knowledge they need to act quickly and wisely in challenging and changing tactical situations. The necessity of situational awareness, mission efficacy, and the quickening pace of contemporary warfare are the driving forces for the urgency of these procedures.

Tools for Mission Planning and Debriefing

Mission planning and debriefing are essential stages of the operational cycle of a Battle Air Management System (BAMS), guaranteeing that missions are meticulously planned, successfully completed, and lessons learned for subsequent operations (Halamek, Cady, & Sterling, 2019). In order of urgency, the following are the instruments frequently used during mission planning and debriefing:

Tools for mission planning: These resources help mission planners and commanders create comprehensive plans for air operations. They could consist of:

- Digital mapping software refers to programs that give planners access to satellite photos and comprehensive maps, making it possible to pinpoint enemy locations, important topographical features, and navigational landmarks.
- Flight Planning Software: Programs that help pilots of participating aircraft with route planning, airspace management, fuel calculations, and other flight planning tasks.
- Tools for modeling and simulating scenarios: computer programs that let planners examine the viability and efficiency of alternative strategies.
- Systems for Communication and Coordination: Instruments that facilitate efficient coordination and cooperation amongst various units and stakeholders during mission planning operations.

The mission's complexity, the threat landscape, and the accessibility of precise and fast information all influence how urgent mission planning tools should be. Mission-planning technologies must support quick decision-making and adaptive planning in time-sensitive missions or scenarios where adversaries may quickly shift tactics.

Debriefing Tools: Following a mission, these tools are used to assess and evaluate mission performance, pinpoint lessons discovered, and offer suggestions for enhancements. These devices could consist of:

- Software programs known as "After Action Review" (AAR) enable participants to submit feedback, talk about important findings, and record lessons learned during the organized assessment of mission events.

- Sensor recordings, communication logs, and performance metrics are just a few examples of the mission data that may be analyzed using data analysis tools to find trends, patterns, and areas that need improvement.
- Collaborative platforms: those that allow participants to discuss and work together while exchanging debriefing materials and lessons learnt.
- Systems used for training and simulation: These give staff members the chance to practice skills, try out new strategies, and improve processes in response to debriefing results by replicating and simulating mission scenarios.

Debriefing technologies are critical because they enable the swift capture and dissemination of mission-specific knowledge. Fast debriefing increases operational performance and mission readiness by enabling organizations to modify and enhance their strategies, methods, and protocols almost instantly. To summarize, the tools for mission planning and debriefing are crucial parts of a Battle Air Management System that help with decision-making, mission execution, and ongoing development. The mission's complexity, the operational environment, and the requirement for prompt feedback and adaptability all influence how urgent these tools are (A. Cheng et al., 2014).

All things considered, the Battle Air Management System is an intricately linked collection of technologies and procedures intended to optimize the efficacy and efficiency of air operations in a conflict zone. In military or defences systems, the term "combat air management systems" refers to a particular or developing concept that describes methods and technologies for controlling and arranging air assets, including aircraft, drones, and other airborne platforms. To guarantee the efficient application of air power in warfare scenarios, these systems may include communications, surveillance, and control techniques (Song, Lee, & Park, 2021; Zhou, Tang, & Zhao, 2019).

Active-Reconfigurable Intelligent Surfaces (A-RIS)

Active-reconfigurable intelligent surfaces (A-RIS) can potentially be applied to UAVs (unmanned aerial vehicles) for communication purposes. Active-reconfigurable intelligent surfaces, sometimes referred to as metamaterials or intelligent reflective surfaces (IRS), are surfaces made up of several electrically actuated components with the ability to modify electromagnetic waves (Zeng & Zhang, 2017). The propagation of electromagnetic signals, such as radio waves, microwaves, or millimetres-wave transmissions, can be dynamically modified on these surfaces. The deployment of RIS on UAV communication systems can support the air warfare management system and provide a number of potential advantages.

- *Signal Enhancement*: By selectively reflecting, refracting, or concentrating communications signals onto the desired recipient, A-RIS can be used to improve signal strength and quality.
- *Interference Mitigation*: By dynamically modifying the phase and amplitude of reflected signals to either raise the signal-to-interference ratio or eliminate interference, RIS can assist reduce interference.
- *Coverage Expansion*: By guiding signals to places with spotty coverage or blocked line-of-sight paths, A-RIS can increase a communications system's coverage area. A-RIS can be used to enable adaptive beamforming in UAV communications systems, allowing the drone to dynamically change the direction of its communications beam in response to shifting communication requirements or environmental changes.

- *Stealth and Security*: By dynamically modifying a UAV's radar cross-section, A-RIS may be able to modify its scattering characteristics, reducing the likelihood that an adversary will discover it or enhancing its security.

Important to keep in mind, though, is that A-RIS implementation in UAV communication systems will present technical difficulties, including power consumption, weight and size restrictions, integration complexity, and the requirement for effective control algorithms to maximize A-RIS performance in real-time communication scenarios (Bejiga, Zeggada, Nouffidj, & Melgani, 2017). Research and development in this field are continuing, and while experimental demonstrations and theoretical concepts show promise, actual implementations of active-reconfigurable intelligent surfaces (A-RIS) in UAV communications systems might still be in the early phases of development (Yang, Meng, Zhang, Hasna, & Di Renzo, 2020).

Signal Enhancement

The following are some ways that A-RIS can improve signal quality:

Beamforming and Directional Control: By directing and concentrating radio waves that transmitters emit in the direction of particular receivers or targeted locations, A-RIS may efficiently reduce interference and boost signal strength in the intended direction. A-RIS can enhance communication range, reduce multipath propagation, and bypass barriers by modifying the signal beam.

Signal Refraction and Reflection: A-RIS can manage the reflection and refraction of incoming signals, directing them toward receivers or improving coverage in low-signal areas. Signal routes can be improved to boost SNR and enhance overall communication reliability by carefully placing A-RIS pieces.

Mitigation of Interference: By manipulating the phase and amplitude properties of interfering signals, A-RIS can selectively cancel or suppress them. In radio frequency (RF) environments that are busy or congested, where several signals may overlap and impede transmission, this interference reduction feature is very helpful.

Dynamic Adaptation: Based on user needs, network dynamics, and environmental factors, A-RIS can optimize signal enhancement in real-time to adapt to changing communication conditions. Even in situations that are unpredictable and dynamic, A-RIS is able to constantly optimize communication lines because to this dynamic adaptation.

Through the utilization of A-RIS's distinct characteristics to control electromagnetic waves, communication systems may effectively enhance signals, resulting in better coverage, dependability, and efficiency (Widodo, Wijayanto, Wijaya, Wisnujati, & Musnansyah, 2023; P. Xu, Chen, Pan, & Di Renzo, 2020). For improving wireless communication in cellular networks, satellite communications, and Internet of Things (IoT) deployments, among other uses, this makes A-RIS a potential technology.

Interference Mitigation

The technique of lessening or decreasing the adverse effects of undesired signals or noise on a communication or sensing system is known as interference mitigation. Interference can cause communication mistakes or decreased reliability by lowering data throughput, deteriorating signal quality, and impairing system performance. The robustness and efficacy of wireless communication systems depend heavily on interference mitigation, particularly in crowded or crowded radio frequency (RF) situations (Nguyen & Do, 2018).

Because of their special qualities, Active-reconfigurable intelligent surfaces (A-RIS) can improve interference mitigation in communication systems. Artificial Resonant Surfaces (A-RIS) are made up of a multitude of passive components, including meta-material patches or antennas, which can dynamically modify their electromagnetic characteristics in reaction to signals that are incident onto them. These surfaces have the ability to control and optimize signal transmission by modifying the amplitude, phase, and direction of electromagnetic waves (Keating, Säily, Hulkkonen, & Karjalainen, 2019).

A-RIS can help reduce interference in the following ways:

- *Selective Signal Reflection*: A-RIS can reduce undesired interference while selectively reflecting or rerouting incoming signals. Certain signal routes can be amplified or suppressed by varying the phase and amplitude of individual A-RIS elements, which efficiently isolates desired signals from interfering sources.
- *Spatial filtering*: The ability of A-RIS to selectively pass desired signals while blocking or attenuating conflicting signals is known as spatial filtering. A-RIS can successfully mitigate the influence of interference sources by creating nulls, or regions of low signal intensity, in the direction of the interference sources by manipulating the spatial distribution of reflected signals.
- *Dynamic Null Steering*: To direct nulls or anti-phase signals toward interference sources and wipe out undesired signals, A-RIS can dynamically modify their reflection qualities. With the help of its dynamic null steering feature, A-RIS can sustain efficient interference mitigation even in dynamic environments by instantly adapting to changing interference conditions.
- *Multi-User Interference Management*: A-RIS can improve signal routes to reduce interference between several users or devices using the same spectrum in multi-user communication environments. A-RIS maximizes spectrum efficiency and total system capacity by minimizing cross-talk and interference by dynamically allocating resources and modifying signal reflections.

Overall, communication systems can achieve good interference reduction, which improves signal quality, dependability, and data throughput by utilizing the special powers of A-RIS to alter electromagnetic waves. For resolving interference issues in a variety of wireless communication applications, such as cellular networks, Wi-Fi networks, and Internet of Things deployments, this makes A-RIS a potential solution.

Coverage Expansion

The term "coverage expansion" describes the process of extending a communication or sensing system's geographic coverage to include previously unreachable or signal-poor places, allowing for connectivity or observation. Expanding coverage is crucial for delivering ubiquitous connection, boosting the efficiency of wireless communication and sensing applications, and ensuring service availability. Because of their special features, Active-reconfigurable intelligent surfaces (A-RIS) can lead to greater coverage expansion in communication systems. Artificial Resonant Surfaces (A-RIS) are made up of a multitude of passive components, including meta-material patches or antennas, which can dynamically modify their electromagnetic characteristics in reaction to signals that are incident onto them. These surfaces have the ability to control and optimize signal transmission by modifying the amplitude, phase, and direction of electromagnetic waves (Long, Liang, Pei, & Larsson, 2021).

A-RIS can aid in the increase of coverage in the following ways:

- *Signal Reflection and Refraction*: A-RIS can extend the coverage area beyond what conventional antennas or base stations can reach by selectively reflecting or refracting incoming signals to steer them into desired places. Signal routes can be tailored to cover certain regions or bridge coverage gaps—such as those seen in remote or indoor spaces—by carefully placing A-RIS elements.
- *Beamforming and Directional Control*: A-RIS may guide and focus radio waves that transmitters emit in the direction that they want to go, hence boosting signal power and coverage. A-RIS can provide connectivity over greater distances or in difficult environments by extending communication range, mitigating signal attenuation, and overcoming obstacles with its shaped signal beam.
- *Enhancement of Multi-Path Propagation*: By generating constructive interference patterns and optimizing signal routes, A-RIS can take advantage of multi-path propagation phenomena to improve coverage extension. A-RIS enhances coverage consistency and reliability by optimizing signal reception at targeted sites and reducing signal fading and attenuation by adjusting the phase and amplitude of reflected signals.
- *Dynamic Adaptation and Optimization*: Based on situational awareness and feedback, A-RIS may optimize coverage extension in real-time in response to shifting network dynamics, user needs, and environmental conditions. Even in dynamic and unpredictable situations, A-RIS is able to continuously modify signal propagation to optimize coverage area due to its dynamic adaptation.

In general, communication systems can accomplish significant coverage expansion by utilizing the special powers of A-RIS to modify electromagnetic waves. This improves service availability, improves connectivity, and increases accessibility in a variety of situations. As a result, A-RIS is a viable technique for expanding coverage in cellular networks, satellite communications, and Internet of Things deployments, among other wireless communication applications.

Stealth and Security

When discussing communication systems, the terms "strength" and "security" relate to the system's resilience to several types of threats, such as cyberattacks, jamming, interference, and eavesdropping. While raising strength means boosting the communication infrastructure's resilience and performance under challenging circumstances, strengthening security entails putting policies in place to safeguard systems, data, and communication channels against illegal access, manipulation, or disruption. Because of its special qualities, Active-reconfigurable intelligent surfaces (A-RIS) have the potential to improve communication systems' strength and security (Ma, Li, Liu, Wu, & Liu, 2022).

- *Mitigation of Interference*: A-RIS has the ability to selectively alter electromagnetic waves in order to reduce interference from outside sources, including nearby networks or electrical equipment. A-RIS has the ability to control and shape signal beams dynamically, hence reducing signal leakage and thwarting unwanted interception by adversaries or eavesdroppers. Moreover, A-RIS can use authentication and encryption techniques to safeguard data sent across communication channels, guaranteeing its integrity and confidentiality.
- *Dynamic Adaptation*: Based on situational awareness and feedback, A-RIS can optimize communication performance and security in real-time to changing environmental circumstances, user requirements, or security threats. In order to maintain reliable and secure communication lines even in dynamic and unpredictable settings, A-RIS is able to continuously modify signal propagation, beamforming parameters, and security measures thanks to this dynamic adaptation.
- *Resilient Communication Infrastructure*: By offering redundant pathways, enhancing signal coverage, and reducing single sources of failure, A-RIS can improve communication infrastructure resilience through the deliberate placement of A-RIS components. Communication networks can become more dependable and resilient by deploying A-RIS elements strategically. This will guarantee that operations continue even in the event of disruptions, attacks, or natural disasters.

All things considered, communication systems can gain more strength and security by utilizing the special powers of A-RIS to control electromagnetic waves and dynamically adapt to changing circumstances. Because of this, A-RIS is a technology that has great promise for improving the security, dependability, and resilience of wireless communication across a range of applications, such as IoT deployments, satellite communications, and cellular networks.

RESOURCE ALLOCATION IN ULTRA-DENSE NETWORKS (UDN) USING A-RIS-ASSISTED ORTHOGONAL MULTIPLE ACCESS (OMA) OR NON-ORTHOGONAL MULTIPLE ACCESS (NOMA)

Resource allocation can be achieved in an effective manner by using either non-orthogonal multiple access (NOMA) or orthogonal multiple access (OMA) in ultra-dense networks (UDNs) equipped with Artificial Intelligence-enabled Active-reconfigurable intelligent surfaces (A-RIS). The decision between OMA and NOMA is influenced by a number of variables, including the desired trade-offs between complexity and spectral efficiency, network conditions, and system needs (Y. Cheng, Li, Liu, Teh, & Poor, 2021; Hemanth, Umamaheswari, Pogaku, Do, & Lee, 2020). An outline of the methods for resource allocation in UDNs using A-RIS-assisted OMA or NOMA is provided below: OMA, or orthogonal multiple access, to prevent interference, OMA assigns orthogonal resources (such as time slots, frequency bands, or codes) to distinct users. The following procedures are involved in resource allocation in A-RIS-assisted OMA UDNs:

- Estimation of Channel State Information (CSI): A-RIS and base stations determine the channel conditions for every network user. This data includes interference levels, fading, and path loss.
- User Grouping: The A-RIS cleverly arranges its reflecting elements to maximize the channels for each group of users, who are grouped according to their channel circumstances.

- Resource Allocation: The A-RIS supports the optimization of signal quality, interference mitigation, and user allocation of orthogonal resources for each group. To increase the signal strength for particular users, this may include modifying the phase shifts of the reflective components.
- Dynamic Adjustment: Based on real-time changes, such as user mobility, interference levels, and environmental conditions, A-RIS dynamically modifies the resource distribution. It does this by continuously monitoring the network conditions.

While, Non-Orthogonal Multiple Access (NOMA) allows several users to share time-frequency resources while employing sophisticated signal processing techniques to isolate each user's signal at the receiver (Chen, Widodo, Lin, Weng, & Do, 2023). In NOMA UDNs supported by A-RIS, resource allocation entails the following steps:

- User Pairing: To optimize the advantages of NOMA, A-RIS and base stations automatically match users with different channel circumstances. Users who experience better channel conditions are linked with those who don't.
- Power Allocation: Based on each user pair's channel circumstances, A-RIS helps to dynamically allocate transmit power to them. Achieving a power balance that maximizes the system's overall performance is the aim.
- Precoding at Transmitters: To help with precoding at the transmitter, the A-RIS ascertains the ideal phase shifts for the reflecting parts. This should improve each user's signal quality and make separation at the receiver easier.
- Dynamic NOMA Grouping: In response to real-time channel conditions, A-RIS continuously modifies the NOMA user grouping. To maximize resource use and system performance, this can entail moving users to separate NOMA groups.

A-RIS-Assisted Resource Allocation Considerations:

- Machine Learning Integration: A-RIS can make better resource allocation decisions by using machine learning algorithms to learn from and adjust to changing network conditions.
- Interference Management: A-RIS is essential to the management of interference in UDNs. In order to reduce interference and improve the signal-to-interference ratio, it is able to dynamically modify the reflective elements.
- Energy Efficiency: Taking power consumption limits and sustainability objectives into account, A-RIS can optimize resource allocation with an emphasis on energy efficiency.
- Security: To ensure the integrity of resource allocation facilitated by A-RIS, security measures should be put in place to thwart unwanted access and interference.

To optimize network performance, intelligent grouping, dynamic adjustment, and resource optimization are all part of A-RIS-assisted resource allocation in UDNs. A-RIS improves spectral efficiency, increases overall system capacity, and allocates resources efficiently, whether it uses OMA or NOMA.

The Coverage Performance and Ergodic Capacity of Cooperative-OMA of BAMS Managed by A-RIS

In a battle air system run by an Active-Reconfigurable Intelligent Surface (A-RIS), analyzing the coverage performance and ergodic capacity of a cooperative orthogonal multiple-access network (OMA) entails taking into account a number of important variables and actions:

- Channel Modelling: In order to describe the wireless communication linkages in the battle air system between the base station, UAVs, and ground users, precise channel models must be developed. This involves using A-RIS to model the multipath propagation, shadowing effects, channel gains, path loss, and multipath propagation for both reflected and direct signals.

- Resource Allocation and Power Management: Create algorithms for resource allocation and power management to maximize the distribution of communication resources across cooperative OMA users, including transmit power levels, time slots, and frequency bands. In order to meet quality-of-service (QoS) criteria and maximize system throughput, the best way to allocate resources is to figure out how to do this.

- Design of a Cooperative OMA Protocol: Provide cooperative OMA protocols to provide effective information sharing and collaboration between UAVs and ground users. To improve coverage, dependability, and spectral efficiency, this may involve power allocation, relay selection, and distributed beamforming strategies.

- A-RIS Deployment and Configuration: Research how the battle air system's A-RIS deployment and configuration affect signal propagation, coverage area, and interference reduction. Analyze how well A-RIS serves cooperative OMA users by improving signal strength, lowering interference, and expanding coverage.

- Coverage Analysis: To determine the likelihood of successful communication between the base station, UAVs, and ground users in various areas of the battle air system, do a coverage study. This entails assessing the outage probability and signal-to-interference-plus-noise ratio (SINR) in order to ascertain the communication links' coverage area and dependability.

- Ergodic Capacity Calculation: To determine the average attainable data rate over fading channels, compute the cooperative OMA network's ergodic capacity. This entails incorporating cooperative OMA protocols and resource allocation algorithms, evaluating fading phenomena like Rayleigh or Nakagami fading, and studying the statistical distribution of channel gains.

- Simulation and Performance Evaluation: To validate the suggested system design and algorithms, conduct comprehensive simulation and performance evaluation studies. Examine the system throughput, ergodic capacity, spectrum efficiency, and coverage performance for a range of user densities, A-RIS configurations, and operational circumstances.

Researchers and engineers can learn more about the coverage performance and ergodic capacity of cooperative OMA networks in A-RIS-managed battle air systems by methodically examining these factors. This makes it possible to create dependable and effective communication systems that can withstand the rigorous demands of military operations in intricately layered settings (Van Chien, Tu, Chatzinotas, & Ottersten, 2020).

The Coverage Performance and Ergodic Capacity of Cooperative-NOMA of a BAMS Managed by A-RIS

In a battle air system controlled by Active-Reconfigurable Intelligent Surface (A-RIS), a cooperative nonorthogonal multiple-access network's (NOMA) coverage performance and ergodic capacity analysis take into account a number of variables, including channel conditions, user distributions, resource allocation strategies, and A-RIS's effect on signal propagation. This is a high-level summary of how ergodic capacity and coverage performance can be assessed in a system and could be explained below (Liu, Ding, Elkashlan, & Poor, 2016; Muhammad, Elhattab, Arfaoui, & Assi, 2023).

Channel Modelling

In order to describe the wireless communication links in the battle air system between the base station, UAVs, and ground users, precise channel models must be developed. This involves using A-RIS to model multipath propagation, shadowing effects, channel gains, path loss, and multipath propagation for both reflected and direct signals. By describing signal propagation, fading effects, and diversity gain in wireless communication systems, channel modelling variables are used to study range performance and ergodic capacity in wireless communication systems (Dai et al., 2015). In order to achieve performance requirements under a variety of channel conditions, this study aids in the optimization of system design, resource allocation, and transmission strategy. Use these variables as follows:

- *Path Loss*: The weakening of a signal during its transmission across a wireless medium is referred to as path loss. It is dependent upon variables like environment, frequency, and distance. Path loss models (such as free space path loss and long-distance path loss) can be combined to estimate the strength of the signal received at various points within the coverage region. Determining the transmitter coverage area and locating probable dead zones or locations with poor signal reception are made easier with the aid of path loss analysis.
- *Shadow*: Shadow considers changes in signal due to obstructions and surrounding conditions (e.g., buildings, topography). A log-normal random variable with a predetermined standard deviation is frequently used to mimic shadowing. Predicting signal variations within a coverage area and evaluating the strength of communications links under different circumstances depend heavily on an understanding of shading effects.
- *Multipath Fading*: Signal fluctuations at the receiver are caused by signal reflection, diffraction, and scattering. These factors contribute to multipath fading. Rayleigh fading (for non-line-of-sight circumstances) and Rician fading (for line-of-sight settings) are common models for multipath fading. The performance and dependability of communication links can be assessed by modeling multipath fading, particularly in settings with strong multipath propagation.
- *Diversity Strategies*: Channel variations are used by diversity techniques, like antenna diversity and frequency diversity, to improve the reliability of communications. One may evaluate how well diversity schemes mitigate fading effects and enhance coverage performance by looking at channel correlation and diversity gain.
- *Channel capacity*: The greatest data rate that can be obtained via a specific communication channel is known as the channel capacity. This is contingent upon variables including channel conditions, signal-to-noise ratio (SNR), and bandwidth. One can estimate the maximum throughput

that can be achieved and assess the system's performance under various channel circumstances by studying the channel capacity using theoretical models (e.g., Shannon's capacity formula) or empirical approaches (e.g., the water filling algorithm).

- *Ergodic capacity*: The average capacity of a channel that decreases over time or space is known as its ergodic capacity. This offers insight into the long-term performance of communications systems and explains statistical fluctuations in channel circumstances. Designing transmission systems that optimize average performance on fading lines and evaluating system sustainability are made easier with the aid of ergodic capacity analysis.

Resource Allocation and Power Management

Create algorithms for resource allocation and power management to maximize the sharing of communication resources among cooperative NOMA users, including transmit power levels, time slots, and frequency bands. This entails figuring out the best user pairing policies and power allocation ratios to improve system throughput while meeting quality-of-service (QoS) standards. In wireless communication systems, the analysis of coverage performance and ergodic capacity is largely dependent on resource allocation and power management variables (X. Huang, 2020). In wireless communication systems, these factors are essential for maximizing ergodic capacity and coverage performance. In order to improve system performance overall, these variables are also used to balance coverage and capacity needs, reduce interference, and make effective use of the resources that are available. This is the usual usage for them:

- *Performance of Coverage*: It is possible to enhance coverage performance by effectively allocating resources like time slots, frequency bands, or codes. When resources are allocated properly, they can be used to service people across a certain geographic area as efficiently as possible. Resource allocation algorithms, for instance, can maximize coverage in cellular networks by dynamically allocating resources to users according to their channel conditions, traffic needs, and Quality of Service (QoS) requirements. In order to maximize coverage, power control mechanisms modify the transmit power levels of base stations and user devices. It is possible to increase a wireless network's coverage area while keeping the signal quality acceptable by adjusting the broadcast power. In order to improve coverage performance, power management algorithms work to reduce interference and lessen the effects of fading, shadowing, and path loss.
- *Ergodic Capacity*: A communication system's ergodic capacity is directly impacted by its best resource allocation practices. The system's capacity can be maximized by intelligently allocating resources, including transmit power and bandwidth, to users based on traffic demands and channel conditions. In order to improve the ergodic capacity of the system, resource allocation algorithms try to make use of both spatial and multi-user variety. By adjusting transmit power levels to strike a balance between capacity and coverage, power control strategies affect ergodic capacity. With aggressive power regulation, interference can be decreased and the signal-to-interference-plus-noise ratio (SINR) can be raised, increasing the system's capacity and spectral efficiency. On the other hand, overly strict power regulation might deteriorate coverage performance and cause coverage gaps. As a result, algorithms for power management must balance capacity and coverage concerns (Jones, Rigling, & Rangaswamy, 2016).

Cooperative NOMA Protocol Design

Create cooperative NOMA protocols to facilitate effective information sharing and collaboration between ground users and UAVs. For the purpose of improving communication reliability and coverage extension, this may involve relay selection, cooperative beamforming, and joint decoding algorithms. Cooperative Non-Orthogonal Multiple Access (NOMA) protocols improve coverage performance and ergodic capacity by leveraging many design variables. In wireless communication systems, cooperative NOMA protocol design factors are utilized to maximize coverage performance and ergodic capacity (Widodo et al., 2023). These variables include power allocation, relay selection, user pairing, relay cooperation method, channel coding, and modulation. These settings are specifically designed to take advantage of NOMA's advantages and user and relay cooperation to improve system performance. The usual usage of these variables is as follows:

- *Allocation of Power*: The distribution of electricity among users in NOMA systems is essential for maximizing capacity and coverage. NOMA uses the power domain to increase spectral efficiency by assigning varying power levels to various users sharing the same time-frequency resources. This idea is expanded upon by cooperative NOMA protocols, which let users work together and share signals they have received. In a cooperative NOMA network, power allocation algorithms are used to distribute transmit powers among users and relays while taking fairness restrictions, channel circumstances, and Quality of Service (QoS) requirements into account.

- *Relay Selection*: To promote user cooperation and improve system performance, cooperative NOMA depends on relays. The best relays to aid in the transmission process are chosen using relay selection algorithms. These algorithms choose relays that can efficiently increase coverage and capacity by taking into account variables including energy efficiency, availability of relays, and channel conditions. Cooperative NOMA protocols can reduce the impacts of channel fading and provide coverage to consumers in locations with weak or shaded signals by strategically choosing relays.

- *User Pairing*: Choosing user pairings that share the same time-frequency resources is known as user pairing in NOMA systems. This idea is expanded upon by cooperative NOMA protocols, which enable users to collaborate and collaboratively decode one another's signals. By pairing users with complimentary channel conditions, user pairing algorithms seek to maximize the ergodic capacity of the system. Cooperative NOMA protocols take use of diversity gains by matching users with different channel characteristics and enhance coverage performance.

- *Relay Cooperation Strategy*: To improve coverage and capacity, cooperative NOMA protocols use a variety of relay cooperation techniques. The amplify-and-forward (AF), decode-and-forward (DF), and compress-and-forward (CF) approaches are some of these methodologies. The selection of relay cooperation strategies is contingent upon the needs of the network, relay capabilities, and channel conditions. Relay cooperation techniques improve the ergodic capacity of the system by minimizing interference, mitigating fading effects, and utilizing cooperative diversity.

- *Modulation and Channel Coding*: For cooperative NOMA systems to maximize coverage and ergodic capacity, channel coding and modulation schemes must be chosen carefully. Cooperative NOMA protocols modify coding and modulation schemes in response to relay capabilities, user demands, and channel conditions. Cooperative NOMA protocols reduce mistakes, guarantee dependable communication, and enhance spectrum efficiency by choosing effective coding and modulation techniques.

A-RIS Deployment and Configuration

Research how the battle air system's A-RIS deployment and configuration affect signal propagation, coverage area, and interference reduction. Analyse how well A-RIS serves cooperative NOMA users by improving signal strength, decreasing interference, and expanding coverage. Coverage performance and ergodic capacity in wireless communication systems are strategically analysed by leveraging A-RIS (Active Reflective Intelligent Surface) deployment and setup variables (Alexandropoulos et al., 2023). In order to strategically assess coverage performance and ergodic capacity in wireless communication systems, A-RIS deployment and configuration variables, such as deployment density, placement optimization, reflection coefficients, and beamforming/signal steering techniques, are employed. The system performance is improved by these factors, which are designed to maximize signal coverage, reduce interference, and improve spectral efficiency. This is the usual usage for them:

- *Density of deployment*: Ergodic capacity and coverage performance are impacted by the number of A-RIS units deployed in a particular area. More exact control over signal reflection and modification is made possible by higher deployment densities, which enhance capacity and coverage. Analysing coverage performance entails examining how various environmental areas' signal coverage is impacted by deployment density. In areas that are shaded, a larger deployment density can improve signal strength and reduce coverage gaps. Ergodic capacity study takes into account the impact of deployment density on system capacity as a whole. Elevated deployment densities have the potential to enhance signal reflection efficiency, resulting in amplified spectral efficiency and ergodic capacity.

- *Placement Optimization*: To maximize ergodic capacity and coverage performance, A-RIS units can be positioned strategically. Algorithms for placement optimization seek to find the best places for A-RIS units, taking into account user distribution, channel properties, and coverage needs. Analysing coverage performance entails determining how various deployment tactics affect signal quality and coverage across the covered region. Placing components optimally can assist reduce interference or blocking of the signal and provide uniform coverage. Ergodic capacity analysis takes into account how placement optimization strategies increase the capacity of the system by minimizing signal attenuation and maximizing reflections of the signal. Increased capacity and better signal-to-noise ratios (SNR) can result from optimal placement.

- *Coefficients of Refraction*: The way that signals are reflected and altered within A-RIS units is determined by the reflection coefficients of particular components. A-RIS units can focus signals, reduce interference, and improve coverage performance and capacity by varying the reflection coefficients. Analysing coverage performance entails examining how various reflection coefficients affect signal quality and coverage. A-RIS units are capable of enhancing uniformity of coverage and signal intensity by optimizing reflection coefficients. Ergodic capacity analysis takes into account the effects of changing reflection coefficients on the system's capacity. A-RIS units are capable of enhancing uniformity of coverage and signal intensity by optimizing reflection coefficients. The impact of changing reflection coefficients on the system's capacity is taken into account in ergodic capacity analysis. A-RIS units raise spectral efficiency and ergodic capacity by optimizing reflection coefficients to enhance signal strength and decrease interference.

- *Signal steering and beamforming*: Real-time optimization of coverage performance and ergodic capacity can be achieved by A-RIS units through dynamic adjustments to beamforming and signal steering. A-RIS units can adjust to shifting channel circumstances and user locations by improving signal reflections or dynamically routing signals towards users. Examining the effects of beamforming and signal steering techniques on signal quality and coverage under various conditions is part of the analysis of coverage performance. In difficult situations, adaptive beamforming can help reduce signal blockage and enhance coverage. Ergodic capacity analysis takes into account how methods such as adaptive beamforming and signal steering, which maximize signal power and reduce interference, can improve the system's capacity. Through dynamic beamforming adjustments, A-RIS devices can improve both ergodic capacity and spectral efficiency (Xiaoxiang Gong, Qian, Ge, & Yan, 2020).

Coverage Analysis

Examine the likelihood of effective communication between the base station, unmanned aerial vehicles, and ground users in various battle air system zones by conducting a coverage analysis. This entails assessing the outage probability and signal-to-interference-plus-noise ratio (SINR) in order to ascertain the communication links' coverage area and dependability. The performance of wireless communication systems, including coverage performance and ergodic capacity, is evaluated using coverage analysis variables, which are crucial metrics (Van Chien et al., 2020). The usual use of these variables is as follows:

- *Path Loss Models*: Signal attenuation is estimated as a function of transmitter and receiver distance using path loss models, such as the Friis free space model or more complex models like the log-distance path loss model. Engineers can forecast signal strength and coverage area, which are essential for planning and optimizing wireless networks, by combining path loss models into coverage analysis.
- *Antenna Gain*: Antenna gain is the capacity of an antenna to direct or focus signals that are received or sent in a specific direction. By focusing signal energy in the intended directions and minimizing interference from other directions, higher antenna gain enhances coverage. Antenna gain is taken into account in coverage analysis to evaluate the coverage area and signal strength.
- *Transmit Power*: Signal strength and coverage area are directly impacted by transmit power. Extending coverage with higher transmit power may result in higher interference levels and power usage. Engineers balance coverage expansion with interference control by analysing the effects of transmit power on coverage performance and ergodic capacity in coverage analysis.
- *The Terrain and Environment*: Buildings, vegetation, topography, and atmospheric conditions are examples of environmental elements that have an impact on signal coverage and propagation. Engineers take into account signal blockage, reflection, diffraction, and attenuation by taking these characteristics into account while doing a coverage analysis. To examine coverage in complicated landscapes, ray tracing simulations and digital elevation models (DEM) are two examples of terrain modelling technologies. 5. Frequency Band: Due to differences in propagation characteristics and regulatory restrictions, the frequency band selected affects coverage performance. Higher frequency bands support denser deployments and give greater capacity, whereas lower frequency

bands are better suited for rural areas and more effective at getting around obstacles. The impact of frequency band selection on coverage area, signal quality, and interference levels is evaluated using coverage analysis.

- *Multipath Fading*: Signal reflections, diffractions, and scattering cause multipath fading, which causes changes in signal strength and phase. Engineers use models of multipath fading phenomena, including Rayleigh or Rician fading, to forecast signal quality metrics like outage likelihood, which affects ergodic capacity, and to examine coverage performance.

- *Noise and Interference*: Noise sources, co-channel interference, and other users' interference reduce coverage capacity and performance. In order to establish coverage boundaries, assess system capacity, and create interference mitigation strategies including frequency reuse, power regulation, and interference cancellation, engineers examine noise levels and interference. Engineers can assess the effectiveness of wireless communication systems, optimize coverage area, enhance signal quality, and increase ergodic capacity while guaranteeing effective resource usage and interference management by taking these coverage analysis variables into account.

Ergodic Capacity Calculation

To determine the average data rate that may be achieved over fading channels, determine the cooperative NOMA network's ergodic capacity. This entails incorporating cooperative NOMA protocols and resource allocation algorithms, evaluating fading effects like Rayleigh or Nakagami fading, and studying the statistical distribution of channel gains. Finding a wireless communication system's average capacity over a large number of channel realizations is known as an ergodic capacity calculation (Li, 2020). To analyze coverage performance and determine the average system capacity, it is essential to understand the variables employed in ergodic capacity calculations. The usual usage of these variables is as follows:

- *Channel State Information (CSI)*: Ergodic capacity computations frequently depend on the channel's statistical characteristics, which are represented by the CSI. Path loss, multipath effects, and channel fading characteristics are all included in this data. Ergodic capacity is computed by averaging capacity over many channel realizations, and the channel's variability over time is modelled by the CSI.

- *SNR, or signal-to-noise ratio*: One important factor in ergodic capacity estimations is the SNR. It is a metric for the quality of the communication link and shows the relationship between the received signal power and noise power. The statistical distribution of SNR, which is impacted by variables like transmit power, route loss, and interference, determines ergodic capacity.

- *Efficiency of Spectrum*: The information rate transmitted per unit of bandwidth is measured by spectral efficiency. Spectral efficiency is a critical variable in ergodic capacity calculations since it directly affects the system's capacity. An improvement in spectral efficiency means that the available bandwidth is used more effectively, which boosts ergodic capacity.

- *Channel Fading Models*: The time-varying characteristics of the wireless channel are represented by channel fading models, such as Rayleigh fading or Nakagami fading. The effects of shadowing, multipath propagation, and other channel changes are captured by these models. A common practice in ergodic capacity estimations is to average capacity over the many fading situations that these models depict.

- *Features of Antenna*: Antenna properties affect a system's ergodic capacity, such as the quantity and spatial arrangement of antennas. Spatial diversity can be achieved by using multiple antennas, which enhances the average capacity over various channel realizations and helps to reduce fading effects.
- *Modulation and Coding Schemes*: A communication system's capacity is impacted by the selection of modulation and coding schemes. Higher data rates are possible with higher-order modulation, but channel defects may cause more problems. Coding techniques are employed to counteract transmission mistakes. The ergodic capacity of the system is dependent on the combination of modulation and coding techniques.
- *Channel Bandwidth*: An essential factor in ergodic capacity computations is the available channel bandwidth. Higher capacity is typically achieved using wider bandwidths. Regulating restrictions and increasing interference are the trade-offs, though.
- *Interference Levels*: A system's ergodic capacity is impacted by interference from other users or sources. The statistical properties of interference are taken into account in ergodic capacity calculations, and methods like interference averaging are used to quantify the effect on the average capacity.

Engineers can assess the effects of many parameters on capacity, design systems that balance coverage, capacity, and dependability, and learn more about the average performance of wireless communication networks by factoring these variables into ergodic capacity estimates (P. Xu et al., 2020).

Performance Evaluation and Simulation

To validate the suggested system design and algorithms, conduct comprehensive simulation and performance evaluation studies. Examine the system throughput, ergodic capacity, spectrum efficiency, and coverage performance for a range of user densities, A-RIS configurations, and operational circumstances.

Through a methodical analysis of these factors, scientists and engineers can learn more about the ergodic capacity and coverage performance of cooperative NOMA networks in A-RIS-managed battle air systems. In order to meet the demanding requirements of military operations in complex and dynamic contexts, this makes it possible to create effective and dependable communication systems (Yang et al., 2020).

IMPLEMENTING UNMANNED AERIAL VEHICLES-AIDED A-RIS TO BAMS

The integration of active-reconfigurable intelligent surfaces (A-RIS) with unmanned aerial vehicles (UAVs) on a Battle-Air Management System (BAMS) presents a number of benefits and opportunities to improve navigation, communication, and system performance.

We provide an example by creating a model system as described below.

Figure 1. Battle-Air management system model

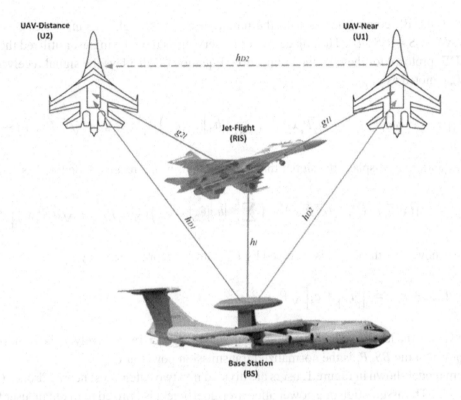

We provide an example by creating a model system as described below. The system consists of a fighter aircraft where the RIS is installed on the fuselage and two combat drones in a position near and a position far from the aircraft. The three vehicles are coordinated by an AWACS aircraft under command. The use of radar and other sensor systems on aircraft for surveillance, command, and control is known as "AWACS" (Airborne Warning and Control System) aviation. Military forces frequently use these aircraft to coordinate air operations, monitor airspace, and identify and track enemy aircraft. The AWACS platform is intended to give military commanders a high degree of situational awareness by providing an aerial platform for remote surveillance and command operations. The BAMS program benefits from AWACS Aviation's capacity to improve surveillance capabilities. AWACS planes can expand their reach by providing aerial surveillance over a greater area, whereas BAMS UAVs offer continuous observation over a small area. In addition to offering command and control capabilities to coordinate surveillance and reaction activities, the AWACS platform can identify and track targets beyond line of sight. In particular, the following are some ways that AWACS flights can benefit the BAMS program: Support for UAV Operations, Wide-Area Surveillance, Command and Control, and Data Fusion and Analysis (Ünal & Başçiftçi, 2020).

The two drones, U_1 (the nearby drone) and U_2 (the distant drone), are the intended recipients of the broadcast signal from the BS. The transmitted signal s(t) indicated in Equation (1) is superposed of the $x_1(t)$ and $x_2(t)$ that are targeted at U_1 and U_2, respectively, because the system uses NOMA (Non-Orthogonal Multiple Access).

$$s(t) = \sqrt{a_1 P_s}\, x_1(t) + \sqrt{a_2 P_s}\, x_2(t) \tag{1}$$

Within a cell, the BS communicates with the remote user U_2 through relaying links U_1 - U_2, and relaying links AWACS – RIS - U_2. The adjacent user U_1 serving as the relaying user utilized the decode and forward (DF) protocol for the relaying connection. Equation (2) displays the signal received by the nearby user U_1, denoted as $y_{S,\,U1}$.

$$y_{S,\,U1} = \left(\left(\left| \hat{h}_{D_1} \right| + e_1 \right) \left(\sqrt{a_1 P_s}\, x_1 + \sqrt{a_2 P_s}\, x_2 \right) + \left(\sum_{l=1}^{L} \left| \hat{h}_{1l} \right| \left\| \hat{g}_{1l} \right| e'_1 \right) \left(\sqrt{a_1 P_s}\, x_1 + \sqrt{a_2 P_s}\, x_2 \right) \right) + n_{U1} \tag{2}$$

Similarly, Equation (3) displays the signal that the remote user, U_2, received, denoted as $y_{S,\,U2}$.

$$y_{S,\,U2} = \left[\left(\left| \hat{h}_{D_2} \right| + e_2 \right) \left(\sqrt{a_1 P_s}\, x_1 + \sqrt{a_2 P_s}\, x_2 \right) + \left(\sum_{l=1}^{L} \left| \hat{h}_{1l} \right| \left\| \hat{g}_{2l} \right| + e'_2 \right) \left(\sqrt{a_1 P_s}\, x_1 + \sqrt{a_2 P_s}\, x_2 \right) \right] + n_{U2} \tag{3}$$

Equation (4) shows that the signal is received by U_2 from U_1, represented as $y_{U1,U2}$.

$$y_{U1,U2} = \left| h_{D_3} \right| \sqrt{P_1}\, x_2 + n_{U2} = \left[\left| \hat{h}_{D_3} \right| + e_3 \right] \sqrt{P_1}\, x_2 + n_{U2} \tag{4}$$

where a_1 and a_2 are the level power coefficient of the signal x_1 and x_2 respectively, P_s is the normalized transmission power at the BS, P_1 is the normalized transmission power at U_1.

In the system model shown in Figure 1, users are divided into two categories: nearby drones (U_1) and distant drones (U_2). The assumption of a power allocation coefficient is utilized to maintain user fairness because the NOMA method allocates less transmit power to the user with the best channel conditions. Sequential interference cancellation (SIC) of the received superposition signal at U_1 is used to decode the U_2 signal since U_2's higher transmit power reduces inter-user interference. Equation (5) below can be used to define the interference and noise ratio (SINR) at U_1 caused by the signal that U_1 received and the signal that U_2 decoded, x_2.

$$\rho_{U_2 \to U_1} = \frac{\left(\left| \sum_{l=1}^{L} \left| h_l \right| \left\| g_{1l} \right| \right|^2 + \left| \hat{h}_{D_1} \right|^2 \right) \alpha_2 \rho_s}{\left(\left| h_{D_1} \right|^2 + \left| \sum_{l=1}^{L} \left| \hat{h}_l \right| \left\| \hat{g}_{1l} \right| \right|^2 \right) \alpha_1 \rho_s + \eta_1 \left(d_{SU_1} \right)^{-x} \rho_s + \eta_2 L \left(d_{SR} \right)^{-x} \left(d_{RU_1} \right)^{-x} \rho_s + 1} \tag{5}$$

where $\rho_s = \dfrac{P_s}{N}$ represents the BS's transmit signal-to-noise ratio (SNR). Moreover, Equation (6) might be used to represent the SINR at U1 resulting from the self-received signal's decoding.

$$\rho_{U_1} = \frac{\left(\left| \sum_{l=1}^{L} \left| \hat{h}_l \right| \left\| \hat{g}_{1l} \right| \right|^2 + \left\| h_{D_1} \right\|^2 \right) \alpha_1 \rho_s}{\eta_1 \left(d_{SU_1} \right)^{-x} \rho_s + \eta_2 L \left(d_{SR} \right)^{-x} \eta_3 \left(d_{RU_1} \right)^{-x} \rho_s + 1} \tag{6}$$

The estimated *Nakagami-m* distribution is used to modify the estimated channel coefficient, or $\hat{h}_k$, which is also the estimated channel gain. Next, by using outage probability analysis, it is gotten Equation (7) below.

$$P_{U_1} = 1 - e^{-\frac{m_1}{L}\left(\frac{d_{SU_1}^{\chi}}{1-\eta_{1_1}} + \frac{d_{SR_1}^{\chi}}{1-\eta_{2_1}} + \frac{d_{RU_1}^{\chi}}{1-\eta_{3_1}}\right)\tau} \sum_{j=0}^{m_1-1} \frac{\left(\frac{m_1}{L}\left(\frac{d_{SU_1}^{\chi}}{1-\eta_{1_1}} + \frac{d_{SR_1}^{\chi}}{1-\eta_{2_1}} + \frac{d_{RU_1}^{\chi}}{1-\eta_{3_1}}\right)\tau\right)^j}{j!} \tag{7}$$

It takes a few steps to get a mathematical statement of the outage probability. To begin with, U_2 can decode its own signal by simply treating the x_1 signal from U_1 as noise. Equation (8) expresses the received SINR for U_2, which translates its own signal from x_2.

$$\rho_{1,U2} = \frac{\left|\|h_{D_2}\|^2 + \left|\sum_{l=1}^{L}|h_l|\|g_{2l}|\right|^2\right|\alpha_2\rho_s}{\left|\|h_{D_2}\|^2 + \left|\sum_{l=1}^{L}|h_l|\|g_{2l}|\right|^2\right|\alpha_1\rho_s + \eta_4\left(d_{SU_2}\right)^{-\chi}\rho_s + \eta_5 L\left(d_{SR}\right)^{-\chi}\rho_s + \eta_6\left(d_{RU_2}\right)^{-\chi}\rho_s + 1} \tag{8}$$

Equation (9) expresses the SINR that U_2 received in order to decode signal x_2 for the relay connection.

$$\rho_{2,U2} = \frac{\|h_{D_2}\|^2\rho}{\eta_7\left(d_{SU_2}\right)^{-\chi}\rho + 1} \tag{9}$$

While, Equation (10) expresses the SINR obtained after combining selection (SC) on U_2.

$$\rho_{U_2}^{SC} = \frac{\|h_{D_2}\|^2\rho}{\eta_7\left(d_{SU_2}\right)^{-\chi}\rho + 1} + \frac{\left|\|h_{D_2}\| + \sum_{l=1}^{L}|h_l|\|g_{2l}|\right|^2\alpha_2\rho_s}{\left|\|h_{D_2}\| + \sum_{l=1}^{L}|h_l|\|g_{2l}|\right|^2\alpha_1\rho_s + \eta_4\left(d_{SU_2}\right)^{-\chi}\rho_s + \eta_5 L\left(d_{SR}\right)^{-\chi}\eta_6\left(d_{RU_2}\right)^{-\chi}\rho_s + 1} \tag{10}$$

Additionally, the coverage probability in Equation (11) is calculated as follows for arbitrary phase shifts:

$$P_{U_2} = 1 - e^{-(\delta_1\tau_2 + \delta_2\tau_3)} \sum_{j=0}^{m_1-1}\sum_{k=0}^{m_2-1} \frac{(\delta_1\tau_2)^j(\delta_2\tau_3)^k}{j!k!} \tag{11}$$

The outage performance of the RIS-aided NOMA network with p-CSI is evaluated numerically in terms of outage probability via *Nakagami-m* fading channels (Xianli Gong, Yue, & Liu, 2020; Lv, Ni, Ding, & Chen, 2016). by establishing sensible settings for the simulation. Using Monte Carlo simulation, the accuracy of the outage probability expression is confirmed. Utilizing location parameters transferred into a Cartesian coordinate system, the aforementioned study is quantitatively validated. In particular, if the locations of AWACS, U_1, and U_2 are on a straight line, the source (BS) is at (0, 0) and the RIS is at (60, 10). To configure the simulation parameters, use Table 1 below to normalize the distance between BS and U2 to one.

Table 1. Simulation parameters

Description	NOMA						
Power allocation coefficient	$\alpha_1 = 0.2, \alpha_2 = 0.8$						
Path loss exponent							
Fading parameter							
Relative channel estimation error							
Distance between two nodes				,			
Transmit SNR							
The antenna gain at transmitter and receiver							
Target Rate							

Additionally, $\beta_k = G_t + G_r + 10\chi_k \log_{10}\left(\frac{d_k}{1m}\right) - 500 + z_k$ models the large-scale fading coefficient [dB] at both adjacent and distant user locations, with $k \in \{SU_1, SR, RU_1, SU_2, SR, RU_2, U_1U_2\}$ and z_k representing shadow fading. Next, the charting of outage probability vs transmit signal-noise-ratio (SNR) is achieved, as shown in Fig. 2, using Eqs. (7) and (11) and $z_k = 0$.

The outage probability for the two users in the non-direct connection scenario is plotted against the transmitted signal strength (SNR) in Figure 2. For U1 and U2, respectively, the goal rates are set at R1 = 3.6 and R2 = 1 bit per channel use (BPCU). Equations are used to plot the precise outage probability curves for the two users. (7) and (11), in that order. It is clear that there is a high degree of consistency between the findings of the Monte Carlo simulation and the precise outage probability curves. In terms of outage performance, RIS-assisted NOMA performs better than traditional NOMA, as illustrated in Figure 2. At low transmission SNR, the outage probability reduces with rising SNR and reaches a stable value at high SNR.

Figure 2. Outage probability vs. transmit SNR

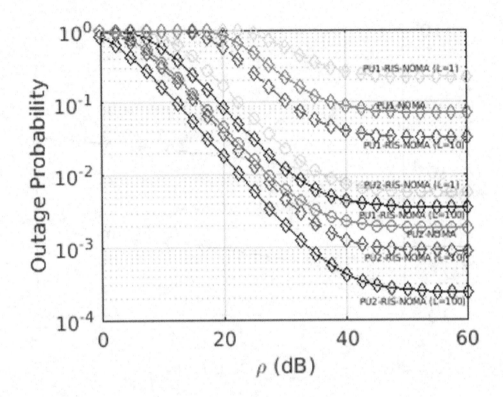

For every RIS element, the approximate outage probability curves for the two users are also plotted for L = 1, 10, and 100, respectively. Moreover, in terms of outage probability, the conventional NOMA network performs better than the RIS-assisted NOMA network when the RIS has just one element. Nonetheless, it becomes clear that the RIS-assisted NOMA network performs better when the element count rises above ten. Channel estimate mistakes produce an error floor at high SNR, which leads to zero diversity order.

Figure 3. Outage probability vs. relative channel estimation error

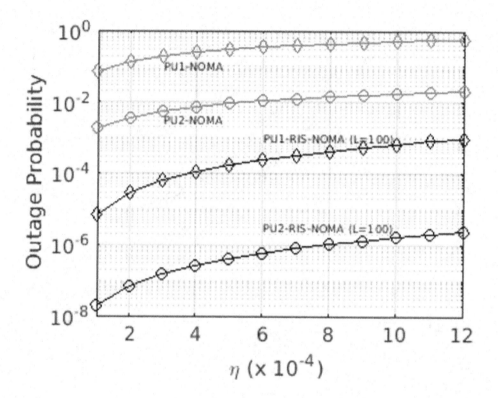

In Figure 3, the relative channel estimation error is displayed against the outage probability of a pair of users in the scenario, assuming that R_1 and R_2 are equal to 3.6 BPCU and 1 BPCU, 50 dB for ρ_s, 100 for L, and 0.5 km for d_{su_1}. The exact outage probability curves and the Monte Carlo simulation results accord remarkably well, as shown in Figure 3. It is evident that in terms of outage performance, RIS-assisted NOMA performs better than traditional NOMA. This is because each RIS piece has a superposition coding scheme. Furthermore, the influence of the error floor is shown to increase the probability of an outage as the percentage channel estimation error grows (Chen et al., 2023).

Figure 4. a) Outage Probability versus normalized distance between BS and U_1 (; b) Outage Probability versus normalized distance between BS and U_1 (

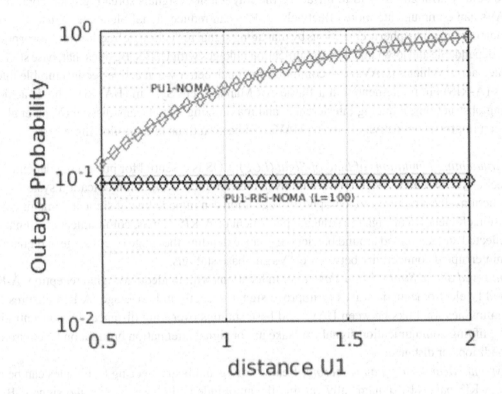

Assuming that $R_1 = 3.6$ dB, R_2 equals 1 dB, L equals 100 units of elements, and equals 0.0001, Figures 4.a and 4.b plot the chance of the pair of users having an outage vs. the normalized distance between AWACS and U_1 for equals 30 dB and 50 dB, respectively, for the two simulations. The exact outage probability curves match the Monte Carlo simulation findings exactly, as shown in Figure 4.a. It's clear that the best place for the user relaying U_1 is nearer the base station than the best place for the user relaying U_2. In order to get a high received SNR at U_1, it is best to locate U_1 closer to the base station because U_1 with better channel conditions receives less transmit power.

Furthermore, it seems that in terms of outage performance, RIS-assisted NOMA performs better than traditional NOMA. The superposition coding system is to blame for this. It is noteworthy that for traditional NOMA networks, the outage performance deteriorates as U_1 gets closer to U_2. This is because U_1, which has superior channel conditions, is given a lower transmit power allocation; as a result, as U_1 gets closer to U_2, the received signal strength at U_1 drops. As a result, the user relaying site for both traditional NOMA networks and those supported by RIS should be near the base station.

For RIS-assisted NOMA, as shown in Figures 4.a and 4.b, the likelihood of an outage does not change with an increase in the user relaying location. The fact that the outage probability equals the error floor at high SNR and is independent of d_{su_1} and d_{su_2} helps to explain this phenomenon.

Essentially, electromagnetic waves, like radio frequencies, can be controlled by dynamically changeable smart surfaces to achieve desired outcomes. The following explains how A-RIS and BAMS work together to control UAVs:

Improved Communication

By carefully arranging A-RIS to influence the way wireless signals spread, ground control stations and UAVs can communicate more effectively. A-RIS can reduce signal blockages caused by barriers, terrain features, or atmospheric conditions by managing signal reflection, refraction, and absorption. This ensures dependable and low-latency communication linkages with UAVs. Besides that, base stations, and unmanned aerial vehicles (UAVs) can communicate much better when active-reconfigurable intelligent surfaces (A-RIS) are implemented in a Battle-Air Management System (BAMS). This is achieved by removing obstructions, reducing interference, and maximizing signal transmission (Yang et al., 2020). A-RIS can improve connectivity between UAVs and base stations in the following ways:

- *Surmounting Limitations of Line-of-Sight (LOS)*: LOS is essential for preserving dependable connections between UAVs and base stations in conventional UAV communication systems. But obstructions like structures, topography, or vegetation can impede line of sight and impair communication efficiency. Through the strategic placement of A-RIS panels, communication signals can be reflected and redirected around barriers, thereby extending the range of coverage and guaranteeing uninterrupted connectivity between UAVs and base stations.
- *Improved Signal Strength and Coverage*: In locations with inadequate signal reception, A-RIS can modify electromagnetic waves to improve signal strength and coverage. A-RIS ensures reliable communication links between UAVs and base stations over long distances by concentrating and amplifying communication signals to make up for signal attenuation brought on by environmental conditions or distance.
- *Dynamic Beamforming and Steering*: Beamforming and beam steering techniques can be applied by A-RIS panels by dynamically varying the amplitude and phase of reflected signals. By doing so, A-RIS is able to focus communication beams on certain UAVs or regions of interest, boosting signal strength and lowering interference. Even in dynamic and unexpected circumstances, A-RIS ensures optimal signal delivery and coverage by constantly responding to UAV motions and operating requirements.
- *Mitigation of Interference*: In crowded or disputed airspace, UAV communication may be interfered with by electromagnetic interference from hostile sources or other wireless equipment. Signal fading, multipath propagation, and jamming efforts can all be lessened by using A-RIS's ability to deliberately modify signal reflections. A-RIS allows clear and dependable communication between UAVs and base stations even in difficult electromagnetic conditions by attenuating undesirable signals and amplifying needed ones.
- *Adaptability to Environmental Changes*: A-RIS is adaptable and capable of adjusting to alterations in the surrounding environment, including differences in the topography, the weather, or atmospheric disturbances. Through constant observation of communication efficiency and surrounding conditions, A-RIS is able to dynamically modify its setup in order to maximize signal transmission and sustain excellent communication connections between unmanned aerial vehicles and base stations.
- *Security and Stealth*: By selectively blocking or rerouting signals to thwart detection or interception by enemies, A-RIS can improve the security and stealth of UAV communication. A-RIS lowers the possibility of adversarial actors intercepting or using signals for malicious purposes by dynamically modifying signal propagation patterns.

A-RIS implementation in BAMS offers a flexible and efficient way to enhance UAV-base station communication, get around obstacles, reduce interference, and guarantee dependable connectivity in a range of operational scenarios.

Increased Effective Range of Network and Coverage

The communication networks' effective range and coverage can be increased by placing A-RIS panels strategically across the battlefield. This improves situational awareness and mission flexibility by allowing unmanned aerial vehicles (UAVs) to operate over greater distances without losing communication with headquarters. Active-reconfigurable intelligent surfaces (A-RIS) can be integrated into a Battle-Air Management System (BAMS) in a number of ways to improve the coverage and effective range of unmanned aerial vehicles (UAVs) relative to base stations (Zeng & Zhang, 2017).

- *Signal Amplification*: To improve communication signals between UAVs and base stations, A-RIS can be placed at key locations. A-RIS strengthens signals by reflecting and focussing them in particular directions, hence increasing the communication link's range. By compensating for long-distance signal attenuation, this amplification allows UAVs to reliably maintain links with base stations over longer distances.
- *Signal Steering and Reflection*: To maximize the signal propagation channels connecting UAVs to base stations, A-RIS panels have the ability to steer and reflect communication signals. Through signal rerouting around obstructions or over difficult terrain, A-RIS reduces signal blockages and guarantees uninterrupted connectivity over extended distances. By changing the directionality of the signal in response to the location of the UAV and its operational needs, dynamic beamforming and steering techniques further improve the coverage of the signals.
- *LOS Extension*: By reflecting signals over obstructions, A-RIS can extend line-of-sight (LOS) communication in situations when it is impeded by structures, topography, or other factors. Indirect line of sight (LOS) connections between UAVs and base stations can be made by carefully placing A-RIS panels along the communication path. This effectively extends the useful range of communication lines.
- *Multi-Hop Communication*: By establishing a network of reflection points to relay signals between UAVs and base stations, A-RIS can enable multi-hop communication. Indirect communication linkages can be established over larger distances than direct LOS connections by bouncing signals off several A-RIS panels. This method improves resilience and coverage, especially in places with poor line-of-sight or difficult terrain.
- *Dynamic Adaptation to Environmental Conditions*: To maximize signal coverage and propagation, A-RIS has the ability to dynamically modify its configuration in response to external factors. A-RIS maintains optimal communication performance throughout a range of operating situations and distances by continuously monitoring elements like signal quality, terrain elevation, and air interference. This allows A-RIS to adjust its steering and reflection patterns.
- *Mitigation of Interference*: A-RIS can limit the effective range and deteriorate communication performance by mitigating interference from other wireless devices or electromagnetic sources. Through deliberate manipulation of signal reflections, A-RIS attenuates harmful interference and amplifies useful signals, guaranteeing dependable and unambiguous communication channels across long distances between UAVs and base stations.

By taking advantage of these features, A-RIS in BAMS expands the communication's effective range and coverage between UAVs and base stations, providing dependable connectivity across a variety of operational situations and longer distances.

Dynamic Path Planning

To maximize signal strength and minimize interference, A-RIS can dynamically modify the propagation environment to optimize UAV flight paths. This function is especially useful in contested or crowded airspace, where UAVs may run across communication-blocking obstructions or jamming attempts (Tang et al., 2020).

Active-reconfigurable intelligent surfaces (A-RIS) can improve dynamic path planning between unmanned aerial vehicles (UAVs) and base stations via the following principles when implemented in a Battle-Air Management System (BAMS):

- *Obstacle Avoidance*: To steer clear of obstructions and terrain features in the UAV's flight path, A-RIS has the ability to dynamically modify signal propagation patterns. It is feasible to reflect signals over obstructions and establish unobstructed communication pathways for UAV navigation by carefully placing A-RIS panels. This makes it possible for dynamic path planning algorithms to produce clear paths in real time, maximizing the efficiency and safety of UAV trajectories.
- *Terrain Adaptation*: A-RIS is capable of modifying signal propagation patterns to take topographical and elevation variations into consideration. A-RIS provides the best possible communication coverage and signal quality over a range of terrain types by modifying signal reflection angles and intensities in response to terrain data. This makes it possible for dynamic path planning algorithms to optimize UAV trajectories for terrain-following or terrain-avoidance operations by incorporating terrain information into route planning decisions.
- *Dynamic Reconfiguration*: Real-time modifications to signal reflection and steering parameters are possible thanks to the changeable nature of A-RIS panels. Because of this versatility, A-RIS designs can be dynamically controlled by dynamic path planning algorithms according to various factors such as UAV position, mission objectives, and environmental variables. It is feasible to adjust communication routes and optimize UAV trajectories in response to shifting operational needs and situational dynamics by dynamically redesigning A-RIS architectures. A-RIS has the ability to reduce interference that could interfere with UAV-base station communication and navigation signals. A-RIS ensures clear and dependable communication lines for dynamic path planning by precisely controlling signal reflections and attenuations. This prevents undesired interference from other wireless devices or electromagnetic sources. Even in crowded or contested airspace conditions, UAVs may navigate with more confidence and precision thanks to their interference reduction feature.
- *Real-Time Feedback and Control*: To enable dynamic path planning, A-RIS implementations in BAMS can incorporate real-time feedback and control techniques. A-RIS systems give dynamic path-planning algorithms with feedback by continuously monitoring communication and navigation performance indicators. This allows the algorithms to adapt UAV trajectories and A-RIS settings to changing conditions. The adaptive and responsive path planning tactics that maximize the effectiveness and efficiency of UAV navigation are guaranteed by this closed-loop control system.

Through the use of these features, A-RIS in BAMS improves dynamic path planning between unmanned aerial vehicles (UAVs) and base stations, allowing for adaptive, interference-resistant, and obstacle-aware trajectory planning for UAV missions in a variety of operational situations.

Stealth and Security

By selectively blocking or rerouting signals to thwart detection or interception by adversaries, A-RIS can also be used to improve the stealth and security of UAV operations. Operational security and mission effectiveness can be preserved with the aid of A-RIS, which can dynamically adapt to shifting threat scenarios (Zhang, Wu, Dai, & He, 2020).

Active-reconfigurable intelligent surfaces (A-RIS) can be integrated into a Battle-Air Management System (BAMS) to improve communication security and safety between unmanned aerial vehicles (UAVs) and base stations in a number of ways.

- *Signal Encryption and Authentication*: By selectively regulating signal propagation pathways and encrypting sent data, A-RIS can be used to improve communication security. A-RIS reduces the possibility of eavesdropping or interception by unauthorized parties by guaranteeing that communication signals are only accessible to authorized parties. Furthermore, in order to confirm the identities of people communicating and guard against spoofing or impersonation attacks, A-RIS can authenticate communication lines between UAVs and base stations.

- *Interference Suppression*: A-RIS can reduce interference originating from outside sources, like electromagnetic interference or jamming devices. A-RIS minimizes unwanted interference and ensures clear and dependable communication links between UAVs and base stations by selectively adjusting signal reflections and attenuations. The ability to suppress interference improves communication resilience and lowers the possibility that hostile acts or external circumstances would disrupt conversation.

- *Dynamic Security Policies*: Based on shifting threat levels and operating circumstances, A-RIS implementations in BAMS can include dynamic security policies to adaptively modify communication security parameters. A-RIS systems dynamically alter signal encryption methods, authentication protocols, and access control mechanisms to prevent evolving security risks and vulnerabilities by continuously monitoring threat indicators and risk factors. By improving communication security efficacy and reactivity, this dynamic security policy management offers a strong defense against constantly changing cyberthreats.

- *Stealth Communication*: In hostile or sensitive circumstances, A-RIS can help achieve stealth communication by limiting signal emissions and carefully managing signal propagation routes. A-RIS improves communication security and lowers the possibility of adversary discovery by guiding communication signals via covert channels and limiting signal exposure. UAVs can operate safely and covertly in contested or hostile airspace thanks to their stealthy communication capabilities, which reduce the possibility of being intercepted or interfered with by hostile forces.

- *Redundancy and Resilience*: By creating several communication channels between UAVs and base stations, A-RIS can improve redundancy and resilience in communication. Replicated communication links that are resistant to interruptions and attacks are ensured by A-RIS through the dynamic modification of signal propagation paths and the utilization of alternate communication channels. In difficult operating circumstances, this redundancy and resilience capacity improves communication availability and dependability, guaranteeing uninterrupted contact between UAVs and base stations.

By offering encryption, authentication, interference suppression, dynamic security rules, stealth communication, and redundancy methods, A-RIS integration in BAMS improves the overall safety and security of communication between UAVs and base stations. Through the use of these features, A-RIS makes it possible to create safe and robust communication networks that are resistant to cyberattacks and guarantee mission success in challenging and dangerous operating circumstances.

Energy Efficiency

A-RIS can help improve overall energy efficiency in UAV operations by minimizing signal attenuation and optimizing communication connectivity. This can decrease the need for large, power-hungry signal amplifiers or bulky communication equipment and increase mission duration (Zhai, Dai, Duo, Wang, & Yuan, 2022).

Active-reconfigurable intelligent surfaces (A-RIS) are a useful tool for improving the energy efficiency of communication between unmanned aerial vehicles (UAVs) and base stations in Battle-Air Management Systems (BAMS) for a number of reasons.

- *Signal Amplification and Directionality*: A-RIS minimizes signal loss and eliminates the need for high-power transmitters on UAVs and base stations by amplifying communication signals and directing them in particular directions. A-RIS guarantees that transmitted signals reach their intended destinations with the least amount of energy consumption by concentrating signals and limiting dispersion.
- *Dynamic Power Adaptation*: In response to communication needs and environmental conditions, A-RIS implementations in BAMS can dynamically modify the parameters of signal reflection and propagation. A-RIS ensures effective use of available power resources by reducing the energy consumption of communication equipment on UAVs and base stations by optimizing signal strength and directionality in real-time.
- *Interference Mitigation*: A-RIS has the ability to reduce interference from outside sources, such as rival wireless networks or electromagnetic disturbances. A-RIS reduces the influence of interference on communication performance by selectively adjusting signal reflections and attenuations. This lessens the requirement for energy-intensive error correcting methods and signal amplification techniques.
- *Adaptive Beamforming*: To direct communication messages toward certain UAVs or regions of interest, A-RIS can employ adaptive beamforming algorithms. A-RIS ensures effective signal delivery with low energy dispersion by dynamically altering the phase and amplitude of the signal, thereby minimizing power consumption in both transmitting and receiving devices.
- *Dynamic Path Optimization*: A-RIS can maximize energy efficiency and minimize signal attenuation by real-time optimizing communication paths between UAVs and base stations. With A-RIS, transmitted signals are guaranteed to reach their intended destinations with the least amount of energy consumption by dynamically adjusting signal propagation channels based on UAV position, environmental conditions, and communication requirements.

- *Energy Harvesting*: By integrating energy harvesting technology, A-RIS panels can use ambient energy sources, including solar or radiofrequency energy, to power A-RIS control systems and communication devices. A-RIS lowers the total energy consumption of communication systems in BAMS by combining renewable energy sources with conventional power sources, improving sustainability, and lowering dependency on external power supplies.

By maximizing signal propagation, reducing interference, and utilizing renewable energy sources, the implementation of A-RIS in BAMS improves the energy efficiency of communication between UAVs and base stations overall. A-RIS guarantees sustainable and effective communication networks that support long-duration UAV operations and lower operating costs by lowering energy consumption and maximizing power utilization.

A-RIS is capable of detecting and reacting to hostile electronic warfare methods, such jamming or spoofing, in dynamic battlefield conditions. This is known as adaptive response to threats. A-RIS can contribute to ensuring the durability and survivability of UAV-based systems by continuously adjusting to mitigate these risks.

The potential for improving communication, navigation, and overall mission success in challenging and disputed areas is considerable when A-RIS is integrated into a Battle-Air Management System for UAV control. While creating and putting into practice such solutions, it's crucial to take into account elements like cost, scalability, and compatibility with current systems. In-depth testing and validation are also required to guarantee the dependability and resilience of BAMS platforms with A-RIS functionality in actual operational situations.

Considerations in Implementing A-RIS-Assisted UAVs in BAMS

Theoretically, unmanned aerial vehicles (UAVs) might be controlled by active-reconfigurable intelligent surfaces (A-RIS) in an air battle management system (BAMS). However, implementing such a system requires combining a number of different technologies and factors. A number of components and potential difficulties are introduced when active-reconfigurable intelligent surfaces (A-RIS) are implemented in a Battle-Air Management System (BAMS) to support unmanned aerial vehicles (UAVs) (Pogaku, Do, Lee, & Nguyen, 2022). Let's dissect the elements and possible difficulties:

- *A-RIS Infrastructure*: To maximize UAV communication and navigation, the A-RIS infrastructure is set up by placing A-RIS panels strategically across the operating region. To obtain the required coverage and performance, meticulous planning is needed to define the positions, orientations, and combinations of A-RIS panels. There may be logistical difficulties in deploying and maintaining the A-RIS infrastructure, such as with site selection, installation, power supply, and maintenance.
- *Optimization of Communication and Navigation*: A-RIS-assisted UAVs in BAMS are primarily concerned with maximizing communication and navigation efficiency through electromagnetic wave manipulation. To guarantee dependable and effective communication between UAVs and base stations, this entails dynamically modifying signal propagation pathways, improving signal strength, reducing interference, and optimizing signal-to-noise ratios. Achieving these goals requires developing algorithms and protocols to manage A-RIS setups efficiently and communicate with base stations and UAVs.

- *Integration with Existing Systems*: Compatibility with other communication, navigation, and control systems is necessary for A-RIS to be integrated into the current BAMS architecture. This could entail overcoming interoperability issues, standardizing protocols, and making sure that A-RIS-enabled components and legacy systems communicate and coordinate seamlessly. For seamless integration and operation inside the BAMS framework, compatibility with current UAV platforms, ground control stations, and network architectures must be taken into account.

- *Dynamic Adaptation and Optimization*: In BAMS, A-RIS-assisted UAVs must be able to adjust dynamically to shifting environmental variables, mission requirements, and operating situations. To adapt to new risk or problems, improve communication channels, and modify A-RIS settings, real-time monitoring, analysis, and decision-making skills are needed. Optimizing performance and effectiveness requires the development of intelligent algorithms and control techniques that can change RIS behaviour independently based on situational information and mission objectives.

- *Security and Reliability*: It is crucial to guarantee the protection and dependability of A-RIS-assisted communication and navigation systems, especially in mission-critical scenarios like military operations. Robust authentication, encryption, and intrusion detection systems are necessary to defend against spoofing attempts, jamming attacks, and other types of interference. Maintaining continuous operation and mission success also depend on building fault-tolerant and resilient A-RIS architectures that can survive environmental dangers, hardware failures, and hostile acts.

- *Regulatory and Ethical Considerations*: The implementation of A-RIS-assisted UAVs in BAMS may give rise to legislative and ethical issues concerning safety, privacy, airspace access, and spectrum management. To guarantee the legal and appropriate use of A-RIS technology in UAV operations, compliance with regulatory requirements, including frequency allocations, licensing, and operational limits, is required. Building public trust and acceptance of A-RIS-assisted UAVs in BAMS also requires resolving ethical issues with surveillance, data privacy, and the misuse of A-RIS-enabled capabilities.

- *Power Consumption*: Since UAVs normally have a limited power supply, any extra technology—such as A-RIS—must be power-efficient in order to maintain the UAV's operating endurance. When integrating active-reconfigurable intelligent surfaces (A-RIS)-assisted unmanned aerial vehicles (UAVs) in a Battle-Air Management System (BAMS), power consumption is a critical component and a source of difficulty. In order to run the electronic components that manipulate electromagnetic waves, A-RIS panels need power. The number of panels, their size, the intricacy of the reconfiguration algorithms, and the frequency of modifications all affect how much power A-RIS uses. The BAMS infrastructure's entire energy budget is directly impacted by A-RIS power consumption. For A-RIS panels to operate continuously, a dependable power supply is required. Power can be supplied by cable connections, batteries, solar panels, or energy harvesting systems, depending on the deployment circumstances. It can be difficult to guarantee A-RIS panels receive a steady and sufficient power supply, especially in isolated or hard-to-reach places where infrastructure could be scarce. Next, power management and compatibility must be carefully considered when integrating A-RIS-assisted UAVs into BAMS. The limited onboard power resources of UAVs are already required to run many systems, including sensors, avionics, propulsion, and communication. The addition of A-RIS-assistance functionality raises power requirements, which may have an effect on the mission capabilities, endurance, and range of UAVs. Furthermore, the key to reducing energy consumption and optimizing performance in A-RIS control and optimization is the development of power-efficient algorithms. The trade-off between attaining the intended communication improvements and energy conservation must be balanced using A-RIS algorithms. Creating effective algorithms that yield the best outcomes while consuming the least amount of power is a difficult technical task. Investigating energy harvesting and renewable power sources can help reduce the power consumption issues that come with using A-RIS-assisted UAVs in BAMS. In order to lessen dependency on external power grids, conventional power sources can be supplemented by solar panels, RF energy scavenging systems, and kinetic energy harvesters. Energy harvesting system integration into A-RIS infrastructure, however, is more difficult and could require more resources for setup and upkeep. In BAMS, power consumption concerns for A-RIS-assisted UAVs are influenced by operational constraints such as mission time, payload capacity, and environmental variables. The power consumption patterns and system dependability of UAV missions may be impacted by the need for extended flight times or by the need to operate in challenging conditions with strong winds or temperatures. Benefit-Cost analysis is crucial to weigh the advantages of A-RIS-assisted communication improvements against the related power and infrastructure requirements. Evaluating the viability and sustainability of A-RIS implementation in BAMS involves completing a comprehensive cost-benefit analysis. It is necessary to carefully consider the initial investment, ongoing costs, upkeep needs, and possible mission benefits when deciding whether or not to use A-RIS technology. When using A-RIS-assisted UAVs in a battle-air management system, power consumption is a key component and possible source of difficulty. To ensure the effective and long-lasting operation of A-RIS-enabled communication systems in BAMS, addressing power-related factors calls for creative solutions in power supply, integration, algorithm design, energy harvesting, and cost-benefit analysis.

- *Testing and Validation*: Implementing active-reconfigurable intelligent surfaces (A-RIS)-assisted UAVs in a Battle-Air Management System (BAMS) requires testing and validation for a number of reasons, including the following: 1) To achieve desired effects, including signal augmentation, interference reduction, and beamforming, A-RIS technology uses complex mechanisms to manipulate electromagnetic waves. A thorough understanding of these mechanisms and how they interact with UAV communication systems is necessary for the implementation of A-RIS in BAMS. To make sure that A-RIS setups and control algorithms work as planned and improve communication performance, testing is essential., 2) Linking A-RIS-assisted UAVs to BAMS necessitates integrating A-RIS parts with current network protocols, UAV control systems, and communication infrastructure. Verifying compatibility, interoperability, and smooth integration between A-RIS and other BAMS components requires testing. Through validation, one may be sure that the integrated system will function dependably and effectively in a range of operational and environmental settings., 3) Communication performance is greatly impacted by A-RIS configuration factors such as phase shifts, amplitude modifications, and the spatial distribution of A-RIS pieces. Optimizing A-RIS configurations for particular UAV missions necessitates testing and validation, taking into account variables like energy efficiency, interference levels, coverage area, and communication range. By going through this iterative process, A-RIS is able to minimize resource consumption and operational expenses while improving communication between UAVs and base stations., 4) To assess A-RIS-assisted UAVs' performance in practical settings, they must be tested in real-world settings. Through field experiments, engineers and researchers can evaluate the durability of the system, communication quality, and A-RIS functionality in erratic and dynamic operational circumstances. Validating A-RIS performance in real-world scenarios offers insightful information about how the system behaves, points out possible problems, and guides future iterations and enhancements to A-RIS-assisted UAVs in BAMS., 5) Compliance with Regulations and Safety: The safety issues and legal obligations related to A-RIS-assisted UAV operations must be taken into consideration during testing and validation procedures. This entails evaluating any threats to the security of airspace, making sure aviation laws are followed, and reducing safety risks associated with the installation and use of A-RIS. Strict validation and testing processes are required to prove that A-RIS-assisted UAVs are safe, dependable, and compliant with BAMS operations.

Implementing A-RIS-assisted UAVs in BAMS requires testing and validation in order to guarantee the integrated system's performance, safety, and compliance with regulations. Through these efforts, possible difficulties are addressed, system design is optimized, and the efficacy of A-RIS technology in improving UAV communication and mission capabilities is validated.

All things considered, there is a lot of room for improvement when it comes to communication, navigation, and mission capabilities when A-RIS-assisted UAVs are used in BAMS. But for the deployment and integration of A-RIS technology within the BAMS framework to be successful, resolving the aforementioned components and difficulties is essential. To fully realize the promise of A-RIS-assisted UAVs in BAMS applications and to overcome technological, operational, and regulatory obstacles, cooperation among researchers, industry stakeholders, regulatory agencies, and end users is important.

REFERENCES

Albarado, K., Coduti, L., Aloisio, D., Robinson, S., Drown, D., & Javorsek, D. (2022). AlphaMosaic: An Artificially Intelligent Battle Management Architecture. *Journal of Aerospace Information Systems*, 19(3), 203–213. DOI: 10.2514/1.I010991

Alexandropoulos, G. C., Phan-Huy, D.-T., Katsanos, K. D., Crozzoli, M., Wymeersch, H., Popovski, P., & Gonzalez, S. H. (2023). RIS-enabled smart wireless environments: Deployment scenarios, network architecture, bandwidth and area of influence. *EURASIP Journal on Wireless Communications and Networking*, 2023(1), 103. DOI: 10.1186/s13638-023-02295-8

Ali, A., Jadoon, Y. K., Changazi, S. A., & Qasim, M. (2020). Military operations: Wireless sensor networks based applications to reinforce future battlefield command system. Paper presented at the *2020 IEEE 23rd International Multitopic Conference (INMIC)*. IEEE. DOI: 10.1109/INMIC50486.2020.9318168

Azcarate, S. M., Ríos-Reina, R., Amigo, J. M., & Goicoechea, H. C. (2021). Data handling in data fusion: Methodologies and applications. *Trends in Analytical Chemistry*, 143, 116355. DOI: 10.1016/j.trac.2021.116355

Bastidas-Puga, E. R., Andrade, Á. G., Galaviz, G., & Covarrubias, D. H. (2019). Handover based on a predictive approach of signal-to-interference-plus-noise ratio for heterogeneous cellular networks. *IET Communications*, 13(6), 672–678. DOI: 10.1049/iet-com.2018.5126

Bejiga, M. B., Zeggada, A., Nouffidj, A., & Melgani, F. (2017). A convolutional neural network approach for assisting avalanche search and rescue operations with UAV imagery. *Remote Sensing (Basel)*, 9(2), 100. DOI: 10.3390/rs9020100

Blasch, E., & Bélanger, M. (2016). *Agile battle management efficiency for command, control, communications, computers and intelligence (C4I)*. Paper presented at the Signal Processing, Sensor/Information Fusion, and Target Recognition XXV.

Bronk, J., Reynolds, N., & Watling, J. (2022). *The Russian air war and Ukrainian requirements for air defence*.

Chen, H.-C. (2019). Collaboration IoT-based RBAC with trust evaluation algorithm model for massive IoT integrated application. *Mobile Networks and Applications*, 24(3), 839–852. DOI: 10.1007/s11036-018-1085-0

Chen, H.-C., Widodo, A. M., Lin, J. C.-W., Weng, C.-E., & Do, D.-T. (2023). Outage behavior of the downlink reconfigurable intelligent surfaces-aided cooperative non-orthogonal multiple access network over Nakagami-m fading channels. *Wireless Networks*, 1–18. DOI: 10.1007/s11276-022-03074-x

Cheng, A., Eppich, W., Grant, V., Sherbino, J., Zendejas, B., & Cook, D. A. (2014). Debriefing for technology-enhanced simulation: A systematic review and meta-analysis. *Medical Education*, 48(7), 657–666. DOI: 10.1111/medu.12432 PMID: 24909527

Cheng, Y., Li, K. H., Liu, Y., Teh, K. C., & Poor, H. V. (2021). Downlink and uplink intelligent reflecting surface aided networks: NOMA and OMA. *IEEE Transactions on Wireless Communications*, 20(6), 3988–4000. DOI: 10.1109/TWC.2021.3054841

Dai, L., Wang, B., Yuan, Y., Han, S., Chih-Lin, I., & Wang, Z. (2015). Non-orthogonal multiple access for 5G: Solutions, challenges, opportunities, and future research trends. *IEEE Communications Magazine*, 53(9), 74–81. DOI: 10.1109/MCOM.2015.7263349

Eisenberg, D. A., Alderson, D. L., Kitsak, M., Ganin, A., & Linkov, I. (2018). Network foundation for command and control (C2) systems: Literature review. *IEEE Access : Practical Innovations, Open Solutions*, 6, 68782–68794. DOI: 10.1109/ACCESS.2018.2873328

Fernandes, S. V., & Ullah, M. S. (2022). A Comprehensive Review on Features Extraction and Features Matching Techniques for Deception Detection. *IEEE Access : Practical Innovations, Open Solutions*, 10, 28233–28246. DOI: 10.1109/ACCESS.2022.3157821

Gao, K., Xiao, H., Qu, L., & Wang, S. (2022). Optimal interception strategy of air defence missile system considering multiple targets and phases. *Proceedings of the Institution of Mechanical Engineers. Part O, Journal of Risk and Reliability*, 236(1), 138–147. DOI: 10.1177/1748006X211022111

Gong, X., Qian, L., Ge, W., & Yan, J. (2020). Research on electronic brake force distribution and anti-lock brake of vehicle based on direct drive electro hydraulic actuator. *International Journal of Automotive Engineering*, 11(2), 22–29. DOI: 10.20485/jsaeijae.11.2_22

Gong, X., Yue, X., & Liu, F. (2020). Performance analysis of cooperative NOMA networks with imperfect CSI over Nakagami-m fading channels. *Sensors (Basel)*, 20(2), 424. DOI: 10.3390/s20020424 PMID: 31940864

Halamek, L. P., Cady, R. A., & Sterling, M. R. (2019). Using briefing, simulation and debriefing to improve human and system performance. *Seminars in Perinatology*. Elsevier. DOI: 10.1053/j.semperi.2019.08.007

Hemanth, A., Umamaheswari, K., Pogaku, A. C., Do, D.-T., & Lee, B. M. (2020). Outage performance analysis of reconfigurable intelligent surfaces-aided NOMA under presence of hardware impairment. *IEEE Access : Practical Innovations, Open Solutions*, 8, 212156–212165. DOI: 10.1109/ACCESS.2020.3039966

Huang, S., Zhang, X., Chen, N., Ma, H., Zeng, J., Fu, P., Nam, W.-H., & Niyogi, D. (2022). Generating high-accuracy and cloud-free surface soil moisture at 1 km resolution by point-surface data fusion over the Southwestern US. *Agricultural and Forest Meteorology*, 321, 108985. DOI: 10.1016/j.agrformet.2022.108985

Huang, X. (2020). Quality of service optimization in wireless transmission of industrial Internet of Things for intelligent manufacturing. *International Journal of Advanced Manufacturing Technology*, 107(3), 1007–1016. DOI: 10.1007/s00170-019-04288-8

Jones, A. M., Rigling, B., & Rangaswamy, M. (2016). Signal-to-interference-plus-noise-ratio analysis for constrained radar waveforms. *IEEE Transactions on Aerospace and Electronic Systems*, 52(5), 2230–2241. DOI: 10.1109/TAES.2016.150511

Keating, R., Säily, M., Hulkkonen, J., & Karjalainen, J. (2019). Overview of positioning in 5G new radio. *Paper presented at the 2019 16th International Symposium on Wireless Communication Systems (ISWCS)*. IEEE. DOI: 10.1109/ISWCS.2019.8877160

Kiohara, P., de Souza, R., Ivo, F. S., Mippo, N. T., Coutinho, O. L., Pérennou, A., & Quintard, V. (2021). Microwave Photonic Approach to Antenna Remote on Airborne Radar Warning Receiver System. *Paper presented at the 2021 SBMO/IEEE MTT-S International Microwave and Optoelectronics Conference (IMOC)*. IEEE. DOI: 10.1109/IMOC53012.2021.9624933

La Porte, T. R. (2019). The United States Air Traffic System: Increasing Reliability in the Midst of Raped Growth 1. *The development of large technical systems*, 215-244.

Li, D. (2020). Ergodic capacity of intelligent reflecting surface-assisted communication systems with phase errors. *IEEE Communications Letters*, 24(8), 1646–1650. DOI: 10.1109/LCOMM.2020.2997027

Liu, Y., Ding, Z., Elkashlan, M., & Poor, H. V. (2016). Cooperative non-orthogonal multiple access with simultaneous wireless information and power transfer. *IEEE Journal on Selected Areas in Communications*, 34(4), 938–953. DOI: 10.1109/JSAC.2016.2549378

Long, R., Liang, Y.-C., Pei, Y., & Larsson, E. G. (2021). Active reconfigurable intelligent surface-aided wireless communications. *IEEE Transactions on Wireless Communications*, 20(8), 4962–4975. DOI: 10.1109/TWC.2021.3064024

Lv, L., Ni, Q., Ding, Z., & Chen, J. (2016). Application of non-orthogonal multiple access in cooperative spectrum-sharing networks over Nakagami-$ m $ fading channels. *IEEE Transactions on Vehicular Technology*, 66(6), 5506–5511. DOI: 10.1109/TVT.2016.2627559

Ma, Y., Li, M., Liu, Y., Wu, Q., & Liu, Q. (2022). Active Reconfigurable Intelligent Surface for Energy Efficiency in MU-MISO Systems. *IEEE Transactions on Vehicular Technology*, 72(3), 4103–4107. DOI: 10.1109/TVT.2022.3221720

Muhammad, A., Elhattab, M., Arfaoui, M. A., & Assi, C. (2023). Optimizing age of information in ris-empowered uplink cooperative noma networks. *IEEE Transactions on Network and Service Management*.

Nguyen, T. L., & Do, D. T. (2018). Power allocation schemes for wireless powered NOMA systems with imperfect CSI: An application in multiple antenna–based relay. *International Journal of Communication Systems*, 31(15), e3789. DOI: 10.1002/dac.3789

Nichols, R. K., Mumm, H. C., Lonstein, W. D., Ryan, J. J., Carter, C., & Hood, J.-P. (2019). *Unmanned Aircraft Systems in the Cyber Domain*. New Prairie Press.

Nichols, R. K., Mumm, H. C., Lonstein, W. D., Ryan, J. J., Carter, C., & Hood, J.-P. (2020). *Counter unmanned aircraft systems technologies and operations*. New Prairie Press.

Park, G., Jeon, G.-Y., Sohn, M., & Kim, J. (2022). A Study of Recommendation Systems for Supporting Command and Control (C2) Workflow. *Journal of Internet Computing and Services*, 23(1), 125–134.

Paucar, C., Morales, L., Pinto, K., Sánchez, M., Rodríguez, R., Gutierrez, M., & Palacios, L. (2018). Use of drones for surveillance and reconnaissance of military areas. *Developments and Advances in Defense and Security: Proceedings of the Multidisciplinary International Conference of Research Applied to Defense and Security (MICRADS 2018)*. Springer. DOI: 10.1007/978-3-319-78605-6_10

Pogaku, A. C., Do, D.-T., Lee, B. M., & Nguyen, N. D. (2022). UAV-assisted RIS for future wireless communications: A survey on optimization and performance analysis. *IEEE Access : Practical Innovations, Open Solutions*, 10, 16320–16336. DOI: 10.1109/ACCESS.2022.3149054

Şandru, V. (2016). *Performances of Air Defence Systems measured with AHP-SWOT analysis.* Paper presented at the Forum Scientiae Oeconomia.

Seizovic, A., Thorpe, D., & Goh, S. (2022). Emergent behavior in the battle management system. *Applied Artificial Intelligence*, 36(1), 2151183. DOI: 10.1080/08839514.2022.2151183

Smagh, N. S. (2020). Intelligence, surveillance, and reconnaissance design for great power competition. *Congressional Research Service, 46389.*

Song, J. G., Lee, J. H., & Park, I. S. (2021). Enhancement of cooling performance of naval combat management system using heat pipe. *Applied Thermal Engineering*, 188, 116657. DOI: 10.1016/j.applthermaleng.2021.116657

Tang, W., Chen, M. Z., Dai, J. Y., Zeng, Y., Zhao, X., Jin, S., Cheng, Q., & Cui, T. J. (2020). Wireless communications with programmable metasurface: New paradigms, opportunities, and challenges on transceiver design. *IEEE Wireless Communications*, 27(2), 180–187. DOI: 10.1109/MWC.001.1900308

Tuncer, O., & Cirpan, H. A. (2022). Target priority based optimisation of radar resources for networked air defence systems. *IET Radar, Sonar & Navigation*, 16(7), 1212–1224. DOI: 10.1049/rsn2.12255

Ünal, H. T., & Başçiftçi, F. (2020). Using Evolutionary Algorithms for the Scheduling of Aircrew on Airborne Early Warning and Control System. *Defence Science Journal*, 70(3), 240–248. DOI: 10.14429/dsj.70.15055

USAE, R. N. (2020). Table Stakes of the Advanced Battle Management System. *Air & Space Power Journal*, 34(3), 81–86.

Van Chien, T., Tu, L. T., Chatzinotas, S., & Ottersten, B. (2020). Coverage probability and ergodic capacity of intelligent reflecting surface-enhanced communication systems. *IEEE Communications Letters*, 25(1), 69–73. DOI: 10.1109/LCOMM.2020.3023759

Widodo, A. M., Wijayanto, H., Wijaya, I. G. P. S., Wisnujati, A., & Musnansyah, A. (2023). Analyzing Coverage Probability of Reconfigurable Intelligence Surface-aided NOMA. *JOIV: International Journal on Informatics Visualization*, 7(3), 839–846. DOI: 10.30630/joiv.7.3.2054

Xu, P., Chen, G., Pan, G., & Di Renzo, M. (2020). Ergodic secrecy rate of RIS-assisted communication systems in the presence of discrete phase shifts and multiple eavesdroppers. *IEEE Wireless Communications Letters*, 10(3), 629–633. DOI: 10.1109/LWC.2020.3044178

Xu, X., Yang, J., & Tang, Z. (2015). *Data virtualization for coupling command and control (C2) and combat simulation systems.* Paper presented at the Advances in Image and Graphics Technologies: 10th Chinese Conference, IGTA 2015, Beijing, China. DOI: 10.1007/978-3-662-47791-5_22

Yang, L., Meng, F., Zhang, J., Hasna, M. O., & Di Renzo, M. (2020). On the performance of RIS-assisted dual-hop UAV communication systems. *IEEE Transactions on Vehicular Technology*, 69(9), 10385–10390. DOI: 10.1109/TVT.2020.3004598

Zeng, Y., & Zhang, R. (2017). Energy-efficient UAV communication with trajectory optimization. *IEEE Transactions on Wireless Communications*, 16(6), 3747–3760. DOI: 10.1109/TWC.2017.2688328

Zhai, Z., Dai, X., Duo, B., Wang, X., & Yuan, X. (2022). Energy-efficient UAV-mounted RIS assisted mobile edge computing. *IEEE Wireless Communications Letters*, 11(12), 2507–2511. DOI: 10.1109/LWC.2022.3206587

Zhang, Z., Wu, J., Dai, J., & He, C. (2020). A novel real-time penetration path planning algorithm for stealth UAV in 3D complex dynamic environment. *IEEE Access : Practical Innovations, Open Solutions*, 8, 122757–122771. DOI: 10.1109/ACCESS.2020.3007496

Zhou, Y., Tang, Y., & Zhao, X. (2019). A novel uncertainty management approach for air combat situation assessment based on improved belief entropy. *Entropy (Basel, Switzerland)*, 21(5), 495. DOI: 10.3390/e21050495 PMID: 33267209

Chapter 10
Innovative Horizons:
The Role of AI and Robotics in Special Education

Princy Pappachan

https://orcid.org/0000-0001-6728-0228

National Chengchi University, Taiwan

Sreerakuvandana

Jain University, India

Siwada Piyakanjana

Asia University, Taiwan

Harlinda Syofyan

Esa Unggul University, Indonesia

Gunawan Nugroho

https://orcid.org/0000-0001-7032-8210

Institut Teknologi Sepuluh Nopember, Indonesia

ABSTRACT

The introduction of robotics and AI represents a paradigm shift in the quickly developing field of special education, providing unprecedented opportunities for improving learning opportunities. The chapter highlights this by examining how these technologies are changing how special education is taught to students with special needs. The transformative role of AI and robotics is evident in how they have moved from auxiliary tools to central elements in fostering adaptive and inclusive learning environments. The chapter delves into personalized learning, highlighting AI's role in customizing educational content to individual styles and enhancing strategy efficacy through data-driven insights. The chapter also explores the role of robotic assistants in interactive teaching and therapy, enhancing physical, sensory, and cognitive skills. Additionally, by addressing ethical concerns and advocating a balanced approach to technology use, the chapter provides educators, researchers, and policymakers with a thorough and forward-looking perspective on AI and robotics in special education.

DOI: 10.4018/979-8-3693-2707-4.ch010

INTRODUCTION

As we move through the post-2020 era that marked the dawn of the Fifth Industrial Revolution (5IR) (Noble et al., 2022), we find ourselves at a pivotal juncture where the combination of human ingenuity and advanced technology in the areas of artificial intelligence and robotics grants us an opportunity to reshape the educational landscape in the field of special education. Unlike its predecessor, the Fourth Industrial Revolution (4IR), which focused on achieving efficiency using technologies, the 5IR emphasizes a harmonious human-machine collaboration focused on improving social well-being, which is a principle that resonates loudly within special education (Fiestas et al., 2024; Papakostas et al., 2021; Gauri & Van Eerden, 2019). In this context, artificial intelligence and robotics are collaborative partners in the educational process rather than just adjuncts.

The context of "special needs education," referred to as "special education," has its roots in the educational reforms of the 20th century, which recognized the need to offer specialized support and resources to students with diverse learning requirements. Prior to this, education for students with disabilities was one of segregation and inequality. Accordingly, the incorporation of special needs education found its way into mainstream educational discourse with the 1975 Education for All Handicapped Children Act (EHA) in the United States, which made free and appropriate education for children with disabilities (Wehmeyer, 2022).

The history of robots dates back to the Greek civilizations and the Middle Ages, when mechanical instruments were meant to carry out certain functions. Though the word "robot" was first used in a play in 1920, it was not until the 20th century that the discipline of robotics started to take shape as advances in computation and technology allowed for the creation of increasingly complex and self-governing robots (Neumann, 2023). The digital revolution further sped these technological advancements and set the stage for the current robotics industry, which is now essential to many industries, from manufacturing to healthcare and education. Figure 1 traces the technological transitions of artificial intelligence and robotics in special education through their initial integrations in education, the digital revolution, the emergence of social robots, and the current advanced applications in special education.

Figure 1. Technological transitions of AI and robotics in special education

In this context, using robots in special education signals an integration of technology advancement and the progressive philosophy of inclusive education. The application of robotics provides new options and paths for students left behind in traditional educational settings (Chu et al., 2022).

This chapter addresses the depth of the role of artificial intelligence and robotics in special education by highlighting its potential and challenges. The following section explores the concept of personalized learning and emphasizes how artificial intelligence may be used to tailor educational content to individual learning styles and needs. It also focuses on data-driven insights and adaptive learning strategies by analyzing how artificial intelligence can be used to improve educational techniques. The following section examines the contribution of artificial intelligence to the advancement of communication enhancement and language development. The following section highlights how robots improve students' cognitive, sensory, and therapy.

The chapter further delves into how artificial intelligence-powered tools and robotics foster the development of social skills, emotional intelligence, and behavioral management. It examines how these technologies support social interaction training, emotional recognition, and behavioral modification. The following section examines how artificial intelligence and robotics create more equitable and inclusive learning environments. The final section addresses the ethical challenges of using artificial intelligence and robotics and presents the different measures that can be taken to protect student data. It also examines the potential advancements to provide educators, researchers, policymakers, stakeholders, and parents with a comprehensive understanding of artificial intelligence and robotics in special education.

ROLE OF AI IN PERSONALIZED LEARNING AND ADAPTIVE EDUCATION

In a conventional educational setting, a one-size-fits-all model is adopted with the assumption that all students will gain equally from a standardized approach. This approach operates under the belief that each student will benefit equally from similar pedagogical strategies regardless of their unique characteristics. However, this approach overlooks the diversity in learning environments where students come from varied backgrounds and have different experiences, abilities, and comprehension levels. Accordingly, this calls for an individualized approach that can address the range of learning styles and abilities found in the classroom. Adaptive learning acknowledges these variations and employs artificial intelligence to analyze students' performance, habits, and preferences (Rane et al., 2023). Central to this educational approach are adaptive learning technologies, platforms, and programs that work in tandem to create personalized educational experiences, as illustrated in Figure 2.

Figure 2. A conceptual mind map depicting the interconnected framework of adaptive learning

Adaptive learning technologies are highlighted as one of the six pivotal educational technology advancements poised to significantly impact higher education during the 2020s (Becker et al., 2017). Under the broad umbrella of adaptive technologies, adaptive learning platforms are stand-alone systems that combine all courseware functionalities into one sturdy working unit. In contrast, adaptive learning programs offer a component that can be plugged into an existing Learning Management Systems (LMS) course to deliver an adaptive and personalized experience (Taylor et al., 2021).

At the core of the adaptive learning process is the application of artificial intelligence algorithms that can sort through large datasets and analyze student data to facilitate the customization and adaptability of instructional material. Adaptive learning platforms employ this to customize learning programs to each student's specific requirements and preferences, along with comprehensive tools such as content management, assessment modules, and real-time feedback mechanisms.

By monitoring how students interact with learning content, the system identifies the student's strengths and weaknesses, preferred learning styles, and areas where more support might be helpful (Tedre et al., 2021; Kolachalama, 2022; Cardona et al., 2023). For instance, the system might prefer video-based content for a student who learns best visually, while it might give more text-centric resources to a student who prefers to learn best by text. This flexibility guarantees that students interact with the content in a way that suits their learning styles, eventually improving understanding and retention. Beyond content type, learning materials can be tailored based on student performance and preferences regarding difficulty levels, pacing, and even assessment format. With the help of AI algorithms, teachers can dynamically modify the difficulty of the problems they give their students, ensuring they are suitably challenged without feeling overburdened. Accordingly, specific curriculums or courses that utilize adaptive learning principles can be offered to provide a customized educational experience. In addition, lesson pacing can be adjusted to suit different learning styles, enabling students to advance at their ideal speed. This personalized approach moves away from the constraints of traditional learning methods, showcasing the ability of artificial intelligence to establish flexible learning settings, or what is known as Adaptive Learning (Kim, 2022; Alda et al., 2020).

Additionally, adaptive learning platforms employ artificial intelligence algorithms that can discern what appeals to each person and suggest additional resources, assignments, or pursuits that complement their educational path, which enhances participation and fosters a feeling of ownership and autonomy when it comes to learning. Figure 3 details the role of adaptive learning platforms, the application of artificial intelligence to cater to individual learning styles, and the advantages of such systems in special education.

Figure 3. A mind map outlining the detailed role of adaptive learning platforms and their benefits in special education scenarios

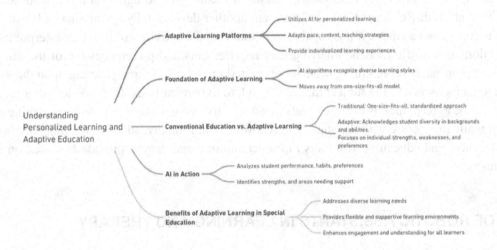

One notable example of an adaptive learning platform is Khan Academy, which leverages artificial intelligence to offer resources and tailored exercises in various subjects, from science to maths. With its adaptive platform, Khan Academy provides students with a personalized learning path by customizing performance-based questions. Another example is Smart Sparrow, which provides a flexible, adaptive learning environment for post-secondary education. Teachers can create interactive, personalized content that changes based on the progress of individual students with the help of this adaptive e-learning platform. This strategy guarantees that students receive focused support while increasing engagement. These systems use real-time data to adjust the pace, content, and how the content is delivered based on the learner's performance. This adaptability ensures that students progress at their own pace and provides additional support if the learner needs it, making education holistic and inclusive. Also, by utilizing AI-powered virtual reality (VR) and augmented reality (AR) technologies, students can experience lifelike and immersive simulations that surpass the limitations of conventional classrooms (Zhang & Wang, 2021; Joo & Jeong, 2020). All these experiences enhance subject comprehension, creativity, and problem-solving skills, which are crucial to success.

Adaptive learning platforms thus offer advantages that go beyond customized teaching by promoting a more inclusive learning environment. Additionally, adaptive learning programs encourage independent learning by letting students move on to more challenging material when ready and offering more help when needed. In these environments, computer systems today are not merely a platform where course content is delivered or taught but continuously identify the needs of every single learner, thus providing learning strategies in real-time.

Despite these developments, the concept of adaptive learning remains ambiguous in academic settings. According to Cavanagh et al. (2020), one obstacle to adopting adaptive learning in higher education is the absence of precise and uniform terminology across the educational field. Although efforts are made to define "what" adaptive learning is, it is difficult even to determine what should be included in the definition due to the wide variety of adaptive learning technologies. According to the 2019 Horizon Report (Alexander et al., 2019), adaptive learning happens when digital tools and systems design unique learning paths for each student according to their learning style, strengths, and weaknesses. Although many people agree with this definition, some people (Cavanagh et al., 2020; Pugliese, 2016) point out that the taxonomy is still fairly ambiguous, making it challenging to agree on a practical definition. Paramythis and Loidl-Reisinger (2004:182) present another definition by noting that "a learning environment is considered adaptive if it is capable of monitoring the activities of its users; interpreting these based on domain-specific models; inferring user requirements and preferences out of the interpreted activities, appropriately representing these in associated models; and, finally, acting upon the available knowledge on its users and the subject matter at hand, to dynamically facilitate the learning process."

What can be agreed upon is the role of advanced adaptive systems that can create a learning path on their own while interacting with the student in real-time. Additionally, adaptive learning environments provide teachers and educators with tools to use technology and data to provide feedback on student performance.

ROLE OF ROBOTIC ASSISTANTS IN LEARNING AND THERAPY

The reshaping of traditional pedagogical approaches to incorporate technology in enhancing the learning experience has extended beyond the mere digitalization of textbooks and incorporating multimedia resources to the creation of interactive, responsive environments. Alongside technologies that allow adaptive learning, technological advancement has opened new opportunities for therapy and support in therapeutic settings, particularly for individuals with special needs. The emergence of robotics in this domain marks a groundbreaking development, progressing from simple bots and other advanced artificial intelligence systems developed by OpenAI to fundamentally alter the delivery and reception of educational materials. Students are, therefore, no longer confined to passive reception of information; instead, they are active participants in a dynamic exchange facilitated by artificial intelligence and robotics.

Conventionally, within education, robots are designed and programmed to play either of the two roles: tutor or peer learner. As tutors, robots not only provide students individualized support and guidance, much like a human teacher would, but also provide focused feedback and tailor the pace of instruction by assessing the student's understanding. On the other hand, as peer learners, robots can engage with students in a reciprocal learning process, which makes the educational experience more interactive and engaging, leading to a deeper understanding and retention of material, as well as the development of empathy and social skills.

Table 1 presents an overview of several studies exploring the role of robotic assistants in learning and therapy, detailing the learning outcomes and theoretical foundations of each study. The studies demonstrate how robotic technology can be used for educational enhancement.

Table 1. Overview of studies on the role of robots in educational settings and their impact on learning outcomes

Study Reference	Role of Robot	Learning Outcome	Theoretical Foundation	Significant Findings
Vogt (2019)	Tutor	Second language learning	Interaction Hypothesis in Second Language Acquisition (SLA)	Children showed increased English comprehension, but overall learning was sluggish
Note: The robot and the young student were seated around a tablet computer featuring short stories in Dutch and English. The children learned various words, from nouns to mathematical concepts, from the robot, which also taught them grammar through play instead of formal instruction.				
Van den Berghe et al. (2019) Belpaeme & Tanaka (2021)	Tutor	Second language learning and fluency	Second Language Acquisition (SLA) Affective Filter Hypothesis (Krashen, 2009)	Positive effect on learning motivation. Robots' target language fluency and interaction also reduced foreign language anxiety.
Note: The robot can support the learner in learning a second language through tutoring and interaction in the target language.				
Tanaka & Kimura (2009) Tanaka & Matsuzoe (2012)	Peer Learner	Enhance learning through 'learning together' or 'through teaching.'	Teachable agent or "protégé" effect (Chase et al.,2009)	Children learn more effectively when they teach the robot
Note: This positive effect is more significant for weaker students. The weakest students are no longer the weakest in the classroom when a peer-like robot is present because the robot is weaker and depends more on instruction, raising the children's status.				
Alemi et al. (2014)	Robot Nima (teacher's sidekick)	Vocabulary acquisition and retention	Vygotsky's Sociocultural Theory (Lindblom & Ziemke, 2002)	The group demonstrated better vocabulary acquisition and retention
Note: The robot assists the 12- to 13-year-old all-female class in honing their English by providing comments on exercises or exemplifying how to correctly pronounce words and phrases in English. With the help of a robot assistant, the teacher instructed a class of thirty students in the official curriculum over five weeks.				
Yun et al. (2011).	Telepresence robot	Improve performance in English-speaking	Telepresence Learning (Wolff et al., 2023)	Students' performance was enhanced by the telepresence robot
Note: The Korea Institute of Science and Technology created the Engkey robot, which is controlled by a computer and faces a distant teacher. The robot's purpose is to aid in the teaching of English in primary schools.				
Bethel et al. (2011).	Socio-emotional support Robot	Open discussion about bullying	Human-Robot Interaction (Spitale et al., 2023)	Kids were more forthcoming about bullying incidents with robots
Note: The study revealed that kids were more forthcoming about bullying incidents at school than when reporting bullying via an anonymous form.				
Caruana et al. (2022).	Reading Companion	Encouraged participation in reading	Affective Computing (Gervasi et al., 2022)	Children preferred reading challenging books to robots rather than alone or with themselves.

continued on following page

Table 1. Continued

Study Reference	Role of Robot	Learning Outcome	Theoretical Foundation	Significant Findings
Note: All of the robots showed the ability to encourage children's participation by responding with nonspeech vocalizations, sounds, and movements (e.g., grunting, head shaking, and tail wagging), even though only one robot (NAO) could have social conversations. Most kids stated in in-depth interviews that the robots provided a pleasant, interesting, "calming," and accepting social context for reading.				
Lubold et al., (2018)	Learning Facilitator	Social rapport	Embodied Cognition (Hoffmann & Pfeifer, 2018)	Children engaged more and learned better when the robot engaged in social dialogue and adjusted to the child's pitch.

Note: Compared to robots that engaged in social dialogue without adaptation or no social dialogue, children experienced more significant learning and social rapport when robots engaged in dialogue and adapted their speech characteristics, showing that supportive and dialogical robots hold much promise for educational interventions.

In robotics, second language learning has been the most promising learning area for innovation. Unlike the immersive, conversational approach used to acquire a first language, a second language is learned through rote learning of grammatical rules and vocabulary lists, making learning a second language tedious. The leading cause of this striking disparity in learning style is largely due to the scarcity of resources. The instructor must use class-based instruction because they cannot communicate in the target language with individual students in the classroom. Furthermore, it is possible that the teacher lacks proficiency in certain areas of the target language or is insufficiently confident when speaking it. In such scenarios, robots have a significant advantage as they can support the learner in learning a second language through tutoring and actual interaction in the target language. In addition to teaching language, the robot can probably speak the target language more fluently than the teacher because contemporary computer voices are virtually indistinguishable from human voices.

Besides teaching, robots can also be used to provide socio-economic support for learners. Given the common perception of robots as impartial and judgmental, people frequently divulge information to them that they would be hesitant to divulge to others. Robots can thus be used to talk about personal matters and give guidance on how to handle issues. If the student grants permission, the robot can also be allowed to share limited data with support or instructional personnel. Furthermore, supportive robots can indirectly support learning by enhancing the child's socioemotional context to encourage engagement in (and lessen avoidance of) learning tasks. This might benefit kids struggling with learning, attention, or literacy. The studies mentioned above indicate that children's socioemotional state may very well be enhanced by the mere presence of "supportive" robots during challenging learning tasks.

Figure 4 illustrates this relationship where robotic assistants serve as intermediaries between students and learning materials, fulfilling roles extending beyond traditional teaching pedagogies by incorporating companionship and emotional support.

Figure 4. Interaction model between students and robotic assistants in educational settings

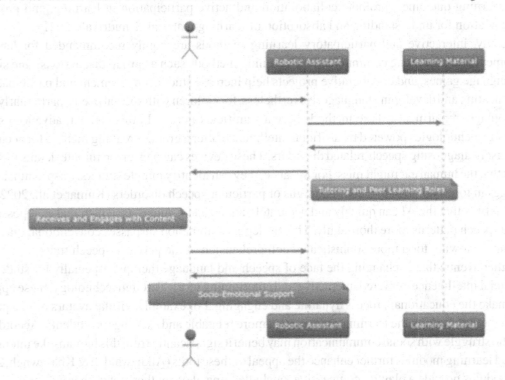

Numerous studies have examined these supportive roles in healthcare settings to increase treatment adherence for long-term conditions (Ferrante et al., 2021). Robotic speech characteristics can influence the environment for learning, especially when the robots can mimic or adjust to the speech characteristics of a human speaker (Kumar et al., 2010). However, autonomous robots must quickly and accurately detect, understand, and react to child speech when adult operators are absent to reach their full potential. However, most robots and speech recognition production systems currently cannot meet these needs.

So, while robotic assistants have shown immense promise in education, primarily in language acquisition and providing socio-emotional support, the future depends on developing autonomous robots capable of intuitive detection, comprehension, and response to learners' needs. With the current pace of technological advances, it can be rightly believed that the role of robotic assistants will transform the learning experience, particularly for students facing challenges in traditional learning environments.

COMMUNICATION ENHANCEMENT AND LANGUAGE DEVELOPMENT

Effective communication skills are crucial to a person's success, whether personal or professional (Darno & Mesiono, 2021). These skills incorporate an excellent understanding of the language used, mastery of various effective communication techniques, knowledge of grammar, a rich vocabulary, and the ability to speak clearly. Good communication skills can assist a person in building solid relationships with others, contribute to more efficient conflict resolution, and also strengthen one's leadership. In addition, good language development can also open doors to wider educational and career opportunities.

In an educational context, effective communication skills and good language development can improve student learning outcomes, promote collaboration and active participation in learning, and provide a solid foundation for understanding and absorption of learning material (Khudriyah, 2021).

Typically, interactive and participatory learning methods are highly recommended for language development and improving communication. Learning methods such as group discussions, role simulations, language games, and collaborative projects help increase student engagement and participation in communicating and developing language. Nevertheless, for children with special needs, particularly those with language impairments, these methods pose a significant hurdle. In this context, advanced speech recognition technologies powered by artificial intelligence offer a groundbreaking method for accurately and precisely diagnosing speech-related disorders. These devices can pick up on minute details in speech patterns that the human ear might miss. For instance, they can identify minute stutters, mispronunciations, or changes in tone and pitch that could be signs of particular speech disorders (Kumar et al., 2022). The main benefit is that the AI can quickly and accurately analyze large datasets, which makes it possible to evaluate speech patterns more thoroughly. This in-depth examination surpasses conventional diagnostic techniques, allowing for a more sophisticated comprehension of the person's speech traits.

Another avenue that is changing the face of speech and language therapy, especially for students, is the artificial intelligence-powered interactive and entertaining tools. Through technology, these apps and games make the educational process dynamic and engaging. For example, virtual avatars offer a personified interaction, making the learning environment more relatable and exciting for students. Accordingly, those who struggle with social communication may benefit significantly from this human-like interaction. Gamified learning modules further enhance the appeal of these tools (Alkhawaldeh & Khasawneh, 2024). These modules provide a playful, competitive, or challenging element that makes therapy sessions more appealing. Students are thus motivated to participate actively, cultivating a constructive and optimistic outlook on their language and speech development (Gallud et al., 2023).

Another significant technological advancement is the Augmentative and Alternative Communications Aids known as AAC aids. These artificial intelligence-enhanced AAC devices, made explicitly for non-verbal students or people with severe speech impairments, mark the beginning of a new era in personalized and adaptive communication support (Yang & Kristensson, 2023; Syriopoulou-Delli & Eleni, 2022). With AI integration, AAC devices can learn from and adjust to users' preferences and capabilities. These aids' machine learning algorithms examine how users communicate, enabling the device to adapt to meet their specific needs over time. This flexibility offers a dynamic and responsive communication solution, essential for people whose communication preferences or abilities change over time (Qian et al., 2022). For instance, in user profiling, artificial intelligence algorithms can generate user profiles by analyzing a user's communication patterns, preferences, and capabilities, which makes it possible for the AAC device to comprehend the person's distinct communication style, vocabulary, and commonly used expressions.

Another feature that lessens the effort needed for non-verbal people to express themselves is predictive text and phrases, where, as the user starts typing, the artificial intelligence can anticipate and recommend words or phrases based on the user profile it has learned. This predictive feature accelerates communication and increases efficiency (Shen et al., 2022; Norré, 2020).

Additionally, integrating artificial intelligence with other therapeutic domains, like occupational therapy or social skills training, amplifies the capacity to offer a more all-encompassing and integrative approach. This integration aims to address a wider range of needs and promote a comprehensive educational experience by utilizing the unique strengths of multiple therapeutic disciplines. Besides, artificial intelligence can also be employed in collaboration with occupational therapy to develop applications and tools explicitly targeting fine motor skill development. For instance, interactive games and simulations can be designed to improve hand-eye coordination, finger dexterity, and precision.

In the educational domain, artificial intelligence systems can analyze student performance data to pinpoint particular motor skill difficulties. Using this data, occupational therapists can create interventions specific to every student's needs, guaranteeing that each receives individualized and efficient support. Artificial intelligence-powered tools can also produce adaptive grammar exercises catering to individual students' skill levels. In order to create a dynamic and exciting learning environment, these exercises can include interactive tests, sentence construction challenges, and real-time feedback.

Figure 5. Sequence diagram depicting how AI-powered tools can address the educational needs of students with special needs to enhance language development

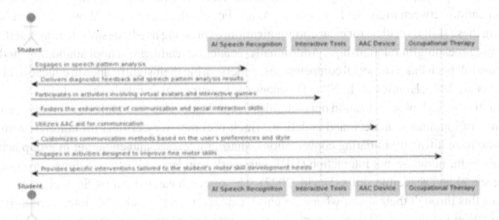

Artificial intelligence interventions thus cover more ground than standard therapeutic methods, as illustrated in Figure 5, improving grammar, expanding vocabulary, and fostering a more sophisticated understanding of language, among other aspects for students with special needs. AI systems also can compile information from various therapeutic approaches, giving a thorough picture of a student's development in all areas. Teachers and therapists can thus make well-informed decisions and adjustments to support a student's overall growth thanks to this holistic tracking. However, incorporating artificial intelligence in therapeutic contexts requires cooperation between education, psychology, and technology experts. Diverse viewpoints need to be considered to ensure the development and execution of comprehensive learning solutions through this interdisciplinary approach.

SOCIAL SKILLS, EMOTIONAL INTELLIGENCE, AND BEHAVIORAL MANAGEMENT

Compared to typical students, students with special educational needs such as attention-deficit disorder (ADHD), intellectual disabilities, learning disabilities, and autism spectrum disorders often encounter challenges that can hinder their social, emotional, and behavioral development. These challenges, ranging from trouble reading social cues and controlling emotions to managing challenging behaviors and navigating complex social interactions, can significantly impact their daily life, well-being, academic performance, ability to forge relationships, and capacity to adapt to their environment.

The development of artificial intelligence (AI) and robotics, in particular, has created new possibilities for assisting the development of children facing these obstacles. These technological advancements have the potential to address their specific needs, promoting empowerment and equality through creative solutions with customized approaches by exploring three essential domains, which are social interaction training, emotional identification, and behavioral modification.

Social Interaction Training

Social interaction training is a pivotal element in fostering societal participation, as it encompasses appropriate language, body language, understanding social norms, and empathy in the exchange of communication between individuals or within groups. For children, especially those with special needs, mastering these skills is critical for clear communication, expressing needs, desires, feelings, self-esteem, and forming meaningful relationships. Additionally, academic readiness, school adaptation, and performance have all been linked to social competencies, underscoring their significance in educational settings (Frogner et al., 2022; Hinkley et al., 2018; Hosokawa & Katsura, 2017; Takahashi et al., 2015; Ziv, 2013). Accordingly, the lack of socialization opportunities due to inadequate peer and adult interaction can lead to feelings of loneliness, anxiety, and isolation, which becomes very critical for children with special needs who have difficulty initiating conversations, sharing interests, and engaging in group activities.

In this light, human-robot interaction (HRI) can work wonders as it offers a unique avenue for enhancing social and academic skills (David, 2022). The research carried out by Scassellati et al. (2018) illustrates this through their study, where the child, caregiver, and social robot interacted during a 30-minute session per day for 30 days as part of a robot-assisted intervention, which allowed the child to interact and share experiences with the caregiver alongside the robot who observed social gaze behaviors throughout the sessions, such as maintaining eye contact and sharing attention. According to the findings, providing joint attention improved the caregiver and their child's social skill performance, including increased initiation of conversation and more frequent eye contact, highlighting the robot's pivotal role in reinforcing social skills.

Another more widely popular humanoid robot in autism therapy and with children with intellectual disabilities, cerebral palsy, and Down syndrome is the NAO robot. As a socially assistive robot, NAO stands out due to its accessibility and multifunctionality. The application of these robots also extends to children with physical disabilities and hearing impairments (Amirova et al., 2023; 2021). Similar to this is the QTrobot, which has been employed as an effective tool for Autism Spectrum Disorders (ASD) to aid the development of cognitive, social, communicative, and emotional skills under the guidance of therapists (Yi. et al., 2023; Kouroupa et al., 2022; Costescu et al., 2014). The QTrobot humanoid robot, equipped with a screen for facial expressions and an easily approachable expressive social appearance, takes on the role of a conversational partner to foster engagement and learning through interactive reading and imitation games (Puglisi et al., 2022). Figure 6 illustrates various other social robots used in special education.

Figure 6. The various social robots available for interactive and assistive technology in special education

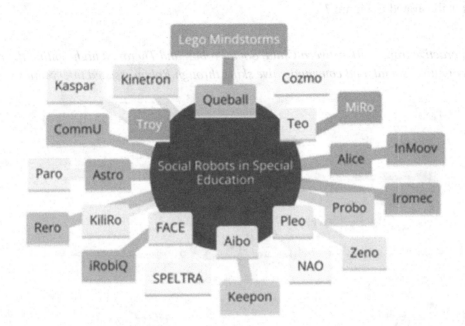

The use of social robots in special education demonstrates great potential; however, it faces significant limitations as existing research focuses on short-term impacts and initial reception of the robotic tools by children and therapists. Though this is necessary to establish acceptance of the new technology, there is a lack of comprehensive understanding of these interventions' long-term benefits and effectiveness. This opens avenues for longitudinal studies to explore how the skills are developed during these sessions and examine the lasting effects of how these skills can transfer to other aspects of life outside therapy.

Emotional Recognition

Emotional recognition is a fundamental skill, as recognizing and interpreting emotions from facial expressions and other non-verbal cues is an essential adaptive skill critical for social integration. It also plays a pivotal role in children's ability to understand, manage, and express emotions effectively (Albayrak et al., 2022). The inability of several children with special needs to identify and understand emotions leads to heightened levels of stress and anxiety, misinterpretation of facial cues, and difficulties in processing emotions (Uphoff, 2023; Pearcey et al., 2020; Corbett et al., 2009). A response to these challenges is the emergence of artificial intelligence and robotics to enhance emotional intelligence via emotional recognition.

In addition to the previously mentioned robots NAO and QTRobot, Kaspar, a humanoid child-sized robot that can mimic human gestures and facial expressions, was created as a social companion for children with special needs (Wood et al., 2019). Also, Keepon, a tiny robot, was created for easy, conversational, and nonverbal communication with children. The appearance and behavior of Keepon were explicitly designed to be as simple as possible for children to communicate effortlessly and comfortably. Research found that children could comprehend the social meaning of the robot's actions without getting bored or overwhelmed because of its straightforward manner (Kozima et al., 2009). A typical session with a social robot is illustrated in Figure 7.

Figure 7. Interactive framework between Child, Social robot, and Therapist highlighting the process of enhancing cognitive, social, and communicative skills through Robot-assisted interventions

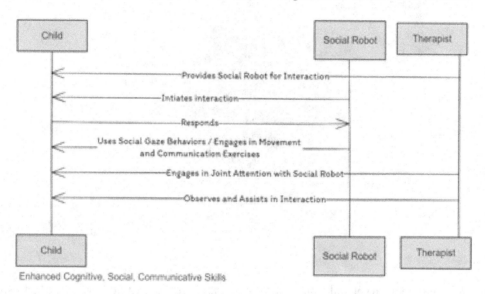

Recent years have witnessed the emergence of artificial intelligence algorithms to predict or identify the emotional states of children with special needs through the use of physiological signals like heart rate variability (HRV) and galvanic skin response (GSR) (Landowska et al., 2022). According to a recent study by Talaat et al. (2024), it may be possible to identify an autistic child's emotions in real-time, such as pain or anger, by using facial expression analysis and deep learning algorithms to discern and categorize diverse emotional states. Such technology could become invaluable tools for families and medical professionals to enhance the quality of life for those who frequently struggle with emotional expression.

Behavioral Modification

Behavioral modification is crucial in supporting children with special needs in managing challenging behaviors effectively. For instance, children with autism spectrum disorder may engage in repetitive behaviors such as hand flapping or rocking, which can interfere with their social and learning activities. In addition, children with attention-deficit/hyperactivity disorder (ADHD) may exhibit impulsivity and difficulty maintaining attention, causing disruptions in the classroom and at home. Furthermore, children with intellectual disabilities might find it challenging to understand and follow instructions, which can lead to frustration and violent behavior.

In these contexts, behavioral modification is required to assist children in learning alternative behaviors, managing their emotions, and adapting to social expectations more effectively. Accordingly, through targeted interventions and support, children with special needs can develop coping strategies and adaptive behaviors that improve their overall functioning and quality of life. Traditional behavioral modification methods involve reward structures, behavioral contracts, and structured interventions, which are lacking in addressing the complex requirements of the present diverse population. In light of this, artificial intelligence (AI) presents innovative methods to assist behavior control and encourage positive outcomes.

Social robots are excellent examples as they offer opportunities for social interaction and skill development by involving kids with special needs in structured activities, games, and conversations. Through these interactions, children practice and learn essential social skills like sharing, taking turns, and listening to instructions. Additionally, since disruptive behaviors frequently interfere with children's access to education, robots may be crucial in removing children's frustrations by helping them to express themselves more thoroughly (Minot, 2021).

Beyond interactive robots, predictive analytics driven by AI can assist teachers in spotting possible learning barriers and offering prompt assistance. A study by Ghafghazi et al. (2021) used the AI-Augmented Learning and Applied Behavior Analytics (AI-ABA) platform to assist therapists in diagnosing and giving feedback to patients by observing audio-visual cues and other physiological data. Furthermore, integrating AI-ABA platforms with multimodal information collection alongside reinforcement-based Virtual reality (VR) and augmented reality (AR) technologies can encourage self-regulatory behavior (Ayşe, 2024; da Silva et al., 2023).

Artificial intelligence and robots have thus emerged as powerful allies to empower children with special needs, offering opportunities to enhance social skills, emotional understanding, and behavior.

ADVANCING ACCESSIBILITY AND INCLUSIVE EDUCATION

The potential of artificial intelligence and robotics in education hinges on its ability to cater to a diverse range of student needs, particularly ensuring that students with different disabilities are not marginalized due to technological limitations or design oversights. This diversity includes different learning styles and speeds and those requiring specialized language and overall development tools for visual, auditory, cognitive, and motor disabilities. However, designing such artificial intelligence tools and ensuring the accessibility of robots can be challenging (Zhang et al., 2020).

The initial hurdle these technological tools face is understanding the spectrum of special needs. Since this spectrum covers a wide range of disabilities, including physical, cognitive, sensory, and emotional, each category presents unique challenges. For instance, students with visual impairments have different needs than those with auditory or mobility impairments. So, students with physical disabilities might struggle with manipulating robotic interfaces, while those with cognitive disabilities might find complex tasks overwhelming. Thus, an artificial intelligence tool effective for one group may be entirely ineffective for another group. Table 2 presents the challenges of making artificial intelligence and robotics accessible in special education.

Table 2. Comparative analysis of accessibility challenges in robotics and AI tools in special education

Aspect	Making Robotics Accessible in Special Education	Making AI Tools Accessible in Special Education
Catering to Different Special Needs	Effective for students with cognitive, physical, and sensory disabilities or autism. Offers structured, predictable interactions.	Beneficial for students with learning disabilities through personalized content. Offers aid to students with visual, auditory, and mobility impairments.
Primary Focus	Physical interaction, manipulation, and sensory feedback.	Data processing, cognitive interaction, and virtual engagement.
Customization Approach	Physical adaptability (modified controls, tactile interfaces).	Software adaptability (personalized learning algorithms, adjustable content complexity).
Technology Interface	Tactile and physical interfaces often require physical manipulation or presence.	Screen-based or voice-activated interfaces require less physical interaction.
Integration with Educational Content	Direct physical or interactive activities (robot-assisted tasks, hands-on experiments).	Indirect or virtual interaction (AI-driven content recommendations, virtual simulations).
User Interaction Style	It often involves direct, tangible interaction with the robot, including movement and touch.	Interactions primarily through digital mediums such as computer interfaces, often without physical contact.
Sensory Engagement	The focus is on physical and kinesthetic learning styles, which include movement and touch.	Often reliant on visual and auditory information, requiring adaptations for sensory impairments.
Training for Educators	Training is required to operate and maintain physical robotics hardware and integrate it into teaching.	Focuses on training in software, data interpretation, and integrating AI tools into curriculum development.
Development and Production Costs	High due to physical components, manufacturing, and hardware development.	It is low since costs are associated mainly with software development, data storage, and computational resources.
Ongoing Maintenance	Involves hardware maintenance, updates, and potential physical repairs.	Primarily software updates, digital troubleshooting, and managing data-driven systems.
Ethical and Safety Concerns	Includes managing robot movements to avoid injury and appropriate robot-student interaction.	It involves data privacy, algorithmic bias, and ensuring AI's cognitive influence is beneficial and ethical.

In addition to these aspects, there are also social and cultural barriers to overcome, particularly in the case of incorporating robotics into special education since it can be viewed as a radical departure from traditional teaching methods. Addressing all these challenges requires considerable efforts from educators, therapists, stakeholders, and policymakers to ensure that students can use the full potential of artificial intelligence tools and robotics.

ETHICAL CONSIDERATIONS AND FUTURE OUTLOOK IN AI AND ROBOTICS

There are several ethical considerations to consider when implementing artificial intelligence technology. One of them is related to data security and privacy. The use of AI often involves collecting and analyzing user data, so it is important to ensure that the data remains secure and user privacy is maintained (Sidiq, 2023; Pappachan et al., 2023; Wang, 2021). Applying ethics in using AI is essential to ensure that this technology provides the most significant benefit to society without violating moral and ethical principles (Panesar & Panesar, 2021).

When using artificial intelligence tools in education, protecting students' data and privacy is paramount. This is particularly necessary in cases where student data like learning patterns, behavioral traits, and academic performance are constantly collected and analyzed to provide personalized learning, predictive analytics, and real-time feedback. Though the ultimate purpose of the data collection is to improve educational outcomes, it becomes a target for potential security breaches. There is also the moral obligation to ensure that the collected data is used in ways that respect the privacy and dignity of the students (Akgun & Greenhow, 2022). In the case of students with special needs, this is all the more important as they may lack the capacity to understand or consent fully to the extent of data collection. The sensitive nature of the data, including information about the student's physical, cognitive, and emotional abilities, also poses significant privacy concerns. There is, thus, the need for robust measures to safeguard student data, which are presented in Figure 8.

Figure 8. Diagram illustrating the different measures that can be taken to protect student data in the educational context

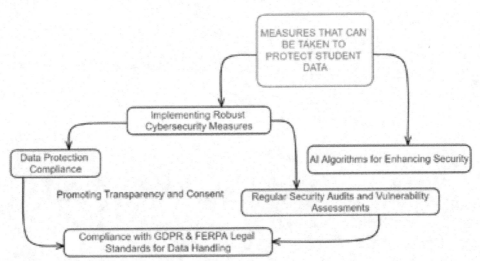

Note: GDPR: General Data Protection Regulation, FERPA: Family Educational Rights and Privacy Act

Adopting the measures illustrated in Figure 7 will foster an inclusive environment that values data privacy and security. In these ethical challenges, collaboration between different parties is critical. Open dialogue and cross-disciplinary discussions must be improved to ensure that the development and implementation of AI in education is technically adequate and considers relevant ethical, social, and environmental aspects (Zhou & Chen, 2023). Thus, through a comprehensive and sustainable approach, the use of AI in education can provide optimal benefits while still paying attention to ethical values important for inclusive and sustainable education development (Celik, 2023; Wu et al., 2018). Ensuring such a culture also promises a positive future by incorporating artificial intelligence and robotics into special education. Additionally, contrary to the notion that teachers will be replaced by artificial intelligence and robots, the future of special education lies in integrating the strength of human educators (expertise, empathy, and human connection) and technological advancements. Also, by overcoming hurdles related to teacher training and the high cost of advanced robotic systems that worsen the digital divide between students with easy access to these tools and those without, educational robotics can emerge as an instrumental tool in classroom settings.

In the ongoing 5IR, Large Language Models (LLMs) are poised to enhance conversational interactions with educational robots significantly. Additionally, integrating the Internet of Things (IoT) in special education promises to create personalized and adaptable learning environments to improve accessibility and engagement for students with special needs (Song et al., 2020). IoT devices are advantageous in improving communication, safety, and behavioural management by enabling real-time monitoring and feedback and providing educators with useful data-driven insights.

The future outlook of artificial intelligence and robotics in special education is thus highly promising, considering the ongoing research and development to enrich the learning experiences and life opportunities for students with special needs.

REFERENCES

Akgun, S., & Greenhow, C. (2022). Artificial intelligence in education: Addressing ethical challenges in K-12 settings. *AI and Ethics*, 2(3), 431–440. DOI: 10.1007/s43681-021-00096-7 PMID: 34790956

Albayrak, Z. S., Kadak, M. T., Akkin Gurbuz, H. G., & Dogangun, B. (2022, November 22). Emotion Recognition Skill in Specific Learning Disorder and Attention-Deficit Hyperactivity Disorder. *Alpha Psychiatry*, 23(6), 268–273. DOI: 10.5152/alphapsychiatry.2022.22219 PMID: 36628377

Alemi, M., Meghdari, A., & Ghazisaedy, M. (2014). Employing humanoid robots for teaching English language in Iranian junior high-schools. *International Journal of HR; Humanoid Robotics*, 11(03), 1450022. DOI: 10.1142/S0219843614500224

Alexander, B., Ashford-Rowe, K., Barajas-Murph, N., Dobbin, G., Knott, J., McCormack, M., & Weber, N. (2019). *Horizon report 2019 higher education edition* (pp. 3–41). EDU19.

Alkhawaldeh, M., & Khasawneh, M. (2024). Designing gamified assistive apps: A novel approach to motivating and supporting students with learning disabilities. *International Journal of Data and Network Science*, 8(1), 53–60. DOI: 10.5267/j.ijdns.2023.10.018

Amirova, A., Rakhymbayeva, N., Yadollahi, E., Sandygulova, A., & Johal, W. (2021). 10 Years of Human-NAO Interaction Research: A Scoping Review. *Frontiers in Robotics and AI*, 8, 744526. DOI: 10.3389/frobt.2021.744526 PMID: 34869613

Amirova, A., Rakhymbayeva, N., Zhanatkyzy, A., Telisheva, Z., & Sandygulova, A. (2023). Effects of parental involvement in robot-assisted autism therapy. *Journal of Autism and Developmental Disorders*, 53(1), 438–455. DOI: 10.1007/s10803-022-05429-x PMID: 35088233

Ayşe, T. U. N. A. (2024). Approaches and Strategies in Applied Behavior Analysis for Children with Autism Spectrum Disorder. *Psikiyatride Güncel Yaklasimlar*, 16(2), 347–357. DOI: 10.18863/pgy.1315911

Becker, S. A., Cummins, M., Davis, A., Freeman, A., Hall, C. G., & Ananthanarayanan, V. (2017). *NMC horizon report: 2017 higher education edition*. The New Media Consortium.

Belpaeme, T., & Tanaka, F. (2021). Social robots as educators. In *OECD Digital Education Outlook 2021 Pushing the Frontiers with Artificial Intelligence, Blockchain and Robots: Pushing the Frontiers with Artificial Intelligence, Blockchain and Robots* (p. 143). OECD Publishing. DOI: 10.1787/1c3b1d56-en

Bethel, C. L., Stevenson, M. R., & Scassellati, B. (2011, October). Secret-sharing: Interactions between a child, robot, and adult. In *2011 IEEE International Conference on systems, man, and cybernetics* (pp. 2489–2494). IEEE. DOI: 10.1109/ICSMC.2011.6084051

Cardona, T., Cudney, E. A., Hoerl, R., & Snyder, J. (2023). Data mining and machine learning retention models in higher education. *Journal of College Student Retention*, 25(1), 51–75. DOI: 10.1177/1521025120964920

Caruana, N., Moffat, R., Miguel-Blanco, A., & Cross, E. S. (2023). Perceptions of intelligence & sentience shape children's interactions with robot reading companions. *Scientific Reports*, 13(1), 7341. DOI: 10.1038/s41598-023-32104-7 PMID: 37147422

Cavanagh, T., Chen, B., Lahcen, R. A. M., & Paradiso, J. R. (2020). Constructing a design framework and pedagogical approach for adaptive learning in higher education: A practitioner's perspective. *International review of research in open and distributed learning, 21*(1), 173–197.

Celik, I. (2023). Towards Intelligent-TPACK: An empirical study on teachers' professional knowledge to ethically integrate artificial intelligence (AI)-based tools into education. *Computers in Human Behavior,* 138, 107468. DOI: 10.1016/j.chb.2022.107468

Chase, C. C., Chin, D. B., Oppezzo, M. A., & Schwartz, D. L. (2009). Teachable agents and the protégé effect: Increasing the effort towards learning. *Journal of Science Education and Technology,* 18(4), 334–352. DOI: 10.1007/s10956-009-9180-4

Chu, S. T., Hwang, G. J., & Tu, Y. F. (2022). Artificial intelligence-based robots in education: A systematic review of selected SSCI publications. *Computers and education. Artificial Intelligence,* ●●●, 100091.

Corbett, B. A., Carmean, V., Ravizza, S., Wendelken, C., Henry, M. L., Carter, C., & Rivera, S. M. (2009, September). A functional and structural study of emotion and face processing in children with autism. *Psychiatry Research: Neuroimaging,* 173(3), 196–205. DOI: 10.1016/j.pscychresns.2008.08.005 PMID: 19665877

da Silva, A. P., Bezerra, I. M. P., Antunes, T. P. C., Cavalcanti, M. P. E., & de Abreu, L. C. (2023). Applied behavioral analysis for the skill performance of children with autism spectrum disorder. *Frontiers in Psychiatry,* 14, 1093252. DOI: 10.3389/fpsyt.2023.1093252 PMID: 37181882

Darno, D., & Mesiono, M. (2021). Strategi Komunikasi Kepala Sekolah Dalam Meningkatkan Efektivitas Manajemen Sekolah Di Mtsn 3 Langkat. *PIONIR: JURNAL PENDIDIKAN, 10*(3).

David, D., Thérouanne, P., & Milhabet, I. (2022). The acceptability of social robots: A scoping review of the recent literature. *Computers in Human Behavior,* 137, 107419. DOI: 10.1016/j.chb.2022.107419

Fiestas Lopez Guido, J. C., Kim, J. W., Popkowski Leszczyc, P. T., Pontes, N., & Tuzovic, S. (2024). Retail robots as sales assistants: How speciesism moderates the effect of robot intelligence on customer perceptions and behaviour. *Journal of Service Theory and Practice,* 34(1), 127–154. DOI: 10.1108/JSTP-04-2023-0123

Frogner, L., Hellfeldt, K., Ångström, A. K., Andershed, A. K., Källström, Å., Fanti, K. A., & Andershed, H. (2022). Stability and change in early social skills development in relation to early school performance: A longitudinal study of a Swedish cohort. *Early Education and Development,* 33(1), 17–37. DOI: 10.1080/10409289.2020.1857989

Gallud, J. A., Carreño, M., Tesoriero, R., Sandoval, A., Lozano, M. D., Durán, I., Penichet, V. M. R., & Cosio, R. (2023). Technology-enhanced and game based learning for children with special needs: A systematic mapping study. *Universal Access in the Information Society,* 22(1), 227–240. DOI: 10.1007/s10209-021-00824-0 PMID: 34248457

Gauri, P., & Van Eerden, J. (2019). *What the Fifth Industrial Revolution is and why it matters.* Europeansting.com.

Gervasi, R., Barravecchia, F., Mastrogiacomo, L., & Franceschini, F. (2023). Applications of affective computing in human-robot interaction: State-of-art and challenges for manufacturing. *Proceedings of the Institution of Mechanical Engineers. Part B, Journal of Engineering Manufacture*, 237(6-7), 815–832. DOI: 10.1177/09544054221121888

Ghafghazi, S., Carnett, A., Neely, L., Das, A., & Rad, P. (2021, October). AI-Augmented Behavior Analysis for Children With Developmental Disabilities: Building Toward Precision Treatment. *IEEE Systems, Man, and Cybernetics Magazine*, 7(4), 4–12. DOI: 10.1109/MSMC.2021.3086989

Hinkley, T., Brown, H., Carson, V., & Teychenne, M. (2018). Cross sectional associations of screen time and outdoor play with social skills in preschool children. *PLoS One*, 13(4), e0193700. DOI: 10.1371/journal.pone.0193700 PMID: 29617366

Hoffmann, M., & Pfeifer, R. (2018). Robots as powerful allies for the study of embodied cognition from the bottom up. *arXiv preprint arXiv:1801.04819*.

Hosokawa, R., & Katsura, T. (2017). A longitudinal study of socio-economic status, family processes, and child adjustment from preschool until early elementary school: The role of social competence. *Child and Adolescent Psychiatry and Mental Health*, 11(1), 1–28. DOI: 10.1186/s13034-017-0206-z

Joo, H. J., & Jeong, H. Y. (2020). A study on eye-tracking-based Interface for VR/AR education platform. *Multimedia Tools and Applications*, 79(23-24), 16719–16730. DOI: 10.1007/s11042-019-08327-0

Kim, J. (2022). The Interconnectivity of Heutagogy and Education 4.0 in Higher Online Education. *Canadian Journal of Learning and Technology*, 48(4), 1–17. DOI: 10.21432/cjlt28257

Kolachalama, V. B. (2022). Machine learning and pre-medical education. *Artificial Intelligence in Medicine*, 129, 102313. DOI: 10.1016/j.artmed.2022.102313 PMID: 35659392

Kouroupa, A., Laws, K. R., Irvine, K., Mengoni, S. E., Baird, A., & Sharma, S. (2022). The use of social robots with children and young people on the autism spectrum: A systematic review and meta-analysis. *PLoS One*, 17(6), e0269800. DOI: 10.1371/journal.pone.0269800 PMID: 35731805

Kozima, H., Michalowski, M. P., & Nakagawa, C. (2009). Keepon: A playful robot for research, therapy, and entertainment. *International Journal of Social Robotics*, 1(1), 3–18. DOI: 10.1007/s12369-008-0009-8

Krashen, S. (2009). *Principles and practice in second language acquisition* (Internet edition). Oxford, UK: Pergamon. https://sdkrashen.com/content/books/principles_and_practice.pdf

Kumar, R., Ai, H., Beuth, J. L., & Rosé, C. P. (2010). Socially capable conversational tutors can be effective in collaborative learning situations. In *Intelligent Tutoring Systems: 10th International Conference*, (pp. 156–164). Springer.

Kumar, T., Mahrishi, M., & Meena, G. (2022). A comprehensive review of recent automatic speech summarization and keyword identification techniques. *Artificial Intelligence in Industrial Applications: Approaches to Solve the Intrinsic Industrial Optimization Problems*, 111-126.

Landowska, A., Karpus, A., Zawadzka, T., Robins, B., Erol Barkana, D., Kose, H., Zorcec, T., & Cummins, N. (2022). Automatic emotion recognition in children with autism: A systematic literature review. *Sensors (Basel)*, 22(4), 1649. DOI: 10.3390/s22041649 PMID: 35214551

Lindblom J, Ziemke T. (2002). Social Situatedness of Natural and Artificial Intelligence: Vygotsky and Beyond. *Adaptive Behavior, 11*(2), 79-96.

Lubold, N., Walker, E., Pon-Barry, H., & Ogan, A. (2018). Automated pitch convergence improves learning in a social, teachable robot for middle school mathematics. In *Artificial Intelligence in Education: 19th International Conference.* Springer.

Minot, D. (2021). *Robot-Assisted Instruction for Children with Autism: How Can Robots Be Used in Special Education?* Autism Spectrum News.

Neumann, M. M. (2023). Bringing Social Robots to Preschool: Transformation or Disruption? *Childhood Education*, 99(4), 62–65. DOI: 10.1080/00094056.2023.2232283

Noble, S. M., Mende, M., Grewal, D., & Parasuraman, A. (2022). The Fifth Industrial Revolution: How harmonious human–machine collaboration is triggering a retail and service [r] evolution. *Journal of Retailing*, 98(2), 199–208. DOI: 10.1016/j.jretai.2022.04.003

Norré, M. (2020). Evaluation of a Word Prediction System in an Augmentative and Alternative Communication for Disabled People. *Journal*, 81(1-4), 49–54. http://iieta. org/journals/mmc_c. DOI: 10.18280/mmc_c.811-409

Panesar, A., & Panesar, A. (2021). Machine learning and AI ethics. *Machine Learning and AI for Healthcare: Big Data for Improved Health Outcomes*, 207-247.

Papakostas, G. A., Sidiropoulos, G. K., Papadopoulou, C. I., Vrochidou, E., Kaburlasos, V. G., Papadopoulou, M. T., Holeva, V., Nikopoulou, V.-A., & Dalivigkas, N. (2021). Social robots in special education: A systematic review. *Electronics (Basel)*, 10(12), 1398. DOI: 10.3390/electronics10121398

Pappachan, P. Sreerakuvandana, & Rahaman, M. (2024). Conceptualizing the Role of Intellectual Property and Ethical Behaviour in Artificial Intelligence. In B. Gupta & F. Colace (Eds.), *Handbook of Research on AI and ML for Intelligent Machines and Systems* (pp. 1-26). IGI Global. https://doi.org/ DOI: 10.4018/978-1-6684-9999-3.ch001

Paramythis, A., & Loidl-Reisinger, S. (2003). Adaptive learning environments and e-learning standards. In *Second European conference on e-learning* (Vol. 1, No. 2003, pp. 369–379).

Pearcey, S., Gordon, K., Chakrabarti, B., Dodd, H., Halldorsson, B., & Creswell, C. (2021). Research Review: The relationship between social anxiety and social cognition in children and adolescents: a systematic review and meta-analysis. *Journal of Child Psychology and Psychiatry, and Allied Disciplines*, 62(7), 805–821. DOI: 10.1111/jcpp.13310 PMID: 32783234

Pearcey, S., Gordon, K., Chakrabarti, B., Dodd, H., Halldorsson, B., & Creswell, C. (2021). Research Review: The relationship between social anxiety and social cognition in children and adolescents: a systematic review and meta-analysis. *Journal of Child Psychology and Psychiatry, and Allied Disciplines*, 62(7), 805–821. DOI: 10.1111/jcpp.13310 PMID: 32783234

Pugliese, L. (2016). Adaptive learning systems: Surviving the storm. *EDUCAUSE Review*, 10(7).

Puglisi, A., Caprì, T., Pignolo, L., Gismondo, S., Chilà, P., Minutoli, R., Marino, F., Failla, C., Arnao, A. A., Tartarisco, G., Cerasa, A., & Pioggia, G. (2022, June 25). Social Humanoid Robots for Children with Autism Spectrum Disorders: A Review of Modalities, Indications, and Pitfalls. *Children (Basel, Switzerland)*, 9(7), 953. DOI: 10.3390/children9070953 PMID: 35883937

Qian, R., Sengan, S., & Juneja, S. (2022). English language teaching based on big data analytics in augmentative and alternative communication system. *International Journal of Speech Technology*, 25(2), 409–420. DOI: 10.1007/s10772-022-09960-1

Rane, N., Choudhary, S., & Rane, J. (2023). Education 4.0 and 5.0: Integrating Artificial Intelligence (AI) for personalized and adaptive learning. *SSRN* 4638365. DOI: 10.2139/ssrn.4638365

Scassellati, B., Boccanfuso, L., Huang, C. M., Mademtzi, M., Qin, M., Salomons, N., Ventola, P., & Shic, F. (2018, August 22). Improving social skills in children with ASD using a long-term, in-home social robot. *Science Robotics*, 3(21), eaat7544. DOI: 10.1126/scirobotics.aat7544 PMID: 33141724

Shen, J., Yang, B., Dudley, J. J., & Kristensson, P. O. (2022, March). Kwickchat: A multi-turn dialogue system for aac using context-aware sentence generation by bag-of-keywords. In *27th International Conference on Intelligent User Interfaces* (pp. 853-867). DOI: 10.1145/3490099.3511145

Sidiq, M. S. (2023, January 23). *The Ethics of Machine Learning: Understanding the Role of Developers and Designers*. HackerNoon. https://hackernoon.com/the-ethics-of-machine-learning-understanding-the-role-of-developers-and-designers

Song, H., Bai, J., Yi, Y., Wu, J., & Liu, L. (2020). Artificial intelligence enabled Internet of Things: Network architecture and spectrum access. *IEEE Computational Intelligence Magazine*, 15(1), 44–51. DOI: 10.1109/MCI.2019.2954643

Spitale, M., Axelsson, M., Kara, N., & Gunes, H. (2023, August). Longitudinal evolution of coachees' behavioural responses to interaction ruptures in robotic positive psychology coaching. In *2023 32nd IEEE International Conference on Robot and Human Interactive Communication (RO-MAN)* (pp. 315-322). IEEE.

Syriopoulou-Delli, C. K., & Eleni, G. (2022). Effectiveness of different types of Augmentative and Alternative Communication (AAC) in improving communication skills and in enhancing the vocabulary of children with ASD: A review. *Review Journal of Autism and Developmental Disorders*, 9(4), 493–506. DOI: 10.1007/s40489-021-00269-4

Takahashi, Y., Okada, K., Hoshino, T., & Anme, T. (2015, August 12). Developmental Trajectories of Social Skills during Early Childhood and Links to Parenting Practices in a Japanese Sample. *PLoS One*, 10(8), e0135357. DOI: 10.1371/journal.pone.0135357 PMID: 26267439

Talaat, F. M., Ali, Z. H., Mostafa, R. R., & El-Rashidy, N. (2024, January 4). Real-time facial emotion recognition model based on kernel autoencoder and convolutional neural network for autism children. *Soft Computing*, 28(9-10), 6695–6708. DOI: 10.1007/s00500-023-09477-y

Tanaka, F., & Matsuzoe, S. (2012). Children teach a care-receiving robot to promote their learning: Field experiments in a classroom for vocabulary learning. *Journal of Human-Robot Interaction*, 1(1), 78–95. DOI: 10.5898/JHRI.1.1.Tanaka

Taylor, D. L., Yeung, M., & Bashet, A. Z. (2021). *Personalized and adaptive learning. Innovative Learning Environments in STEM Higher Education: Opportunities*. Challenges, and Looking Forward.

Tedre, M., Toivonen, T., Kahila, J., Vartiainen, H., Valtonen, T., Jormanainen, I., & Pears, A. (2021). Teaching machine learning in K–12 classroom: Pedagogical and technological trajectories for artificial intelligence education. *IEEE Access : Practical Innovations, Open Solutions*, 9, 110558–110572. DOI: 10.1109/ACCESS.2021.3097962

Uphoff, M. K. (2023). *Social emotional learning and its needs and benefits for students with down syndrome, autism spectrum disorder, and emotional behavioral disorder* [Master's thesis, Bethel University]. Spark Repository.

Van den Berghe, R., Verhagen, J., Oudgenoeg-Paz, O., Van der Ven, S., & Leseman, P. (2019). Social robots for language learning: A review. *Review of Educational Research*, 89(2), 259–295. DOI: 10.3102/0034654318821286

Vogt, P., van den Berghe, R., De Haas, M., Hoffman, L., Kanero, J., Mamus, E., . . . Pandey, A. K. (2019, March). Second language tutoring using social robots: a large-scale study. In *2019 14th ACM/IEEE International Conference on Human-Robot Interaction (HRI)* (pp. 497-505). Ieee. DOI: 10.1109/HRI.2019.8673077

Wang, Y. (2021). When artificial intelligence meets educational leaders' data-informed decision-making: A cautionary tale. *Studies in Educational Evaluation*, 69, 100872. DOI: 10.1016/j.stueduc.2020.100872

Wehmeyer, M. L. (2022). From segregation to strengths: A personal history of special education. *Phi Delta Kappan*, 103(6), 8–13. DOI: 10.1177/00317217221082792

Wolff, F., Wickord, L. C., Rahe, M., & Quaiser-Pohl, C. M. (2023). Effects of an intercultural seminar using telepresence robots on students' cultural intelligence. *Computers & Education: X Reality*, 2, 100007.

Wood, L. J., Robins, B., Lakatos, G., Syrdal, D. S., Zaraki, A., & Dautenhahn, K. (2019, March 1). Developing a protocol and experimental setup for using a humanoid robot to assist children with autism to develop visual perspective taking skills. *Paladyn : Journal of Behavioral Robotics*, 10(1), 167–179. DOI: 10.1515/pjbr-2019-0013

Wu, J., Guo, S., Huang, H., Liu, W., & Xiang, Y. (2018). Information and communications technologies for sustainable development goals: State-of-the-art, needs and perspectives. *IEEE Communications Surveys and Tutorials*, 20(3), 2389–2406. DOI: 10.1109/COMST.2018.2812301

Yang, B., & Kristensson, P. O. (2023, September). Designing, Developing, and Evaluating AI-driven Text Entry Systems for Augmentative and Alternative Communication Users and Researchers. In *Proceedings of the 25th International Conference on Mobile Human-Computer Interaction* (pp. 1-4). ACM. DOI: 10.1145/3565066.3609738

Yi, H., Liu, T., & Lan, G. (2024). The key artificial intelligence technologies in early childhood education: A review. *Artificial Intelligence Review*, 57(1), 12. DOI: 10.1007/s10462-023-10637-7

Yun, S., Shin, J., Kim, D., Kim, C. G., Kim, M., & Choi, M. T. (2011). Engkey: Tele-education robot. In *Social Robotics: Third International Conference*. Springer.

Zhang, W., & Wang, Z. (2021). Theory and practice of VR/AR in K-12 science education—A systematic review. *Sustainability (Basel)*, 13(22), 12646. DOI: 10.3390/su132212646

Zhang, X., Tlili, A., Nascimbeni, F., Burgos, D., Huang, R., Chang, T. W., Jemni, M., & Khribi, M. K. (2020). Accessibility within open educational resources and practices for disabled learners: A systematic literature review. *Smart Learning Environments*, 7(1), 1–19. DOI: 10.1186/s40561-019-0113-2

Zhou, J., & Chen, F. (2023). AI ethics: From principles to practice. *AI & Society*, 38(6), 2693–2703. DOI: 10.1007/s00146-022-01602-z

Ziv, Y. (2013, February). Social information processing patterns, social skills, and school readiness in preschool children. *Journal of Experimental Child Psychology*, 114(2), 306–320. DOI: 10.1016/j.jecp.2012.08.009 PMID: 23046690

Chapter 11
Drones and Unmanned Aerial Vehicles Automation Using Reinforcement Learning

Jaskirat Kaur

(iD) https://orcid.org/0009-0001-1665-6763

Chandigarh College of Engineering and Technology, Chandigarh, India

Sudhakar Kumar

(iD) https://orcid.org/0000-0001-7928-4234

Chandigarh College of Engineering and Technology, Chandigarh, India

Sunil K. Singh

(iD) https://orcid.org/0000-0003-4876-7190

Chandigarh College of Engineering and Technology, Chandigarh, India

Ruchika Thakur

Chandigarh College of Engineering and Technology, Chandigarh, India

Shavi Bansal

Insights2Techinfo, India

Varsha Arya

Asia University, Taiwan, & Hong Kong Metropolitan University, Hong Kong

ABSTRACT

This chapter provides an in-depth study at the latest developments in deep reinforcement learning (DRL) as applied to drones and unmanned aerial vehicles (UAVs), with a focus on safety standards and overcoming challenges like object detection and cybersecurity. Real-world instances across sectors demonstrate how DRL significantly improves efficiency and adaptability in autonomous systems. By contrasting various DRL techniques, the chapter highlights their effectiveness and potential to advance drone and UAV capabilities. The practical implications and advantages of DRL applications are emphasized, showcasing their transformative influence on industries. Moreover, the chapter explores future research paths to stimulate innovation and enhance self-piloting vehicle technology. In essence, it provides a thorough overview of DRL's role in drone and UAV operations, underscoring the importance of safety standards, DRL's capacity to boost efficiency and safety in autonomous systems, and its contribution to the evolving autonomous technology landscape.

DOI: 10.4018/979-8-3693-2707-4.ch011

INTRODUCTION

The seamless integration of Unmanned Aerial Vehicles (UAVs) across diverse industries has ignited a wave of unprecedented advancements in autonomous robotics, fundamentally reshaping the landscape of industrial automation (Soori et al., 2023). This comprehensive introduction serves as a launching pad for a meticulous exploration into the intricate domain of UAV and drone technologies, intricately intertwined with edge computing and artificial intelligence (AI). Our journey commences with a detailed examination of UAVs, looking closely at their varied hardware and software setups, detailed performance features, and complex communication systems (Dwivedi et al., 2023). Moreover, the exploration includes a dedicated examination of the role of Deep Reinforcement Learning (DRL) within the context of autonomous vehicles, elucidating its significance in optimising decision-making processes and enhancing the overall performance of manufacturing UAVs (Shuford, 2024).

In our exploration of UAV applications, we uncover their diverse uses in engineering geology, precision agriculture, and future transportation (Giordan et al., 2020). Autonomous robotics' evolution and UAVs' transformative potential are revealed, addressing challenges and promising prospects (Mohsan et al., 2023). Synthesising insights from various sources, we aim for a comprehensive narrative to drive innovation in autonomous vehicles, especially UAVs, across industries (Muslimov, 2023). Our goal is to unlock their full potential and propel them to new heights in various sectors. All the acronyms used in this chapter are shown Table 3.

Automation in UAVs and Drones

The historical development of industrial drones spans decades, evolving from military origins to diverse industrial applications. Initially focused on reconnaissance and surveillance, drones saw significant advancements in miniaturisation, sensors, and autonomy in the late 20th and early 21st centuries. Milestones include the introduction of commercial drones, advanced sensor integration, and improved control algorithms. Today, industrial drones are integral to sectors like agriculture, construction, and logistics, enhancing efficiency and safety and paving the way towards achieving full automation in UAV operations.

The integration of automation in Unmanned Aerial Vehicles (UAVs) follows distinct levels delineated by the Society of Automotive Engineers (SAE) International, progressing from Manual Control (Level 0) to Full Automation (Level 5). This advancement enhances operational efficiency by automating tasks traditionally performed manually, allowing for scalable deployment of UAV fleets across various applications like surveillance and delivery services. Automation reduces reliance on human pilots, thereby improving safety through mitigating human error and ensuring reliable operation even in challenging environments. Automated UAV systems comprise essential components such as sensors for real-time data, actuators for controlling movement, onboard processors as cognitive centres, communication systems for data exchange, and power systems for energy provision, as elaborated further in subsequent sections (Mengi et al., 2023).

Architecture and Components of UAVs

Figure 1. Architecture and components of UAVs

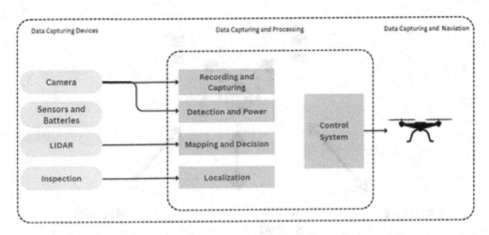

Unmanned Aerial Vehicles (UAVs), commonly referred to as drones, come in various sizes and serve different purposes, categorized as Macro Aerial Vehicles, Micro Aerial Vehicles, and Nano Aerial Vehicles. Studies have addressed challenges like flight time for quad-copters, essential for tasks such as path planning and inspection (Wen et al., 2024). Architecture and components of UAVs is shown in Figure 1.

UAVs typically consist of three main components: data collection, processing, and actuation. The data collection system integrates onboard sensors, cameras, batteries, and smart devices to gather pertinent information. The central core unit processes this data for tasks like path planning, surveillance, and localization.

The hardware components onboard UAVs include a variety of sensors tailored for different applications, such as path planning, collision avoidance, and inspection during flight operations. Essential sensors like LIDAR and infrared devices play crucial roles in collision avoidance and mapping, while cameras and GPS facilitate surveillance and navigation along predefined routes (Javaid et al., 2024). The central core system acts as the nerve centre, coordinating UAV actions in real-world environments (Bayomi & Fernandez, 2023).

The conventional methods of controlling Unmanned Aerial Vehicles (UAVs) typically involve ground control stations (GCS) utilized in both manual and autonomous missions. With the growing drone industry, many countries have established regulations governing UAV operations. To operate a UAV effectively, three primary components are essential: the ground control room (GCR), the communication infrastructure—which may include satellite links, radio frequency systems, or internet connectivity—and, of course, the UAV itself. Communication with the UAV can be facilitated through various means, including satellite links, radio signals, and internet connections. These communication methods enable operators to command and receive data from the UAV during its mission. This framework underscores the interconnectedness of ground control, communication, and UAV technology in the effective operation of unmanned aerial systems. Figure 2 illustrates the communication channels mostly used to control UAVs.

Figure 2. Communication channels mostly used to control UAVs

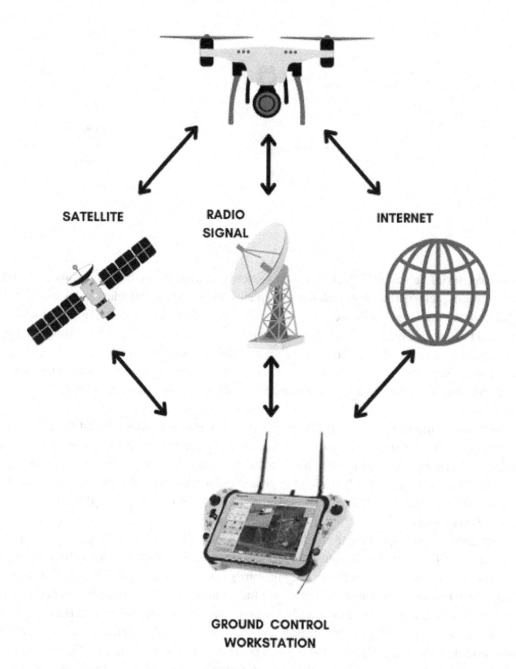

To understand UAV architectures, we utilise classifications like Macro Aerial Vehicles, Micro Aerial Vehicles, and Nano Aerial Vehicles, each tailored to diverse applications. Expanding on this, we delve into the essential components defining UAVs, highlighting their structural framework involving data collection, processing, and actuation (Jacobsen et al., 2023). These components are pivotal for executing tasks like path planning, inspection, and evaluation.

Applications of Reinforcement Learning in Robotics

Reinforcement Learning, a subset of machine learning, enables agents to learn optimal decision-making through interaction with their environment to maximise rewards. In robotics and UAVs, it finds applications in autonomous navigation, mission planning, and adaptive control (Aboueleneen et al., 2023). These algorithms enhance collision avoidance, object manipulation, and surveillance capabilities. In precision agriculture, Reinforcement Learning-equipped UAVs optimise tasks like crop monitoring and pest control. It aids search and rescue missions by enabling efficient search patterns and adaptability. Its versatility enhances autonomy and adaptability across scenarios, unlocking new possibilities in robotics and UAVs. These applications will be discussed in detail in further sections.

This paper makes a significant contribution to understanding and implementing reinforcement learning (RL) in drone and UAV automation. It explores fundamental RL concepts like Markov Decision Processes and Deep Q-Networks, highlighting their role in advancing UAV autonomy. Practical applications of RL in obstacle avoidance, exploration, target tracking, and precision agriculture are showcased, demonstrating the versatility and effectiveness of RL-enabled UAVs. Additionally, the integration of edge computing and AI technologies is discussed, offering insights into enhancing real-time processing capabilities in UAV systems. The paper not only emphasises the benefits of automation but also identifies key challenges and research directions, providing valuable contributions to autonomous aerial systems.

Our paper is structured for efficiency, featuring numbered sections to facilitate a thorough understanding of reinforcement learning integration in drone and UAV automation. Section 1 introduces drones and UAVs, followed by a historical perspective on automation. Section 2 delves into reinforcement learning fundamentals, covering its definition, components, and methods. Section 3 examines reinforcement learning's application in UAV automation, discussing benefits, system components, and real-world case studies. In Section 4, we explore edge computing and AI integration for UAVs, highlighting opportunities and challenges. Section 5 addresses current challenges, potential advances, and future research directions, concluding with a comprehensive overview of findings and contributions in Section 6.

FUNDAMENTALS OF REINFORCEMENT LEARNING

Reinforcement learning (RL) is a way machines learn by trying things out and getting feedback. The aim is to figure out the best way to do something by getting the most reward over time. RL works by agents doing actions, seeing what happens, and getting rewards. It's useful for solving problems like robotics, gaming, and decision-making where things happen in order and need to adapt.

Definition and Components of Reinforcement Learning

Reinforcement Learning is a type of machine learning where an intelligent agent learns to make decisions by interacting with its environment. Unlike supervised learning, RL doesn't rely on labelled datasets. Instead, it learns through trial and error, receiving feedback in the form of rewards or penalties. The components of reinforcement learning is explain in Table 1.

Table 1. Components of reinforcement learning

Component	Definition	Role
Agent	The entity responsible for decision-making and interaction with the environment.	The agent observes the present state of the environment, makes choices in actions, and then receives feedback.
Environment	The external system in which the agent operates and learns.	The environment responds to the agent's actions, transitioning to new states and providing feedback through rewards.
State (S)	A portrayal of the current state or arrangement of the surroundings.	The set of States, represented by S, assists the agent in comprehending the context and making knowledgeable decisions grounded in its observations.
Action (A)	The decisions or moves made by the agent in response to the observed state.	Actions, denoted by the set A, determine the agent's influence on the environment, leading to state transitions.
Reward (R)	immediate environmental feedback expressing the agent's action's desirability.	Rewards, specified by the reward function R, guide the learning process, encouraging the agent to take actions that lead to positive outcomes.
Policy	An approach, or state-to-action mapping, that the agent gradually picks up.	The policy guides the decision-making process, helping the agent choose actions that maximise cumulative rewards.
Transition Probability (P)	The possibility of changing from one state to another, taking into account a particular activity.	Transition function P specifies the likelihood of state transitions, providing a probabilistic aspect to the agent's decision-making.
Discount Factor (γ)	A factor that influences the extent to which the agent prioritizes future rewards over immediate ones.	The discount factor γ impacts the agent's evaluation of future outcomes, striking a balance between the significance of immediate and long-term rewards.

The agent's primary objective is to devise a plan, a set of rules dictating actions in various situations, to maximise long-term outcomes. It computes this outcome by summing expected rewards, discounted based on their temporal distance. The discount factor determines the balance between immediate and future rewards; a higher factor prioritises long-term rewards. By interacting with the environment, the agent adjusts its estimations of situation quality, termed the value function. This function indicates the expected reward for each situation or situation-action pair, aiding plan evaluation and generation. Value functions vary based on whether they consider only situations, actions, or both.

The agent employs diverse learning methods like dynamic programming, temporal difference learning, Monte Carlo methods, Q-learning, policy gradient methods, and deep reinforcement learning to acquire the value function and plan. These methods differ in information utilisation, value function and plan updates, and exploration-exploitation trade-offs. Balancing exploration and exploitation is crucial; exploration involves trying new strategies, while exploitation sticks with known ones to maximise reward. An effective agent finds this balance to avoid suboptimal plans or missed opportunities for better ones.

Markov Decision Processes (MDPs)

Within Markov Decision Processes (MDPs), the system progresses through discrete time steps, and the agent engages with the environment by making decisions informed by the observed states (Han et al., 2021). The mathematical framework for MDPs involves several key components:

State (S): - Assume that S is the collection of states, encompassing every potential scenario the agent could face. In this context, $(s_t) \in S$ denotes the state observed at time step t.

Action (A): - Let A be the collection of all potential decisions the agent may make in each state; the action the agent took at time step t is indicated by (a_t) in A.

Transition Probability (P): - The transition function P defines the likelihood of moving from one state to another when a specific action is taken. It can be denoted as $P(s_{t+1} \mid s_t, a_t)$, representing the probability of transitioning to state s_{t+1} based on the current state s_t and action (a_t) (Garcia & Rachelson, 2013).

Reward (R): - Each state-action pair is given a numerical value by the reward function R, which is represented as $R(s_t, a_t)$. It stands for the prompt feedback or punishment that the agent gets for acting in a certain way in a certain condition (Garcia & Rachelson, 2013).

Discount Factor (γ): - The discount factor ($0 <= \gamma <= 1$) establishes the agent's inclination towards immediate rewards over future ones. The total discounted reward G_t at time step t is computed by summing up future rewards as follows (Garcia & Rachelson, 2013):

$$G_t = R_{t+1} + \gamma R_{t+2} + \gamma^2 R_{t+3} + \ldots \ldots (i)$$

Value Function (V): - The value function $V(s_t)$ signifies the anticipated cumulative reward associated with being in a specific state (s_t) and adhering to a particular policy (Garcia & Rachelson, 2013). It is defined recursively as:

$$V(S_t) = \sum_{t=0}^{\infty} \gamma t . R_{t+1} \ (ii)$$

Policy (π): - A policy π is basically a mapping from states to actions. It defines the decision-making strategy of the agent. A deterministic policy is denoted as $\pi(st) \rightarrow at$, and a stochastic policy is denoted as $\pi(a_t \mid s_t)$ (Garcia & Rachelson, 2013).

Bellman Equation: - The Bellman equation shows how the value of one state is connected to the values of the states that come after it:

$$V(S_t) = R_{(S_t, a_t)} + \gamma \sum S_{t+1 \in S} P(S_{t+1} \mid a_t) . V(S_{t+1}) \ (iii)$$

Finding the solution to the Bellman equation helps figure out the best plan that gets the most expected total reward in a Markov Decision Process

Q-Learning and Deep Q-Networks (DQN)

Q-Learning is a basic model-free method in reinforcement learning, playing a crucial role in tackling complex decision-making problems. Its importance is evident in the field of Unmanned Aerial Vehicles (UAVs), particularly for finding routes in changing environments. This algorithm learns from experience using a value function called Q(s, a), which predicts the total rewards when taking action (a) in state (s) and following the best strategy.

The iterative update rule for Q-Learning is expressed through the Bellman equation (Watkins & Dayan, 1992):

$$Q_{(s,a)} \leftarrow Q_{(s,a)} + \alpha \left[r + \gamma_{max} a' Q(s', a') - Q(s,a) \right] \ (iv)$$

where,

$Q_{(s,a)}$ is the Q-value for state s and action a,

α is the learning rate,

r is the immediate reward,

γ is the discount factor,

and s′ is the next state. Q-Learning is effective for discrete state and action spaces; however, it becomes impractical for larger or continuous spaces due to the necessity of maintaining a vast table of Q-values.

Deep Q-Learning (DQL) blends Q-Learning with deep neural networks to overcome the limitations of Q-Learning when working with large or continuous regions. These neural networks function as approximations for the Q-values connected to every state-action pair in DQL. The network generates Q-values for each possible action after receiving the state as input. Training aims to reduce the mean squared error between the target and predicted Q-values, as determined by the Bellman equation (Watkins & Dayan, 1992).

The loss function for DQL is represented as:

$$\mathbf{L}(\mathbf{\theta}) = \mathbf{E}_s, \mathbf{a}, \mathbf{r}, \mathbf{s}'\left[\left(\mathbf{r} + \gamma_{max}\mathbf{a}'\mathbf{Q}(\mathbf{s}', \mathbf{a}'\ ; \mathbf{\theta}^-) - \mathbf{Q}(\mathbf{s}, \mathbf{a}; \mathbf{\theta})\right)^2\right] (v)$$

where,

L(θ) is the loss function with respect to the neural network parameters θ.

$θ^-$ represents the parameters of a target network (periodically updated).

Deep Q-Network (DQN), introduced by DeepMind in 2013, is a specific implementation of DQL designed to learn from raw pixels. A neural network with convolutional layers and fully linked layers is used by DQN (Watkins & Dayan, 1992).

Several enhancements contribute to the performance and stability of DQN in UAV path planning:

1. **Experience Replay**: Storing experiences in a replay buffer and sampling mini batches for training, reducing sample correlation and enhancing data efficiency.
2. **Frame Skipping**: Repeating actions for several frames and updating the network at regular intervals, reducing computational costs, and enhancing robustness to game dynamics.
3. **Reward Clipping**: Clipping rewards to a range [-1, 1] to prevent dominance of large rewards in gradients, ensuring algorithm generalizability.
4. **Double Q-Learning**: A Q-Learning variation that separates action selection from assessment to reduce overestimation bias.
5. **Dueling Network**: Utilising a duelling network architecture, which separates state value estimation and state-dependent action advantage estimation, improving learning efficiency. (Watkins & Dayan, 1992)

Policy Gradient Methods and Deep Deterministic Policy Gradient

Policy gradient methods optimize policies in reinforcement learning to maximize long-term reward. They adjust policy settings to improve reward over time, suitable for complex scenarios with many options or continuous actions, encouraging exploration. Two types exist: actor-only and actor-critic methods, the latter incorporating a value function for stability. These methods employ function estimation techniques like neural networks to handle large state-action spaces effectively.

Policy function: π(a|s;θ) is the probability of taking action an in states with policy parameters θ.

Objective function: J(θ) is the expected sum of discounted rewards, indicating the overall performance of the policy.

Policy Gradient Theorem (Silver et al., 2014):

$$\nabla_\theta \mathbf{J}(\theta) = \mathbf{E}_\pi \left[\sum_{t=0}^T \nabla_\theta \log_\pi(\mathbf{a}_t | \mathbf{s}_t ; \theta) \, \mathbf{Q}^\pi \left(\mathbf{s}_t, \mathbf{a}_t \right) \right] \text{ (vi)}$$

providing the gradient of the expected return.

Actor-only method update: θ ← θ + α∇_θ J(θ), where α is the learning rate.

Actor-critic method modification (Watkins & Dayan, 1992):

$$\nabla_\theta \mathbf{J}(\theta) = \mathbf{E}_\pi \left[\sum_{t=0}^T \nabla_\theta \log_\pi(\mathbf{a}_t | \mathbf{s}_t ; \theta) \, \mathbf{Q}^\pi(\mathbf{s}_t, \mathbf{a}_t) - \mathbf{b} \left(\mathbf{s}_t \right) \right] \text{ (vii)}$$

integrating a baseline b(s).

UAV path planning is complex, requiring drones to navigate safely while avoiding obstacles and adapting to dynamic environments. Traditional methods like A*, D*, or RRT may falter due to predefined maps and models. Policy gradient methods offer a data-driven solution, enabling drones to learn path planning policies autonomously. With onboard sensors, drones perceive critical environmental aspects, facilitating comprehensive state representation. Action spaces can be discrete or continuous, and the reward function guides drones to destinations while avoiding obstacles and threats. Leveraging neural networks, drones learn complex policies for robust navigation. Challenges include data and computational requirements, the exploration-exploitation dilemma, and interactions with other agents. Addressing these challenges offers opportunities for research in enhancing exploration strategies and developing cooperative multi-agent reinforcement learning approaches for UAV path planning (Yan et al., 2020).

AUTOMATION IN DRONES AND UAVS

Unmanned Aerial Vehicles (UAVs), colloquially known as drones, have rapidly evolved from niche military applications to ubiquitous tools across numerous industries, ranging from agriculture to cinematography. Central to this evolution is the incorporation of automation, which has profoundly transformed the capabilities and applications of UAVs. The evaluation of automatic parallelization algorithms may have relevance to the automation processes in drones and UAVs. Automation often involves complex algorithms, and understanding parallelization methods could contribute to the efficiency and optimization of processes within UAV systems (S S. Kumar et al., 2021) (S. Kumar et al., 2022).

Various Levels of Automation

Automation in UAVs is often conceptualised across different levels, each representing varying degrees of autonomy and human involvement. The Society of Automotive Engineers (SAE) International has identified these stages/levels (as depicted in Figure 3), which offer a framework for comprehending the evolution from manual to completely autonomous operation:

i) Level 0 (No Automation):

At this level, all aspects of UAV operation, including navigation, flight control, and mission execution, are entirely under the direct control of a human operator. While manual control offers precise oversight and decision-making, it limits the scalability and efficiency of UAV operations, particularly in complex or dynamic environments.

ii) Level 1 (Human Assisted):

Assisted Control introduces partial automation into UAV operations, augmenting human decision-making with automated features such as stability assistance, waypoint navigation, and altitude hold. At Level 1, the drones provide suggestions or warnings to the human operator, who retains ultimate authority over decision-making.

iii) Level 2 (Partial Automation):

Partial automation represents a significant step towards autonomy, with the UAV assuming greater responsibility for executing predefined tasks or missions. At Level 2, the UAV can perform more advanced features such as GPS-guided navigation and automated takeoff and landing, reducing the cognitive load on the operator while maintaining human oversight.

iv) Level 3 (Conditional Automation):

At this level, the UAV can perform specific tasks autonomously, such as following a predefined flight path or maintaining a fixed altitude, but still requires human intervention for higher-level decision-making and supervision. Level 3 automation enables the UAV to operate within predefined operational boundaries, with human intervention required only in exceptional circumstances.

v) Level 4 (High Automation):

High automation enables the UAV to operate autonomously within predefined operational boundaries or mission parameters, with human intervention required only in exceptional circumstances. At this level, the UAV can handle complex missions with minimal human oversight, adapting to changing conditions and unforeseen events.

vi) Level 5 (Full Automation):

Full automation represents the pinnacle of autonomy, where the UAV is capable of completely autonomous operation across all aspects of its mission, from takeoff to landing. At this level, human intervention is optional and typically reserved for strategic oversight or unforeseen circumstances beyond the capabilities of the UAVs automated systems.

Figure 3. Various levels of automation in UAVs

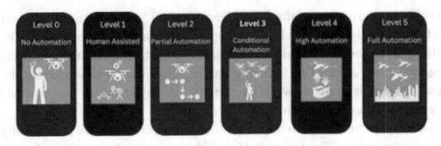

Benefits of Automation in UAVs

Automation offers a multitude of advantages for UAVs, reshaping the landscape of aerial operations. One of the primary benefits lies in its efficiency. By automating tasks that were previously performed manually, UAVs can execute missions with unparalleled precision and speed, leading to significant improvements in overall efficiency (S. Kumar et al., 2023). Moreover, the scalability of automated UAV systems enables the deployment of fleets of vehicles for large-scale operations without the need for individual human operators for each vehicle. This scalability is particularly advantageous in applications such as surveillance, monitoring, and delivery services, where multiple UAVs can work collaboratively to cover expansive areas efficiently (S. K. Sharma et al., 2021). Additionally, automation enhances safety by reducing reliance on human pilots in certain tasks, mitigating the risk of human error and ensuring reliable operation, even in hazardous or challenging environments.

Components of Automated Drone Systems

Automated drone systems consist of several key components that work together to enable autonomous operation:

1. **Sensors:** Provide real-time data on surroundings, including position, orientation, and environment, using GPS, IMUs, cameras, LiDAR, and radar.
2. **Actuators:** Control movement and behavior based on input commands and environmental feedback, including propulsion systems, control surfaces, and payload deployment mechanisms.
3. **Onboard Processors:** Serve as the UAV's brain, processing sensor data, executing algorithms, and making real-time decisions via microcontrollers and embedded computers.
4. **Communication Systems:** Facilitate data exchange between the UAV and external entities like ground control stations, utilizing radio frequency, satellite, cellular networks, or ad-hoc mesh networks.

5. **Power Systems**: Power systems provide the necessary energy to operate the UAVs components and subsystems. This includes batteries or fuel cells for electric-powered UAVs, as well as fuel tanks and engines for combustion-powered UAVs. Efficient power management is essential to ensure long-endurance flights and mission sustainability (Kumar et al., 2023).

Integration of Deep Reinforcement Learning Into UAV Automation

The seamless integration of deep reinforcement learning (DRL) into unmanned aerial vehicle (UAV) automation has fundamentally transformed the landscape of autonomous aerial operations. DRL harnesses the power of advanced deep neural networks for intricate feature extraction and nuanced decision-making processes (Aggarwal et al., 2022). This technological leap empowers UAVs to iteratively refine their control strategies by learning from real-world interactions, thereby unlocking the potential for unparalleled efficiency and autonomy in mission execution.

UAVs with DRL algorithms excel at navigating complex environments, quickly adjusting to changes, and optimizing mission success. This fusion of AI and aerial robotics marks a groundbreaking advancement in autonomous systems. This integration is depicted in Figure 4, which illustrates the process of DRL-based feature extraction in a UAV system.

Figure 4. Deep learning feature extraction in a UAV system

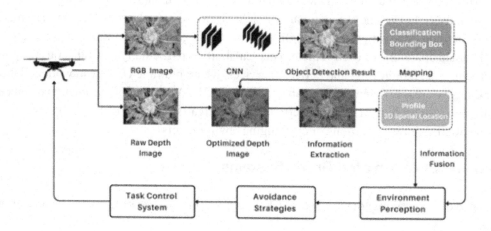

The following Table 2 outlines key benefits of integrating multi-modal sensors with CRNNs for industrial drones, highlighting enhanced perception, real-time decision-making, adaptability to dynamic environments, robustness to sensor variability, and facilitation of autonomous operation.

Table 2. Enhanced performance using multi-modal sensors with CRNNs

Benefits	Description
Enhanced Perception	Multi-modal sensor fusion provides diverse sources of environmental data, including depth perception, object detection, and terrain mapping, enabling more comprehensive understanding.
Real-time Decision Making	CRNNs process sequential sensor data efficiently, enabling rapid interpretation and informed decision-making, crucial for dynamic tasks like path planning and collision avoidance.
Adaptability to Dynamic Environments	Integration allows drones to adapt to rapidly changing conditions effectively, analysing sensor data to maintain optimal performance even in challenging environments.
Robustness to Sensor Variability	Fusing data from multiple sensors mitigates individual sensor limitations, improving overall accuracy and reliability by leveraging CRNNs' ability to learn complex patterns.

CASE STUDIES AND APPLICATIONS

Navigation and Path Planning

Navigation and path planning are key for UAVs, shaping their ability to navigate complex areas and achieve goals efficiently. Reinforcement learning helps UAVs adjust strategies and optimize paths for smart, real-time decisions.

Reinforcement Learning for Dynamic Obstacle Avoidance

UAVs face a tough time navigating through dynamic obstacles, which traditional systems struggle with, leading to higher collision risks. Reinforcement learning (RL) offers a solution by letting UAVs learn avoidance strategies through continuous interactions. This process involves feedback-driven learning where UAVs refine navigation policies by receiving rewards or penalties based on their actions near obstacles. Deep reinforcement learning (DRL) enhances this by using deep neural networks to extract complex features from sensor data, improving real-time obstacle perception and response.

In industrial drone operations, challenges like object detection and SLAM are crucial for mission safety and efficacy. Object detection issues include varied environmental conditions and complex backgrounds, while SLAM challenges involve accurate perception and autonomous navigation. Overcoming these hurdles demands robust algorithms, sensor fusion techniques, and real-time processing for reliable detection and precise localization, ensuring successful drone operations across diverse industries.

Autonomous Exploration using Reinforcement Learning

Autonomous exploration is crucial for UAVs navigating unknown or inaccessible areas like disaster zones or dense forests. Traditional methods often result in incomplete coverage or missed opportunities. However, reinforcement learning (RL) empowers UAVs to autonomously navigate and map uncharted territories, optimising exploration while balancing known information. RL-based exploration involves refining strategies through interaction with the environment, dynamically adjusting trajectories based on the value of unexplored areas. Leveraging uncertainty-aware strategies, RL-enabled UAVs prioritise exploration of regions with high informational content, ensuring flexibility and robustness in changing conditions (Li et al., 2023). Continuously updating their strategies through online learning and adaptation, RL-equipped UAVs enhance coverage and efficiency over time, showcasing the potential of RL in autonomous aerial operation

Surveillance and Monitoring

Advanced UAVs transform surveillance in security and conservation. Industrial drones cut costs in data gathering and hazardous site inspections. They excel with high-res imaging, but face flight limits and regulations. Reinforcement learning (RL) techniques enhance UAV surveillance, particularly in target tracking, improving monitoring capabilities in dynamic environments.

Target Tracking with Reinforcement Learning

Target tracking in UAV surveillance is essential for continuous monitoring of moving objects within designated areas. While traditional methods may struggle in dynamic environments, reinforcement learning (RL) presents a promising solution. RL enables UAVs to learn optimal tracking strategies through iterative feedback-driven processes. RL-equipped UAVs refine their tracking policies based on rewards or penalties, ensuring robust tracking even in challenging conditions. Deep reinforcement learning (DRL) techniques further enhance tracking precision by processing complex features from sensor data. RL-based target tracking improves surveillance effectiveness and efficiency, enabling adaptation to dynamic scenarios such as tracking multiple targets simultaneously or reacquiring lost targets. This ensures accurate and reliable surveillance in diverse environments.

Anomaly Detection in Aerial Surveillance and Engineering Geology

Unmanned aerial vehicles (UAVs) play a pivotal role in engineering geology, offering diverse applications from terrain mapping to hazard identification. Their extensive coverage and high-resolution data collection capabilities enable comprehensive geological assessments, particularly in challenging-to-access areas. UAV photogrammetry, especially in hazard-prone regions, accurately maps earth cracks and monitors landslides, providing vital insights for disaster management (Giordan et al., 2020; Koutalakis et al., 2021). These applications highlight UAVs' significant contribution to engineering geology, fostering a deeper understanding of geological processes and promoting sustainable development practices (Chhabra et al., 2024).

Delivery and Logistics

Delivery and logistics play pivotal roles in commerce and trade, encompassing the intricate processes involved in transporting goods and services from suppliers to customers. These operations are conducted through diverse modes of transportation such as road, rail, air, water, or pipelines (Wang et al., 2023). With the dynamic landscape shaped by technology, e-commerce, and globalisation, the delivery and logistics sector is experiencing notable transformations and innovations. Emerging trends and challenges include the increasing demand for faster and more convenient delivery services, heightened complexity and uncertainty in operations, and a growing awareness of the environmental and social impacts. To address these, the industry is embracing technologies like unmanned aerial vehicles (UAVs) or drones, artificial intelligence (AI), machine learning (ML), blockchain, smart contracts, and the Internet of Things (IoT) (S. Kumar et al., 2019). These advancements aim to enhance efficiency, sustainability, and responsibility in delivery and logistics practices.

Optimising Delivery Routes with Reinforcement Learning

Efficiently optimising delivery routes is critical for delivery and logistics, impacting cost, time, and service quality. This involves sequencing locations and actions while considering constraints like time windows and capacity limits (Eu & Phang, 2023). Traditional methods struggle with the complexity and variability of real-world scenarios, relying on fixed models and assumptions.

To address these challenges, reinforcement learning (RL) emerges as a practical solution for optimising delivery routes. RL, a subset of machine learning, allows the system to learn from its experiences and interactions without relying on preset models or assumptions. By maximising long-term rewards, RL is effective in managing extensive or continuous state and action spaces, using methods like neural networks to understand complex and nonlinear strategies.

Using RL to optimise delivery routes means representing the problem as a Markov decision process (MDP), which includes the following components:

1. A set of states representing possible situations of the delivery system, including the location, load, and status of the vehicle, as well as the demand, supply, and condition of customers.
2. A set of actions representing possible decisions of the delivery system, such as moving to a different location, picking up or dropping off a package, or waiting at the current location (Eu & Phang, 2023).
3. A transition function reflecting the probability of moving from one state to another given an action, capturing the dynamics and uncertainty of the environment, such as traffic, weather, or customer behaviour.
4. A reward function offering immediate feedback on taking an action in a state, reflecting the objectives and preferences of the delivery system, such as minimising travel distance, time, or cost, or maximising customer satisfaction, service quality, or profit.

Through RL, delivery systems learn policies mapping states to actions, dynamically adapting to the environment without predefined models (R. R. Kumar et al., 2023). RL offers advantages over traditional methods, enabling adaptability to changes and uncertainties (Yadav et al., 2023). UAVs have seen increased use, especially during the COVID-19 pandemic, for delivering medical supplies and parcels, with advancements in technology improving their efficiency and safety (Eu & Phang, 2023). The drop-box mechanism, designed using SOLIDWORKS software, facilitates contactless delivery by automatically dropping off parcels (Eu & Phang, 2023).

RL enables the management of large state and action spaces, fostering exploration and robustness in delivery systems (Vats et al., 2023; Gupta et al., 2023). However, challenges such as data requirements and the exploration-exploitation dilemma remain (R. R. Kumar et al., 2023). Furthermore, RL encounters the exploration-exploitation dilemma, necessitating a delicate balance between attempting new actions for improved solutions and exploiting the current best action for immediate rewards.

Automated Payload Management in UAVs

Efficient payload management in unmanned aerial vehicles (UAVs) is a pivotal aspect of ensuring the safety, efficiency, and quality of delivery services (Dwivedi et al., 2023). The complexity of managing variable and diverse payloads, coupled with the need for constant adaptation to changes and uncertainties, necessitates innovative solutions. Traditional methods involving human intervention or mechanical arms may fall short, prompting the adoption of automated methods. These leverage technologies like smart packaging, smart landing gears, and smart algorithms to perform payload operations autonomously. Such advancements contribute to streamlining the loading and unloading processes, adhering to weight limits, ensuring stability, and aligning with customer preferences and delivery standards.

UAVs in smart farming

Smart farming revolutionises agriculture by integrating advanced technologies and data-driven methodologies to enhance productivity, profitability, and sustainability. Unmanned Aerial Vehicles (UAVs), equipped with cameras and sensors, play a vital role in collecting high-resolution data of farms in near-real-time, capturing crucial parameters like plant height, biomass, and nutrient levels (Peñalvo et al., 2022). This data, integrated and processed using artificial intelligence and machine learning, creates a dynamic digital representation of the farm, reflecting spatial and temporal variations (Peñalvo et al., 2022). By leveraging mixed cloud-edge computing paradigms, smart farming optimizes coordination strategies, facilitating faster decision-making and adaptive resource allocation (Wu et al., 2018).

UAVs utilise digital farm representations for simulations and analyses, aiding in crop growth modelling, yield prediction, and pest detection, among others. Mathematical models, optimization algorithms, and reinforcement learning provide valuable insights and recommendations, guiding actions like pesticide spraying and crop harvesting. Smart actuators and controllers enable optimal farm strategies. UAVs contribute significantly to precision agriculture, pest management, and yield estimation, enhanced by integration with IoT and big data (S. Kr. Singh et al., 2022)(Sharma et al., 2023).

However, creating and updating digital farm representations for UAVs may require substantial data and computation, posing challenges in storage, processing, and privacy. Capturing the complexity of agricultural systems, including climate change and market demand, may affect representation accuracy (Peñalvo et al., 2022). Technical and regulatory hurdles in drone operation, such as battery life and airspace management, can limit data collection and intervention effectiveness (Javaid et al., 2024).

UAVs as Digital Twins

UAVs play a crucial role in creating and maintaining digital twins of farms, crops, and animals in smart farming. These virtual replicas offer real-time access, simulations, and interventions based on comprehensive digital representations, leveraging sensors, data, and algorithms (Vats et al., 2023). Digital twins provide high-resolution, near-real-time data and images, capturing farm features and parameters (Bai et al., 2022). However, challenges such as data storage, transmission, and processing, along with issues of data quality, security, and privacy (R. Kumar et al., 2024), may arise. Capturing the complexity of agricultural systems and addressing technical and regulatory hurdles like battery life and airspace management remain critical (Javaid et al., 2024) (Lawrence et al., 2023).

AI-Empowered UAV Perception Systems for Precision Agriculture

AI is crucial in improving UAV perception systems, which is essential for precision agriculture (Lawrence et al., 2023). AI enables UAVs to gather, analyse, and understand data and information from the environment through different sensors, data sources, and algorithms. Analysis and utilisation of large amounts of data and information, improving knowledge and understanding of agricultural problems and solutions. Automation and optimization of tasks and processes, enhancing the efficiency and effectiveness of operations and outcomes. Adaptation and innovation of methods and tools, improving the flexibility and creativity of operations and outcomes.

UAVs, empowered by AI, enhance their perception capabilities, enabling them to perform various tasks and functions in smart farming. UAV perception systems, defined as systems enabling UAVs to acquire, process, and interpret data, are crucial for precision agriculture (Lawrence et al., 2023). Collaboration between AI and UAVs in perception systems contributes to more efficient and effective operations in smart farming. These systems facilitate the acquisition, processing, and interpretation of data from the environment, ensuring UAVs can perform tasks with enhanced capabilities.

UAV Swarm Management for Smart Farming

UAV swarm management introduces a collaborative approach to data collection and analysis in smart farming. Coordinated efforts of multiple UAVs operating as a swarm cover larger areas, improve data accuracy, and enhance overall farm management. Swarm intelligence, distributed control, and game theory become key considerations for efficiently coordinating multiple drones (Muslimov, 2023).

UAV swarm management offers various benefits:

1. Enhanced coverage through collaborative efforts of multiple UAVs in a swarm, ensuring comprehensive data collection across agricultural areas.
2. Increased accuracy due to the coordinated actions of a UAV swarm, leading to more reliable insights for farmers.
3. Efficient data processing facilitated by swarm intelligence, enabling quicker analysis and decision-making in smart farming operations.

Ongoing research focuses on refining multi-UAV systems to address challenges like communication range, collision avoidance, and airspace management, aiming to enhance precision agriculture. Embracing technologies like IoT and Human Digital Twin (HDT), this research introduces an innovative framework for predicting cardiovascular disease (CVD) early. Leveraging wearable sensors and machine learning, the framework utilizes data from 70,000 patient records to achieve highly accurate predictions, with XG Boost reaching 88.91% accuracy. The incorporation of HDT technology enhances healthcare system operations, improving predictive accuracy and enabling a more effective approach to patient care (Vats et al., 2023).

The development of swarm intelligence and distributed control mechanisms has the potential to revolutionise smart farming operations, enhancing scalability, efficiency, and adaptability to agricultural needs (Ming et al., 2023).

INTEGRATION OF EDGE COMPUTING AND AI FOR UAVS

Advancements in edge computing and artificial intelligence (AI) have paved the way for transformative applications in unmanned aerial vehicles (UAVs). As compared to cloud computing, edge computing processes near the data source, typically at the "edge" of the network, thereby providing better latency and reduces bandwidth usage (Kumar et al., 2021). The comparative architecture is shown in Figure 5. This section explores the integration of edge computing and AI technologies in UAV systems, highlighting the opportunities they present and the challenges they pose (McEnroe et al., 2022).

Figure 5. Edge computing vs. cloud computing

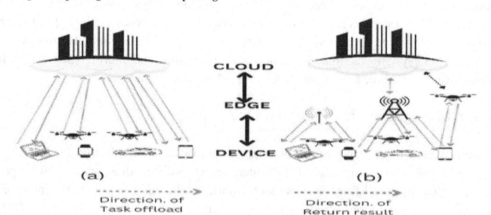

Opportunities and Challenges in Implementing Edge AI in UAVs

Implementing edge AI in UAVs offers a myriad of opportunities for enhancing their autonomy, efficiency, and capabilities. By deploying AI algorithms directly on-board UAVs or on edge computing devices situated close to the UAVs, real-time decision-making can be achieved without relying on centralised processing or communication with ground stations (R. Singh et al., 2022). This enables UAVs to execute tasks autonomously, such as object detection, navigation, and obstacle avoidance, even in environments with limited or unreliable connectivity. Additionally, edge AI enables UAVs to process large volumes of sensor data locally, reducing latency and bandwidth requirements while preserving privacy and security (M. Singh et al., 2023). However, implementing edge AI in UAVs also presents significant challenges, including limited computational resources, power constraints, and the need for robust algorithms capable of operating in resource-constrained environments (Ma et al., 2021). Furthermore, ensuring the reliability and safety of AI-driven UAV systems is paramount, requiring rigorous testing, validation, and certification processes. The research on joint UAV position optimization and resource scheduling in space-air-ground integrated networks with mixed cloud-edge computing intersects with discussions on Deep Reinforcement Learning (DRL) for drones and UAVs in their mutual concern for efficient resource utilization and optimization. Both domains strive to enhance performance and reduce delays in complex networked environments, reflecting a shared focus on advancing autonomous systems through innovative technological solutions (Mao et al., 2020).

Impact of Edge AI on UAV-Based IoT Services

The integration of edge AI in UAVs revolutionizes UAV-based Internet of Things (IoT) services, enhancing data collection, analysis, and dissemination. By leveraging edge computing and AI, UAVs become intelligent data collection and processing platforms, enabling real-time monitoring, analysis, and response in various applications like environmental monitoring and disaster management (Aldaej et al., 2022) (A. Singh et al., 2022). Edge AI-equipped UAVs autonomously process sensor data on board, reducing bandwidth requirements and response times (Ahmed et al., 2022). This improves the scalability, efficiency, and responsiveness of UAV-based IoT services, facilitating data-driven decision-making and optimization in diverse domains. However, integrating edge AI raises concerns about data privacy, security, and regulatory compliance, necessitating robust encryption, authentication, and access control mechanisms to safeguard sensitive information and ensure compliance with data protection regulations (Yazid et al., 2021) (R. Kumar et al., 2023).

CHALLENGES AND FUTURE DIRECTIONS

As unmanned aerial vehicles (UAVs) continue to evolve and integrate advanced technologies, they face a variety of challenges and opportunities for future development.

For the future outlook of industrial drones, it's crucial to anticipate both challenges and opportunities that will shape their trajectory. Regulatory hurdles, including evolving rules on airspace access and privacy, pose significant challenges for operators and manufacturers alike. Meanwhile, advancements in AI and machine learning hold promise for enhancing drone capabilities, but integrating these technologies while ensuring safety remains a challenge. Addressing scalability, interoperability, and data security concerns is crucial as demand for drone services continues to rise. Additionally, emerging market trends such as drone-as-a-service models and niche applications offer both opportunities and challenges for industry players. Essentially, industrial drones must navigate regulatory, technological, and market-driven challenges while seizing opportunities for innovation and growth. The challenges encountered in the various sectors in the drone industry have been depicted in Figure 6.

Figure 6. Challenges faced in UAV's in various sectors

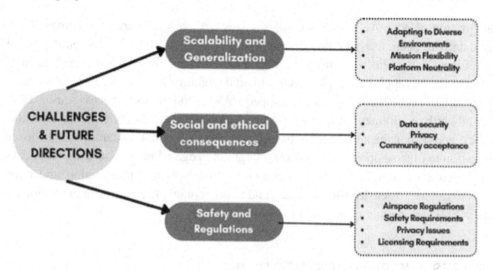

This section explores key challenges in scalability, safety, regulations, deployment, and outlines potential advances and research directions to address these challenges.

Scalability and Generalization

Scalability and generalisation are critical challenges in UAV technology, particularly concerning the deployment of autonomous systems in diverse environments and applications (Bathla et al., 2022). While advancements in machine learning and reinforcement learning have enabled UAVs to perform complex tasks autonomously, scaling these capabilities across different missions, environments, and platforms remains a significant challenge. UAV systems must be capable of adapting and generalising their learned behaviours to new scenarios without extensive retraining (A. Singh et al., 2022).

● Adapting to Diverse Environments: UAV technology must overcome the challenge of operating in various environments, such as urban, rural, and adverse weather conditions, requiring robust algorithms capable of adapting to diverse scenarios.
● Mission Flexibility: Ensuring scalability involves enabling UAV systems to perform a wide range of missions, from surveillance and reconnaissance to package delivery, while maintaining efficiency and effectiveness across different tasks.
● Platform Neutrality: Achieving generalization necessitates developing UAV systems that can seamlessly integrate with different platforms and hardware configurations, allowing for interoperability and flexibility in deployment.

Safety and Regulations

Safety and regulatory compliance are paramount considerations in the deployment of UAVs, particularly in urban and populated areas. Ensuring the safe operation of UAVs involves mitigating risks associated with collisions, airspace congestion, and human-machine interaction. Additionally, compliance with regulations governing UAV operations, such as flight restrictions, privacy laws, and data protection regulations, is essential to prevent accidents and legal issues. Future research should focus on developing advanced sense-and-avoid systems, collision avoidance algorithms, and human-machine interface designs to enhance the safety and usability of UAVs (A. Sharma et al., 2023). The PPDTSA framework's privacy-preserving object detection aligns with discussions on Deep Reinforcement Learning (DRL) for drones and UAVs, as both prioritise object detection, attention mechanisms, and privacy preservation within advanced technological integration, fostering safer and more efficient autonomous systems (Ma et al., 2021). Legal regulations governing industrial drone usage encompass a range of considerations:

1. Airspace Regulations: These rules dictate drone flight locations and altitudes, often including restrictions near airports, crowds, or specific zones. Compliance ensures aviation safety and prevents collisions with manned aircraft.
2. Safety Requirements: Regulations outline operational standards to mitigate drone-related risks, covering maintenance, pilot training, flight planning, and emergencies. Adherence minimises accidents and property damage.
3. Privacy Issues: Concerns arise regarding privacy violations from drone surveillance and data collection. Regulations restrict such activities, prohibiting filming in private spaces without consent and governing data storage and use, maintaining public trust.
4. Licensing Requirements: Many jurisdictions mandate licences or permits for drone operation, requiring knowledge of aviation regulations, passing exams, and obtaining insurance. Compliance ensures operators' competence and accountability.

Overall, adherence to legal regulations is vital for responsible drone operations. Complying with airspace rules, safety standards, privacy laws, and licensing ensures ethical and effective drone use, mitigating risks and liabilities.

Social and ethical consequences

Drone technology presents significant social and ethical impacts, notably concerning data security, privacy, and community acceptance. Concerns regarding data security arise from the vulnerability of data transmission channels and storage systems to hacking or unauthorized access. Privacy issues stem from drones' ability to capture images and footage without consent, raising questions about surveillance boundaries and personal privacy rights. Moreover, community acceptance varies depending on factors such as cultural norms and perceptions of safety and security, with concerns about noise pollution, invasion of privacy, and potential misuse leading to opposition in some areas (A. Sharma et al., 2023). Addressing these concerns requires robust data security measures, transparent and accountable drone operations to respect privacy rights, and community engagement to foster understanding and acceptance. Policymakers also play a crucial role in developing regulations that balance the benefits of drone technology with privacy and community interests.

Real-world Deployment Challenges

Real-world deployment presents various challenges for UAVs, including environmental factors, infrastructure limitations, and operational constraints. UAVs must contend with adverse weather conditions, GPS signal interference, and limited battery life, which can affect their performance and reliability. Furthermore, integrating UAV operations into existing airspace management systems and infrastructure requires coordination with aviation authorities, air traffic control, and other stakeholders (Singh et al., 2022)

Potential Advances and Future Research Directions

Despite the challenges, UAV technology holds immense potential for future advancements and innovations. Areas of future research may include the development of autonomous swarming capabilities, collaborative UAV networks, and adaptive mission planning algorithms. Additionally, advancements in sensor technology, materials science, and propulsion systems may enable the design of more efficient and versatile UAV platforms. Furthermore, interdisciplinary research efforts that integrate UAV technology with fields such as artificial intelligence, robotics, and materials science could unlock new possibilities for UAV applications in areas such as environmental monitoring, disaster response, and infrastructure inspection. Continuing to invest in research and development is crucial for driving innovation and fully realising the potential of UAV technology in addressing various needs (Ghamari et al., 2022).

Examining technological developments and emerging trends in the industrial drone sector offers valuable insights into the current landscape and future directions of the industry. Key areas of focus include advancements in data analytics, artificial intelligence (AI), autonomy, and sensor technology. Data analytics techniques enable the extraction of actionable insights from the vast amounts of data generated by drones, optimising workflows and facilitating predictive analytics. AI empowers drones to process sensor data, recognize objects, and make real-time decisions autonomously, enhancing capabilities in navigation, object detection, and more. Increasing autonomy in drones allows them to plan and execute missions independently, navigate complex environments, and adapt to changing conditions with minimal human intervention, utilising technologies such as simultaneous localization and mapping (SLAM) and obstacle avoidance systems. Moreover, continuous advancements in sensor technology, such as high-resolution cameras and LiDAR scanners, enable drones to capture detailed and specialised data for various applications, including mapping, surveying, and infrastructure inspection. These developments collectively drive innovation and progress in the industrial drone sector, opening up new possibilities for diverse applications across industries.

CONCLUSION

A thorough overview of how reinforcement learning (RL) can be integrated into the automation of drones and unmanned aerial vehicles (UAVs). Through a systematic examination of fundamental RL concepts, such as Markov Decision Processes (MDPs), Q-Learning, Deep Q-Networks (DQN), and policy gradient methods, coupled with an exploration of their application in robotics, we have explained the significance of RL in advancing UAV autonomy. The benefits of automation in UAVs are multiple, ranging from enhanced operational efficiency and safety to increased mission capabilities. By automating critical functions such as navigation, path planning, surveillance, monitoring, delivery, and logistics, RL algorithms empower UAVs to adapt to dynamic environments and perform complex tasks autonomously with precision and reliability. Moreover, combining edge computing and AI presents fresh opportunities and challenges for UAV systems. While edge AI offers the potential to enhance real-time processing capabilities and optimise resource utilisation, it also raises concerns regarding latency, bandwidth, and security, which must be carefully addressed in the design and deployment of UAV-based IoT services. In summary, our goal was to shed light on how reinforcement learning (RL) can strengthen UAV autonomy and highlight the merging prospects and hurdles of combining edge computing with AI. Looking forward, the focus lies on advancing autonomous features, tackling scalability and regulatory hurdles, and fostering interdisciplinary endeavours to unlock UAV technology's full potential.

Table 3. Acronyms used

Acronyms	Description
UAVs	Unmanned Aerial Vehicles
DQN	Deep Q-Network
RL	Reinforcement Learning
AI	Artificial Intelligence
LiDAR	Light Detection and Ranging
GPS	Global Positioning System
ESCs	Electronic Speed Controllers
MDPs	Markov Decision Processes
A*, D*, or RRT	Path Planning Algorithms
SAE	Society of Automotive Engineers
IMUs	Inertial Measurement Units
DRL	Deep Reinforcement Learning
IoT	Internet of Things

REFERENCES

Aboueleneen, N., Alwarafy, A., & Abdallah, M. (2023). Deep Reinforcement Learning for Internet of Drones Networks: Issues and Research Directions. *IEEE Open Journal of the Communications Society*, 4, 671–683. DOI: 10.1109/OJCOMS.2023.3251855

Aggarwal, K., Singh, S. K., Chopra, M., Kumar, S., & Colace, F. (2022). Deep Learning in Robotics for Strengthening Industry 4.0.: Opportunities, Challenges and Future Directions. In Nedjah, N., Abd El-Latif, A. A., Gupta, B. B., & Mourelle, L. M. (Eds.), *Robotics and AI for Cybersecurity and Critical Infrastructure in Smart Cities* (pp. 1–19). Springer International Publishing. DOI: 10.1007/978-3-030-96737-6_1

Ahmed, F., Mohanta, J. C., Keshari, A., & Yadav, P. S. (2022). Recent Advances in Unmanned Aerial Vehicles: A Review. *Arabian Journal for Science and Engineering*, 47(7), 7963–7984. DOI: 10.1007/s13369-022-06738-0 PMID: 35492958

Aldaej, A., Ahanger, T. A., Atiquzzaman, M., Ullah, I., & Yousufudin, M. (2022). Smart Cybersecurity Framework for IoT-Empowered Drones: Machine Learning Perspective. *Sensors (Basel)*, 22(7), 7. DOI: 10.3390/s22072630 PMID: 35408244

Bai, S., Song, S., Liang, S., Wang, J., Li, B., & Neretin, E. (2022). UAV Maneuvering Decision-Making Algorithm Based on Twin Delayed Deep Deterministic Policy Gradient Algorithm. *Journal of Artificial Intelligence and Technology*, 2(1), 1. DOI: 10.37965/jait.2021.12003

Bathla, G., Bhadane, K., Singh, R. K., Kumar, R., Aluvalu, R., Krishnamurthi, R., Kumar, A., Thakur, R. N., & Basheer, S. (2022). Autonomous Vehicles and Intelligent Automation: Applications, Challenges, and Opportunities. *Mobile Information Systems*, 2022, e7632892. DOI: 10.1155/2022/7632892

Bayomi, N., & Fernandez, J. E. (2023). Eyes in the Sky: Drones Applications in the Built Environment under Climate Change Challenges. *Drones (Basel)*, 7(10), 10. DOI: 10.3390/drones7100637

Chhabra, A., Singh, S. K., Sharma, A., Kumar, S., Gupta, B. B., Arya, V., & Chui, K. T. (2024). Sustainable and intelligent time-series models for epidemic disease forecasting and analysis. *Sustainable Technology and Entrepreneurship*, 3(2), 100064. DOI: 10.1016/j.stae.2023.100064

Dwivedi, K., Govindarajan, P., Srinivasan, D., Keerthi Sanjana, A., Selvanambi, R., & Karuppiah, M. (2023). Intelligent Autonomous Drones in Industry 4.0. In Sarveshwaran, V., Chen, J. I.-Z., & Pelusi, D. (Eds.), *Artificial Intelligence and Cyber Security in Industry 4.0* (pp. 133–163). Springer Nature. DOI: 10.1007/978-981-99-2115-7_6

Eu, K. Q., & Phang, S. K. (2023). Automated Parcel Loading-Unloading Mechanism Design for Delivery UAV. *Journal of Physics: Conference Series*, 2523(1), 012016. DOI: 10.1088/1742-6596/2523/1/012016

Garcia, F., & Rachelson, E. (2013). Markov Decision Processes. In *Markov Decision Processes in Artificial Intelligence* (pp. 1–38). John Wiley & Sons, Ltd. DOI: 10.1002/9781118557426.ch1

Ghamari, M., Rangel, P., Mehrubeoglu, M., Tewolde, G. S., & Sherratt, R. S. (2022). Unmanned Aerial Vehicle Communications for Civil Applications: A Review. *IEEE Access : Practical Innovations, Open Solutions*, 10, 102492–102531. DOI: 10.1109/ACCESS.2022.3208571

Giordan, D., Adams, M. S., Aicardi, I., Alicandro, M., Allasia, P., Baldo, M., De Berardinis, P., Dominici, D., Godone, D., Hobbs, P., Lechner, V., Niedzielski, T., Piras, M., Rotilio, M., Salvini, R., Segor, V., Sotier, B., & Troilo, F. (2020). The use of unmanned aerial vehicles (UAVs) for engineering geology applications. *Bulletin of Engineering Geology and the Environment*, 79(7), 3437–3481. DOI: 10.1007/s10064-020-01766-2

Gupta, A., Singh, S. K., Gupta, B. B., Chopra, M., & Gill, S. S. (2023). Evaluating the Sustainable COVID-19 Vaccination Framework of India Using Recurrent Neural Networks. *Wireless Personal Communications*, 133(1), 73–91. DOI: 10.1007/s11277-023-10751-3

Han, D., Chen, W., & Liu, J. (2021). Energy-Efficient UAV Communications Under Stochastic Trajectory: A Markov Decision Process Approach. *IEEE Transactions on Green Communications and Networking*, 5(1), 106–118. DOI: 10.1109/TGCN.2020.3016266

Jacobsen, R. H., Matlekovic, L., Shi, L., Malle, N., Ayoub, N., Hageman, K., Hansen, S., Nyboe, F. F., & Ebeid, E. (2023). Design of an Autonomous Cooperative Drone Swarm for Inspections of Safety Critical Infrastructure. *Applied Sciences (Basel, Switzerland)*, 13(3), 3. DOI: 10.3390/app13031256

Javaid, A., Alduais, A., Shullar, M. H., Baroudi, U., & Alnaser, M. (2024). Monocular-based collision avoidance system for unmanned aerial vehicle. *IET Smart Cities*, 6(1), 1–9. DOI: 10.1049/smc2.12067

Koutalakis, P. D., Tzoraki, O. A., Prazioutis, G. I., Gkiatas, G. T., & Zaimes, G. N. (2021). Can Drones Map Earth Cracks? Landslide Measurements in North Greece Using UAV Photogrammetry for Nature-Based Solutions. *Sustainability (Basel)*, 13(9), 9. DOI: 10.3390/su13094697

Kumar, R., Singh, S. K., & Lobiyal, D. K. (2023). Communication structure for Vehicular Internet of Things (VIoTs) and review for vehicular networks. In *Automation and Computation*. CRC Press. DOI: 10.1201/9781003333500-34

Kumar, R., Singh, S. K., Lobiyal, D. K., Kumar, S., & Jawla, S. (2024). Security Metrics and Authentication-based RouTing (SMART) Protocol for Vehicular IoT Networks. *SN Computer Science*, 5(2), 236. DOI: 10.1007/s42979-023-02566-7

Kumar, S., Karnani, G., Gaur, M. S., & Mishra, A. (2021). Cloud Security using Hybrid Cryptography Algorithms. *2021 2nd International Conference on Intelligent Engineering and Management (ICIEM)*, (pp. 599–604). IEEE. DOI: 10.1109/ICIEM51511.2021.9445377

Kumar, S., Singh, S. K., & Aggarwal, N. (2023). Speculative Parallelism on Multicore Chip Architecture Strengthen Green Computing Concept: A Survey. In *Advanced Computer Science Applications*. Apple Academic Press. DOI: 10.1201/9781003369066-2

Kumar, S., Singh, S. K., & Aggarwal, N. (2023b). Sustainable Data Dependency Resolution Architectural Framework to Achieve Energy Efficiency Using Speculative Parallelization. *2023 3rd International Conference on Innovative Sustainable Computational Technologies (CISCT)*, (pp. 1–6). IEEE. DOI: 10.1109/CISCT57197.2023.10351343

Kumar, S., Singh, S. K., Aggarwal, N., Gupta, B. B., Alhalabi, W., & Band, S. S. (2022). An efficient hardware supported and parallelization architecture for intelligent systems to overcome speculative overheads. *International Journal of Intelligent Systems*, 37(12), 11764–11790. DOI: 10.1002/int.23062

Kumar, S., Tiwari, P., & Zymbler, M. (2019). Internet of Things is a revolutionary approach for future technology enhancement: A review. *Journal of Big Data*, 6(1), 111. DOI: 10.1186/s40537-019-0268-2

Kumar, S. S., Singh, S., Aggarwal, N., & Aggarwal, K. (2021). Efficient speculative parallelization architecture for overcoming speculation overheads. *International Conference on Smart Systems and Advanced Computing (Syscom-2021)*, (pp. 132–138). IEEE.

Lawrence, I. D., Vijayakumar, R., & Agnishwar, J. (2023). Dynamic Application of Unmanned Aerial Vehicles for Analyzing the Growth of Crops and Weeds for Precision Agriculture. In *Artificial Intelligence Tools and Technologies for Smart Farming and Agriculture Practices* (pp. 115–132). IGI Global. DOI: 10.4018/978-1-6684-8516-3.ch007

Li, L., Li, W., Wang, J., Chen, X., Peng, Q., & Huang, W. (2023). UAV Trajectory Optimization for Spectrum Cartography: A PPO Approach. *IEEE Communications Letters*, 27(6), 1575–1579. DOI: 10.1109/LCOMM.2023.3265214

Ma, B., Wu, J., Lai, E., & Hu, S. (2021, December). PPDTSA: Privacy-preserving deep transformation self-attention framework for object detection. In *2021 IEEE Global Communications Conference (GLOBECOM)* (pp. 1-5). IEEE. DOI: 10.1109/GLOBECOM46510.2021.9685855

Mao, S., He, S., & Wu, J. (2020). Joint UAV position optimization and resource scheduling in space-air-ground integrated networks with mixed cloud-edge computing. *IEEE Systems Journal*, 15(3), 3992–4002. DOI: 10.1109/JSYST.2020.3041706

McEnroe, P., Wang, S., & Liyanage, M. (2022). A Survey on the Convergence of Edge Computing and AI for UAVs: Opportunities and Challenges. *IEEE Internet of Things Journal*, 9(17), 15435–15459. DOI: 10.1109/JIOT.2022.3176400

Mengi, G., Singh, S. K., Kumar, S., Mahto, D., & Sharma, A. (2023). Automated Machine Learning (AutoML): The Future of Computational Intelligence. In Nedjah, N., Martínez Pérez, G., & Gupta, B. B. (Eds.), *International Conference on Cyber Security, Privacy and Networking (ICSPN 2022)* (pp. 309–317). Springer International Publishing. DOI: 10.1007/978-3-031-22018-0_28

Ming, R., Jiang, R., Luo, H., Lai, T., Guo, E., & Zhou, Z. (2023). Comparative Analysis of Different UAV Swarm Control Methods on Unmanned Farms. *Agronomy (Basel)*, 13(10), 10. DOI: 10.3390/agronomy13102499

Mohsan, S. A. H., Othman, N. Q. H., Li, Y., Alsharif, M. H., & Khan, M. A. (2023). Unmanned aerial vehicles (UAVs): Practical aspects, applications, open challenges, security issues, and future trends. *Intelligent Service Robotics*, 16(1), 109–137. DOI: 10.1007/s11370-022-00452-4 PMID: 36687780

Muslimov, T. (2023). Curl-Free Vector Field for Collision Avoidance in a Swarm of Autonomous Drones. In Ronzhin, A., Sadigov, A., & Meshcheryakov, R. (Eds.), *Interactive Collaborative Robotics* (pp. 369–379). Springer Nature Switzerland. DOI: 10.1007/978-3-031-43111-1_33

Peñalvo, F. J. G., Maan, T., Singh, S. K., Kumar, S., Arya, V., Chui, K. T., & Singh, G. P. (2022). Sustainable Stock Market Prediction Framework Using Machine Learning Models. [IJSSCI]. *International Journal of Software Science and Computational Intelligence*, 14(1), 1–15. DOI: 10.4018/IJSSCI.313593

Peñalvo, F. J. G., Sharma, A., Chhabra, A., Singh, S. K., Kumar, S., Arya, V., & Gaurav, A. (2022). Mobile Cloud Computing and Sustainable Development: Opportunities, Challenges, and Future Directions. [IJCAC]. *International Journal of Cloud Applications and Computing*, 12(1), 1–20. DOI: 10.4018/IJCAC.312583

Sharma, A., Singh, S. K., Badwal, E., Kumar, S., Gupta, B. B., Arya, V., Chui, K. T., & Santaniello, D. (2023). Fuzzy Based Clustering of Consumers' Big Data in Industrial Applications. *2023 IEEE International Conference on Consumer Electronics (ICCE)*, (pp. 01–03). IEEE. DOI: 10.1109/ICCE56470.2023.10043451

Sharma, A., Singh, S. K., Chhabra, A., Kumar, S., Arya, V., & Moslehpour, M. (2023). A Novel Deep Federated Learning-Based Model to Enhance Privacy in Critical Infrastructure Systems. [IJSSCI]. *International Journal of Software Science and Computational Intelligence*, 15(1), 1–23. DOI: 10.4018/IJSSCI.334711

Sharma, A., Singh, S. K., Kumar, S., Chhabra, A., & Gupta, S. (2023). Security of Android Banking Mobile Apps: Challenges and Opportunities. In Nedjah, N., Martínez Pérez, G., & Gupta, B. B. (Eds.), *International Conference on Cyber Security, Privacy and Networking (ICSPN 2022)* (pp. 406–416). Springer International Publishing. DOI: 10.1007/978-3-031-22018-0_39

Sharma, S. K., Singh, S. K., & Panja, S. C. (2021). Human Factors of Vehicle Automation. In *Autonomous Driving and Advanced Driver-Assistance Systems (ADAS)*. CRC Press. DOI: 10.1201/9781003048381-17

Shuford, J. (2024). Deep Reinforcement Learning Unleashing the Power of AI in Decision-Making. *Journal of Artificial Intelligence General Science (JAIGS), 1*(1). DOI: 10.60087/jaigs.v1i1.36

Silver, D., Lever, G., Heess, N., Degris, T., Wierstra, D., & Riedmiller, M. (2014). Deterministic Policy Gradient Algorithms. *Proceedings of the 31st International Conference on Machine Learning*, (pp. 387–395). MLR. https://proceedings.mlr.press/v32/silver14.html

Singh, A., & Singh, S. Kr., & Mittal, A. (2022). A Review on Dataset Acquisition Techniques in Gesture Recognition from Indian Sign Language. In P. Verma, C. Charan, X. Fernando, & S. Ganesan (Eds.), *Advances in Data Computing, Communication and Security* (pp. 305–313). Springer Nature. DOI: 10.1007/978-981-16-8403-6_27

Singh, I., & Singh, S. Kr., Kumar, S., & Aggarwal, K. (2022). Dropout-VGG Based Convolutional Neural Network for Traffic Sign Categorization. In M. Saraswat, H. Sharma, K. Balachandran, J. H. Kim, & J. C. Bansal (Eds.), *Congress on Intelligent Systems* (pp. 247–261). Springer Nature. https://doi.org/DOI: 10.1007/978-981-16-9416-5_18\

Singh, M., Singh, S. K., Kumar, S., Madan, U., & Maan, T. (2023). Sustainable Framework for Metaverse Security and Privacy: Opportunities and Challenges. In Nedjah, N., Martínez Pérez, G., & Gupta, B. B. (Eds.), *International Conference on Cyber Security, Privacy and Networking (ICSPN 2022)* (pp. 329–340). Springer International Publishing. DOI: 10.1007/978-3-031-22018-0_30

Singh, R., Singh, S. K., Kumar, S., & Gill, S. S. (2022). SDN-Aided Edge Computing-Enabled AI for IoT and Smart Cities. In *SDN-Supported Edge-Cloud Interplay for Next Generation Internet of Things*. Chapman and Hall/CRC. DOI: 10.1201/9781003213871-3

Singh, S. Kr., Sharma, S. K., Singla, D., & Gill, S. S. (2022). Evolving Requirements and Application of SDN and IoT in the Context of Industry 4.0, Blockchain and Artificial Intelligence. In *Software Defined Networks* (pp. 427–496). John Wiley & Sons, Ltd. https://doi.org/DOI: 10.1002/9781119857921.ch13

Soori, M., Arezoo, B., & Dastres, R. (2023). Artificial intelligence, machine learning and deep learning in advanced robotics, a review. *Cognitive Robotics*, 3, 54–70. DOI: 10.1016/j.cogr.2023.04.001

Vats, T., Singh, S. K., Kumar, S., Gupta, B. B., Gill, S. S., Arya, V., & Alhalabi, W. (2023). Explainable context-aware IoT framework using human digital twin for healthcare. *Multimedia Tools and Applications*, 83(22), 62489–62490. DOI: 10.1007/s11042-023-16922-5

Wang, X., Yang, Z., Chen, G., & Liu, Y. (2023). A Reinforcement Learning Method of Solving Markov Decision Processes: An Adaptive Exploration Model Based on Temporal Difference Error. *Electronics (Basel)*, 12(19), 19. DOI: 10.3390/electronics12194176

Watkins, C. J. C. H., & Dayan, P. (1992). Q-learning. *Machine Learning*, 8(3), 279–292. DOI: 10.1007/BF00992698

Wen, H., Xie, Z., Wu, Z., Lin, Y., & Feng, W. (2024). Exploring the future application of UAVs: Face image privacy protection scheme based on chaos and DNA cryptography. *Journal of King Saud University. Computer and Information Sciences*, 36(1), 101871. DOI: 10.1016/j.jksuci.2023.101871

Wu, J., Guo, S., Huang, H., Liu, W., & Xiang, Y. (2018). Information and communications technologies for sustainable development goals: State-of-the-art, needs and perspectives. *IEEE Communications Surveys and Tutorials*, 20(3), 2389–2406. DOI: 10.1109/COMST.2018.2812301

Yadav, N., Singh, S. K., & Sharma, D. (2023). Forecasting Air Pollution for Environment and Good Health Using Artificial Intelligence. *2023 3rd International Conference on Innovative Sustainable Computational Technologies (CISCT)*, (pp. 1–5). IEEE. DOI: 10.1109/CISCT57197.2023.10351334

Yan, C., Xiang, X., & Wang, C. (2020). Towards Real-Time Path Planning through Deep Reinforcement Learning for a UAV in Dynamic Environments. *Journal of Intelligent & Robotic Systems*, 98(2), 297–309. DOI: 10.1007/s10846-019-01073-3

Yazid, Y., Ez-Zazi, I., Guerrero-González, A., El Oualkadi, A., & Arioua, M. (2021). UAV-Enabled Mobile Edge-Computing for IoT Based on AI: A Comprehensive Review. *Drones (Basel)*, 5(4), 4. DOI: 10.3390/drones5040148

Chapter 12
Multi–Modal Sensor Fusion With CRNNs for Robust Object Detection and Simultaneous Localization and Mapping (SLAM) in Agile Industrial Drones

Ujjwal Thakur

Chandigarh College of Engineering and Technology, Chandigarh, India

Sunil K. Singh

ⓘ https://orcid.org/0000-0003-4876-7190

Chandigarh College of Engineering and Technology, Chandigarh, India

Sudhakar Kumar

ⓘ https://orcid.org/0000-0001-7928-4234

Chandigarh College of Engineering and Technology, Chandigarh, India

Anubhav Singh

Chandigarh College of Engineering and Technology, Chandigarh, India

Varsha Arya

Asia University, Taiwan, & Hong Kong Metropolitan University, Hong Kong

Kwok Tai Chui

Hong Kong Metropolitan University, Hong Kong

ABSTRACT

The integration of multi-modal sensor fusion with CRNNs in agile industrial drones addresses the need for improved object detection and SLAM capabilities. This technology enhances the drone's ability to navigate and map complex industrial environments with greater accuracy and reliability. It is crucial for tasks such as inspection, monitoring, and autonomous navigation in dynamic and challenging industrial settings. By integrating visual, LiDAR, and inertial measurement unit (IMU) sensor information, the approach enhances situational awareness, facilitating safer navigation and intelligent interaction amidst challenging conditions. The integration of LiDAR, IMU, and optical sensors provides awareness of the environment, enabling the drone to adjust to changing circumstances instantly. The suggested method

DOI: 10.4018/979-8-3693-2707-4.ch012

meets the needs of industrial applications, which demand dependable and durable solutions. It achieves this by offering improved precision, dependability, and flexibility that eventually expands the potential of industrial drones in a variety of operational contexts.

INTRODUCTION

Background and Motivation

The evolution of industrial-grade drones (Floreano & Wood, 2015) is now opening up formerly inaccessible areas to a variety of fields like manufacturing, inspection of infrastructure, agriculture and emergency services among others. Compared with other aircraft or similar technology, these drones can provide sensing, monitoring, and data collection in places and situations which were considered unconquerable in the past. It is only possible to harness their power after being provided with the right "scanning" technique by which they are able to identify things correctly and drive on their own. The classical methods of robot design often have problems catching up with the demands of production environments which are characterized by changeable lighting, moving objects, and varied layouts. Therefore, innovative development of contemporary perception techniques is needed to allow drones to operate safely and effectively in difficult situations.

Objectives of the Chapter

The essential objective of this chapter is to propose an approach for upgrading question discovery and Concurrent Localization and Mapping in dexterous mechanical rambles through the integration of multi-modal sensor combination with convolutional repetitive neural systems (CRNNs)(Liebl & Burghardt, 2020). By leveraging information from visual, LiDAR(Zhou et al., 2024) (Liu, 2008), and inertial estimation unit (IMU)(Höflinger et al., 2013) sensors, our system points to increase the recognition capabilities of mechanical rambles, empowering them to explore, identify objects, and outline their environment with expanded exactness and unwavering quality. By exploring further into multi-modal combination techniques(Lingenfelser et al., 2011) and implementing CRNNs into execution, to address the problems that come with mechanical rambling operations(Zatsiorsky & Duarte, 1999) and open the door to improvements in efficiency and autonomy in a range of mechanical applications.

Overview of Industrial Drone Applications

Mechanical rambles have found far reaching applications over different businesses. These rambles offer unparalleled flexibility and proficiency, empowering assignments that were already time-consuming, expensive, or perilous to be performed with ease and exactness. From assessing basic frameworks such as bridges, control lines, and pipelines to looking over agrarian areas and evaluating natural risks, mechanical rambles have illustrated potential in revolutionizing various divisions. In any case, to completely capitalize on their capabilities, it is fundamental to prepare them with progressed recognition frameworks capable of vigorous question location and precise in challenging situations. Parallelization algorithms(Kumar et al., 2021b, 2021a, 2022, 2023b; I. Singh, Singh, Singh, et al., 2022) play a crucial role in efficiently processing and fusing data from various sensors concurrently.

Challenges in Object Detection and SLAM for Agile Industrial Drones

Despite their potential, industrial drones face a number of challenges in object detection and SLAM(Bailey & Durrant-Whyte, 2006) (Frese et al., 2010) (Kümmerle et al., 2009), especially in the flexible and dynamic environments commonly encountered in industrial environments. Traditional methods often have difficulty handling factors such as varying lighting conditions, occlusions, dynamic obstacles, and complex structures, which can significantly reduce accuracy and reliability of the cognitive system. Additionally, the need for real-time decision making further exacerbates these challenges, requiring the development of advanced techniques capable of rapidly processing and integrating data from multiple sensors. Addressing these challenges is critical to exploiting the full potential of industrial drones and enabling them to operate safely and efficiently in various industrial applications(S. Singh, Sharma, Singla, et al., 2022).

The contribution of this chapter covers Multi-Modal Sensor Fusion with CRNNs for Robust Object Detection and Simultaneous Localization and Mapping (SLAM) in Agile Industrial Drones.

- The brief overview of multi modal sensor fusion which includes principles of sensor fusion, types of sensor and its importance in perception enhancement are covered.
- Overview of CNNs and RNNs,role of CNNs and RNNs in CRNNs, CRNNs architecture, decision making with CRNNs and advantages of it were discussed .
- Further explores the multi modal fusion framework which explores integration of LIDAR, visual and IMU sensors, design of fusion pipeline and dataflow of fusion pipeline.
- The overview of object detection in the environment, challenges in object detection for drones, object detection techniques,algorithms and role of Multi-Modal Fusion in improving object detection accuracy.
- Simultaneous Localization And Mapping (SLAM) importance in drone application, traditional SLAM methods and limitations and enhancing the SLAM performance through Multi-Modal Sensor Fusion is discussed.

The organization of this chapter is as follows: Section 2 covers literature overview, history was discussed in section 3. The fundamentals of multimodal sensor fusion are covered in section 4, and CRNNs and their design are explored in section 5. The multimodal sensor fusion framework was explored in Section 6, industrial object detection was discovered in Section 7, SLAM was discussed in Section 8, real-world applications were discussed in Section 9, problems and future directions were discussed in Section 10, and a conclusion was provided in Section 11.

LITERATURE OVERVIEW

The integration of multi-modal sensor combination with Convolutional Repetitive Neural Systems (CRNNs) in spry mechanical rambles speaks to a critical progression within the field of mechanical autonomy, especially in protest location and Synchronous Localization and Mapping (Hammer) capabilities. This segment gives a comprehensive survey of important writing, centering on the standards of sensor combination, the design of CRNNs, and the significance of multi-modal combination for recognition improvement in mechanical ramble applications.

Sensor combination may be a basic preparation in mechanical rambles, pointed at joining information from different sensors to improve discernment accuracy and unwavering quality. Elmenreich and Pitzek (2001) characterize sensor fusion as the method of combining information from different sensors to make a more total and exact representation of the environment. By leveraging the qualities of sensors such as visual cameras, LiDAR, and IMUs, rambles can overcome the confinements of personal sensors and accomplish a more vigorous discernment framework. This approach moves forward accuracy, unwavering quality, and flexibility, fundamental for secure and productive ramble operations in energetic mechanical situations.

Mechanical rambles utilize an assortment of sensors, counting visual cameras, LiDAR, and IMUs, to gather data around their environment. Visual cameras capture high-resolution pictures, LiDAR sensors give exact remote estimations and make nitty gritty 3D point clouds, whereas IMUs collect information related to ramble developments. Understanding the capacities and commitments of these sensors is significant for planning compelling multimodal combination calculations that misuse their qualities . Convolutional Neural Systems (CNNs) and Repetitive Neural Systems (RNNs) are crucial structures in profound learning, each with interesting characteristics and applications. CNNs exceed expectations in picture acknowledgment errands by extracting spatial highlights from input pictures, whereas RNNs specialize in capturing transient conditions inside successive information (Yin et al., 2017). CRNNs coordinated both CNNs and RNNs, making them well-suited for dealing with multi-modal sensor data in real-time applications. By combining spatial and transient data, CRNNs empower rambles to form educated choices based on both inactive and energetic perspectives of the environment. The design of CRNNs ordinarily comprises convolutional layers for spatial highlight extraction taken after by repetitive layers, such as LSTM or GRU layers, for capturing transient conditions (Cheng et al., 2020). This various leveled design permits CRNNs to successfully coordinate data from different sensor modalities, such as visual, LiDAR, and IMU information, for moved forward discernment capabilities. Furthermore, CRNNs can handle variable-length arrangements of sensor information, empowering real-time handling of multi-modal inputs. Multi-modal sensor combination plays a crucial part in improving the cognitive capabilities of mechanical rambles by coordination information from different sensors. By combining data at different levels of reflection, rambles can make educated choices based on a comprehensive understanding of their environment. Multi-modal combination empowers rambles to require advantage of the complementary data given by diverse sensor modalities, making a more vigorous and adaptable discernment framework.

SHORT HISTORY

A story of technical advancement and convergence is the history of Multi-Modal Sensor Fusion with Convolutional Recurrent Neural Networks (CRNNs) for Robust Object Detection and Simultaneous Localization and Mapping (SLAM) in Agile Industrial Drones. This journey started with the awareness that the dynamic and complicated industrial settings that drones operate in could not be fully handled by standard techniques to object identification and SLAM. A system that could map the environment in real time, follow items' motions, and recognize things with robustness was what engineers and researchers were aiming for. The combination of Convolutional Neural Networks (CNNs) and Recurrent Neural

Networks (RNNs), which brought together the sequential processing powers of RNNs and the spatial knowledge of CNNs, was revolutionary. In order to provide drones a thorough awareness of their environment, early studies concentrated on improving the architecture and training techniques of CRNNs. These techniques were intended to efficiently integrate data from many sensors, including cameras, LiDAR, and IMUs. CRNNs acquired popularity for its application in object identification and SLAM tasks in agile industrial drones as they demonstrated effectiveness in processing multi-modal sensor data. The drone industry adopted CRNN-based solutions at a faster pace over time due to developments in both hardware and software. Industrial drones with vision systems driven by CRNNs might find their way through congested areas, recognize and avoid obstacles, and maintain precise localization even in difficult circumstances. Numerous industrial applications, such as agricultural monitoring, warehouse management, and infrastructure inspection, have been transformed by this technology.

Anticipating future developments, current research endeavors to expand the frontiers of Multi-Modal Sensor Fusion using CRNNs, with the goal of augmenting the resilience, effectiveness, and expandability of object identification and self-localization in maneuverable industrial drones. In the future, autonomous drones will likely play an ever-more-important part in our everyday lives as these technologies develop and open up new opportunities for automation, safety, and productivity in a variety of industrial contexts.

FUNDAMENTALS OF MULTI-MODAL SENSOR FUSION

Principles of Sensor Fusion

Sensor fusion(Elmenreich & Pitzek, 2001) is the process of integrating data from multiple sensors to create a more complete and accurate representation of the environment than is possible with a single sensor. This subsection explores the principles behind sensor fusion in industrial drones.

By combining data from sensors such as visual cameras, LiDAR, and IMUs, drones can overcome the limitations of individual sensors and achieve a more powerful perception system. Sensor fusion aims to leverage the strengths of each sensor while compensating for their weaknesses, thereby improving the drone's perception and interaction with the surrounding world. This approach improves precision, reliability, and resilience to environmental changes, which are essential for safe and efficient drone operations in dynamic industrial environments.

Types of Sensors in Industrial Drones

Industrial drones use a variety of sensors to collect information about their surroundings. The visual camera captures high-resolution images, providing valuable visual data for tasks such as object detection and scene understanding. LiDAR sensors excel at providing precise distance measurements and creating detailed 3D point clouds of the environment, allowing drones to accurately perceive the shape and structure of objects nearby. Inertial measurement units (IMUs) collect data related to drone movements, including acceleration, rotation, and orientation, which are essential for navigation and control (Lobiyal, 2023).Understanding the functions and contributions of these sensors is critical to designing effective multimodal fusion algorithms(Sharma et al., 2024) that take advantage of their strengths and improve overall cognitive abilities.

Importance of Multi-Modal Fusion for Perception Enhancement

Multimodal sensor(Yang et al., 2022) fusion plays an important role in improving the cognitive capabilities of industrial drones. By integrating data from multiple sensors, drones can better understand their surroundings and be more accurate. This subsection emphasizes the importance of multimodal fusion to improve accuracy, reliability, and adaptability under different operating conditions.

Multi-modal fusion allows drones to take advantage of the complementary information provided by different sensor modalities, creating a more robust and flexible perception system. Additionally, by fusing data at multiple levels of abstraction, drones can make informed decisions based on a comprehensive understanding of their environment, improving autonomy and efficiency in various industrial applications.

CONVOLUTIONAL RECURRENT NEURAL NETWORKS (CRNNs)

Overview of CNNs and RNNs

Convolutional Neural Systems (CNNs) (Li et al., 2022) (Lindsay, 2021) and Repetitive Neural Systems (RNNs) (Yin et al., 2017) (Li et al., 2022) are two crucial structures in profound learning, each with interesting characteristics and applications. In this subsection, we offer a brief diagram of both models and their significance to multi-modal sensor combinations in mechanical rambles. Computer vision has been enhanced by convolutional neural networks, particularly in image recognition applications. Characterized by their hierarchical structure, CNNs (Girshick, 2015) employ convolutional layers to extract spatial features from input images. These layers consist of learnable filters that convolve across the input, capturing local patterns and progressively building a hierarchy of complex features. By minimizing spatial dimensions even further, pooling layers enhance computational efficiency while maintaining important information. CNNs(Sharma, Singh, Chhabra, et al., 2023) are successful due to the fact that it can automatically learn hierarchical representations, which allows them to detect complex patterns in visual data. In relation to the agile drone CNNs (Shewalkar et al., 2019) are broadly utilized for preparing visual information, such as pictures and video outlines, due to their capacity to capture spatial pecking orders of highlights through convolutional layers. They exceed expectations at assignments like picture classification, protest location, and semantic division.

On the other hand, Recurrent Neural Networks (Elmenreich & Pitzek, 2001) (Ilci & Toth, 2020) specialize in capturing temporal dependencies within sequential data. In relation to drone technology, where movements unfold over time, RNNs become instrumental. Unlike traditional networks, RNNs possess connections that form loops, allowing them to maintain a memory of previous inputs. This architecture is well-suited for tasks requiring an understanding of context and temporal patterns. RNNs are planned to handle successive information, making them reasonable for handling transient data such as time-series information or arrangements of sensor readings. They are commonly utilized in assignments like characteristic dialect handling, discourse acknowledgment, and time-series expectation. Understanding the capabilities and impediments of CNNs and RNNs is fundamental for planning successful multi-modal combination structures that use the qualities of both models for progressed discernment in mechanical rambles.

Architecture of CRNNs

Within the interest of improving the capabilities of industrial drones, the design of Convolutional Repetitive Neural Systems (CRNNs) rises as a basic component. This subsection explores into the complexities of CRNN engineering, investigating the consistent integration of Convolutional Neural Systems (CNNs) and Recurrent Neural Systems (RNNs) making it well-suited for handling multi-modal sensor information in real-time. Discussing the design of CRNNs, outlining how convolutional layers extricate spatial highlights from visual information whereas repetitive layers capture worldly conditions in successive information. The CRNN engineering regularly comprises a few convolutional layers taken after by repetitive layers, such as Long Short-Term Memory (LSTM)(Ali-Löytty et al., 2005) or Gated Repetitive Unit (GRU) layers(Cheng et al., 2020). These repetitive layers empower the demonstration to preserve memory of past perceptions and make expectations based on both spatial and transient settings. The various leveled nature of the design permits CRNNs to viably coordinate data from different modalities, such as visual, LiDAR, and IMU information, for improved discernment capabilities. A methodical procedure is used by the Convolutional-Recurrent Neural Network (CRNN) (I. Singh, Singh, Kumar, et al., 2022) to evaluate input images and provide predictions. The system initiates with the input image then utilizes convolutional layers for feature extraction. It additionally uses feature reuse networks in order to make maximum use of relevant characteristics and asymmetric convolutional layers to achieve various receptive fields. A recurrent layer, that is often used as a Deep Bi-Directional Long Short-Term Memory (LSTM) network, receives the output from these convolutional layers. This recurrent layer allows for the complete comprehension of sequential patterns by capturing temporal dependencies in both forward and backward directions.The final stage involves a transcription layer, where pre-frame prediction and sequence generation take place based on the learned representations as depicted in Figure 1.

Figure 1. Architecture of CRNNs

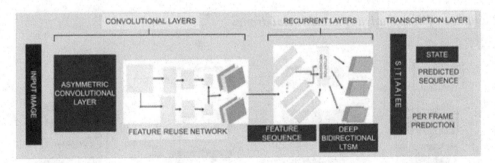

Advantages of CRNNs in Multi-Modal Fusion

CRNNs offer a few preferences for multi-modal sensor combinations in mechanical rambles. This subsection highlights the key benefits of utilizing CRNNs for coordination information from diverse sensor modalities. One advantage is the capacity to consistently combine spatial and transient data inside a bound together engineering, permitting rambles to create educated choices based on both inactive and energetic perspectives of the environment. Object detection in industrial environments demands precision and versatility taking into the account of factors like changing light conditions and complex structures. Furthermore, CRNNs can handle variable-length arrangements of sensor information, obliging the nonconcurrent nature of sensor readings and empowering real-time preparation of multi-modal inputs. Besides, the various leveled nature of CRNNs encourages highlight learning at numerous levels of reflection, capturing both low-level points of interest and high-level semantic data from the sensor information. In general, CRNNs give a capable system for multi-modal combination, empowering mechanical rambles to see and explore their environment more successfully in complex and energetic situations.

MULTI-MODAL SENSOR FUSION FRAMEWORK

Integration of Visual, LiDAR, and IMU Sensors

This subsection traces the method of joining information from visual cameras, Light Discovery and Extending (LiDAR) sensors, and Inertial Estimation Units (IMUs) into a bound together system(Chopra, 2023) for multi-modal sensor combination. Visual cameras give high-resolution pictures, capturing nitty gritty visual data around the environment(S. Singh, Sharma, Singla, et al., 2022). LiDAR sensors, on the other hand, offer exact remote estimations and create 3D point clouds, giving profitable spatial information. IMUs capture information related to the drone's movement, counting increasing speed, turn, and introduction. By combining information from these sensors, the multi-modal combination system makes a comprehensive representation of the environment, empowering the ramble to see and explore its environment viably. This integration preparation includes synchronizing information from diverse sensors, adjusting facilitated frameworks, and preprocessing crude sensor information to guarantee compatibility and consistency over modalities.

Figure 2. Outline of data analytics in industrial drones

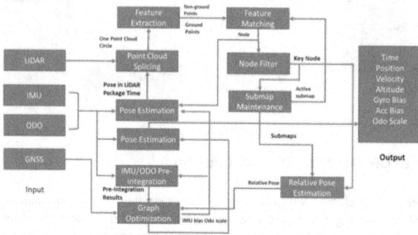

The process involved in data analytics are as follows:

Information Procurement: This is often the primary step in any information analytics preparation. Here, data is collected from different sources. The sorts of data mentioned within the flowchart are:

(a) **Tabular/Structured Information:** Typically information that's organized in an unthinkable shape with columns and columns. Cases incorporate spreadsheets and SQL databases.
(b) **Signals Information:** This alludes to information that's captured over time, regularly from sensors or other gadgets. Cases incorporate sound signals, temperature readings over time, etc.

Preprocessing: Once the information is procured, it ought to be preprocessed to form it appropriate to encourage investigation. This seems to include cleaning the information (expelling or rectifying wrong information), normalizing the data (scaling the information to a little, indicated extent), and changing the information (changing over the information into an appropriate organization for examination).

Feature/Data Fusion: After preprocessing, the diverse sorts of information are combined or intertwined together. This is often done to make a more comprehensive set of information that gives a more full picture of the circumstance or issue at hand.

Dimensionality Reduction: High-dimensional data can be troublesome to work with and can lead to issues like overfitting in machine learning models (Peñalvo, Maan, et al., 2022). Dimensionality reduction techniques are utilized to decrease the number of features in the dataset while holding as much data as conceivable. Methods incorporate Central Component Examination (PCA), t-Distributed Stochastic Neighbor Implanting (t-SNE), and others.

Predictive Analytics: This is the ultimate step where the preprocessed and reduced data is utilized to create forecasts or choices. This seems to include utilizing machine learning models (Khade et al., 2012) to foresee future patterns, classify information, or indeed distinguish peculiarities.

The flowchart moreover mentions an elective way where a Dimensionality Diminishment Strategy (DRT) is connected straightforwardly to the combined highlights, bypassing the Dimensionality Diminishment step. This recommends that in some cases, the dimensionality diminishment can be performed prior within the preparation, depending on the particular necessities of the investigation. The ultimate comes about at that point gotten from an ML Model.

Design of the Fusion Pipeline

In this subsection, we examine the plan and engineering of the combination pipeline, which forms the multi-modal sensor information and extricates important highlights for recognition assignments such as protest location and Synchronous Localization and Mapping (Pummel). The combination pipeline regularly comprises a few stages, counting information preprocessing, highlight extraction, combination, and decision-making (Alibabaei et al., 2022; Sharma, Singh, Badwal, et al., 2023; Sharma, Singh, Chhabra, et al., 2023). Amid information preprocessing, crude sensor information is sifted, calibrated, and synchronized to guarantee transient arrangement and consistency (depicted in Figure 3). Highlight extraction includes extricating instructive highlights from the sensor information utilizing methods such as convolutional neural systems (CNNs) and repetitive neural systems (RNNs). The extricated highlights capture spatial and worldly designs within the sensor information, giving wealthy representations for consequent combination and decision-making stages. Combination procedures such as feature-level combination, decision-level combination, and late combination are utilized to combine data from diverse modalities and produce a bound together discernment yield.

Figure 3. Dataflow in fusion pipeline

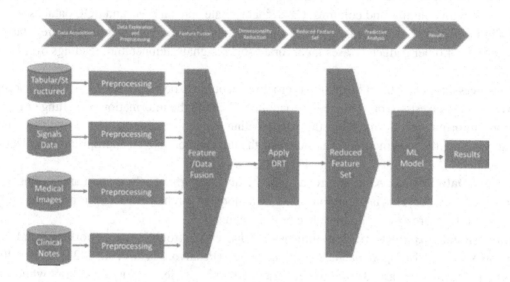

Role of CRNNs in Data Fusion and Decision Making

This subsection investigates the particular part of Convolutional Repetitive Neural Systems (CRNNs) within the combination pipeline for multi-modal sensor combination. CRNNs are well-suited for handling multi-modal sensor information due to their capacity to capture both spatial and transient conditions within a bound together engineering. Within the combination pipeline, CRNNs are utilized to meld highlights extricated from visual, LiDAR, and IMU information, empowering the ramble to create educated choices based on coordinated multi-modal data. CRNNs can successfully learn to extract discriminative highlights from multi-modal sensor information and make forecasts in real-time, encouraging assignments such as protest discovery and Hammer. By joining CRNNs into the combination pipeline, mechanical rambles can accomplish vigorous discernment capabilities and explore complex situations with made strides precision and unwavering quality.

OBJECT DETECTION IN ENVIRONMENTS

Challenges in Object Detection for Industrial Drones

Recognizing and understanding the challenges in question location inside mechanical situations is fundamental for creating viable arrangements. Mechanical rambles experience various impediments, counting fluctuating lighting conditions, occlusions, energetic deterrents, and complicated structures.

These challenges posture noteworthy obstacles for conventional question location strategies, frequently driving to diminished precision and unwavering quality in down to earth applications. In addition, the nearness of cluttered and congested situations complicates the assignment of recognizing objects of interest. Recognizing and tending to these challenges are crucial steps in formulating vigorous protest location calculations custom fitted to mechanical ramble operations.

Object Detection Techniques and Algorithms

A comprehensive investigation of different question location procedures and calculations underscores the assorted procedures utilized in mechanical ramble applications. Conventional strategies such as Haar cascades, Histogram of Arranged Slopes (Hoard), and format coordinating have served as foundational approaches. Be that as it may, their confinements in strength and versatility, especially in complex mechanical situations, require the investigation of more advanced techniques. Profound learning-based approaches, counting Single Shot MultiBox Locator (SSD), You Simply See Once (YOLO), and Region-based Convolutional Neural Systems (R-CNN), have developed as promising choices. SSD is a real-time item identification approach that uses multiple bounding boxes and class probabilities to be predicted simultaneously through a single deep neural network. SSD efficiently detects objects across different spatial resolutions with a collection of default bounding boxes with variable scales and aspect ratios. This enables SSD to balance speed and accuracy in object localization tasks. The groundbreaking object detection technique You Only Look Once (YOLO) was developed for real-time applications. The input image is divided into a grid by YOLO, which then forecasts bounding boxes and class probabilities for every grid cell. YOLO's one-pass technique allows it to get remarkably fast results, which makes it especially useful in situations where accurate and quick object recognition (D. Zhang & Chang, 2023) is required, such as industrial drone operations. R-CNN created region suggestions and classified them with a CNN, establishing a region-based approach to object recognition. Further enhancements in speed and accuracy were made by its variants, Fast R-CNN and Faster R-CNN. R-CNN enhanced object detection by using region proposals to concentrate computation on possible object locations. This made object detection more effective and appropriate for intricate situations such as those found in industrial drone applications. Privacy-preserving Deep Transformation Self-attention guarantees the protection and privacy of inference results. (Ma et al., 2021) Leveraging convolutional neural systems (CNNs), these strategies independently extricate discriminative highlights from input information, empowering real-time predictions—a basic necessity in mechanical ramble operations requesting quick and precise protest location.

Role of Multi-Modal Fusion in Improving Object Detection Accuracy

Analyzing the centrality of multi-modal combination divulges its essential part in lifting question discovery precision in mechanical settings. By coordinating information from different sensors such as visual cameras, LiDAR, and Inertial Estimation Units (IMUs), rambles pick up a comprehensive understanding of their environment. This all encompassing approach leads to improved question discovery execution by leveraging complementary data over distinctive sensor modalities. Multi-modal combination strategies enable rambles to compensate for person sensor impediments, subsequently supporting general strength. Moreover, combination at numerous levels of reflection empowers rambles to capture both fine-grained points of interest and high-level semantic data, in this manner refining protest discovery precision. Tackling the control of multi-modal combination prepares mechanical rambles with the capabilities required to explore and work securely and proficiently inside challenging mechanical situations.

SIMULTANEOUS LOCALIZATION AND MAPPING (SLAM)

Importance of SLAM in Industrial Drone Applications

Understanding the central role of simultaneous localization and mapping (SLAM) in industrial drone applications is essential. SLAM constitutes the core technology that enables drones to navigate autonomously and map their environment in real time (Aggarwal et al., 2022).Especially in industrial environments, where accurate positioning and mapping are critical for safe and efficient operations, SLAM becomes indispensable. By continuously updating the location and simultaneously building a map of the environment, drones equipped with SLAM can skillfully maneuver in complex and dynamic industrial environments with precision and reliability. This capability is especially important in environments where GPS signals are unreliable or unavailable, such as indoor facilities or areas with signal interference.

Traditional SLAM Methods and Limitations

Diving into conventional Slam methods discloses their characteristic limitations. Routine Pummel procedures regularly depend on probabilistic approaches like Expanded Kalman Channels (EKF)(Ali-Löytty et al., 2005) or Molecule Channels to appraise the drone`s posture (position and introduction) and outline the environment at the same time. Whereas viable in numerous scenarios, these strategies show a few disadvantages. They may battle to handle large-scale situations or energetic scenes with moving objects, driving to float or irregularity within the evaluated posture and outline. Additionally, conventional Hammer strategies can be computationally serious, constraining their real-time pertinence in resource-constrained situations. Recognizing these confinements underscores the need for investigating elective approaches that address these challenges and empower more strong and proficient routes in mechanical ramble applications.

Enhancing SLAM Performance Through Multi-Modal Sensor Fusion

Investigating the potential of multi-modal sensor combination in upgrading SLAM execution sheds light on inventive arrangements. By coordinating information from numerous sensors such as visual cameras, LiDAR, and Inertial Estimation Units (IMUs), rambles can increase conventional Pummel strategies with wealthier and more assorted data around the environment. Multi-modal combination strategies engage rambles to compensate for person sensor impediments and progress by and large Pummel strength and precision. For occasion, LiDAR sensors offer exact profundity estimations, complementing visual information and upgrading highlight extraction and mapping precision. IMUs give profitable movement data that can help in making strides, posture estimation and diminishing float within the Pummel framework. Leveraging data from numerous sensor modalities prepares mechanical drones with Hammer capabilities to realize more precise and solid localization and mapping in challenging and energetic mechanical situations, in this manner encouraging more secure and more effective operations. The method of Simultaneous Localization and Mapping, or SLAM, is an essential technique in robotics and autonomous systems, and Figure 4 depicts its main steps. The system begins collecting raw sensor data from various sources, such as cameras, lidar, or Inertial Measurement Units (IMUs), throughout the localization phase. This data is then processed by algorithms (Mengi et al., 2023) to identify the robot's or device's current location with respect to the surroundings. Additional measurements are provided for navigational reasons by the Measurement Navigation Sensor, which can consist of GPS or lidar. Based on the present estimated position, the system predicts what the sensors should detect in the Predicted Observation step. Perception Lidar along with additional cameras are examples of sensors that provide useful information about the environment (Kumar et al., 2021a).

Figure 4. Implementation of SLAM

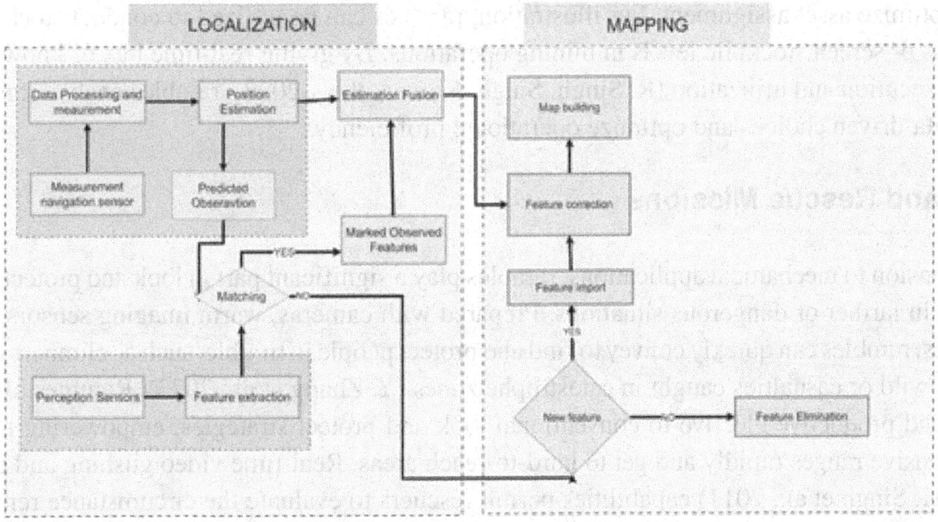

REAL-WORLD APPLICATIONS

Inspection and Monitoring in Industrial Settings

One of the essential real-world applications of mechanical rambles lies in assessment and checking errands inside mechanical settings. Rambles prepared with cameras, LiDAR sensors, and other specialized sensors can independently explore through complex mechanical situations to examine framework, gear, and offices. For example, drones can be used to examine pipelines, control lines, and mechanical apparatus for signs of harm, erosion, or wear and tear, forest fire monitoring, disaster rescue (Mao et al., 2021) and sustainable development. (Wu et al., 2018) (Kumar et al., 2023c) (Peñalvo, Sharma, et al., 2022) By capturing high-resolution pictures and 3D point clouds of the foundation, rambles empower reviewers to distinguish potential issues and inconsistencies rapidly and precisely(Kumar et al., 2023a). In addition, ramblers can get to hard-to-reach or unsafe regions that are challenging for human auditors to reach securely, subsequently upgrading work environment security and proficiency. Real-time information gushing capabilities encourage proactive support and decision-making, permitting mechanical offices to address potential issues some time recently they heighten into exorbitant disappointments or mischances.

Maintenance Operations and Asset Management

Another vital application of mechanical rambles is in support operations and resource administration. Rambles prepared with sensors and cameras can perform schedule reviews of mechanical resources, such as capacity tanks, distribution centers, and fabricating offices, to evaluate their condition and screen for any signs of disintegration or breakdown. By robotizing review assignments, rambles streamline upkeep operations, decrease downtime, and drag out the life expectancy of basic resources. Moreover, rambles can be coordinated with resource administration frameworks to track resource areas, screen stock levels, and optimize asset assignment. For illustration, rambles can be utilized to conduct stock checks in stockrooms or screen stockpile levels in mining operations. By giving real-time bits of knowledge into resource execution and utilization (R. Singh, Singh, Kumar, et al., 2022), rambles enable organizations to form data-driven choices and optimize operational proficiency.

Search and Rescue Missions

In expansion to mechanical applications, rambles play a significant part in look and protect missions, especially in farther or dangerous situations. Prepared with cameras, warm imaging sensors, and GPS capabilities, rambles can quickly convey to find and protect people in trouble, such as climbers misplaced within the wild or casualties caught in catastrophe zones (Y. Zhang et al., 2023). Rambles offer a cost-effective and productive elective to conventional look and protect strategies, empowering rescuers to cover expansive ranges rapidly and get to hard-to-reach areas. Real-time video gushing and communication(S. K. Singh et al., 2011) capabilities permit rescuers to evaluate the circumstance remotely and give help or direction to the people in need. Besides, rambles prepared with warm imaging sensors can distinguish body warm marks, indeed in low-light or antagonistic climate conditions, making strides the chances of finding survivors. By leveraging the nimbleness and flexibility of rambles, look and protect groups can improve their capabilities and spare lives in basic circumstances.

CHALLENGES AND FUTURE DIRECTIONS

Remaining Challenges in Multi-Modal Fusion

In spite of noteworthy progressions, a few challenges hold on within the space of multi-modal sensor combination for mechanical rambles. One major challenge is the integration of heterogeneous sensor information, each with its own characteristics and clamor designs. Harmonizing information from visual cameras, LiDAR sensors, and IMUs whereas bookkeeping for contrasts in determination, inspecting rates, and estimation vulnerabilities remains a complex errand. Also, guaranteeing vigor to natural inconstancy, such as changes in lighting conditions, climate, and territory, postures a noteworthy challenge for multi-modal combination calculations. Creating strategies that can adaptively combine sensor information in real-time whereas keeping up precision and unwavering quality in energetic situations is fundamental for overcoming these challenges.

Emerging Technologies and Opportunities

Rising innovations offer promising roads for tending to existing challenges and opening unused openings in multi-modal sensor combinations for mechanical rambles. Propels in machine learning, especially profound learning, have revolutionized the field by empowering more modern combination calculations capable of learning complex designs and connections from multi-modal sensor information. Procedures such as profound neural systems, convolutional repetitive neural systems (CRNNs), and consideration instruments have appeared to guarantee in moving forward the strength and productivity of multi-modal combination calculations. Additionally, progressions in sensor innovation, such as the advancement of low-cost, miniaturized sensors with made strides precision and unwavering quality, advance the possibility and versatility of multi-modal combination for mechanical ramble applications.

Directions for Future Research and Development

Looking ahead, a few key ranges warrant advancements to progress the state-of-the-art in multi-modal sensor combination for mechanical rambles. One critical course is the investigation of versatile combination calculations that can powerfully alter combination parameters based on natural conditions and errand necessities. Versatile combination calculations can upgrade the versatility and versatility of mechanical rambles, empowering them to function viably in differing and unusual situations. More-over, investigate endeavors ought to center on the improvement of standardized assessment systems and benchmark datasets for dispassionately comparing the execution of distinctive combination calculations. Setting up common assessment measurements and datasets will encourage collaboration and quicken advance within the field by empowering analysts to construct upon each other's work and approve modern methods in real-world scenarios.

CONCLUSION

The integration of multi-modal sensor fusion and Convolutional Recurrent Neural Networks (CRNNs) to provide robust object detection and Simultaneous Localization and Mapping (SLAM) in agile industrial drones is explored. The aim was to enhance the drone's proficiency in effectively detecting objects and conducting operations in complex industrial settings. The use of LiDAR and inertial measurement unit (IMU) sensors, providing a complete understanding of the environment and facilitating intelligent interaction in challenging conditions. The fundamentals of multi-modal sensor fusion is important for perception enhancement in industrial drones as it allows for a more robust and reliable perception system. Moreover, focus on Convolutional Recurrent Neural Networks providing insights into the architecture, advantages and role of CRNNs in multi-modal fusion. Therefore, CRNNs is an effective tool for integrating geographical and temporal data in a smooth manner, which is ideal for handling the complexities of industrial drone operations. The importance of SLAM in industrial drone applications, traditional SLAM methods and their limitations were discussed, leading to the need of enhancing SLAM performance through multi-modal sensor fusion. This allows drones to navigate autonomously and map their environment in real time for safe and efficient operations. The real world applications showcase the practical implications of proposed technology. The challenges and future directions of multi-modal fusion technology are showcased.

REFERENCES

Aggarwal, K., Singh, S. K., Chopra, M., Kumar, S., & Colace, F. (2022). Deep Learning in Robotics for Strengthening Industry 4.0.: Opportunities, Challenges and Future Directions. In Nedjah, N., Abd El-Latif, A. A., Gupta, B. B., & Mourelle, L. M. (Eds.), *Robotics and AI for Cybersecurity and Critical Infrastructure in Smart Cities* (pp. 1–19). Springer International Publishing. DOI: 10.1007/978-3-030-96737-6_1

Ali-Löytty, S., Sirola, N., & Piché, R. (2005). *Consistency of three Kalman filter extensions in hybrid navigation.*

Alibabaei, K., Gaspar, P. D., Lima, T. M., Campos, R. M., Girão, I., Monteiro, J., & Lopes, C. M. (2022). A Review of the Challenges of Using Deep Learning Algorithms to Support Decision-Making in Agricultural Activities. *Remote Sensing (Basel)*, 14(3), 3. DOI: 10.3390/rs14030638

Bailey, T., & Durrant-Whyte, H. (2006). Simultaneous localization and mapping (SLAM): Part II. *IEEE Robotics & Automation Magazine*, 13(3), 108–117. DOI: 10.1109/MRA.2006.1678144

Cheng, Z., Xu, Y., Cheng, M., Qiao, Y., Pu, S., Niu, Y., & Wu, F. (2020). *Refined Gate: A Simple and Effective Gating Mechanism for Recurrent Units* (arXiv:2002.11338). arXiv. https://doi.org//arXiv.2002.11338DOI: 10.48550

Chopra, A. G. (2023). Impact of Artificial Intelligence and the Internet of Things in Modern Times and Hereafter: An Investigative Analysis. In *Advanced Computer Science Applications*. Apple Academic Press.

Elmenreich, W., & Pitzek, S. (2001). Using sensor fusion in a time-triggered network. *IECON'01. 27th Annual Conference of the IEEE Industrial Electronics Society (Cat. No.37243)*. IEEE. DOI: 10.1109/IECON.2001.976510

Floreano, D., & Wood, R. J. (2015). Science, technology and the future of small autonomous drones. *Nature*, 521(7553), 460–466. DOI: 10.1038/nature14542 PMID: 26017445

Frese, U., Wagner, R., & Röfer, T. (2010). A SLAM Overview from a User's Perspective. *KI - Künstliche Intelligenz, 24*(3), 191–198. DOI: 10.1007/s13218-010-0040-4

Girshick, R. (2015). *Fast R-CNN*. 1440–1448. https://openaccess.thecvf.com/content_iccv_2015/html/Girshick_Fast_R-CNN_ICCV_2015_paper.html

Höflinger, F., Müller, J., Zhang, R., Reindl, L. M., & Burgard, W. (2013). A Wireless Micro Inertial Measurement Unit (IMU). *IEEE Transactions on Instrumentation and Measurement*, 62(9), 2583–2595. DOI: 10.1109/TIM.2013.2255977

Khade, G., Kumar, S., & Bhattacharya, S. (2012). Classification of web pages on attractiveness: A supervised learning approach. *2012 4th International Conference on Intelligent Human Computer Interaction (IHCI)*. IEEE. DOI: 10.1109/IHCI.2012.6481867

Kumar, S., Singh, S., Aggarwal, N., & Aggarwal, K. (2021a, December 26). *Efficient speculative parallelization architecture for overcoming speculation overheads.*

Kumar, S., Singh, S., Aggarwal, N., & Aggarwal, K. (2021b). Evaluation of automatic parallelization algorithms to minimize speculative parallelism overheads: An experiment. *Journal of Discrete Mathematical Sciences and Cryptography*, 24(5), 1517–1528. DOI: 10.1080/09720529.2021.1951435

Kumar, S., Singh, S. K., & Aggarwal, N. (2023a). Speculative Parallelism on Multicore Chip Architecture Strengthen Green Computing Concept: A Survey. In *Advanced Computer Science Applications*. Apple Academic Press. DOI: 10.1201/9781003369066-2

Kumar, S., Singh, S. K., & Aggarwal, N. (2023b). Sustainable Data Dependency Resolution Architectural Framework to Achieve Energy Efficiency Using Speculative Parallelization. *2023 3rd International Conference on Innovative Sustainable Computational Technologies (CISCT)*, (pp. 1–6). IEEE. DOI: 10.1109/CISCT57197.2023.10351343

Kumar, S., Singh, S. K., & Aggarwal, N. (2023c). Sustainable Data Dependency Resolution Architectural Framework to Achieve Energy Efficiency Using Speculative Parallelization. *2023 3rd International Conference on Innovative Sustainable Computational Technologies (CISCT)*, (pp. 1–6). IEEE. DOI: 10.1109/CISCT57197.2023.10351343

Kumar, S., Singh, S. K., Aggarwal, N., Gupta, B. B., Alhalabi, W., & Band, S. S. (2022). An efficient hardware supported and parallelization architecture for intelligent systems to overcome speculative overheads. *International Journal of Intelligent Systems*, 37(12), 11764–11790. DOI: 10.1002/int.23062

Kümmerle, R., Steder, B., Dornhege, C., Ruhnke, M., Grisetti, G., Stachniss, C., & Kleiner, A. (2009). On measuring the accuracy of SLAM algorithms. *Autonomous Robots*, 27(4), 387–407. DOI: 10.1007/s10514-009-9155-6

Li, Z., Liu, F., Yang, W., Peng, S., & Zhou, J. (2022). A Survey of Convolutional Neural Networks: Analysis, Applications, and Prospects. *IEEE Transactions on Neural Networks and Learning Systems*, 33(12), 6999–7019. DOI: 10.1109/TNNLS.2021.3084827 PMID: 34111009

Liebl, B., & Burghardt, M. (2020). *On the Accuracy of CRNNs for Line-Based OCR: A Multi-Parameter Evaluation* (arXiv:2008.02777). arXiv. https://doi.org//arXiv.2008.02777DOI: 10.48550

Lindsay, G. W. (2021). Convolutional Neural Networks as a Model of the Visual System: Past, Present, and Future. *Journal of Cognitive Neuroscience*, 33(10), 2017–2031. DOI: 10.1162/jocn_a_01544 PMID: 32027584

Lingenfelser, F., Wagner, J., & André, E. (2011). A systematic discussion of fusion techniques for multi-modal affect recognition tasks. *Proceedings of the 13th International Conference on Multimodal Interfaces*, (pp. 19–26). IEEE. DOI: 10.1145/2070481.2070487

Liu, X. (2008). Airborne LiDAR for DEM generation: Some critical issues. *Progress in Physical Geography*, 32(1), 31–49. DOI: 10.1177/0309133308089496

Lobiyal, R. K., Sunil, K., & Singh, D. K. (2023). Communication structure for Vehicular Internet of Things (VIoTs) and review for vehicular networks. In *Automation and Computation*. CRC Press.

Ma, B., Wu, J., Lai, E., & Hu, S. (2021). PPDTSA: Privacy-preserving Deep Transformation Self-attention Framework For Object Detection. *2021 IEEE Global Communications Conference (GLOBECOM)*, (pp. 1–5). IEEE. DOI: 10.1109/GLOBECOM46510.2021.9685855

Mao, S., He, S., & Wu, J. (2021). Joint UAV Position Optimization and Resource Scheduling in Space-Air-Ground Integrated Networks With Mixed Cloud-Edge Computing. *IEEE Systems Journal*, 15(3), 3992–4002. DOI: 10.1109/JSYST.2020.3041706

Mengi, G., Singh, S. K., Kumar, S., Mahto, D., & Sharma, A. (2023). Automated Machine Learning (AutoML): The Future of Computational Intelligence. In Nedjah, N., Martínez Pérez, G., & Gupta, B. B. (Eds.), *International Conference on Cyber Security, Privacy and Networking (ICSPN 2022)* (pp. 309–317). Springer International Publishing. DOI: 10.1007/978-3-031-22018-0_28

Peñalvo, F. J. G., Maan, T., Singh, S. K., Kumar, S., Arya, V., Chui, K. T., & Singh, G. P. (2022). Sustainable Stock Market Prediction Framework Using Machine Learning Models. [IJSSCI]. *International Journal of Software Science and Computational Intelligence*, 14(1), 1–15. DOI: 10.4018/IJSSCI.313593

Peñalvo, F. J. G., Sharma, A., Chhabra, A., Singh, S. K., Kumar, S., Arya, V., & Gaurav, A. (2022). Mobile Cloud Computing and Sustainable Development: Opportunities, Challenges, and Future Directions. [IJCAC]. *International Journal of Cloud Applications and Computing*, 12(1), 1–20. DOI: 10.4018/IJCAC.312583

Sharma, A., Singh, S. K., Badwal, E., Kumar, S., Gupta, B. B., Arya, V., Chui, K. T., & Santaniello, D. (2023). Fuzzy Based Clustering of Consumers' Big Data in Industrial Applications. *2023 IEEE International Conference on Consumer Electronics (ICCE)*, (pp. 01–03). IEEE. DOI: 10.1109/ICCE56470.2023.10043451

Sharma, A., Singh, S. K., Chhabra, A., Kumar, S., Arya, V., & Moslehpour, M. (2023). A Novel Deep Federated Learning-Based Model to Enhance Privacy in Critical Infrastructure Systems. [IJSSCI]. *International Journal of Software Science and Computational Intelligence*, 15(1), 1–23. DOI: 10.4018/IJSSCI.334711

Sharma, A., Singh, S. K., Kumar, S., Preet, M., Gupta, B. B., Arya, V., & Chui, K. T. (2024). Revolutionizing Healthcare Systems: Synergistic Multimodal Ensemble Learning & Knowledge Transfer for Lung Cancer Delineation & Taxonomy. *2024 IEEE International Conference on Consumer Electronics (ICCE)*, (pp. 1–6). IEEE. DOI: 10.1109/ICCE59016.2024.10444476

Singh, I., Singh, S. Kumar, S., & Aggarwal, K. (2022). Dropout-VGG Based Convolutional Neural Network for Traffic Sign Categorization. In M. Saraswat, H. Sharma, K. Balachandran, J. H. Kim, & J. C. Bansal (Eds.), *Congress on Intelligent Systems* (pp. 247–261). Springer Nature. DOI: 10.1007/978-981-16-9416-5_18

Singh, I., Singh, S. K., Singh, R., & Kumar, S. (2022). Efficient Loop Unrolling Factor Prediction Algorithm using Machine Learning Models. *2022 3rd International Conference for Emerging Technology (INCET)*, (pp. 1–8). IEEE. DOI: 10.1109/INCET54531.2022.9825092

Singh, R., Singh, S. K., Kumar, S., & Gill, S. S. (2022). SDN-Aided Edge Computing-Enabled AI for IoT and Smart Cities. In *SDN-Supported Edge-Cloud Interplay for Next Generation Internet of Things*. Chapman and Hall/CRC. DOI: 10.1201/9781003213871-3

Singh, S., Sharma, S., Singla, D., & Gill, S. S. (2022). *Evolving Requirements and Application of SDN and IoT in the Context of Industry 4.0*. Blockchain and Artificial Intelligence. DOI: 10.1002/9781119857921. ch13

Singh, S. K., Kumar, A., Gupta, S., & Madan, R. (2011). Architectural performance of WiMAX over WiFi with reliable QoS over wireless communication. *International Journal of Advanced Networking and Applications*, 3(1), 1017.

Wu, J., Guo, S., Huang, H., Liu, W., & Xiang, Y. (2018). Information and Communications Technologies for Sustainable Development Goals: State-of-the-Art, Needs and Perspectives. *IEEE Communications Surveys and Tutorials*, 20(3), 2389–2406. DOI: 10.1109/COMST.2018.2812301

Yang, R., Zhang, W., Tiwari, N., Yan, H., Li, T., & Cheng, H. (2022). Multimodal Sensors with Decoupled Sensing Mechanisms. *Advancement of Science*, 9(26), 2202470. DOI: 10.1002/advs.202202470 PMID: 35835946

Yin, W., Kann, K., Yu, M., & Schütze, H. (2017). *Comparative Study of CNN and RNN for Natural Language Processing* (arXiv:1702.01923). arXiv. https://doi.org//arXiv.1702.01923DOI: 10.48550

Zatsiorsky, V. M., & Duarte, M. (1999). Instant equilibrium point and its migration in standing tasks: Rambling and trembling components of the stabilogram. *Motor Control*, 3(1), 28–38. DOI: 10.1123/mcj.3.1.28 PMID: 9924099

Zhang, D., & Chang, W. (2023). A Novel Semantic Segmentation Approach Using Improved SegNet and DSC in Remote Sensing Images. *International Journal on Semantic Web and Information Systems*, 19(1), 1–17. DOI: 10.4018/IJSWIS.332769

Zhang, Y., Liu, M., Guo, J., Wang, Z., Wang, Y., Liang, T., & Singh, S. K. (2023). Optimal Revenue Analysis of the Stubborn Mining Based on Markov Decision Process. In Xu, Y., Yan, H., Teng, H., Cai, J., & Li, J. (Eds.), *Machine Learning for Cyber Security* (pp. 299–308). Springer Nature Switzerland. DOI: 10.1007/978-3-031-20099-1_25

Zhou, S., Li, L., Zhang, X., Zhang, B., Bai, S., Sun, M., Zhao, Z., Lu, X., & Chu, X. (2024). *LiDAR-PTQ: Post-Training Quantization for Point Cloud 3D Object Detection* (arXiv:2401.15865). arXiv. https://doi.org//arXiv.2401.15865DOI: 10.48550

Chapter 13
Climbing Human–Machine Interaction and Wireless Tele–Operation in Smart Autonomous Robots:
Exploring the Future of Space Robotics and Autonomous Systems

Bhupinder Singh
https://orcid.org/0009-0006-4779-2553
Sharda University, India

Christian Kaunert
https://orcid.org/0000-0002-4493-2235
Dublin City University, Ireland

ABSTRACT

The future of space exploration is defined by a close partnership between intelligent autonomous robots and humans. These robots serve as extensions in space which do tasks in dangerous or harsh environments where direct human participation is impossible. In recent years, researchers, commercial firms and space organizations have made considerable advances in the creation of autonomous and remotely controlled space robots. These robots serve an important role in space exploration, assisting people on the International Space Station (ISS) and exploring distant celestial bodies. These robots require highly developed tele-operation interfaces and HMI designs to bridge the knowledge gap between human competency and machine execution. This chapter provides a thorough analysis of the dynamic interactions between intelligent autonomous robots and humans. It also offers insights for space agencies, academics, engineers and everyone else interested in the fascinating field of space technology and exploration which lays out a vision for the future of autonomous systems and space robots.

DOI: 10.4018/979-8-3693-2707-4.ch013

INTRODUCTION AND BACKGROUND

The frontiers of technology innovation and human understanding have continuously been pushed by space travel. The need for advanced robotic systems that can navigate and perform activities in difficult alien conditions is growing as this expands our knowledge of the universe. The ability to conduct research in areas that are uninhabitable or inaccessible to human astronauts has been greatly aided by the use of autonomous robots into space travel. In this context, the combination of wireless teleoperation and human-machine interaction (HMI) becomes a key paradigm in the development of intelligent autonomous robots, providing a technique to improve these robots' overall efficiency, flexibility and adaptability in traversing the intricacies of space (Gao, 2021).

This work addresses the combination of climbing mechanics, HMI, and wireless teleoperation to create intelligent autonomous robots for space exploration. As the historical timeline of space robotics shows, the development of intelligent and autonomous robotic platforms has been a gradual evolution from fundamental mechanical systems. These devices are now essential instruments for anything from maintenance and repair missions to interplanetary surface investigation. But given the needs of space travel, new developments are needed all the time to get beyond obstacles like unreliable terrain, very high temperatures, and delayed communication. Given this context, climbing robots represent a potentially exciting new frontier in exploration, as they can traverse uneven and steep surfaces, opening up hitherto unexplored areas (Mahmud et al., 2020).

The smooth incorporation of human operators into autonomous robots via advanced human-machine interface (HMI) technologies represents a noteworthy advancement in the cooperation between humans and machines. Human-machine interface (HMI) design and implementation focuses on creating user-friendly interfaces that allow human operators and robotic systems to communicate intuitively. The goal is to develop an interface structure that bridges the gap between human thought and machine execution, allowing people to cooperate, supervise, and direct robots with effectiveness. The significance of HMI in providing accurate control and real-time decision-making becomes more and more crucial as space missions get more complex. Concurrently, the wireless teleoperation becomes a fundamental component of space robotics development, especially in situations where direct human presence is impractical. The ability to remotely control autonomous robots wirelessly enhances the reach and flexibility of these machines, enabling them to navigate vast distances while maintaining a constant communication link with human operators (Shukla & Karki, 2016).

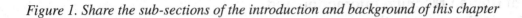

Figure 1. Share the sub-sections of the introduction and background of this chapter

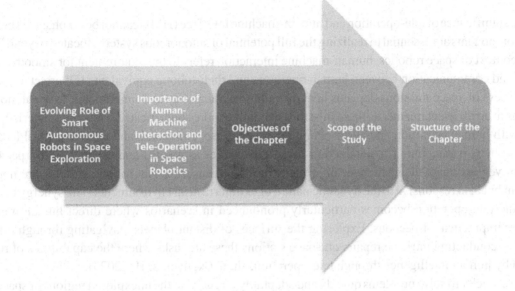

Evolving Role of Smart Autonomous Robots in Space Exploration

During space missions, intelligent autonomous robots are essential for enhancing human skills. Although human astronauts are incredibly adaptive and skilled problem solvers, they are also susceptible to the severe circumstances of space travel, such as radiation exposure, extremely high temperatures, and prolonged solitude. Because autonomous robots can do dangerous and repetitive jobs, human astronauts may concentrate on more strategic and complicated tasks. This human-robot cooperative strategy improves mission safety and efficiency (Shukla & Karki, 2016).

The incorporation of artificial intelligence (AI) into intelligent self-governing robots enhances their efficacy in space exploration. Robots can now assess data in real time, decide wisely, and adjust to new situations thanks to AI algorithms. This capacity is especially helpful in situations where it is impossible to immediately intervene humanly due to communication delays. These robots may adjust their performance according to previous experiences and continually enhance their capacity to travel and function in various interplanetary situations by utilizing machine learning and complex algorithms. With space agencies and commercial companies aiming for ambitious exploration missions, intelligent autonomous robots are becoming more important than just data collectors. These robots have the potential to be essential parts of future interplanetary expeditions, space habitats and lunar colonies. Their ability to do building, maintenance, and repair activities in zero gravity creates new opportunities for human presence that is sustainable beyond Earth. The idea of autonomous robots working alongside human astronauts on challenging missions has the potential to promote a mutually beneficial partnership that optimizes the abilities of both parties (Ha et al., 2019).

Importance of Human-Machine Interaction and Tele-Operation in Space Robotics

The significance of tele-operation and human-machine interface (HMI) cannot be emphasized because these components are essential to realizing the full potential of autonomous systems located beyond Earth. In the context of space robotics, human-machine interaction refers to the requirement for smooth coordination and communication between human operators and their robotic equivalents (Ollero et al., 2019). The success of space missions depends heavily on humans' capacity to communicate with autonomous systems in an efficient manner as these missions becoming more sophisticated. Designing interfaces that align with human cognition and allow operators to supervise, command, and receive critical feedback in real-time is paramount. This interaction paradigm enables astronauts and ground control personnel to intuitively navigate the challenges of space exploration, ensuring that robotic systems are not only efficient but also responsive to the dynamic nature of extraterrestrial environments. The synergy between HMI and tele-operation becomes particularly pronounced in scenarios where direct human presence is either impractical or perilous. Exploring the surfaces of distant planets, navigating through asteroid fields, or conducting intricate repairs on space stations these are tasks where the capabilities of robots, guided by human intelligence through tele-operation, shine (Nguyen & Ha, 2023).

The capacity to solve problems quickly and adaptably is crucial in the unexplored regions of space, and it is made possible by the combination of human supervision that is intuitive and the accuracy provided by tele-operation. Beyond the initial operational stage of space missions, HMI and tele-operation are crucial. Remote control and interaction with autonomous systems is becoming essential for long-term exploration and living on Mars, especially as space agencies and commercial companies aim to establish long-term missions, lunar outposts, or possibly human settlement there. The combination of these components improves regular work efficiency while acting as a safety net that allows human operators to take over and intervene when the autonomy of robotic systems proves inadequate. These elements are not merely technological conveniences but represent the nexus between human ingenuity and the capabilities of autonomous systems (Lee et al., 2022).

Objectives of the Chapter

This chapter has the following objectives to-
- analyze the current state of space robotics and identify limitations.
- study climbing mechanisms for enhanced mobility in microgravity environments.
- explore the role of HMI in improving human-robot collaboration.
- examine the implementation of wireless teleoperation for remote control and data transmission in space.

Figure 2. Shows the objectives of this chapter

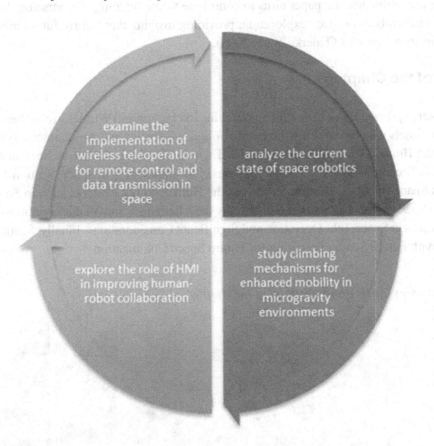

Scope of the Study

The scope of this chapter encompasses a comprehensive exploration of the dynamic interplay between human-machine interaction (HMI) and wireless teleoperation in the context of smart autonomous robots, with a specific focus on their application in space robotics. As this stand at the forefront of a new era in space exploration, marked by ambitious missions to celestial bodies and the establishment of human presence beyond Earth, understanding and optimizing the capabilities of autonomous systems become imperative (Zhang et al., 2017).

The study explores the subtleties of human-machine interaction and clarifies its importance in promoting natural language exchange and cooperation between human operators and intelligent autonomous robots. The research attempts to offer a thorough overview of the state-of-the-art in HMI and its implications for space travel by reviewing the body of current literature and technology advancements (Ollero et al., 2005). The scope includes investigating new technologies and how they may revolutionize space robots. The goal of the article is to pinpoint possible uses and situations in which the combination of wireless teleoperation, HMI, and climbing robots might completely alter the field of space exploration. Furthermore, a critical examination of the theoretical and technological obstacles is conducted to shed light on the directions that future research and development in the subject should follow. By elucidating the prospects and obstacles, the paper aims to contribute to the ongoing discourse on the evolution of smart autonomous robots in space exploration, providing insights that inform future missions and advancements in space robotics (Tiderko et al., 2008).

Structure of the Chapter

This chapter expressly dives in the Human-Machine Interaction and Wireless Tele-Operation in Smart Autonomous Robots which Exploring the Future of Space Robotics and Autonomous Systems. Section 2 elaborates the Human-Machine Interaction (HMI) Design for Space Robotics. Section 3 specifies the Tele-Operation Technologies- Low-Latency Wireless Communication in Space. Section 4 lays down the Autonomous Space Robotics. Section 5 explores the Human-Robot Collaboration in Space: Trust and Teamwork in Space Missions. Section 6 highlights the Challenges and Viable Solutions- Security and Privacy Concerns in Space Tele-Operation and Significant Considerations. Finally, Section 7 Conclude the Chapter with Future Scope- Preparing for Future Space Colonization

Figure 3. Expresses the flow/organization of the chapter

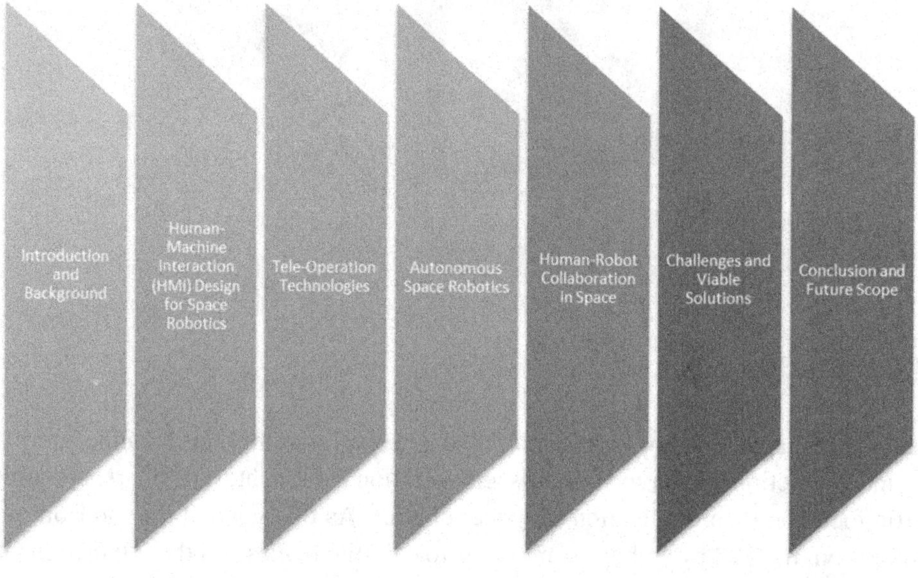

HUMAN-MACHINE INTERACTION (HMI) DESIGN FOR SPACE ROBOTICS

Human-Machine Interaction (HMI) design for space robotics represents a critical facet in the quest to enhance the efficacy and adaptability of autonomous systems operating beyond Earth. Sensor readouts, augmented reality displays, and 3D maps are examples of visualization tools that are essential to HMI design because they provide operators with the data they need to make choices quickly and intelligently.

The ease of use and intuitiveness of autonomous systems for human operators are greatly dependent on the ergonomic and cognitive components of HMI design. The design needs to take into consideration the difficulties presented by microgravity, small areas, and the possible mental strain on operators over long missions (Haidegger et al., 2019).

Design features like gesture controls, voice commands, and ergonomically optimized interfaces are a few instances of how to lessen the cognitive load on operators so they may concentrate on making higher-level decisions rather than figuring out complicated interfaces. The inclusivity of HMI design extends beyond traditional control rooms, considering the potential for remote and even collaborative control of robots by astronauts during spacewalks or extravehicular activities. Portable and wearable HMI devices equipped with haptic feedback and responsive interfaces become indispensable tools, ensuring that astronauts can maintain control and situational awareness even when physically distant from the central control hub (Mourtzis et al., 2021).

The design of Human-Machine Interaction for space robotics is a multidimensional challenge that demands a fusion of technological innovation, ergonomic considerations, and cognitive science. As space missions become more intricate and extend to destinations such as Mars and beyond, the importance of intuitive, adaptive, and user-friendly HMI designs cannot be overstated. Through thoughtful and anticipatory design principles, HMI becomes the linchpin that connects human intellect with the capabilities of smart autonomous robots, enabling a harmonious collaboration that propels space exploration into new frontiers.

Figure 4. Conveys the HMI design for space robotics

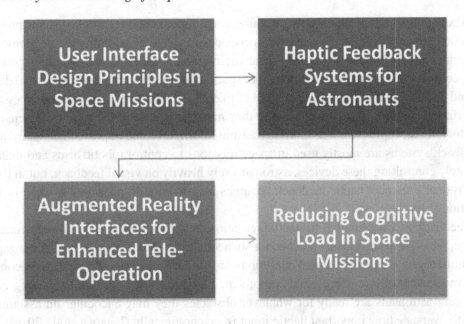

User Interface Design Principles in Space Missions

User Interface (UI) design principles take on heightened significance in the context of space missions, where the efficiency and clarity of communication between human operators and complex systems are paramount. The effectiveness of space exploration missions depends heavily on the design of user interfaces, especially as these missions get more ambitious and reach further celestial bodies. The guidelines address the particular difficulties presented by the extraterrestrial environment and the requirement for smooth communication with intelligent autonomous systems in order to create user interfaces (UIs) that are successful for space missions (Dufourd & Dalgalarrondo, 2006).

First and foremost, the fundamentals of UI design for space missions are simplicity and clarity. Given that operators may be handling complicated tasks and dynamic situations, the user interface has to provide information in a clear and succinct way to minimize cognitive burden. Clear representations combined with easy-to-use controls improve astronauts' and ground control staff's capacity to quickly analyze data and make educated decisions (Handa, 2015). Adaptability is a fundamental tenet of UI design, given the particular difficulties associated with space habitats. Unexpected difficulties, alterations in the mission context, and environmental factors must all be taken into account while designing the interface. When data varies, adaptive user interfaces (UIs) may react to it intelligently, facilitating a smooth transition between mission phases and giving astronauts and operators the flexibility they need to react to changing circumstances (Tiddi et al., 2020).

Haptic Feedback Systems for Astronauts

At the nexus of technology and human-machine interaction, haptic feedback systems provide novel approaches that enhance astronauts' sensory experiences in space by adding a tactile element. Haptic feedback devices offer astronauts an essential means of interacting with and understanding their surroundings in the peculiar and demanding environment of space, where communication delays and dependence on visual and auditory signals may be hampered (Yin et al., 2023). These systems are an essential part of user interface design for space travel because they make advantage of the astronauts' sense of touch, using vibratory, force, or pressure sensations to transmit information and improve their spatial awareness. Haptic feedback systems are mostly used in space missions to control robotic arms and manipulators. When remotely controlling these devices, astronauts rely heavily on visual feedback, but in the absence of direct physical presence, haptic feedback becomes instrumental in providing a sense of touch and proprioception (Elbasheer et al., 2023).

Haptic feedback systems simulate the tactile sensations related to jobs and situations astronauts would face in space, which helps astronauts practice for such situations. These devices bridge the gap between training simulators on Earth and the real conditions of space by allowing astronauts to become used to the feel of various controls, surfaces, and equipment. This proactive approach to training is essential to guaranteeing astronauts are ready for whatever obstacles they may encounter on assignment. It is impossible to overstate how important haptic input is psychologically (Ivanova et al., 2016).

Augmented Reality Interfaces for Enhanced Tele-Operation

In the field of improved teleoperation, augmented reality (AR) interfaces are a revolutionary development, especially when it comes to space exploration. The use of augmented reality interfaces provides a previously unattainable means for astronauts and operators to engage with distant robotic systems in a natural and engaging way as mankind continues its exploration of space. Digital and physical information are seamlessly combined in augmented reality to give users instructions, visualizations, and contextual data overlayed right in their field of vision. AR interfaces provide a bridge between human intellect and autonomous system capabilities in the context of improved teleoperation, enhancing the teleoperation experience in space missions (Liang et al., 2021). The principal benefit of AR interfaces is their capacity to improve situational awareness in the context of teleoperation. With AR headsets or smart glasses, astronauts and operators may see vital information superimposed on the physical world, including navigation routes, sensor data, and equipment status. With providing a more thorough awareness of the remote environment, this visual augmentation lessens cognitive strain and gives users the ability to make wise judgments instantly (Kuru, 2021).

This instantaneous and contextually relevant information becomes indispensable for precision control and navigation of robotic systems in space missions, where communication delays might be substantial. The capacity of AR interfaces to deliver contextual information in real-time is one of its main benefits for teleoperation. When donning AR-capable gear, including helmets or smart glasses, astronauts and operators may see visual overlays that superimpose vital information, navigational cues, and status updates right over their field of sight. This feature allows operators to easily combine digital information with the real environment around them, which is crucial in situations where quick decisions are needed. AR interfaces can improve astronauts' situational awareness in space exploration by displaying navigation markers, diagnostic data, or even holographic representations of far-off celestial entities (Kubota & Kunii, 2009).

Reducing Cognitive Load in Space Missions

Reduced cognitive load is a result of improved training programs that make use of simulation and virtual reality (VR) technology. Astronauts can rehearse emergency situations, acclimate to the alien environment, and hone their decision-making abilities in a safe and authentic environment by using immersive training settings. This type of training not only increases their readiness for mission-specific difficulties but also develops muscle memory, which lessens the mental strain of carrying out activities in the strange and possibly dangerous environment of space. The adoption of simple and user-friendly user interfaces is a crucial tactic to reduce cognitive burden in space missions. When interfaces are designed with clarity, consistency, and ease of use in mind, it takes less mental strain for ground control staff and astronauts to evaluate data and give orders. Easy-to-understand graphics combined with uncomplicated controls facilitate a more efficient cognitive process, freeing operators to concentrate on making high-level decisions instead of figuring out the intricate interface (Robinson et al., 2023).

TELE-OPERATION TECHNOLOGIES- LOW-LATENCY WIRELESS COMMUNICATION IN SPACE

Teleoperation technologies are pivotal in enabling human control of robotic systems in space, and the effectiveness of these technologies is profoundly influenced by the latency of the wireless communication channels involved. Low-latency wireless communication represents a cornerstone in the quest for seamless teleoperation, especially in the extraterrestrial environment where vast distances and signal propagation delays are inherent challenges. As it extend reach to distant planets, moons, and asteroids, the necessity for real-time or near-real-time communication becomes increasingly critical for precise and responsive control of robotic systems (Benaoumeur et al., 2015).

The developments in communication protocols and technology are directly related to the deployment of low-latency wireless communication. Reducing signal travel times can be achieved in part by using high-frequency radio waves or laser communication systems in conjunction with protocols that prioritize data transfer speed and reliability. These technology advancements are essential to making sure that teleoperation experiences closely resemble real-time situations and allow operators to precisely manage robotic equipment quickly. Low-latency wireless communication has applications outside of Earth-based teleoperation. The capacity to remotely manage robotic devices in almost real-time becomes a mission-critical skill when it considers crewed trips to Mars and beyond. Low-latency communication systems are essential to sustaining a responsive link between humans and autonomous robotic assets in these deep-space missions. At the vanguard of teleoperation technology, low-latency wireless communication is reshaping the possibilities for human control over robotic systems in space. The integration of these technologies becomes essential as this push the limits of space exploration in order to overcome the difficulties presented by signal propagation delays. Low-latency wireless technologies enable us to negotiate the complexity of space with agility and accuracy by emphasizing real-time or near-real-time communication. This opens up new possibilities for scientific research and advances in our knowledge of the universe (Das et al., 2017).

Telerobotics and Remote Control of Space Probes

In space exploration, telerobotics asthe fusion of robotics and telecommunication—is essential, especially for the remote operation of space missions. The immensity of the universe poses problems that call for creative solutions as humankind expands beyond Earth, and telerobotics emerges as a crucial enabler for human operators to communicate with and operate space probes in far-off and frequently hostile settings. This can carry out complex activities, collect data, and carry out scientific experiments in places that are inaccessible to human astronauts by using space probes that are controlled remotely (Basanez et al., 2023).

Telerobotics is one way that space research overcomes the constraints placed on it by the great distances and communication lags that come with interplanetary missions. Given the size of space and the amount of time needed for travel, it is unfeasible, if not impossible, to send human astronauts to investigate every area of our solar system. By enabling operators on Earth to remotely operate space probes, rovers, and landers, telerobotics acts as a stand-in for human presence and expands our capacity to investigate and engage with far-off celestial worlds. Telerobotic-equipped space probes extend human capabilities by enabling precise and complex operations on far-off planets, moons, or asteroids. Telerobotic technology enables the execution of scientific investigations, the collecting of samples, and the performance of movements that demand human-like dexterity, from the Martian rovers to probes investigating the ice moons of Jupiter. This capacity not only adds to our scientific understanding but also establishes the foundation for possible future human expeditions by giving vital information about the requirements and difficulties of particular alien settings (Amador et al., 2022).

The advancement of cutting-edge communication technology is critical to the success of telerobotics in space exploration. For control to be precise and immediate, dependable low-latency communication channels between Earth and the distant probes must be established. Telerobotic systems gain from increased data transfer speeds, less signal propagation delays, and better overall responsiveness as communication technology advance; these benefits all add to the efficacy of remote control in space missions (Ebrahimzadeh & Maier, 2019).

Challenges in Long-Distance Teleoperation

In order to guarantee the effectiveness and success of remote activities, long-distance teleoperation presents a number of major obstacles, particularly in the context of space exploration. These difficulties result from the great distances that must be covered, the delays in signal transmission, the bandwidth constraints, and the complexity of working in extraterrestrial conditions (Aijaz & Sooriyabandara, 2018). The following are some significant obstacles related to long-distance teleoperation as-

Latency in Communications: There will always be a delay in signal propagation between Earth and distant robotic systems due to the limited speed of light. This delay might vary from several minutes to tens of minutes in deep-space missions, such those to Mars or beyond. Real-time control is made more difficult by the delay, particularly in situations when quick decisions or exact modifications are needed.

Restricted Bandwidth: There is a limit to the bandwidth that can be used to send data from Earth to far-off space missions. The restricted bandwidth creates a bottleneck when teleoperation activities get more complicated, including transferring huge datasets or high-definition video streams. Making the most of the communication resources at hand requires effective data compression, prioritizing, and optimization strategies.

Autonomy and Decision-Making: The delicate balance between human control and autonomous decision-making by robotic systems is essential for long-distance teleoperation. It is not feasible to manage every component of an operation in real time due to the communication latency. Therefore, while human operators concentrate on making high-level decisions, autonomous systems need to be able to manage repetitive tasks and adapt to unforeseen circumstances.

Complexity of Operations: The tasks performed in long-distance teleoperation are often complex and intricate, requiring precise control and coordination. For example, manipulating robotic arms, conducting scientific experiments, or navigating challenging terrains on other planets demand a level of dexterity and precision that poses a significant challenge when operated from a remote location.

Variability in the Environment: Extraterrestrial settings might pose unforeseen difficulties due to their intrinsic unpredictability. Systems for long-distance teleoperation need to be able to adjust to changing circumstances, including variations in the environment, topography, or the robotic system's capabilities. This flexibility is essential to the accomplishment of tasks that take place in a variety of environments and over long periods of time.

Human-System Interaction: Considering the cognitive strain on human operators, long-distance teleoperation control methods and user interfaces need to be simple and easy to use. One major problem is designing interfaces that allow for efficient control and communication even in the presence of communication delay. The operator's tactile feedback and situational awareness can be improved by using augmented reality interfaces with haptic feedback.

Protection and Dependability: It is crucial to guarantee the protection and dependability of long-range teleoperation systems. To protect the communication channels from any threats or unwanted access, strong encryption techniques must be used. Maintaining control in the case of system breakdowns requires redundant communication channels and fail-safe methods.

A multidisciplinary strategy incorporating developments in robotics, artificial intelligence, communication technologies, and human-machine interaction is needed to address these difficulties (Nof et al., 2015).

Figure 5. Highlights the challenges in long-distance teleoperation

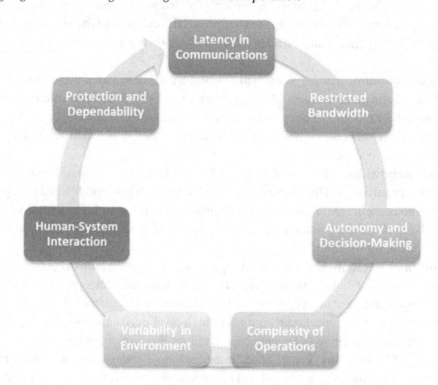

Real-Time Feedback Systems for Astronauts

When it comes to improving astronauts' decision-making, performance, and general well-being during space missions, real-time feedback systems are essential because they give them access to timely and useful information. Real-time feedback becomes an invaluable instrument for preserving situational awareness and guaranteeing the efficiency and safety of astronauts in the demanding and dynamic environment of space, where decisions made in a matter of seconds can make a big difference. Real-time feedback systems are used in the physiological parameter monitoring of astronauts. Vital indications including body temperature, oxygen saturation, and heart rate are continuously monitored, and this information is critical to understanding the astronauts' health. Any departures from the usual can be quickly identified, enabling timely action in the event of any health problems (Mathiassen et al., 2021).

Real-time feedback is essential in space robotics and teleoperation to guarantee accurate and responsive control. The ability of robotic devices or vehicles to respond instantly to orders is crucial for astronauts operating them remotely on other celestial worlds. Long-distance teleoperation's communication delays are avoided by providing astronauts with real-time input on the location, orientation, and condition of the robotic equipment (Agnisarman et al., 2019).

Technologically, sophisticated sensors, telemetry systems, and data processing power are necessary for the deployment of real-time feedback systems. These feedback systems rely heavily on wearables with sensors, augmented reality interfaces, and communication networks. Real-time data analysis using machine learning algorithms may also be used to provide adaptive and customized feedback based on the needs of each individual astronaut and the demands of the mission. Astronauts' real-time feedback systems are vital resources that support their performance, safety, and general well-being in the harsh environment of space. These technologies are essential to maximizing the human experience on space missions, whether they are tracking physiological data, enabling teleoperation, improving situational awareness, or honing training methods (Benos et al., 2023).

AUTONOMOUS SPACE ROBOTICS

Using artificial intelligence (AI) and robotics to enable spacecraft and robotic systems to function with a high degree of autonomy in the vast and difficult expanse of space, autonomous space robotics represents a cutting-edge frontier in space exploration. Autonomous space robots have the ability to make judgments, adapt to changing situations, and carry out tasks independently, which reduces the need for ongoing human involvement. Their independence in carrying out regular duties frees up important human resources for more intricate and strategic work (Rojas, 2015).

This is especially important in situations where great distances or communication lag make real-time teleoperation impracticable. By using AI algorithms to make judgments or carry out preprogrammed instructions, autonomous systems may function autonomously, simplifying mission operations and making the most use of their resources. Traditional robotic systems rely significantly on human control. This is particularly important for deep-space missions, because real-time control might be hampered by communication delays caused by the great distances involved. The complex control algorithms, computer vision, and machine learning are just a few of the cutting-edge technologies that must be integrated in the creation of autonomous space robots. Robots with integrated cameras and sensors can see and comprehend their environment thanks to computer vision, and machine learning enables them to gain insight from their experiences and improve with time (Lacava et al., 2021).

The implementation of autonomous space robots is not without difficulties, though. Thorough testing and validation are necessary to guarantee the resilience and dependability of the autonomous systems, particularly in the hostile environment of space. Also, the ethical considerations of autonomous decision-making in space exploration, such as the potential for unintended consequences or conflicts with mission objectives, necessitate careful consideration and oversight. These intelligent systems not only enhance the efficiency and autonomy of space missions but also open new possibilities for scientific discovery and exploration beyond the limitations of direct human presence. As technology continues to advance, the role of autonomous space robotics is poised to expand, reshaping the future of space exploration and our understanding of the universe (Sebbane, 2018).

Figure 6. Travels the autonomous space robotics

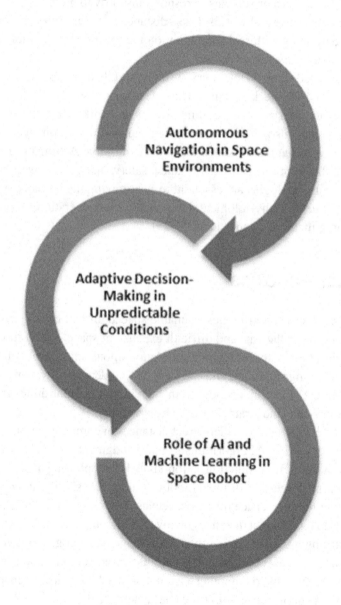

Autonomous Navigation in Space Environments

Autonomous navigation in space settings is a significant technical advancement, allowing spacecraft and robotic systems to navigate and explore celestial bodies with more freedom and adaptability. Autonomous systems grow increasingly skilled at traversing the complexity of space by continually refining their awareness of the environment and adjusting to changing situations, promoting a type of intelligence that transcends pre-programmed reactions and enables adaptive decision-making. However, implementing autonomous navigation in space is not without difficulties. Navigating without a GPS-like system, dealing with limited processing resources, and assuring algorithm resilience under harsh environmental conditions are all continuing technological challenges (Sauer, 2010).

Adaptive Decision-Making in Unpredictable Conditions

Adaptive decision-making in unpredictable conditions stands as a paramount challenge in various domains, ranging from space exploration to disaster response on Earth. In dynamic and uncertain environments, the ability to make swift and contextually informed decisions becomes a critical factor for success. Adaptive decision-making involves continuously assessing the evolving situation, adjusting strategies in real-time, and leveraging a flexible cognitive approach that accommodates unexpected challenges.

Role of AI and Machine Learning in Space Robot

AI helps to improve the planning and scheduling of space missions. Machine learning algorithms can examine mission characteristics, historical data, and external factors in order to create ideal schedules that take into account resource restrictions, energy efficiency, and scientific aims. This intelligent planning ensures that space missions are executed with maximum efficiency and that resources are utilized judiciously, ultimately enhancing the return on investment for space exploration endeavors. In the space robotics, AI and ML enable adaptive and autonomous control systems. These systems can learn from the performance of robotic platforms, adjusting their control strategies to optimize energy consumption, improve precision in operations, and respond effectively to unforeseen challenges. This adaptability is particularly crucial for long-duration missions, where space robots must operate autonomously for extended periods without constant human supervision (Cremer, 2017).

HUMAN-ROBOT COLLABORATION IN SPACE: TRUST AND TEAMWORK IN SPACE MISSIONS

Human-robot collaboration in space missions heralds a transformative era where intelligent machines work synergistically with astronauts, combining their unique strengths to overcome the challenges of the cosmos. The building of trust and productive partnership between humans and their robotic counterparts is just as important to the success of these joint efforts as the technical prowess of the robots. The foundation of human-robot cooperation is trust which reflects astronauts' faith in the skills and dependability of the robotic systems that support them. The creating of robots that exhibit dependability, openness in their decision-making, and the ability to clearly convey their goals are all necessary to gaining this confidence. In order to maximize mission success, teamwork in space missions entails the seamless integration of human and robotic capabilities. Astronauts can focus on more intricate and strategic duties by delegating risky or time-consuming jobs to robots. Human operators provide their intuition, inventiveness, and contextual knowledge to the team, which helps them to make critical judgments in unstructured contexts and quickly adjust to changing circumstances. Human-robot collaboration and communication are facilitated by interfaces that are easy to use and intuitive, which is necessary for effective teamwork. The most important components of effective human-robot collaborations in space is the development of shared autonomy, in which people and robots work together to make decisions and carry out tasks (Paik et al., 2022).

In the special environment of space missions, the psychological component of human-robot teamwork is equally important. In the cramped and lonely confines of spacecraft or space stations, astronauts develop close relationships with their robotic companions, depending on them for aid, company, and support. The sense of connection and collaboration is improved when robots are designed with anthropomorphic qualities or social skills, such as the capacity to communicate emotions or comprehend human gestures. By creating a sense of friendship between humans and robots, these social cues improve the psychological climate and increase astronauts' mental toughness and general well-being throughout protracted missions. The difficulties associated with human-robot cooperation in space go beyond technological ones and include social, cultural, and ethical issues. Achieving a successful partnership requires figuring out how autonomous robots should be, dealing with responsibility concerns when mistakes are made, and taking cultural sensitivities about robots and human contact into consideration (Kiencke et al., 2006).

Figure 7. Carries trust and teamwork in space missions

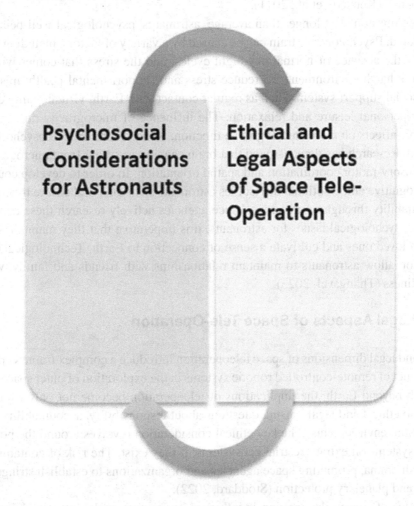

| Psychosocial Considerations for Astronauts | Ethical and Legal Aspects of Space Tele-Operation |

Psychosocial Considerations for Astronauts

Because extended missions to the International Space Station (ISS) or future interplanetary travel require careful attention to the psychological and social aspects of human experience in confined and isolated environments, astronauts' psychosocial well-being is a critical concern in the context of space exploration. A thorough understanding of psychosocial considerations is necessary to protect astronauts' mental health and resilience due to the unique challenges of space missions, which include extended separation from loved ones, the lack of natural environmental stimuli, and confinement of spacecraft or space stations. Feelings of boredom and loneliness might arise from the cramped quarters of spacecraft or space stations, as well as the restricted social contacts with a small crew. In order to overcome these obstacles, space agencies give careful consideration to a person's capacity for managing loneliness, fostering positive interpersonal interactions, and fostering a cohesive team environment in addition to technical proficiency (Kennedy et al., 2011).

Because space missions last longer than average, astronauts' psychological well-being needs to be carefully considered. Psychological strain can be caused by a variety of factors, including the monotony of daily routines, the absence of normal day-night cycles, and the stress that comes with completing difficult tasks in a harsh environment. To reduce stress and support mental health, mission planning include psychosocial support systems such as regular contact with Earth, virtual reality entertainment, and chances for personal leisure and relaxation. The influence of microgravity on the human body, including possible affects on mood and cognitive function, is the second crucial psychosocial factor to take into account. Research has demonstrated that brain changes resulting from microgravity exposure might impact sensory-motor coordination and spatial orientation. In order to develop countermeasures that lessen the cognitive and emotional difficulties astronauts encounter and ensure their mental clarity and emotional stability throughout missions, space agencies actively research these consequences. In order to address psychological issues for astronauts, it is imperative that they maintain frequent communication with loved ones and cultivate a sense of connection to Earth. Technological developments in communication allow astronauts to maintain relationships with friends and family, which reduces feelings of loneliness (Thangavel, 2023).

Ethical and Legal Aspects of Space Tele-Operation

The ethical and legal dimensions of space teleoperation introduce a complex framework that governs the use and conduct of remote-controlled robotic systems in the exploration of outer space. As humanity extends its reach beyond Earth, the implications of teleoperation become not only technological and scientific but also ethical and legal, raising questions about responsibility, accountability, and the preservation of celestial environments. The key ethical consideration revolves around the potential impact of teleoperated systems on extraterrestrial ecosystems, if they exist. The risk of contamination or disruption poses a dilemma, prompting space agencies and organizations to establish stringent guidelines for sterilization and planetary protection (Stoddard, 2022).

The governance of space teleoperation is informed by international treaties and agreements. The Outer Space Treaty, established in 1967, serves as a foundational document, outlining the principles for the peaceful use of outer space and prohibiting the placement of nuclear weapons or other weapons of mass destruction in orbit. The question of space resource utilization further complicates the legal aspects of teleoperation. As humans contemplate the extraction of resources from celestial bodies, legal frameworks must address issues of ownership, property rights, and environmental preservation. The prospect of exploiting extraterrestrial resources through teleoperation demands ethical and legal considerations that balance the potential benefits for humanity with the imperative to minimize harm and preserve the integrity of celestial environments. The ethical and legal dimensions of space teleoperation underscore the need for comprehensive frameworks that navigate the intricacies of remote-controlled missions beyond Earth. As teleoperation technologies advance, addressing questions of responsibility, environmental impact, data security, and resource utilization becomes imperative to ensure that humanity's foray into space aligns with ethical principles, international law, and a commitment to the responsible exploration and stewardship of the cosmos (Vanhee, 2021).

CHALLENGES AND VIABLE SOLUTIONS- SECURITY AND PRIVACY CONCERNS IN SPACE TELE-OPERATION AND SIGNIFICANT CONSIDERATIONS

The burgeoning field of space teleoperation introduces a spectrum of challenges related to security and privacy, demanding vigilant consideration as humanity extends its reach into the cosmos. The remote control of spacecraft and robotic systems from Earth raises unique concerns regarding the safeguarding of sensitive data, the potential for unauthorized access, and the overall security of teleoperation activities. The primary obstacle lies in the susceptibility of Earth-to-space robotic systems communication links. Since teleoperation entails sending data, telemetry, and orders across great distances, it is critical to make sure that information is sent securely and secured. The integrity of teleoperation operations is threatened by the possibility of interception, manipulation, or illegal access to communication signals. Implementing robust encryption protocols and secure communication frameworks is imperative to protect the confidentiality and authenticity of the data exchanged during space teleoperation (Xing & Marwala, 2018).

The increased reliance on artificial intelligence (AI) and machine learning (ML) in space teleoperation introduces a new dimension of security concerns. Malicious actors could potentially exploit vulnerabilities in AI algorithms, leading to unauthorized manipulation of robotic systems or the extraction of sensitive information. Securing the algorithms and ensuring the resilience of AI-based systems against adversarial attacks is an ongoing challenge that requires continuous advancements in cybersecurity.

Data privacy emerges as a significant concern in space teleoperation, especially as the volume of information transmitted between Earth and space-based systems continues to grow. The telemetry data, scientific observations, and operational details collected during teleoperation activities may contain sensitive information that requires protection. Establishing protocols for the responsible collection, storage, and transmission of data, with a focus on minimizing privacy risks, becomes essential to uphold ethical standards and legal obligations.

The international and collaborative nature of space exploration further complicates security considerations. As multiple entities from different countries engage in teleoperation activities, the coordination of security measures and the establishment of standardized protocols become imperative. Ensuring that all stakeholders adhere to uniform cybersecurity standards and cooperate in addressing potential security threats fosters a collective approach to maintaining the integrity and security of space teleoperation. An additional challenge lies in the evolving threat landscape of space activities. With an increasing number of countries and private entities participating in space exploration, the risk of deliberate interference, cyberattacks, or industrial espionage targeting teleoperation systems grows. Developing adaptive and resilient security frameworks capable of mitigating emerging threats and vulnerabilities is crucial to sustaining the long-term viability of space teleoperation (Tahir et al., 2023).

The security and privacy concerns in space teleoperation underscore the need for a comprehensive and proactive approach to cybersecurity. As space exploration activities become more interconnected and reliant on teleoperation technologies, addressing these challenges becomes paramount to safeguarding the confidentiality, integrity, and availability of data exchanged between Earth and space-based systems. Ethical considerations, legal frameworks, and international collaboration must be integral components of efforts to fortify the security and privacy of space teleoperation and ensure the responsible and secure exploration of the space (Song et al., 2016).

Viable Solutions Envisioning the Future of Space Robotics and Autonomous Systems

Envisioning the future of space robotics and autonomous systems unveils a landscape marked by unprecedented exploration capabilities and sustained human presence in the cosmos. As technology advances, we foresee intelligent robotic companions working collaboratively with astronauts, seamlessly navigating extraterrestrial landscapes, and conducting intricate scientific tasks. The convergence of artificial intelligence, machine learning, and autonomous navigation will empower spacecraft and robotic explorers to adapt in real-time to the challenges of deep-space missions, minimizing the reliance on constant human intervention. Future space missions will likely witness the deployment of swarms of miniaturized robots, leveraging swarm intelligence for distributed sensing and exploration. The incorporation of advanced human-machine interfaces, such as augmented reality and haptic feedback systems, will enhance the teleoperation of robots in remote environments (Ebrahimzadeh, 2019).

Measures Implications for Space Agencies, Researchers, and Engineers

The development of autonomous systems and space robotics will have a significant impact on space agencies, scientists, and engineers. A paradigm change in mission capabilities will be seen by space organizations like NASA and ESA, allowing for more ambitious and adaptable exploration projects. By integrating autonomous technologies, agencies may perform deep-space missions more efficiently since there is less need for real-time human control. As robotic explorers outfitted with cutting-edge artificial intelligence (AI) travel far-off planets and moons, examine geological characteristics, and expand our grasp of the universe, researchers will discover hitherto unheard-of prospects for scientific discovery.

The technology that power these developments will be designed and improved by engineers, who will be important in building robust, flexible, and sustainable robotic systems. The future of space exploration will be shaped by the combined efforts of agencies, researchers, and engineers as the synergy between human control and autonomous capabilities becomes increasingly crucial, pushing the frontiers of what is possible in the vastness of the space (Beniiche, 2023).

CONCLUSION AND FUTURE SCOPE- PREPARING FOR FUTURE SPACE COLONIZATION

A key component of being ready for future space colonization is the development of space robots and related technologies. Our capacity for space exploration and interaction has been greatly expanded by the fusion of artificial intelligence, autonomous navigation, and creative robotic designs. These developments, which range from cooperative human-robot expeditions to autonomous rovers on far-off planets, are essential steps toward establishing a long-term human presence beyond Earth. The scope of space colonization presents both exciting possibilities and complex challenges. The collaborative efforts between humans and robotic systems will be paramount in overcoming the obstacles that arise in extraterrestrial environments. The utilization of emerging technologies such as 3D printing with in-situ resources and the implementation of green propulsion systems, will be crucial for ensuring the self-sufficiency and sustainability of future colonies. The trend toward miniaturization, swarm robotics, and reusable platforms underscores the need for versatile and adaptable solutions. These developments not only enhance the efficiency of space missions but also contribute to cost-effectiveness, making long-term colonization endeavors more feasible. The collaborative nature of human-robot teams, with humans providing intuition, creativity, and adaptability while robots handle routine or hazardous tasks, will be a hallmark of successful space colonization efforts. The journey toward space colonization is not without its ethical, legal, and psychosocial considerations. Striking a balance between scientific exploration, responsible resource utilization, and the preservation of celestial environments requires thoughtful ethical frameworks and international collaboration. The mental well-being of astronauts and the development of psychological support systems will be essential as humans face extended periods of isolation and confinement during colonization missions.

In the future, as it move beyond our own planetary confines, the lessons learned from robotic exploration and human-robot collaboration in space will inform the blueprint for sustainable and responsible colonization efforts. The integration of diverse disciplines, from robotics and artificial intelligence to psychology and environmental science, will be instrumental in addressing the multifaceted challenges that arise in extraterrestrial habitats. This journey toward space colonization is a testament to human ingenuity and the collaborative synergy between humans and advanced robotic systems. The future holds the promise of establishing viable human habitats on other celestial bodies, expanding our understanding of the universe, and ensuring the long-term survival of humanity beyond the confines of Earth. As it navigates the challenges and opportunities that lie ahead, the trajectory toward space colonization marks a profound and transformative chapter in the collective human quest for exploration and discovery.

REFERENCES

Agnisarman, S., Lopes, S., Madathil, K. C., Piratla, K., & Gramopadhye, A. (2019). A survey of automation-enabled human-in-the-loop systems for infrastructure visual inspection. *Automation in Construction*, 97, 52–76. DOI: 10.1016/j.autcon.2018.10.019

Aijaz, A., & Sooriyabandara, M. (2018). The tactile internet for industries: A review. *Proceedings of the IEEE*, 107(2), 414–435. DOI: 10.1109/JPROC.2018.2878265

Amador, O., Aramrattana, M., & Vinel, A. (2022). A Survey on Remote Operation of Road Vehicles. *IEEE Access : Practical Innovations, Open Solutions*, 10, 130135–130154. DOI: 10.1109/ACCESS.2022.3229168

Basañez, L., Nuño, E., & Aldana, C. I. (2023). Teleoperation and Level of Automation. In *Springer Handbook of Automation* (pp. 457–482). Springer International Publishing. DOI: 10.1007/978-3-030-96729-1_20

Benaoumeur, I., Zoubir, A. F., & Reda, H. E. A. (2015). Remote Control of Mobile Robot using the Virtual Reality. *International Journal of Electrical & Computer Engineering (2088-8708)*, 5(5). IEEE.

Beniiche, A. (2023). *6G and Next-Generation Internet: Under Blockchain Web3 Economy*. CRC Press. DOI: 10.1201/9781003427322

Benos, L., Moysiadis, V., Kateris, D., Tagarakis, A. C., Busato, P., Pearson, S., & Bochtis, D. (2023). Human–robot interaction in agriculture: A systematic review. *Sensors (Basel)*, 23(15), 6776. DOI: 10.3390/s23156776 PMID: 37571559

Cremer, S. (2017). *Neuroadaptive Human-Machine Interfaces for Collaborative Robots* [Doctoral dissertation, The University of Texas at Arlington].

Das, S. K., Sahu, A., & Popa, D. O. (2017, May). Mobile app for human-interaction with sitter robots. In *Smart Biomedical and Physiological Sensor Technology XIV* (Vol. 10216, p. 85). SPIE. DOI: 10.1117/12.2262792

Dai, C. (2022, June Distributed User Association with Grouping in Satellite-Terrestrial Integrated Networks. *IEEE Internet of Things Journal*, 9(12), 10244–10256. DOI: 10.1109/JIOT.2021.3122939

Dufourd, D., & Dalgalarrondo, A. (2006, April). Integrating human/robot interaction into robot control architectures for defense applications. In *1th National Conference on Control Architecture of Robots, April* (pp. 6-7). IEEE.

Ebrahimzadeh, A. (2019). *Tactile Internet over Fiber-Wireless Enhanced HetNets using Edge Intelligence* [Doctoral dissertation, Institut National de la Recherche Scientifique, Canada].

Ebrahimzadeh, A., & Maier, M. (2019). Delay-constrained teleoperation task scheduling and assignment for human+ machine hybrid activities over FiWi enhanced networks. *IEEE Transactions on Network and Service Management*, 16(4), 1840–1854. DOI: 10.1109/TNSM.2019.2937020

Elbasheer, M., Longo, F., Mirabelli, G., Nicoletti, L., Padovano, A., & Solina, V. (2023). Shaping the role of the digital twins for human-robot dyad: Connotations, scenarios, and future perspectives. *IET Collaborative Intelligent Manufacturing*, 5(1), e12066. DOI: 10.1049/cim2.12066

Gao, Y. (2021). *Space robotics and autonomous systems: technologies, advances and applications*. Institution of Engineering and Technology.

Ha, Q. P., Yen, L., & Balaguer, C. (2019). Robotic autonomous systems for earthmoving in military applications. *Automation in Construction*, 107, 102934. DOI: 10.1016/j.autcon.2019.102934

Haidegger, T., Galambos, P., & Rudas, I. J. (2019, April). Robotics 4.0–Are we there yet? In *2019 IEEE 23rd International Conference on Intelligent Engineering Systems (INES)* (pp. 000117-000124). IEEE.

Handa, S. (2015). *Human-Machine Interaction for Unmanned Surface Systems* [Doctoral dissertation, University of Illinois at Urbana-Champaign].

Ivanova, K., Gallasch, G. E., & Jordans, J. (2016). *Automated and autonomous systems for combat service support: scoping study and technology prioritisation*. Defence Science and Technology Group Edinburgh SA Australia.

Kennedy, K., Alexander, L., Alexander, L., Landis, R., Landis, R., Linne, D., & Sims, J. (2011, September). NASA Technology Area 07 Human Exploration Destination Systems Roadmap. In *AIAA SPACE 2011 Conference & Exposition* (p. 7255). IEEE.

Kiencke, U., Nielsen, L., Sutton, R., Schilling, K., Papageorgiou, M., & Asama, H. (2006). The impact of automatic control on recent developments in transportation and vehicle systems. *Annual Reviews in Control*, 30(1), 81–89. DOI: 10.1016/j.arcontrol.2006.02.001

Kubota, T., & Kunii, Y. (2009, April). Intelligent guidance of mobile explorer for Planetary Robotic Exploration. In *2009 IEEE International Conference on Mechatronics* (pp. 1-6). IEEE. DOI: 10.1109/ICMECH.2009.4957151

Kuru, K. (2021). Conceptualisation of human-on-the-loop haptic teleoperation with fully autonomous self-driving vehicles in the urban environment. *IEEE Open Journal of Intelligent Transportation Systems*, 2, 448–469. DOI: 10.1109/OJITS.2021.3132725

Lacava, G., Marotta, A., Martinelli, F., Saracino, A., La Marra, A., Gil-Uriarte, E., & Vilches, V. M. (2021). Cybersecurity Issues in Robotics. *Journal of Wireless Mobile Networks, Ubiquitous Computing and Dependable Applications*, 12(3), 1–28.

Lee, J. S., Ham, Y., Park, H., & Kim, J. (2022). Challenges, tasks, and opportunities in teleoperation of excavator toward human-in-the-loop construction automation. *Automation in Construction*, 135, 104119. DOI: 10.1016/j.autcon.2021.104119

Liang, C. J., Wang, X., Kamat, V. R., & Menassa, C. C. (2021). Human–robot collaboration in construction: Classification and research trends. *Journal of Construction Engineering and Management*, 147(10), 03121006. DOI: 10.1061/(ASCE)CO.1943-7862.0002154

Mahmud, S., Lin, X., & Kim, J. H. (2020, January). Interface for human machine interaction for assistant devices: A review. In *2020 10th Annual computing and communication workshop and conference (CCWC)* (pp. 0768-0773). IEEE. DOI: 10.1109/CCWC47524.2020.9031244

Mathiassen, K., Schneider, F. E., Bounker, P., Tiderko, A., Cubber, G. D., Baksaas, M., Główka, J., Kozik, R., Nussbaumer, T., Röning, J., Pellenz, J., & Volk, A. (2021). Demonstrating interoperability between unmanned ground systems and command and control systems. *International Journal of Intelligent Defence Support Systems*, 6(2), 100–129. DOI: 10.1504/IJIDSS.2021.115236

Mourtzis, D., Angelopoulos, J., & Panopoulos, N. (2021). Smart manufacturing and tactile internet based on 5G in industry 4.0: Challenges, applications and new trends. *Electronics (Basel)*, 10(24), 3175. DOI: 10.3390/electronics10243175

Nguyen, H. A., & Ha, Q. P. (2023). Robotic autonomous systems for earthmoving equipment operating in volatile conditions and teaming capacity: A survey. *Robotica*, 41(2), 486–510. DOI: 10.1017/S0263574722000339

Nof, S. Y., Ceroni, J., Jeong, W., & Moghaddam, M. (2015). *Revolutionizing Collaboration through e-Work, e-Business, and e-Service* (Vol. 2). Springer. DOI: 10.1007/978-3-662-45777-1

Ollero, A., Boverie, S., Goodall, R., Sasiadek, J., Erbe, H., & Zuehlke, D. (2005). MECHATRONICS, ROBOTICS AND COMPONENTS FOR AUTOMATION AND CONTROL IFAC CC MILESTONE REPORT. *IFAC Proceedings Volumes, 38*(1), 1-13.

Ollero, A., Boverie, S., Goodall, R., Sasiadek, J., Erbe, H., & Zühlke, D. (2006). Mechatronics, robotics and components for automation and control: IFAC milestone report. *Annual Reviews in Control*, 30(1), 41–54. DOI: 10.1016/j.arcontrol.2006.02.002

Paik, P., Thudi, S., & Atashzar, S. F. (2022). Power-based velocity-domain variable structure passivity signature control for physical human-(tele) robot interaction. *IEEE Transactions on Robotics*, 39(1), 386–398. DOI: 10.1109/TRO.2022.3197932

Robinson, N., Tidd, B., Campbell, D., Kulić, D., & Corke, P. (2023). Robotic vision for human-robot interaction and collaboration: A survey and systematic review. *ACM Transactions on Human-Robot Interaction*, 12(1), 1–66. DOI: 10.1145/3570731

Rojas, L. C. V. (2015). *Temporarily Distributed Hierarchy in Unmanned Vehicles Swarms*.

Sauer, M. (2010). *Mixed-reality for enhanced robot teleoperation*. Universität Würzburg.

Sebbane, Y. B. (2018). *Intelligent autonomy of UAVs: advanced missions and future use*. CRC Press. DOI: 10.1201/b22485

Shukla, A., & Karki, H. (2016). Application of robotics in onshore oil and gas industry—A review Part I. *Robotics and Autonomous Systems*, 75, 490–507. DOI: 10.1016/j.robot.2015.09.012

Shukla, A., & Karki, H. (2016). Application of robotics in offshore oil and gas industry—A review Part II. *Robotics and Autonomous Systems*, 75, 508–524. DOI: 10.1016/j.robot.2015.09.013

Song, D., Goldberg, K., & Chong, N. Y. (2016). *Springer Handbook of Robotics: Networked Robots.* Springer.

Stoddard, B. (2022). *Designing and Evaluating a User Interface for Multi-Robot Furniture.*

Tahir, N., & Parasuraman, R. (2023). Mobile Robot Control and Autonomy Through Collaborative Simulation Twin. *arXiv preprint arXiv:2303.06172.*

Thangavel, K. (2023). *Trusted autonomous operations of distributed satellite systems for earth observation missions* [Doctoral dissertation, RMIT University].

Tiddi, I., Bastianelli, E., Daga, E., d'Aquin, M., & Motta, E. (2020). Robot–city interaction: Mapping the research landscape—a survey of the interactions between robots and modern cities. *International Journal of Social Robotics*, 12(2), 299–324. DOI: 10.1007/s12369-019-00534-x

Tiderko, A., Bachran, T., Hoeller, F., & Schulz, D. (2008). RoSe—A framework for multicast communication via unreliable networks in multi-robot systems. *Robotics and Autonomous Systems*, 56(12), 1017–1026. DOI: 10.1016/j.robot.2008.09.004

Vanhée, L., Jeanpierre, L., & Mouaddib, A. I. (2021, September). Optimizing Requests for Support in Context-Restricted Autonomy. In *2021 IEEE/RSJ International Conference on Intelligent Robots and Systems (IROS)* (pp. 6434-6440). IEEE. DOI: 10.1109/IROS51168.2021.9636240

Wu, J., Guo, S., Huang, H., Liu, W., & Xiang, Y.Information and Communications Technologies for Sustainable Development Goals. (2018, March). State-of-the-Art, Needs and Perspectives. *IEEE Communications Surveys and Tutorials*, 20(3), 2389–2406. DOI: 10.1109/COMST.2018.2812301

Xing, B., & Marwala, T. (2018). Smart maintenance for human–robot interaction. *Studies in Systems, Decision and Control.* Springer.

Yin, Y., Zheng, P., Li, C., & Wang, L. (2023). A state-of-the-art survey on Augmented Reality-assisted Digital Twin for futuristic human-centric industry transformation. *Robotics and Computer-integrated Manufacturing*, 81, 102515. DOI: 10.1016/j.rcim.2022.102515

Zhang, T., Li, Q., Zhang, C. S., Liang, H. W., Li, P., Wang, T. M., & Wu, C. (2017). Current trends in the development of intelligent unmanned autonomous systems. *Frontiers of information technology & electronic engineering, 18*, 68-85.

Chapter 14
Safety, Ethics, and Regulation in Intelligent Drones

Akshat Gaurav
Ronin Institute, USA

Brij B. Gupta
Asia University, Taiwan

Varsha Arya
Asia University, Taiwan, & Hong Kong Metropolitan University, Hong Kong

Arcangelo Castiglione
University of Salerno, Fisciano, Italy

ABSTRACT

This chapter addresses the critical issues of safety, ethics, and regulation surrounding the deployment of AI in industrial robotics and intelligent drones. It highlights the importance of establishing robust safety standards to mitigate risks associated with mechanical and software failures. Ethical considerations are explored, focusing on accountability, privacy, and the socio-economic impacts of automation. The chapter also navigates the complex regulatory landscape, underscoring the need for adaptive frameworks that balance innovation with public safety. Through case studies and discussions, it calls for a multi-disciplinary approach to develop ethical guidelines and effective legislation. The chapter emphasizes the collective responsibility of stakeholders to ensure the responsible advancement of AI technologies, advocating for policies that protect human welfare while fostering technological growth.

INTRODUCTION

Intelligent drones have indeed become a significant area of interest due to their poten- tial applications in various fields (Figure 1). The integration of advanced technologies such as artificial intelligence, machine learning, and the Internet of Things has led to the development of intelligent drones with enhanced capabilities. These drones are being utilized in diverse sectors such as agriculture, cybersecurity, healthcare, envi- ronmental conservation, and smart city applications. The use of intelligent drones is

DOI: 10.4018/979-8-3693-2707-4.ch014

not only revolutionizing traditional practices but also opening up new possibilities for innovation and efficiency. In agriculture, drones are being used for tasks such as crop monitoring, pesticide spraying, and soil analysis (Waqas 2023; Chui et al. 2023; ?). The integration of artificial intelligence and machine learning in drone technology has implications for cybersecurity, where advancements in drone technology are being examined for their impact on security measures (Sindiramutty 2024; Xie et al. 2023). Additionally, the use of intelligent drones in healthcare, particularly in the context of pandemic management, has been highlighted as a significant contribution to public health efforts (Chandra et al. 2022)(Figure 2).

Furthermore, the concept of smart cities is being revolutionized by the integration of collaborative drones and the Internet of Things (IoT) (Alsamhi et al. 2019; Casillo et al. 2024). This integration aims to enhance real-time applications in smart cities, thereby improving efficiency and resource management. The potential for intelligent drones to contribute to the sustainability and ecological well-being of the environment is also being explored, with applications in conservation efforts and sustainable devel-

Figure 1. Drone market size

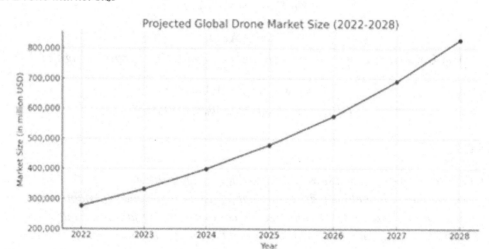

opment (Santangeli et al. 2020; Gupta et al. 2023; Mercier-Laurent 2021). The use of artificial intelligence and machine learning in drone technology has enabled advanced capabilities such as real-time detection, monitoring, and tracking. These capabilities have implications for various industries, including industrial inspection, surveillance, and emergency response systems (Zhang, Fan, and Hou 2022; Yao, Kao, and Lin 2023; Peng 2024; Gupta et al. 2024). Moreover, the integration of intelligent drones with 5G communication networks has the potential to revolutionize wireless systems, enabling applications such as package delivery, surveillance, and remote sensing (Mozaffari et al. 2019)(Zhang et al. 2019).

The ethical implications of using intelligent drones are also being considered, par- ticularly in the context of public healthcare and sustainability. The development, im- plementation, and assessment of drones used in public healthcare are being guided by ethical frameworks that consider the autonomy, operationalization, and explicability of intelligent drone systems (Cawthorne and Wynsberghe 2020).

Figure 2. Drone applications

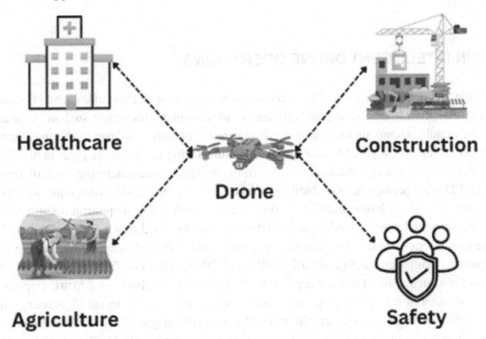

The development and deployment of drones have brought about significant advancements in various fields, but they also raise important considerations related to safety, ethics, and regulation. These aspects are crucial for ensuring the responsible and ben- eficial use of drone technology across different domains.Safety is a paramount concern in the operation of drones, particularly in areas such as aviation, healthcare, and con- struction. The integration of drones into these sectors necessitates adherence to safety regulations and standards to mitigate potential risks and hazards (Umar 2020; Behera et al. 2023; Plioutsias, Karanikas, and Chatzimihailidou 2017; Tran and Nguyen 2022). Furthermore, the use of drones in urban environments requires specific safety measures to ensure the protection of people, property, and the environment (Chen et al. 2020; Molina et al. 2018a). The implementation of safety requirements and standard operat- ing procedures is essential to guarantee safe drone missions (Umar 2020; Kumar et al. 2022).

Ethical considerations also play a significant role in the development and deploy- ment of drones. Issues related to privacy, civil liberties, and public perception need to be carefully addressed. The use of drones raises concerns about privacy infringe- ment and civil liberties, particularly in the context of surveillance and data collection (Guo, Ahmad, and Chang 2020; C¸ etin et al. 2022; Kiran, Pasupuleti, and Eswari 2022). Additionally, public perception of the risks associated with drones influences the ac- ceptance and adoption of this technology, highlighting the importance of ethical frameworks in shaping societal attitudes towards drones (Ayamga, Tekinerdogan, and Rambaldi 2020) (Allouch et al. 2019).

Regulation is fundamental in ensuring the responsible and lawful use of drones. The absence of clear policies, standards, and guidelines governing the safe operation of civilian drones poses a barrier to research and development in this field (Ayamga and Tekinerdo˘gan 2021; Khanam, Tanweer, and Khalid 2022). The development of a comprehensive regulatory framework is essential to address safety, privacy, and security concerns associated with drone technology (Rubagotti et al. 2021; Hu et al. 2022; Cain et al. 2021).

SAFETY IN INTELLIGENT DRONE OPERATIONS

The design and manufacturing of drones require adherence to specific standards and considerations to ensure optimal performance, safety, and efficiency. Additive man- ufacturing, such as 3D printing, has emerged as a valuable technique for producing specialized multirotor UAV parts, offering advantages in time and cost savings com- pared to conventional manufacturing technologies Petar et al. (2022). Additionally, the analysis and optimization of UAV frame design for manufacturing from thermo- plastic materials on FDM 3D printers have enabled the production of drone bodies with unique designs tailored for specific purposes (Palinkas et al. 2022). Furthermore, the exploration of opportunities for research and practice in drone manufacturing has highlighted the potential for cloud-based design and manufacturing to increase pro- ductivity in the design, prototype, and production of drones, enabling faster responses to market needs (Maghazei and Netland 2019) (Wu et al. 2015). The classification and use of composite materials in the structure of unmanned aerial vehicles (UAVs) have also been a focus, emphasizing the importance of material selection and manufacturing techniques in drone design (S¨onmez et al. 2022).

The regulation of the design and manufacturing of drones varies significantly across different countries and regions, impacting the global drone industry in various ways (Table 1). The regulatory landscape influences the development, production, and use.

Table 1. Comparison of drone regulations

Region	Manufacturing Requirements	Operational Limits	Sector-Specific Rules	Notes
EU	Standardized manufacturing processes as per EASA stan- dards	Maximum al- titude and distance from operator re- strictions	Specific guide- lines for com- mercial vs. recreational use	Moving towards a unified regula- tory framework
USA	Compliance with FAA manufacturing standards	Line-of-sight operation, alti- tude limits	Different rules for commercial, recreational, and governmen- tal use	State laws may also apply
Africa	Varies widely, some regions with minimal regulations	Often restric- tive, focusing on safety and privacy	Agricultural use heavily regu- lated in some countries	Regulatory en- vironment still developing
OECD Coun- tries	Diverse, de- pending on the country	Generally align with ICAO guidelines	Often have detailed privacy regulations	Significant variations in approach

of drones, shaping the industry's growth, innovation, and market dynamics.In the Eu- ropean Union, there has been a concerted effort to consolidate fragmented regulations concerning the manufacture and operation of civil drones into a common framework. The aim is to harmonize regulations across European countries, promoting standard- ization and facilitating the free movement of drones within the EU. This initiative is expected to streamline the design and manufacturing processes, enabling manufac- turers to adhere to consistent regulatory requirements and certifications, thereby fos- tering a more integrated

European drone market Molina and Montagud (2017)(Cam- pos, Molina, and Kr¨oner 2017). Similarly, a comparative analysis of the legislation evolution for drone use in OECD countries has revealed sig- nificant diversity among national legal frameworks. While 22 OECD countries belong to the European Union, the observed variations in national legal frameworks underscore the complexity and challenges associated with harmonizing regulations across different countries. This di- versity can impact the design and manufacturing of drones, as manufacturers may need to navigate distinct regulatory requirements when producing drones for different markets (Tsiamis, Efthymiou, and Tsagarakis 2019). In the United States, the regula- tory and legal schemes for drone operation within the construction industry have been reviewed, highlighting the impact of regulations on specific sectors. Understanding the regulatory landscape is crucial for manufacturers, as compliance with industry-specific regulations is essential for designing and producing drones tailored to specific applica- tions, such as construction (Agapiou 2021). Moreover, the impact of regulations on the global drone industry extends beyond manufacturing and design considerations. The regulatory environment influences market dynamics, trade, and international collab- oration. An analysis of the impact of Industry 4.0 innovations in international trade has high- lighted the role of regulations in shaping the functioning of business models and consumer behavior in the context of drone technology (Dumanska et al. 2021). In Africa, the challenges posed by regulations for the use of drones in agriculture have been explored, revealing either very restrictive regulations or a lack of proper regulation in many African countries. This regulatory landscape can significantly impact the adoption and manufacturing of drones for agricultural purposes, potentially hin- dering the develop- ment and deployment of drone technology in the agricultural sector (Ayamga and Tekinerdo˘gan 2021).

Furthermore, the impact of regulations on the global drone industry is evident in the context of healthcare provision. Early adopters in sub-Saharan Africa, such as Madagascar, Malawi, and Senegal, have piloted the use of bi-directional transport drones for health systems. The regulatory environment in these countries can influence the design and manufacturing of drones tailored for healthcare applications, reflecting the intersection of regulatory frameworks and industry-specific needs (Knoblauch et al. 2019).

The impact of regulations on the global drone industry is also evident in the context of safety con- siderations. The review of the current state of UAV regulations has high- lighted the need for effective control and management of the drone revolution. The regulatory impact is expected to manage the substantial increase in drone flights, em- phasizing the role of regulations in addressing safety risks and operational challenges (Merkert and Bushell 2020).

In conclusion, the regulation of the design and manufacturing of drones varies across different coun- tries and regions, impacting the global drone industry in terms of market integration, industry-specific applications, trade, and safety considerations. Harmoniz- ing regulations, addressing industry-specific needs, and managing safety risks are es- sential for fostering a conducive regulatory environment that supports the responsible and innovative development of drone technology on a global scale.

ETHICAL CONSIDERATIONS IN DRONE TECHNOLOGY

Privacy Concerns

Surveillance and data collection ethics related to drones are critical considerations in the deployment and use of drone technology (Table 2). The ethical implications of drone surveillance and data collec- tion have been the subject of extensive analysis and debate, with a focus on privacy, accountability, and the impact on individuals and com- munities.The ethical dimensions of drone surveillance and data collection have been examined in the context of conservation efforts. The use of drones for monitoring and surveillance in conservation raises questions about the balance between targeted data collection for monitoring purposes and the broader implications of omnibus surveil- lance. This distinction is crucial for ensuring that data collection is conducted with specific conservation goals in mind, rather than for indiscriminate surveillance Millner (2024).

Furthermore, the domestic use of drones for surveillance has prompted ethical analysis, emphasizing the need to consider the ethical ramifications of drone surveillance. The potential for privacy infringement and the oversight of surveillance activities are central ethical concerns that need to be addressed in the design and deployment of drones for surveillance purposes (West and Bowman 2016). The deployment of drones for data collection and surveillance has also been explored in the context of border management and policing. The use of drone technology for border surveillance raises challenges related to data pro- tection, human rights, and the potential securitization of border surveillance. The ethical considerations surrounding the impact of drone technology on human rights and privacy in border management high- light the need for robust ethical frameworks and regulatory safeguards (Marin and Krajˇcíková 2015).

In the agricultural sector, the use of drones for data collection and surveillance has become increas- ingly prevalent. Typical drone applications in agriculture include soil sampling, pesticide spraying, and animal population surveillance. However, the ethical implications of data collection and surveillance in agriculture extend to con- siderations of data integrity, confidentiality, and the responsible use of surveillance technologies in farming practices. The ethical implications of drone surveillance and data collection also intersect with security considerations. The efficient and secure transfer of surveillance data collected by drones is essential for disaster management and security applications. Ensuring the integrity and confidentiality of sensed data collected through surveillance drones is a critical ethical consideration in the design and operation of drone communication networks for security purposes (Ab- dallah et al. 2019).

Moreover, the use of digital surveillance technologies in conservation has raised social concerns related to privacy and pervasive surveillance. While these technologies offer technical capabilities for conservation, they also prompt ethical reflections on the balance between conservation goals and the potential impact on privacy and surveil- lance in natural environments (Simlai and Sandbrook 2021). The ethical assessment of drone surveillance has also extended to the spatial and technological dimensions of drone operations. The spatial relationship between drone operators and their targets, as well as the intimate knowledge gained through surveillance, raises ethical questions about the impact of distance and technology on fulfilling ethical requirements in drone operations (Williams 2015).

In the context of collective adaptive systems of drones, the ethical implications of intelligence, surveillance, and reconnaissance (ISR) scenarios have been examined. The ethical considerations of surveillance and data collection in ISR scenarios emphasize the need for responsible and accountable use of surveillance technologies in complex operational environments (Riley, McKinney, and Gamble 2022). The ethical implications of drone surveillance and data collection also intersect with journalism and immersive experiences. Practical considerations regarding technical expertise, ethical issues, and privacy implications are central to the ethical framework for drone journal- ism and the responsible generation of immersive experiences through drone technology (Ntalakas et al. 2017).

Environmental Ethics

The impact of drones on wildlife and ecosystems has been a subject of extensive research and analysis, with a focus on understanding the implications of drone technology on biodiversity, conservation, and environmental management. Drones have been increasingly utilized in ecological research and wildlife monitoring, offering new opportunities and challenges in understanding and managing ecosystems. Drones have proven to be valuable tools for multi-species research in various ecosystems, including coastal and montane forests in the tropics. Their use in wildlife surveys has shown promising results, enabling researchers to gather data in different ecosystems, con- tributing to a better understanding of species distribution and habitat characteristics Rahman, Sitorus, and Condro (2021).

The application of drones in ecological research has extended to approaching sensi- tive wildlife in inaccessible areas. Ethical guidelines have been developed to ensure that drone use minimizes disturbance to wildlife, particularly in the context of approaching.

Table 2. Summary of ethical considerations by use case for drone surveillance

Application	Primary Ethical Concerns	Suggested Mitigations
Conservation	Privacy, potential over-surveillance of protected areas	Implement strict data protocols, limit surveil- lance to necessary ar- eas
Border Security	Human rights, privacy, data protection	Adhere to interna- tional laws, use data minimization tech- niques
Urban Policing	Privacy, civil liberties, community trust	Community engage- ment, transparent policies, oversight mechanisms
Agriculture	Data integrity, confi- dentiality	Use encrypted data storage, ensure data is only accessible to authorized personnel

birds and other sensitive species. This approach has opened up new possibilities for studying wildlife behavior and ecology in remote and challenging environments (Vas et al. 2015).

However, the use of drones in wildlife research and conservation has raised concerns about their po- tential impact on wildlife. Scientific papers specifically evaluating the effects of drones on wildlife are scarce but increasing, highlighting the need for a better understanding of the potential disturbances caused by drone operations on wildlife behavior and ecosystems (Rebolo-Ifr´an, Grilli, and Lambertucci 2019).

In the context of human-wildlife conflict, drones have been deployed to mitigate conflicts, such as human-elephant conflict in Tanzanian parks. Unmanned aerial ve- hicles (UAVs) have been used to monitor and manage wildlife, demonstrating their potential in addressing conservation challenges and minimizing human-wildlife con- flicts in protected areas (Hahn et al. 2016). The behavioral responses of wildlife to drone-based monitoring have been a subject of interest, particularly in understanding how different species, such as kangaroos, respond to the presence of drones. As drone technology becomes more accessible, its use in wildlife ecology has increased, offering new insights into wildlife behavior and responses to environmental stimuli (Brunton et al. 2019).

The use of drones in scientific research and by the tourist industry in Antarctica has raised legal implications and concerns about the potential environmental impacts of drone operations. The need to balance the scientific and operational benefits of drones with their environmental impacts, including potential disturbance to Antarctic wildlife, has been highlighted as a critical consideration in the use of drones in fragile environments (Bollard et al. 2022). Drones have also been used to survey and monitor African elephants, providing guidelines for the effective and ethical use of drones in wildlife surveys. The potential of drones in wildlife sciences has been recognized, offering opportunities to observe and monitor wildlife in inaccessible habitats, contributing to conservation and management efforts (Hartmann, Fishlock, and Leslie 2021). In the context of beach safety and marine conservation, drones have been explored as a plat-

form for sighting sharks, emphasizing the role of drones in environmental protection and the conservation of marine ecosystems. The use of drones in wildlife monitoring and environmental protection has become increasingly important, particularly in addressing the conservation needs of marine species (Butcher et al. 2019).

Furthermore, drones have been utilized to assess canopy cover and analyze tree mortality in oil palm agroforests, highlighting their potential in ecological research and environmental monitoring. The impact of digital transformation on local ecosys- tems and biodiversity has been examined, emphasizing the role of modern technologies, including drones, in understanding and managing environmental changes (Khokthong et al. 2019)(Gavrilovi´c 2024). The use of drones in detecting evidence of winter sports activities in protected mountain areas has demonstrated their potential in environmental monitoring and conservation. The low impact of drone flights on wildlife has been reported, highlighting the potential of drones as a monitoring tool in protected areas (Weber and Knaus 2017).

The impact of drones on biodiversity and ecosystems has been a subject of interest, particularly in understanding their potential effects on invasive species detection and monitoring. The use of drones in ecological research has provided new opportunities for assessing and managing biodiversity, contributing to the conservation and protection of natural habitats (Lahoz-Monfort and Magrath 2021) (Damijani´c 2021).

Ethical Use in Various Domains

The use of drones in humanitarian operations has indeed become increasingly prevalent, offering new opportunities for disaster response, healthcare delivery, and human- itarian aid. However, the ethical implications of employing drones in humanitarian contexts have been the subject of extensive research and analysis. The use of drones in humanitarian operations raises complex ethical considerations related to privacy, equity, sovereignty, and the potential impact on at-risk populations in disaster zones and remote areas lacking sufficient healthcare services.The deployment of drones in humanitarian logistics has been hindered by barriers related to inadequate govern- ment regulations. This highlights the need for improved regulatory frameworks to facilitate the effective use of drones in humanitarian operations Kamat et al. (2022). The turn to technology in humanitarian action has raised ethical concerns regarding the implications for humanitarianism, sovereignty, and access to healthcare services in disaster-affected or remote areas (Wang 2020). Debates have emerged about the eth- ical, legal, and social implications of humanitarian drones, emphasizing the need for a comprehensive understanding of the potential risks and challenges associated with their use (Wang, Christen, and Hunt 2021). The growing use of drones in the human- itarian space has prompted the need for ethical reflection to understand the potential impact of this technology on humanitarian care (Wynsberghe 2019). The ethical impli- cations of using drones to fulfill humanitarian functions, particularly in the aftermath of disasters, have been a subject of scrutiny, emphasizing the importance of deploying this technology in non-violent and ethically desirable ways as part of the humanitarian response (Tatsidou et al. 2019). The promises and perils of using drones and satellite imagery analysis in humanitarian operations have been highlighted, underscoring the need for careful ethical consideration of the consequences of their use (Lichtman and Nair 2015).

The intention of logistics service providers to adopt delivery drones for humani- tarian operations has been a subject of empirical study, shedding light on the ethical considerations and adoption barriers in the context of humanitarian logistics (Edwards et al. 2023). The use of drones in humanitarian logis- tics, particularly in the context of delivering medical supplies, has been explored, emphasizing the need to address orga- nizational challenges and ethical considerations related to the use of drones (Comes, Sandvik, and Walle 2018).

The ethical considerations associated with the development use of drones for delivery in Malawi have been examined, highlighting the need for critical engagement with the social, political, and ethical implications of using drones for humanitarian purposes (Wang 2021). The ethical implications of using drones for healthcare delivery have been the focus of a multi-site qualitative study, contributing to the critical engagement with the social, political, and ethical meanings and implications of using drones in global health problem-solving (Jeyabalan et al. 2020).

REGULATION OF INTELLIGENT DRONES

The regulatory frameworks governing the use of drones vary significantly across dif- ferent countries and regions, reflecting diverse approaches to managing the operation and application of drone technol- ogy. The regulatory landscape for drones encompasses a wide range of considerations, including safety, privacy, airspace management, and the specific applications of drones in various sectors. The following references provide insights into the key regulatory frameworks governing the use of drones in different countries worldwide:

Gallardo-Camacho and Breijo (2020) highlight the Royal Decree 1036/2017 in Spain, which defines the regulatory framework for drones, allowing flights over urban areas, crowds, and non-controlled air- space, subject to specific conditions such as visual line of sight and weight limitations. Zwickle, Farber, and Hamm (2018) discuss the regulatory structure governing the use of drones in the US, emphasizing its alignment with pub- lic safety needs while acknowledging variations in legal protection for privacy based on geographical location. Plioutsias, Karanikas, and Chatzimihailidou (2017) point out the absence of a common regulatory framework based on systemic risk assessment for small drone operations, in- dicating the need for a comprehensive approach to safety requirements. The Organisation for Economic Co-operation and Development (OECD) publication (2018) notes that existing regulatory frameworks for the aviation sector tend to apply to commercial drones, reflecting the integration of drones into regulated airspace. Nugraha, Jeyakodi, and Mahem (2016) compare the regulatory measures for civilian drone usage in Indonesia, India, and Thailand, providing an overview of the impact of existing legal frameworks in these countries. Agapiou (2021) reviews the legal and regulatory landscape for drones in construction, discussing national and interna- tional regulatory bodies, risk-based approaches, safety classifications, and efforts to harmonize drone legislation. St¨ocker et al. (2017) highlight the significant barriers to research and development posed by current legal frameworks regulating UAVs, in- dicating the need for regulatory frameworks that facilitate innovation. Molina et al. (2018b) discuss the increasing awareness in governments about inadequate drone us- age, particularly in terms of security and terrorism, underscoring the evolving regu- latory responses to address security concerns associated with drone technology. These references provide valuable insights into the diverse regulatory frameworks governing the use of drones in different countries worldwide, reflecting the complex interplay of safety, privacy, security, and innovation in the management of drone technology.

CONCLUSION

The intricate balance of safety, ethics, and regulation forms the backbone of advancing drone technology responsibly. This chapter has highlighted how incorporating robust safety protocols, considering ethical implications, and adhering to evolving regulations are essential in harnessing the full potential of drones across various domains. As the application of drones expands, it becomes imperative for stakeholders to engage in continuous dialogue, ensuring that innovations align with societal values and legal frameworks. Looking ahead, the focus must remain on developing adaptive regulatory mechanisms that not only foster innovation but also protect public interests, paving the way for sustainable and ethical advancement in drone technology.

ACKNOWLEDGMENT

This research work is supported by National Science and Technology Council (NSTC), Taiwan Grant No. NSTC112-2221-E-468-008-MY3.

REFERENCES

Abdallah, A., Ali, M., Misic, J., & Misic, V. (2019). Efficient security scheme for disaster surveillance uav communication networks. *Information (Basel)*, 10(2), 43. DOI: 10.3390/info10020043

Agapiou, A. (2021). Drones in construction: an international review of the legal and regulatory landscape. *Proceedings of the Institution of Civil Engineers - Management Procurement and Law*. IEEE. DOI: 10.1680/jmapl.19.00041

Allouch, A., Koubaa, A., Khalgui, M., & Abbes, T. (2019). Qualitative and quantitative risk analysis and safety assessment of unmanned aerial vehicles missions over the internet. *IEEE Access : Practical Innovations, Open Solutions*, 7, 53392–53410. DOI: 10.1109/ACCESS.2019.2911980

Alsamhi, S., Ma, O., Ansari, M., & Almalki, F. (2019). Survey on collaborative smart drones and internet of things for improving smartness of smart cities. *IEEE Access : Practical Innovations, Open Solutions*, 7, 128125–128152. DOI: 10.1109/ACCESS.2019.2934998

Ayamga, M., Tekinerdo˘gan, B., & Kassahun, A. (2021). Exploring the challenges posed by regulations for the use of drones in agriculture in the african context. *Land (Basel)*, 10(2), 164. DOI: 10.3390/land10020164

Ayamga, M., Tekinerdogan, B., & Rambaldi, G. (2020). Developing a policy framework for adoption and management of drones for agriculture in africa. *Technology Analysis and Strategic Management*, 33(8), 970–987. DOI: 10.1080/09537325.2020.1858047

Behera, T. K., Bakshi, S., Sa, P. K., Nappi, M., Castiglione, A., Vijayakumar, P., & Gupta, B. B. (2023). The NITRDrone dataset to address the challenges for road extraction from aerial images. *Journal of Signal Processing Systems for Signal, Image, and Video Technology*, 95(2), 197–209. DOI: 10.1007/s11265-022-01777-0

Bollard, B., Doshi, A., Gilbert, N., Poirot, C., & Gillman, L. (2022). Drone technol- ogy for monitoring protected areas in remote and fragile environments. *Drones (Basel)*, 6(2), 42. DOI: 10.3390/drones6020042

Brunton, E., Bolin, J., Leon, J., & Burnett, S. (2019). Fright or flight? be- havioural responses of kangaroos to drone-based monitoring. *Drones (Basel)*, 3(2), 41. DOI: 10.3390/drones3020041

Butcher, P., Piddocke, T., Colefax, A., Hoade, B., Peddemors, V., Borg, L., & Cullis, B. (2019). Beach safety: Can drones provide a platform for sighting sharks? *Wildlife Research*, 46(8), 701. DOI: 10.1071/WR18119

Cain, S. (2021). *Standards for uas - acceptable means of compliance for low risk sora operations*. ARC. .DOI: 10.2514/6.2021-1177

Campos, V., Molina, M., & Kr̈oner, S. (2017). Ethics and Civil Drones. *Introduction.*, 1–5. DOI: 10.1007/978-3-319-71087-7

Casillo, M., & Colace, F. (2024). Securing Digital Ecosystems: Harnessing the Power of Intelligent Machines in a Secure and Sustainable Environment. In *Handbook of Research on AI and ML for Intelligent Machines and Systems*, (pp. 50–74). IGI Global.

Cawthorne, D., & Wynsberghe, A. (2020). An ethical framework for the design, develop- ment, implementation, and assessment of drones used in public healthcare. *Science and Engineering Ethics*, 26(5), 2867–2891. DOI: 10.1007/s11948-020-00233-1 PMID: 32578062

Chandra, M., Kumar, K., Thakur, P., Chattopadhyaya, S., Alam, F., & Kumar, S. (2022). Dig- ital technologies, healthcare and covid-19: Insights from developing and emerging nations. *Health and Technology*, 12(2), 547–568. DOI: 10.1007/s12553-022-00650-1 PMID: 35284203

Chen, Y. (2020). *Efficient drone mobility support using reinforcement learning*. WCNC. .DOI: 10.1109/WCNC45663.2020.9120595

Chui, K. T., Gupta, B. B., Liu, J., Arya, V., Nedjah, N., Almomani, A., & Chaurasia, P. (2023). A survey of internet of things and cyber-physical systems: Standards, algorithms, applications, security, challenges, and future directions. *Information (Basel)*, 14(7), 388. DOI: 10.3390/info14070388

Comes, T. (2018). Cold chains, interrupted. *Journal of Humani- tarian Logistics and Supply Chain Management, 8*, 49–69. .DOI: 10.1108/JHLSCM-03-2017-0006

Damijani'c, D. (2021). Remote sensing in invasive species detection and monitor- ing. *International Journal of Environmental Sciences & Natural Resources*, 29(1). DOI: 10.19080/IJESNR.2021.29.556255

Dumanska, I., Vasylkivskyi, D., Zhurba, I., Pukhalska, Y., Matviiets, O., & Goncharuk, A. (2021). Dronology and 3d printing as a catalyst for international trade in industry 4.0. *WSEAS Transactions on Environment and Development*, 17, 740–757. DOI: 10.37394/232015.2021.17.71

Edwards, D., Subramanian, N., Chaudhuri, A., Morlacchi, P., & Zeng, W. (2023). Use of delivery drones for humanitarian operations: Analysis of adoption barriers among logistics service providers from the technology acceptance model perspective. *Annals of Operations Research*, 335(3), 1645–1667. DOI: 10.1007/s10479-023-05307-4 PMID: 37361062

Gallardo-Camacho, J., & Breijo, V. (2020). Relationships between law enforcement authorities and drone journalists in spain. *Media and Communication*, 8(3), 112–122. DOI: 10.17645/mac.v8i3.3097

Gavrilovi'c, M., Muhović, A., & Pavlović, N. (2024). Analysis of the application of modern technologies in agriculture in three balkan countries and the impact on biodiversity. *Romanian Agricultural Research*, 41, 79–90. DOI: 10.59665/rar4108

Guo, J., Ahmad, I., & Chang, K. (2020). Classification, positioning, and tracking of drones by hmm using acoustic circular microphone array beamforming. *EURASIP Journal on Wireless Communications and Networking*, 2020(1), 9. DOI: 10.1186/s13638-019-1632-9

Gupta, B. B., Gaurav, A., Chui, K. T., & Arya, V. (2023). Optimized Edge- cCCN Based Model for the Detection of DDoS Attack in IoT Environment. In *International Conference on Edge Computing*. Springer.

Gupta, B. B., Gaurav, A., Chui, K. T., Arya, V., & Choi, C. (2024). Au- toencoders Based Optimized Deep Learning Model for the Detection of Cyber Attack in IoT Environment. In *2024 IEEE International Conference on Consumer Electronics (ICCE)*, (pp. 1–6). IEEE.

Hahn, N., Mwakatobe, A., Konuche, J., Souza, N., Keyyu, J., Goss, M., Chang'a, A., Palminteri, S., Dinerstein, E., & Olson, D. (2016). Unmanned aerial vehicles mitigate hu- man–elephant conflict on the borders of tanzanian parks: A case study. *Oryx*, 51(3), 513–516. DOI: 10.1017/S0030605316000946

Hartmann, W., Fishlock, V., & Leslie, A. (2021). First guidelines and suggested best pro-tocol for surveying african elephants (loxodonta africana) using a drone. *Koedoe*, 63(1). DOI: 10.4102/koedoe.v63i1.1687

Hu, B., Gaurav, A., Choi, C., & Almomani, A. (2022). Evaluation and com- parative analysis of semantic web-based strategies for enhancing educational system devel- opment. [IJSWIS]. *International Journal on Semantic Web and Information Systems*, 18(1), 1–14. DOI: 10.4018/IJSWIS.302895

Jeyabalan, V., Nouvet, E., Meier, P., & Donelle, L. (2020). Context-specific chal- lenges, opportunities, and ethics of drones for healthcare delivery in the eyes of program managers and field staff: A multi-site qualitative study. *Drones (Basel)*, 4(3), 44. DOI: 10.3390/drones4030044

Kamat, A. (2022). Uncovering in- terrelationships between barriers to unmanned aerial vehicles in human- itarian logistics. *Operations Management Research, 15*, 1134–1160. .DOI: 10.1007/s12063-021-00235-7

Khokthong, W., Zemp, D., Irawan, B., Sundawati, L., Kreft, H., & H¨olscher, D. (2019). Drone- based assessment of canopy cover for analyzing tree mortality in an oil palm agroforest. *Frontiers in Forests and Global Change*, 2, 12. DOI: 10.3389/ffgc.2019.00012

Kiran, M. A., Pasupuleti, S. K., & Eswari, R. (2022). Efficient pairing-free identity-based signcryption scheme for cloud-assisted iot. [IJCAC]. *International Journal of Cloud Applications and Computing*, 12(1), 1–15. DOI: 10.4018/IJCAC.305216

Knoblauch, A., Rosa, S., Sherman, J., Blauvelt, C., Matemba, C., & Maxim, L. (2019). Bi-directional drones to strengthen healthcare provision: Experiences and lessons from madagascar, malawi and senegal. *BMJ Global Health*, 4(4), e001541. DOI: 10.1136/bmjgh-2019-001541 PMID: 31413873

Kumar, R., Singh, S. K., Lobiyal, D. K., Chui, K. T., Santaniello, D., & Rafsanjani, M. K. (2022). A novel decentralized group key management scheme for cloud-based vehicular IoT networks. [IJCAC]. *International Journal of Cloud Applications and Computing*, 12(1), 1–34. DOI: 10.4018/IJCAC.311037

Lahoz-Monfort, J., & Magrath, M. (2021). A comprehensive overview of technolo- gies for species and habitat monitoring and conservation. *Bioscience*, 71(10), 1038–1062. DOI: 10.1093/biosci/biab073 PMID: 34616236

Lichtman, A., & Nair, M. (2015). Humanitarian uses of drones and satellite im- agery analysis: The promises and perils. *AMA Journal of Ethics*, 17(10), 931–937. DOI: 10.1001/journalofethics.2015.17.10. stas1-1510 PMID: 26496056

Maghazei, O., & Netland, T. (2019). Drones in manufacturing: Exploring opportunities for research and practice. *Journal of Manufacturing Technology Management*, 31(6), 1237–1259. DOI: 10.1108/ JMTM-03-2019-0099

Marin, L. (2015). *Deploying drones in policing southern european borders: constraints and challenges for data protection and human rights.* Springer. ₆.DOI: 10.1007/978-3-319-23760-2

Mercier-Laurent, E. (2021). *Can artificial intelligence effectively support sustainable development?* Springer. ₁0.DOI: 10.1007/978-3-030-80847-1

Merkert, R., & Bushell, J. (2020). Managing the drone revolution: A systematic lit- erature review into the current use of airborne drones and future strategic direc- tions for their effective control. *Journal of Air Transport Management*, 89, 101929. DOI: 10.1016/j.jairtraman.2020.101929 PMID: 32952321

Millner, N., Newport, B., Sandbrook, C., & Simlai, T. (2024). Between monitoring and surveillance: Geographies of emerging drone tech- nologies in contemporary conservation. *Progress in Environmental Geography*, 3(1), 17–39. DOI: 10.1177/27539687241229739

Molina, M., Campos, V., Montagud, M., & Molina, B. (2018a). Ethics for civil indoor drones: A qualitative analysis. *International Journal of Micro Air Vehicles*, 10(4), 340–351. DOI: 10.1177/1756829318794004

Molina, M. (2017). *Legal and ethical recommendations.* Springer. ₅.DOI: 10.1007/978-3-319-71087-7

Molina, N. (2018b). *A wireless method for drone identification and monitoring using ais technology.* IEEE. .DOI: 10.23919/URSI-AT-RASC.2018.8471616

Mozaffari, M., Kasgari, A., Saad, W., Bennis, M., & Debbah, M. (2019). Beyond 5g with uavs: Founda- tions of a 3d wireless cellular network. *IEEE Transactions on Wireless Communications*, 18(1), 357–372. DOI: 10.1109/TWC.2018.2879940

Ntalakas, A., Dimoulas, C., Kalliris, G., & Veglis, A. (2017). Drone journal- ism: Generating immersive experiences. *Journal of Media Critiques*, 3(11), 187–199. DOI: 10.17349/jmc117317

Nugraha, R., Jeyakodi, D., & Mahem, T. (2016). Urgency for legal framework on drones: Lessons for indonesia, india, and thailand. *Indonesia Law Review*, 6(2), 137. DOI: 10.15742/ilrev.v6n2.229

Palinkas, I., Pekez, J., Desnica, E., Rajic, A., & Nedelcu, D. (2022). Analysis and optimization of uav frame design for manufacturing from thermoplastic materials on fdm 3d printer. *Materiale Plastice*, 58(4), 238–249. DOI: 10.37358/MP.21.4.5549

Plioutsias, A., Karanikas, N., & Chatzimihailidou, M. (2017). Hazard analysis and safety requirements for small drone operations: To what extent do popular drones embed safety? *Risk Analysis*, 38(3), 562–584. DOI: 10.1111/risa.12867 PMID: 28768049

Rahman, D., Sitorus, A., & Condro, A. (2021). From coastal to montane forest ecosystems, using drones for multi-species research in the tropics. *Drones (Basel)*, 6(1), 6. DOI: 10.3390/drones6010006

Rebolo-Ifr'an, N., Grilli, M., & Lambertucci, S. (2019). Drones as a threat to wildlife: Youtube comple- ments science in providing evidence about their effect. *Environmental Conservation*, 46(3), 205–210. DOI: 10.1017/S0376892919000080

Riley, I. (2022). *Improving the expected performance of self- organization in a collective adaptive system of drones using stochastic multiplayer games.* IEEE. .DOI: 10.24251/HICSS.2022.918

Rubagotti, M., Tusseyeva, I., Baltabayeva, S., Summers, D., & Sandygulova, A. (2021). Perceived safety in physical human robot interaction – a survey. https://doi.org//arxiv.2105.14499.DOI: 10.48550

Santangeli, A., Chen, Y., Kluen, E., Chirumamilla, R., Tiainen, J., & Loehr, J. (2020). In- tegrating drone-borne thermal imaging with artificial intelligence to locate bird nests on agricultural land. *Scientific Reports*, 10(1), 10993. DOI: 10.1038/s41598-020-67898-3 PMID: 32665596

Simlai, T. (2021). *Digital surveillance technologies in conservation and their social implications.* Springer. .DOI: 10.1093/oso/9780198850243.003.0012

Sindiramutty, S. (2024). *Future trends and emerging threats in drone cybersecurity.* Springer. .DOI: 10.4018/979-8-3693-0774-8.ch007

S¨onmez, M. (2022). *Unmanned aerial vehicles – classification, types of composite materials used in their structure and applications.* ICAMS. .DOI: 10.24264/icams-2022.I.11

St¨ocker, C., Bennett, R., Nex, F., Gerke, M., & Zevenbergen, J. (2017). Review of the current state of uav regulations. *Remote Sensing (Basel)*, 9(5), 459. DOI: 10.3390/rs9050459

Tatsidou, E., Tsiamis, C., Karamagioli, E., Boudouris, G., Pikoulis, A., Kakalou, E., & Pikoulis, E. (2019). Reflecting upon the humanitarian use of unmanned aerial vehicles (drones). *Swiss Medical Weekly*. DOI: 10.4414/smw.2019.20065 PMID: 30950503

Tran, T. (2022). Management and regulation of drone operation in urban en- vironment: a case study. *Social Sciences, 11.* . https://doi.org/.DOI: 10.3390/socsci11100474

Umar, T. (2020). Applications of drones for safety inspection in the gulf cooperation coun- cil construction. *Engineering, Construction, and Architectural Management*, 28(9), 2337–2360. DOI: 10.1108/ECAM-05-2020-0369

Vas, E., Lescro¨el, A., Duriez, O., Boguszewski, G., & Gr'emillet, D. (2015). Approaching birds with drones: First experiments and ethical guidelines. *Biology Letters*, 11(2), 20140754. DOI: 10.1098/rsbl.2014.0754 PMID: 25652220

Wang, N. (2020). "we live on hope...": Ethical considerations of humanitarian use of drones in post-disaster nepal. *IEEE Technology and Society Magazine*, 39(3), 76–85. DOI: 10.1109/MTS.2020.3012332

Wang, N. (2021). "as it is africa, it is ok"? ethical considerations of development use of drones for delivery in malawi. *IEEE Transactions on Technology and Society*, 2(1), 20–30. DOI: 10.1109/TTS.2021.3058669

Wang, N., Christen, M., & Hunt, M. (2021). Ethical considerations associated with "hu- manitarian drones": A scoping literature review. *Science and Engineering Ethics*, 27(4), 51. DOI: 10.1007/s11948-021-00327-4 PMID: 34342721

Waqas, M. (2023). *Unmanned aerial vehicles (uavs) in modern agriculture.* Springer. .DOI: 10.4018/978-1-6684-9231-4.ch006

Weber, S., & Knaus, F. (2017). Using drones as a monitoring tool to detect evidence of winter sports activities in a protected mountain area. *Eco.Mont*, 9(1), 30–34. DOI: 10.1553/eco.mont-9-1s30

West, J., & Bowman, J. (2016). The domestic use of drones: An ethical analysis of surveillance issues. *Public Administration Review*, 76(4), 649–659. DOI: 10.1111/puar.12506

Williams, J. (2015). Distant intimacy: Space, drones, and just war. *Ethics & International Affairs*, 29(1), 93–110. DOI: 10.1017/S0892679414000793

Wu, D., Rosen, D., Wang, L., & Schaefer, D. (2015). Cloud-based design and manufacturing: A new paradigm in digital manufacturing and design innovation. *Computer Aided Design*, 59, 1–14. DOI: 10.1016/j.cad.2014.07.006

Wynsberghe, A. (2019). Drones in humanitarian contexts, robot ethics, and the human–robot interaction. *Ethics and Information Technology*, 22(1), 43–53. DOI: 10.1007/s10676-019-09514-1

Xie, Y., Li, P., Nedjah, N., Gupta, B. B., Taniar, D., & Zhang, J. (2023). Pri- vacy protection framework for face recognition in edge-based Internet of Things. *Cluster Computing*, 26(5), 3017–3035. DOI: 10.1007/s10586-022-03808-8

Yao, C., Kao, C., & Lin, J. (2023). Drone for dynamic monitoring and tracking with intelligent image analysis. *Intelligent Automation & Soft Computing*, 36(2), 2233–2252. DOI: 10.32604/iasc.2023.034488

Zhang, L., Zhao, H., Hou, S., Zhao, Z., Xu, H., Wu, X., Wu, Q., & Zhang, R. (2019). A survey on 5g millimeter wave communications for uav-assisted wireless networks. *IEEE Access : Practical Innovations, Open Solutions*, 7, 117460–117504. DOI: 10.1109/ACCESS.2019.2929241

Zhang, X., Fan, K., Hou, H., & Liu, C. (2022). Real-time detection of drones using channel and layer pruning, based on the yolov3-spp3 deep learning algorithm. *Micromachines*, 13(12), 2199. DOI: 10.3390/mi13122199 PMID: 36557498

Zwickle, A., Farber, H., & Hamm, J. (2018). Comparing public concern and support for drone regulation to the current legal framework. *Behavioral Sciences & the Law*, 37(1), 109–124. DOI: 10.1002/bsl.2357 PMID: 30004141

Compilation of References

Abdallah, A., Ali, M., Misic, J., & Misic, V. (2019). Efficient security scheme for disaster surveillance uav communication networks. *Information (Basel)*, 10(2), 43. DOI: 10.3390/info10020043

Abdel-Malek, M., Akkaya, K., Bhuyan, A., & Ibrahim, A. (2022). A proxy signature-based swarm drone authentication with leader selection in 5g networks. *IEEE Access : Practical Innovations, Open Solutions*, 10, 57485–57498. DOI: 10.1109/ACCESS.2022.3178121

Abedini, M., & Al-Anbagi, I. (2024). Enhanced Active Eavesdroppers Detection System for Multihop WSNs in Tactical IoT Applications. *IEEE Internet of Things Journal*, 11(4), 6748–6760. DOI: 10.1109/JIOT.2023.3313048

Abid, S., Sulaiman, N., Chan, S., Nazir, U., Abid, M., Han, H., Ariza-Montes, A., & Vega-Muñoz, A. (2021). Toward an integrated disaster management approach: How artificial intelligence can boost disaster management. *Sustainability (Basel)*, 13(22), 12560. DOI: 10.3390/su132212560

Abir, T. A., Kuantama, E., Han, R., Dawes, J., Mildren, R., & Nguyen, P. (2023). Towards Robust Lidar-based 3D Detection and Tracking of UAVs. *Proceedings of the Ninth Workshop on Micro Aerial Vehicle Networks, Systems, and Applications*, (pp. 1–7). IEEE. DOI: 10.1145/3597060.3597236

Aboueleneen, N., Alwarafy, A., & Abdallah, M. (2023). Deep Reinforcement Learning for Internet of Drones Networks: Issues and Research Directions. *IEEE Open Journal of the Communications Society*, 4, 671–683. DOI: 10.1109/OJCOMS.2023.3251855

Abulaish, M., Wasi, N. A., & Sharma, S. (n.d.). The role of lifelong machine learning in bridging the gap between human and machine learning: A scientometric analysis. *WIREs Data Mining and Knowledge Discovery, n/a*(n/a), e1526. DOI: 10.1002/widm.1526

Adu-Gyamfi, S., Gyasi, R., & Darkwa, B. (2021). Historicizing medical drones in africa: A focus on ghana. *History of Science and Technology*, 11(1), 103–125. DOI: 10.32703/2415-7422-2021-11-1-103-125

Agapiou, A. (2021). Drones in construction: an international review of the legal and regulatory landscape. *Proceedings of the Institution of Civil Engineers - Management Procurement and Law*. IEEE. DOI: 10.1680/jmapl.19.00041

Aggarwal, K., Singh, S. K., Chopra, M., & Kumar, S. (2022). Role of Social Media in the COVID-19 Pandemic: A Literature Review. In *Data Mining Approaches for Big Data and Sentiment Analysis in Social Media* (pp. 91–115). IGI Global. DOI: 10.4018/978-1-7998-8413-2.ch004

Aggarwal, K., Singh, S. K., Chopra, M., Kumar, S., & Colace, F. (2022). Deep Learning in Robotics for Strengthening Industry 4.0.: Opportunities, Challenges and Future Directions. In Nedjah, N., Abd El-Latif, A. A., Gupta, B. B., & Mourelle, L. M. (Eds.), *Robotics and AI for Cybersecurity and Critical Infrastructure in Smart Cities* (pp. 1–19). Springer International Publishing. DOI: 10.1007/978-3-030-96737-6_1

Agnisarman, S., Lopes, S., Madathil, K. C., Piratla, K., & Gramopadhye, A. (2019). A survey of automation-enabled human-in-the-loop systems for infrastructure visual inspection. *Automation in Construction*, 97, 52–76. DOI: 10.1016/j.autcon.2018.10.019

Agrawal, A. (2020). *The next generation of human-drone partnerships: co-designing an emergency response system*. ACM. .DOI: 10.1145/3313831.3376825

Agustina, N. P., & Darwito, P. A. (2023). Autonomous Quadcopter Trajectory Tracking and Stabilization Using Control System Based on Sliding Mode Control and Kalman Filter. *2023 International Seminar on Intelligent Technology and Its Applications: Leveraging Intelligent Systems to Achieve Sustainable Development Goals, ISITIA 2023 - Proceeding*, (pp. 489–493). IEEE. DOI: 10.1109/ISITIA59021.2023.10221176

Ahmed, F., Mohanta, J. C., Keshari, A., & Yadav, P. S. (2022). Recent Advances in Unmanned Aerial Vehicles: A Review. *Arabian Journal for Science and Engineering*, 47(7), 7963–7984. DOI: 10.1007/s13369-022-06738-0 PMID: 35492958

Aijaz, A., & Sooriyabandara, M. (2018). The tactile internet for industries: A review. *Proceedings of the IEEE*, 107(2), 414–435. DOI: 10.1109/JPROC.2018.2878265

Akgun, S., & Greenhow, C. (2022). Artificial intelligence in education: Addressing ethical challenges in K-12 settings. *AI and Ethics*, 2(3), 431–440. DOI: 10.1007/s43681-021-00096-7 PMID: 34790956

Alanezi, M., Shahriar, M., Hasan, M., Ahmed, S., Sha'aban, Y., & Bouchekara, H. (2022). Live- stock management with unmanned aerial vehicles: A review. *IEEE Access : Practical Innovations, Open Solutions*, 10, 45001–45028. DOI: 10.1109/ACCESS.2022.3168295

Alatabani, L. (2022). *Robotics architectures based machine learning and deep learning approaches*. CrossRef. .DOI: 10.1049/icp.2022.2274

Al-Bahri, M. (2021). Using augmented reality and drones in tandem to serve smart cities. *Artificial Intelligence & Robotics Development Journal*, 147–157. .DOI: 10.52098/airdj.202144

Albarado, K., Coduti, L., Aloisio, D., Robinson, S., Drown, D., & Javorsek, D. (2022). AlphaMosaic: An Artificially Intelligent Battle Management Architecture. *Journal of Aerospace Information Systems*, 19(3), 203–213. DOI: 10.2514/1.I010991

Albayrak, Z. S., Kadak, M. T., Akkin Gurbuz, H. G., & Dogangun, B. (2022, November 22). Emotion Recognition Skill in Specific Learning Disorder and Attention-Deficit Hyperactivity Disorder. *Alpha Psychiatry*, 23(6), 268–273. DOI: 10.5152/alphapsychiatry.2022.22219 PMID: 36628377

Aldaej, A., Ahanger, T. A., Atiquzzaman, M., Ullah, I., & Yousufudin, M. (2022). Smart Cybersecurity Framework for IoT-Empowered Drones: Machine Learning Perspective. *Sensors (Basel)*, 22(7), 7. DOI: 10.3390/s22072630 PMID: 35408244

Aldeen, Y., & Abdulhadi, H. (2021). Data communication for drone-enabled internet of things. *Indonesian Journal of Electrical Engineering and Computer Science*, 22, 1216. DOI: 10.11591/ijeecs.v22.i2.pp1216-1222

Alemi, M., Meghdari, A., & Ghazisaedy, M. (2014). Employing humanoid robots for teaching English language in Iranian junior high-schools. *International Journal of HR; Humanoid Robotics*, 11(03), 1450022. DOI: 10.1142/S0219843614500224

Alexander, B., Ashford-Rowe, K., Barajas-Murph, N., Dobbin, G., Knott, J., McCormack, M., & Weber, N. (2019). *Horizon report 2019 higher education edition* (pp. 3–41). EDU19.

Alexandropoulos, G. C., Phan-Huy, D.-T., Katsanos, K. D., Crozzoli, M., Wymeersch, H., Popovski, P., & Gonzalez, S. H. (2023). RIS-enabled smart wireless environments: Deployment scenarios, network architecture, bandwidth and area of influence. *EURASIP Journal on Wireless Communications and Networking*, 2023(1), 103. DOI: 10.1186/s13638-023-02295-8

Al-Ghaithi, R., Hamid, A., & Slimi, Z. (2021). Drone delivery efficiency, challenges, and poten- tial in oman during covid-19. *Journal of University of Shanghai for Science and Technology*, 23(7), 811–830. DOI: 10.51201/JUSST/21/07214

Alhussan, A., Al-Dhaqm, A., Yafooz, W., Razak, S., Emara, A., & Khafaga, D. (2022). Towards development of a high abstract model for drone forensic domain. *Electronics (Basel)*, 11(8), 1168. DOI: 10.3390/electronics11081168

Ali, A., Jadoon, Y. K., Changazi, S. A., & Qasim, M. (2020). Military operations: Wireless sensor networks based applications to reinforce future battlefield command system. Paper presented at the *2020 IEEE 23rd International Multitopic Conference (INMIC)*. IEEE. DOI: 10.1109/INMIC50486.2020.9318168

Alibabaei, K., Gaspar, P. D., Lima, T. M., Campos, R. M., Girão, I., Monteiro, J., & Lopes, C. M. (2022). A Review of the Challenges of Using Deep Learning Algorithms to Support Decision-Making in Agricultural Activities. *Remote Sensing (Basel)*, 14(3), 3. DOI: 10.3390/rs14030638

Ali-Löytty, S., Sirola, N., & Piché, R. (2005). *Consistency of three Kalman filter extensions in hybrid navigation*.

Ali, M., Jamaludin, J., Ahmedy, I., & Awalin, L. J. (2023). Energy performance review of battery-powered drones for search and rescue (sar) operations. *IOP Conference Series. Earth and Environmental Science*, 1261(1), 012021. DOI: 10.1088/1755-1315/1261/1/012021

Ali, S., Khan, S., Fatma, N., Ozel, C., & Hussain, A. (2023). Utilisation of drones in achiev-ing various applications in smart warehouse management. *Benchmarking*, 31(3), 920–954. DOI: 10.1108/BIJ-01-2023-0039

Alkhawaldeh, M., & Khasawneh, M. (2024). Designing gamified assistive apps: A novel approach to motivating and supporting students with learning disabilities. *International Journal of Data and Network Science*, 8(1), 53–60. DOI: 10.5267/j.ijdns.2023.10.018

Allouch, A. (2019). *Mavsec: securing the mavlink protocol for ardupilot/px4 unmanned aerial systems.* IEEE. .DOI: 10.1109/IWCMC.2019.8766667

Allouch, A., Koubaa, A., Khalgui, M., & Abbes, T. (2019). Qualitative and quantitative risk analysis and safety assessment of unmanned aerial vehicles missions over the internet. *IEEE Access : Practical Innovations, Open Solutions*, 7, 53392–53410. DOI: 10.1109/ACCESS.2019.2911980

Almalki, F., Alotaibi, A., & Angelides, M. (2021). Coupling multifunction drones with ai in the fight against the coronavirus pandemic. *Computing*, 104(5), 1033–1059. DOI: 10.1007/s00607-021-01022-9

Alsaedi, A., Moustafa, N., Tari, Z., Mahmood, A., & Anwar, A. (2020). Ton iot telemetry dataset: A new generation dataset of iot and iiot for data-driven intrusion detection systems. *IEEE Access : Practical Innovations, Open Solutions*, 8, 165130–165150. DOI: 10.1109/ACCESS.2020.3022862

Alsamhi, S., Ma, O., & Ansari, M. (2019). Survey on artificial intelligence based tech-niques for emerging robotic communication. *Telecommunication Systems*, 72(3), 483–503. DOI: 10.1007/s11235-019-00561-z

Alsamhi, S., Ma, O., Ansari, M., & Almalki, F. (2019a). Survey on collaborative smart drones and internet of things for improving smartness of smart cities. *IEEE Access : Practical Innovations, Open Solutions*, 7, 128125–128152. DOI: 10.1109/ACCESS.2019.2934998

Alsamhi, S., Ma, O., Ansari, M., & Meng, Q. (2019b). Greening internet of things for greener and smart-er cities: A survey and future prospects. *Telecommunication Systems*, 72(4), 609–632. DOI: 10.1007/s11235-019-00597-1

Alsifiany, F. (2023). *Use of ai to diversify and improve the performance of rf sensors drone detection mechanism.* IEEE. .DOI: 10.5121/csit.2023.130504

Alvarez-Dionisi, L., Mittra, M., & Balza, R. (2019). Teaching artificial intelligence and robotics to un-dergraduate systems engineering students. *International Journal of Modern Education and Computer Science*, 11(7), 54–63. DOI: 10.5815/ijmecs.2019.07.06

Amador, O., Aramrattana, M., & Vinel, A. (2022). A Survey on Remote Operation of Road Vehi-cles. *IEEE Access : Practical Innovations, Open Solutions*, 10, 130135–130154. DOI: 10.1109/AC-CESS.2022.3229168

Aman, A., Yadegaridehkordi, E., Attarbashi, Z., Hassan, R., & Park, Y. (2020). A survey on trend and classification of internet of things reviews. *IEEE Access : Practical Innovations, Open Solutions*, 8, 111763–111782. DOI: 10.1109/ACCESS.2020.3002932

Amiri, M., & Ramli, R. (2022). Visual navigation system for autonomous drone using fiducial marker detection. *International Journal of Advanced Computer Science and Applications*, 13(9). DOI: 10.14569/IJACSA.2022.0130981

Amirova, A., Rakhymbayeva, N., Yadollahi, E., Sandygulova, A., & Johal, W. (2021). 10 Years of Human-NAO Interaction Research: A Scoping Review. *Frontiers in Robotics and AI*, 8, 744526. DOI: 10.3389/frobt.2021.744526 PMID: 34869613

Amirova, A., Rakhymbayeva, N., Zhanatkyzy, A., Telisheva, Z., & Sandygulova, A. (2023). Effects of parental involvement in robot-assisted autism therapy. *Journal of Autism and Developmental Disorders*, 53(1), 438–455. DOI: 10.1007/s10803-022-05429-x PMID: 35088233

Anwar, M. (2018). *Navren-rl: learning to fly in real en- vironment via end-to-end deep reinforcement learning using monocular images*. IEEE. .DOI: 10.1109/M2VIP.2018.8600838

Anwar, S., U. (2022). *Artificial intelligence in healthcare: an overview*. InTechOpen. .DOI: 10.5772/intechopen.102768

Araar, O., & Aouf, N. (2014). Full linear control of a quadrotor UAV, LQ vs H∞. *2014 UKACC International Conference on Control, CONTROL 2014*, (pp. 133–138). IEEE. DOI: 10.1109/CONTROL.2014.6915128

Arkhipov, V. (2022). Definition of artificial intelligence in the context of the russian legal system: A critical approach. *Государство и право*, 168(1), 168. DOI: 10.31857/S102694520018288-7

Ashraf, S. (2023). *Iot empowered smart cybersecurity framework for intrusion detection in internet of drones*. Research Square. .DOI: 10.21203/rs.3.rs-3047663/v1

Auflem, M., Kohtala, S., Jung, M., & Steinert, M. (2022). Facing the facs—Using ai to evaluate and control facial action units in humanoid robot face development. *Frontiers in Robotics and AI*, 9, 887645. DOI: 10.3389/frobt.2022.887645 PMID: 35774595

Ayamga, M., Tekinerdo˘gan, B., & Kassahun, A. (2021). Exploring the challenges posed by regulations for the use of drones in agriculture in the african context. *Land (Basel)*, 10(2), 164. DOI: 10.3390/land10020164

Ayamga, M., Tekinerdogan, B., & Rambaldi, G. (2020). Developing a policy framework for adoption and management of drones for agriculture in africa. *Technology Analysis and Strategic Management*, 33(8), 970–987. DOI: 10.1080/09537325.2020.1858047

Ayşe, T. U. N. A. (2024). Approaches and Strategies in Applied Behavior Analysis for Children with Autism Spectrum Disorder. *Psikiyatride Güncel Yaklasimlar*, 16(2), 347–357. DOI: 10.18863/pgy.1315911

Azcarate, S. M., Ríos-Reina, R., Amigo, J. M., & Goicoechea, H. C. (2021). Data handling in data fusion: Methodologies and applications. *Trends in Analytical Chemistry*, 143, 116355. DOI: 10.1016/j.trac.2021.116355

Babko-Malaya, O. (2023). *Assigning semantic meaning to machine derived competency controlling topics*. SPIE. .DOI: 10.1117/12.2663821

Baig, Z., Syed, N., & Mohammad, N. (2022). Securing the smart city airspace: Drone cyber attack detection through machine learning. *Future Internet*, 14(7), 205. DOI: 10.3390/fi14070205

Bailey, T., & Durrant-Whyte, H. (2006). Simultaneous localization and mapping (SLAM): Part II. *IEEE Robotics & Automation Magazine*, 13(3), 108–117. DOI: 10.1109/MRA.2006.1678144

Bai, Q., Li, S., Yang, J., Song, Q., Li, Z., & Zhang, X. (2020). Object detection recognition and robot grasping based on machine learning: A survey. *IEEE Access : Practical Innovations, Open Solutions*, 8, 181855–181879. DOI: 10.1109/ACCESS.2020.3028740

Bai, S., Song, S., Liang, S., Wang, J., Li, B., & Neretin, E. (2022). UAV Maneuvering Decision-Making Algorithm Based on Twin Delayed Deep Deterministic Policy Gradient Algorithm. *Journal of Artificial Intelligence and Technology*, 2(1), 1. DOI: 10.37965/jait.2021.12003

Bajpai, P., Sa, P., & Tewari, R. (2017). Greedy algorithm for image compression in image processing. *International Journal of Computer Applications*, 166(8), 34–37. DOI: 10.5120/ijca2017914118

Banaeian Far, S., & Imani Rad, A. (2024). Internet of Artificial Intelligence (IoAI): The emergence of an autonomous, generative, and fully human-disconnected community. *Discover Applied Sciences*, 6(3), 91. DOI: 10.1007/s42452-024-05726-3

Barbaria, S., Mont, M., Ghadafi, E., Mahjoubi Machraoui, H., & Rahmouni, H. B. (2022). Leveraging patient information sharing using blockchain-based distributed networks. *IEEE Access : Practical Innovations, Open Solutions*, 10, 106334–106351. DOI: 10.1109/ACCESS.2022.3206046

Barnas, A., Chabot, D., Hodgson, A., Johnston, D., Bird, D., & Ellis-Felege, S. (2020). A stan- dardized protocol for reporting methods when using drones for wildlife research. *Journal of Unmanned Vehicle Systems*, 8(2), 89–98. DOI: 10.1139/juvs-2019-0011

Barriga, J., Clemente, P., Hern'andez, J., & P'erez-Toledano, M. (2022). Simulateiot-fiware: Domain specific language to design, code generation and execute iot simulation environments on fiware. *IEEE Access : Practical Innovations, Open Solutions*, 10, 7800–7822. DOI: 10.1109/ACCESS.2022.3142894

Barthelm´e, S. (2019). Imager: an r package for image processing based on cimg. *The Journal of Open Source Software*, 4. DOI: 10.21105/joss.01012

Basañez, L., Nuño, E., & Aldana, C. I. (2023). Teleoperation and Level of Automation. In *Springer Handbook of Automation* (pp. 457–482). Springer International Publishing. DOI: 10.1007/978-3-030-96729-1_20

Bassyouni, Z., & Elhajj, I. (2021). Augmented reality meets artificial intel- ligence in robotics: A systematic review. *Frontiers in Robotics and AI*, 8, 724798. DOI: 10.3389/frobt.2021.724798 PMID: 34631805

Bastidas-Puga, E. R., Andrade, Á. G., Galaviz, G., & Covarrubias, D. H. (2019). Handover based on a predictive approach of signal-to-interference-plus-noise ratio for heterogeneous cellular networks. *IET Communications*, 13(6), 672–678. DOI: 10.1049/iet-com.2018.5126

Bathla, G., Bhadane, K., Singh, R. K., Kumar, R., Aluvalu, R., Krishnamurthi, R., Kumar, A., Thakur, R. N., & Basheer, S. (2022). Autonomous Vehicles and Intelligent Automation: Applications, Challenges, and Opportunities. *Mobile Information Systems*, 2022, e7632892. DOI: 10.1155/2022/7632892

Bayomi, N., & Fernandez, J. E. (2023). Eyes in the Sky: Drones Applications in the Built Environment under Climate Change Challenges. *Drones (Basel)*, 7(10), 10. DOI: 10.3390/drones7100637

Beam, A. (2020). Challenges to the reproducibility of machine learning models in health care. *JAMA, 323*. . DOI: 10.1001/jama.2019.20866

Becker, S. A., Cummins, M., Davis, A., Freeman, A., Hall, C. G., & Ananthanarayanan, V. (2017). *NMC horizon report: 2017 higher education edition.* The New Media Consortium.

Beck, H., Lesueur, J., Charland-Arcand, G., Akhrif, O., Gagne, S., Gagnon, F., & Couillard, D. (2016). Autonomous takeoff and landing of a quadcopter. *2016 International Conference on Unmanned Aircraft Systems, ICUAS 2016,* (pp. 475–484). IEEE. DOI: 10.1109/ICUAS.2016.7502614

Behera, T. (2023). The NITRDrone dataset to address the challenges for road extraction from aerial images. *Journal of Signal Process-ing Systems,95*(2).

Behera, T. K., Bakshi, S., Sa, P. K., Nappi, M., Castiglione, A., Vijayakumar, P., & Gupta, B. B. (2023). The NITRDrone dataset to address the challenges for road extraction from aerial images. *Journal of Signal Processing Systems for Signal, Image, and Video Technology,* 95(2), 197–209. DOI: 10.1007/s11265-022-01777-0

Bejiga, M. B., Zeggada, A., Nouffidj, A., & Melgani, F. (2017). A convolutional neural network approach for assisting avalanche search and rescue operations with UAV imagery. *Remote Sensing (Basel),* 9(2), 100. DOI: 10.3390/rs9020100

Belpaeme, T., & Tanaka, F. (2021). Social robots as educators. In *OECD Digital Education Outlook 2021 Pushing the Frontiers with Artificial Intelligence, Blockchain and Robots: Pushing the Frontiers with Artificial Intelligence, Blockchain and Robots* (p. 143). OECD Publishing. DOI: 10.1787/1c3b1d56-en

Benaoumeur, I., Zoubir, A. F., & Reda, H. E. A. (2015). Remote Control of Mobile Robot using the Virtual Reality. *International Journal of Electrical & Computer Engineering (2088-8708),* 5(5). IEEE.

Ben-Ari, M. (2017). *Image processing.* Springer. ₁2.DOI: 10.1007/978-3-319-62533-1

Beniiche, A. (2023). *6G and Next-Generation Internet: Under Blockchain Web3 Economy.* CRC Press. DOI: 10.1201/9781003427322

Bennett, C. (2020). *Use of symmetrical peak extraction in drone micro-doppler classification for staring radar.* IEEE. .DOI: 10.1109/RadarConf2043947.2020.9266702

Benos, L., Moysiadis, V., Kateris, D., Tagarakis, A. C., Busato, P., Pearson, S., & Bochtis, D. (2023). Human–robot interaction in agriculture: A systematic review. *Sensors (Basel),* 23(15), 6776. DOI: 10.3390/s23156776 PMID: 37571559

Benotsmane, R., Dud'as, L., & Kov'acs, G. (2020). Survey on artificial intelligence al- gorithms used in industrial robotics. *Multidiszciplin'aris Tudom'anyok,* 10(4), 194–205. DOI: 10.35925/j.multi.2020.4.23

Bernard-Cooper, J. (2022). *Multiple drone type classi-fication using machine learning techniques based on fmcw radar micro-doppler data.* SPIE. .DOI: 10.1117/12.2618026

Besada, J. A., Bergesio, L., Campaña, I., Vaquero-Melchor, D., López-Araquistain, J., Bernardos, A. M., & Casar, J. R. (2018). Drone Mission Definition and Implementation for Automated Infrastructure Inspection Using Airborne Sensors. *Sensors (Basel),* 18(4), 4. DOI: 10.3390/s18041170 PMID: 29641506

Bethel, C. L., Stevenson, M. R., & Scassellati, B. (2011, October). Secret-sharing: Interactions between a child, robot, and adult. In *2011 IEEE International Conference on systems, man, and cybernetics* (pp. 2489–2494). IEEE. DOI: 10.1109/ICSMC.2011.6084051

Bisht, J., & Vampugani, V. S. (2022). Load and cost-aware min-min workflow scheduling algorithm for heterogeneous resources in fog, cloud, and edge scenarios. *International Journal of Cloud Applications and Computing*, 12(1), 1–20. DOI: 10.4018/IJCAC.2022010105

Bithas, P., Michailidis, E., Nomikos, N., Vouyioukas, D., & Kanatas, A. G. (2019). A survey on machine-learning techniques for uav-based communications. *Sensors (Basel)*, 19(23), 5170. DOI: 10.3390/s19235170 PMID: 31779133

Bj¨orklund, S., Petersson, H., & Hendeby, G. (2015). Features for micro-doppler based activity classification. *IET Radar, Sonar & Navigation*, 9(9), 1181–1187. DOI: 10.1049/iet-rsn.2015.0084

Bl¨ocher, K., & Alt, R. (2020). Ai and robotics in the european restaurant sector: Assessing potentials for process innovation in a high-contact service industry. *Electronic Markets*, 31(3), 529–551. DOI: 10.1007/s12525-020-00443-2

Blasch, E., & Bélanger, M. (2016). *Agile battle management efficiency for command, control, communications, computers and intelligence (C4I)*. Paper presented at the Signal Processing, Sensor/Information Fusion, and Target Recognition XXV.

Blight, L., Bertram, D., & Kroc, E. (2019). Evaluating uav-based techniques to census an urban-nesting gull population on canada's pacific coast. *Journal of Unmanned Vehicle Systems*, 7(4), 312–324. DOI: 10.1139/juvs-2019-0005

Blockchain for Data Science. (2021, October 16). *Insights2Techinfo*. https://insights2techinfo.com/blockchain-for-data-science/

Bogue, R. (2023). The role of robots in environmental monitoring. *The Industrial Robot*, 50(3), 369–375. DOI: 10.1108/IR-12-2022-0316

Bollard, B., Doshi, A., Gilbert, N., Poirot, C., & Gillman, L. (2022). Drone technol- ogy for monitoring protected areas in remote and fragile environments. *Drones (Basel)*, 6(2), 42. DOI: 10.3390/drones6020042

Borgmann, T. (2015). *Image guided phase unwrapping for real-time 3d-scanning*. IEEE. .DOI: 10.1109/PCS.2015.7170058

Borikar, G., Gharat, C., & Deshmukh, S. (2022). Application of drone systems for spraying pesticides in advanced agriculture: A review. *IOP Conference Series. Materials Science and Engineering*, 1259(1), 012015. DOI: 10.1088/1757-899X/1259/1/012015

Bouadi, H., Bouchoucha, M., & Tadjine, M. (2007). Sliding mode control based on backstepping approach for an UAV type-quadrotor. *International Journal of Mechanical, Aerospace, Industrial, Mechatronic and Manufacturing Engineering, 1*(2), 39–44. http://www.waset.org/publications/11524

Brock, D. (2018). Learning from artificial intelligence's previous awakenings: The history of expert systems. *AI Magazine*, 39(3), 3–15. DOI: 10.1609/aimag.v39i3.2809

Bronk, J., Reynolds, N., & Watling, J. (2022). *The Russian air war and Ukrainian requirements for air defence.*

Brunton, E., Bolin, J., Leon, J., & Burnett, S. (2019). Fright or flight? be- havioural responses of kangaroos to drone-based monitoring. *Drones (Basel)*, 3(2), 41. DOI: 10.3390/drones3020041

Brunton, E., Leon, J., & Burnett, S. (2020). Evaluating the efficacy and optimal deployment of thermal infrared and true-colour imaging when using drones for monitoring kangaroos. *Drones (Basel)*, 4(2), 20. DOI: 10.3390/drones4020020

Bruzzone, A. G., Massei, M., Di Matteo, R., & Kutej, L. (2019). Introducing Intelligence and Autonomy into Industrial Robots to Address Operations into Dangerous Area. In Mazal, J. (Ed.), *Modelling and Simulation for Autonomous Systems* (pp. 433–444). Springer International Publishing. DOI: 10.1007/978-3-030-14984-0_32

Buchelt, A., Adrowitzer, A., Kieseberg, P., Gollob, C., Nothdurft, A., Eresheim, S., Tschiatschek, S., Stampfer, K., & Holzinger, A. (2024). Exploring artificial intelligence for applications of drones in forest ecology and management. *Forest Ecology and Management*, 551, 121530. DOI: 10.1016/j.foreco.2023.121530

Bulusu, N. (2021). *Towards adaptive, self-configuring networked unmanned aerial vehicles.* ACM. .DOI: 10.1145/3469259.3470488

Burke, C., Rashman, M., Wich, S., Symons, A., Theron, C., & Longmore, S. (2019). Optimizing observing strategies for monitoring animals using drone-mounted thermal infrared cameras. *International Journal of Remote Sensing*, 40(2), 439–467. DOI: 10.1080/01431161.2018.1558372

Butcher, P., Piddocke, T., Colefax, A., Hoade, B., Peddemors, V., Borg, L., & Cullis, B. (2019). Beach safety: Can drones provide a platform for sighting sharks? *Wildlife Research*, 46(8), 701. DOI: 10.1071/WR18119

Cai´c, M. (2019). Value of social robots in services: social cognition perspective. *Journal of Services Marketing,33.* .DOI: 10.1108/JSM-02-2018-0080

Cain, S. (2021). *Standards for uas - acceptable means of compliance for low risk sora operations.* ARC. .DOI: 10.2514/6.2021-1177

Cain, L., Thomas, J., & Alonso, M.Jr. (2019). From sci-fi to sci-fact: The state of robotics and ai in the hospitality industry. *Journal of Hospitality and Tourism Technology*, 10(4), 624–650. DOI: 10.1108/JHTT-07-2018-0066

Callaos, N., Cowin, J., Erkollar, A., & Oberer, B. (2023). Cybernetic Rela- tionships between Technological Innovations, Ethics, and the Law. Touro Scholar.

Campos, V., Molina, M., & Kr¨oner, S. (2017). Ethics and Civil Drones. *Introduction.*, 1–5. DOI: 10.1007/978-3-319-71087-7

Cangelosi, A., & Schlesinger, M. (2018). From babies to robots: The contribution of develop- mental robotics to developmental psychology. *Child Development Perspectives*, 12(3), 183–188. DOI: 10.1111/cdep.12282

Cao, Y. (2021). Technical composition and cre-ation of interactive installation art works under the background of artificial intelligence. *Mathematical Problems in Engineering*, 1–11. .DOI: 10.1155/2021/7227416

Cardona, T., Cudney, E. A., Hoerl, R., & Snyder, J. (2023). Data mining and machine learning retention models in higher education. *Journal of College Student Retention*, 25(1), 51–75. DOI: 10.1177/1521025120964920

Caruana, N., Moffat, R., Miguel-Blanco, A., & Cross, E. S. (2023). Perceptions of intelligence & sentience shape children's interactions with robot reading companions. *Scientific Reports*, 13(1), 7341. DOI: 10.1038/s41598-023-32104-7 PMID: 37147422

Carvalho, D., Pereira, E., & Cardoso, J. (2019). Machine learning interpretability: A survey on methods and metrics. *Electronics (Basel)*, 8(8), 832. DOI: 10.3390/electronics8080832

Casillo, M., & Colace, F. (2024). Securing Digital Ecosystems: Harnessing the Power of Intelligent Machines in a Secure and Sustainable Environment. In *Handbook of Research on AI and ML for Intelligent Machines and Systems*, (50–74). IGI Global.

Casillo, M., & Colace, F. (2024). Securing Digital Ecosystems: Harnessing the Power of Intelligent Machines in a Secure and Sustainable Environment. In *Handbook of Research on AI and ML for Intelligent Machines and Systems*, (pp. 50–74). IGI Global.

Casillo, M., & Colace, F. (2024). Securing Digital Ecosystems: Harnessing the Power of Intelligent Machines in a Secure and Sustainable Environment. In *Handbook of Research on AI and ML for Intelligent Machines and Systems*. IGI Global.

Cavanagh, T., Chen, B., Lahcen, R. A. M., & Paradiso, J. R. (2020). Constructing a design framework and pedagogical approach for adaptive learning in higher education: A practitioner's perspective. *International review of research in open and distributed learning, 21*(1), 173–197.

Cawthorne, D., & Wynsberghe, A. (2020). An ethical framework for the design, develop-ment, implementation, and assessment of drones used in public healthcare. *Science and Engineering Ethics*, 26(5), 2867–2891. DOI: 10.1007/s11948-020-00233-1 PMID: 32578062

Celik, I. (2023). Towards Intelligent-TPACK: An empirical study on teachers' professional knowledge to ethically integrate artificial intelligence (AI)-based tools into education. *Computers in Human Behavior*, 138, 107468. DOI: 10.1016/j.chb.2022.107468

Chabot, D., & Bird, D. (2015). Wildlife research and management methods in the 21st century: Where do unmanned aircraft fit in? *Journal of Unmanned Vehicle Systems*, 3(4), 137–155. DOI: 10.1139/juvs-2015-0021

Chai, J., Zeng, H., Li, A., & Ngai, E. W. T. (2021). Deep learning in computer vision: A critical review of emerging techniques and application scenarios. *Machine Learning with Applications*, 6, 100134. DOI: 10.1016/j.mlwa.2021.100134

Chamoso, P. (2015). *Swarm agent-based architecture suitable for internet of things and smartcities*. Springer. .DOI: 10.1007/978-3-319-19638-1

Chanda, S., & Banerjee, D. (2022). Omission and commission errors underlying ai failures. *AI & Society*. DOI: 10.1007/s00146-022-01585-x PMID: 36415822

Chandhar, P., & Larsson, E. (2019). Massive mimo for connectivity with drones: Case studies and future directions. *IEEE Access : Practical Innovations, Open Solutions*, 7, 94676–94691. DOI: 10.1109/ACCESS.2019.2928764

Chandra, M., Kumar, K., Thakur, P., Chattopadhyaya, S., Alam, F., & Kumar, S. (2022). Dig- ital technologies, healthcare and covid-19: Insights from developing and emerging nations. *Health and Technology*, 12(2), 547–568. DOI: 10.1007/s12553-022-00650-1 PMID: 35284203

Chase, C. C., Chin, D. B., Oppezzo, M. A., & Schwartz, D. L. (2009). Teachable agents and the protégé effect: Increasing the effort towards learning. *Journal of Science Education and Technology*, 18(4), 334–352. DOI: 10.1007/s10956-009-9180-4

Chaves, T., Martins, M., Martins, K., & Macedo, A. (2022). Development of an automated distribution grid with the application of new technologies. *IEEE Access : Practical Innovations, Open Solutions*, 10, 9431–9445. DOI: 10.1109/ACCESS.2022.3142683

Chen, Y. (2020). *Efficient drone mobility support using reinforcement learning*. WCNC. .DOI: 10.1109/WCNC45663.2020.9120595

Cheng, A., Eppich, W., Grant, V., Sherbino, J., Zendejas, B., & Cook, D. A. (2014). Debriefing for technology-enhanced simulation: A systematic review and meta-analysis. *Medical Education*, 48(7), 657–666. DOI: 10.1111/medu.12432 PMID: 24909527

Cheng, Y., Li, K. H., Liu, Y., Teh, K. C., & Poor, H. V. (2021). Downlink and uplink intelligent reflecting surface aided networks: NOMA and OMA. *IEEE Transactions on Wireless Communications*, 20(6), 3988–4000. DOI: 10.1109/TWC.2021.3054841

Chen, H.-C. (2019). Collaboration IoT-based RBAC with trust evaluation algorithm model for massive IoT integrated application. *Mobile Networks and Applications*, 24(3), 839–852. DOI: 10.1007/s11036-018-1085-0

Chen, H.-C., Widodo, A. M., Lin, J. C.-W., Weng, C.-E., & Do, D.-T. (2023). Outage behavior of the downlink reconfigurable intelligent surfaces-aided cooperative non-orthogonal multiple access network over Nakagami-m fading channels. *Wireless Networks*, 1–18. DOI: 10.1007/s11276-022-03074-x

Chen, J., & Hengjinda, P. (2019). Applying ai technology to the operation of smart farm robot. *Sensors and Materials*, 31(5), 1777. DOI: 10.18494/SAM.2019.2389

Chen, J., Li, S., Liu, D., & Li, X. (2020). AiRobSim: Simulating a Multisensor Aerial Robot for Urban Search and Rescue Operation and Training. *Sensors (Basel)*, 20(18), 5223. DOI: 10.3390/s20185223 PMID: 32933186

Chen, K., Lai, Y., & Hu, S. (2015). 3d indoor scene modeling from rgb-d data: A survey. *Computational Visual Media*, 1(4), 267–278. DOI: 10.1007/s41095-015-0029-x

Chen, Y., & Luca, G. (2021). Technologies supporting artificial intelligence and robotics application development. *JAIT*, 1(1), 1–8. DOI: 10.37965/jait.2020.0065

Chhabra, A., Singh, S. K., Sharma, A., Kumar, S., Gupta, B. B., Arya, V., & Chui, K. T. (2024). Sustainable and intelligent time-series models for epidemic disease forecasting and analysis. *Sustainable Technology and Entrepreneurship*, 3(2), 100064. DOI: 10.1016/j.stae.2023.100064

Chong, Z., Low, C., Mohammad, U., Rahman, R., & Shaari, M. (2018). Conception of logistics management system for smart factory. *IACSIT International Journal of Engineering and Technology*, 7(4.27), 126. DOI: 10.14419/ijet.v7i4.27.22499

Chopra, A. G. (2023). Impact of Artificial Intelligence and the Internet of Things in Modern Times and Hereafter: An Investigative Analysis. In *Advanced Computer Science Applications*. Apple Academic Press.

Christie, D. (2016). *3d reconstruc- tion of dynamic vehicles using sparse 3d-laser-scanner and 2d image fusion*. IEEE. .DOI: 10.1109/IAC.2016.7905690

Christou, A. G., Stergiou, C. L., Memos, V. A., Ishibashi, Y., & Psannis, K. E. (2023). Revolutionizing Connectivity: The Power of AI, IoT, and Edge Computing for Smart and Autonomous Systems. *2023 6th World Symposium on Communication Engineering (WSCE)*, 56–60. DOI: 10.1109/WSCE59557.2023.10365771

Chui, K. T., Gupta, B. B., Arya, V., & Torres-Ruiz, M. (2024). Selective and Adaptive Incremental Transfer Learning with Multiple Datasets for Machine Fault Diagno- sis. *Computers, Materials & Continua*, 78(1), 1363–1379. DOI: 10.32604/cmc.2023.046762

Chui, K. T., Gupta, B. B., Liu, J., Arya, V., Nedjah, N., Almomani, A., & Chaurasia, P. (2023). A survey of internet of things and cyber-physical systems: Standards, algorithms, applications, security, challenges, and future directions. *Information (Basel)*, 14(7), 388. DOI: 10.3390/info14070388

Chu, S. T., Hwang, G. J., & Tu, Y. F. (2022). Artificial intelligence-based robots in education: A systematic review of selected SSCI publications. *Computers and education. Artificial Intelligence*, ●●●, 100091.

Citron, R., Jenniskens, P., Watkins, C., Sinha, S., Shah, A., Ra¨ıssi, C., Devillepoix, H., & Albers, J. (2021). Recovery of meteorites using an autonomous drone and machine learning. *Meteoritics & Planetary Science*, 56(6), 1073–1085. DOI: 10.1111/maps.13663

Clemente, C., L. Pallotta, C. Ilioudis, F. Fioranelli, G. Giunta, and A. Farina. 2021. "Chebychev moments based drone classification, recognition and fingerprinting." .DOI: 10.23919/IRS51887.2021.9466211

Coindreau, M.-A., Gallay, O., & Zufferey, N. (2021). Parcel delivery cost minimization with time window constraints using trucks and drones. *Networks*, 78(4), 400–420. DOI: 10.1002/net.22019

Colace, F., Guida, C. G., Gupta, B., Lorusso, A., Marongiu, F., & Santaniello, D. (2022). A BIM-based approach for decision support system in smart buildings. In *Proceedings of Seventh International Congress on Information and Communication Technology: ICICT 2022*, (pp. 471–481). Springer.

Colace, F., Guida, C. G., Gupta, B., Lorusso, A., Marongiu, F., & Santaniello, D. (2022). A BIM-based approach for decision support system in smart buildings. In *Proceedings of Seventh International Congress on Information and Communication Technology: ICICT 2022*. Springer.

Coluccia, A. (2020). Detection and classification of multirotor drones in radar sensor networks: a review. *Sensors,20*, 4172. .DOI: 10.3390/s20154172

Comes, T. (2018). Cold chains, interrupted. *Journal of Humani- tarian Logistics and Supply Chain Management, 8,* 49–69. .DOI: 10.1108/JHLSCM-03-2017-0006

Corbett, B. A., Carmean, V., Ravizza, S., Wendelken, C., Henry, M. L., Carter, C., & Rivera, S. M. (2009, September). A functional and structural study of emotion and face processing in children with autism. *Psychiatry Research: Neuroimaging,* 173(3), 196–205. DOI: 10.1016/j.pscychresns.2008.08.005 PMID: 19665877

Cremer, S. (2017). *Neuroadaptive Human-Machine Interfaces for Collaborative Robots* [Doctoral dissertation, The University of Texas at Arlington].

Croon, G., Dupeyroux, J., Fuller, S., & Marshall, J. (2022). Insect-inspired ai for autonomous robots. *Science Robotics,* 7(67), eabl6334. DOI: 10.1126/scirobotics.abl6334 PMID: 35704608

Cross, E., Hortensius, R., & Wykowska, A. (2019). From social brains to social robots: applying neurocognitive insights to human–robot interaction. *Philosophical Transactions of the Royal Society B Biological Sciences, 374.* ..DOI: 10.1098/rstb.2018.0024

Cui, Y., Wang, L., Duan, L., & Suiqing, C. (2020). Quality evaluation based on color grading - relationship between chemical susbtances and commercial grades by machine version in corni fructus. *Tropical Journal of Pharmaceutical Research,* 19(7), 1495–1501. DOI: 10.4314/tjpr.v19i7.23

da Silva, A. P., Bezerra, I. M. P., Antunes, T. P. C., Cavalcanti, M. P. E., & de Abreu, L. C. (2023). Applied behavioral analysis for the skill performance of children with autism spectrum disorder. *Frontiers in Psychiatry,* 14, 1093252. DOI: 10.3389/fpsyt.2023.1093252 PMID: 37181882

Dai, C. (2022, June Distributed User Association with Grouping in Satellite-Terrestrial Integrated Networks. *IEEE Internet of Things Journal,* 9(12), 10244–10256. DOI: 10.1109/JIOT.2021.3122939

Dai, N. (2021). *Drone application in smart cities: the general overview of security vulnerabilities and countermeasures for data communication.* Springer. DOI: 10.1007/978-3-030-63339-4

Dai, L., Wang, B., Yuan, Y., Han, S., Chih-Lin, I., & Wang, Z. (2015). Non-orthogonal multiple access for 5G: Solutions, challenges, opportunities, and future research trends. *IEEE Communications Magazine,* 53(9), 74–81. DOI: 10.1109/MCOM.2015.7263349

Damijani'c, D. (2021). Remote sensing in invasive species detection and monitor- ing. *International Journal of Environmental Sciences & Natural Resources,* 29(1). DOI: 10.19080/IJESNR.2021.29.556255

Daponte, P., Vito, L., Glielmo, L., Iannelli, L., Liuzza, D., Picariello, F., & Silano, G. (2019). A review on the use of drones for precision agriculture. *IOP Conference Series. Earth and Environmental Science,* 275(1), 012022. DOI: 10.1088/1755-1315/275/1/012022

Darapureddy, N. (2021). A comprehensive study on artificial intelligence and robotics for machine intelligence. In *Methodologies and Applica-tions of Computational Statistics for Machine Intelligence,* (203–222). IGI Global.

Darno, D., & Mesiono, M. (2021). Strategi Komunikasi Kepala Sekolah Dalam Meningkatkan Efektivitas Manajemen Sekolah Di Mtsn 3 Langkat. *PIONIR: JURNAL PENDIDIKAN,* 10(3).

Darwito, P. A., & Wahyuadnyana, K. D. (2022). Performance Examinations of Quadrotor with Sliding Mode Control-Neural Network on Various Trajectory and Conditions. *Mathematical Modelling of Engineering Problems*, 9(3), 707–714. DOI: 10.18280/mmep.090317

Das, S. K., Sahu, A., & Popa, D. O. (2017, May). Mobile app for human-interaction with sitter robots. In *Smart Biomedical and Physiological Sensor Technology XIV* (Vol. 10216, p. 85). SPIE. DOI: 10.1117/12.2262792

Daudi, J. (2015). An overview of application of artificial immune system in swarm robotic systems. *Automation Control and Intelligent Systems*, 3(2), 11. DOI: 10.11648/j.acis.20150302.11

David, D., Thérouanne, P., & Milhabet, I. (2022). The acceptability of social robots: A scoping review of the recent literature. *Computers in Human Behavior*, 137, 107419. DOI: 10.1016/j.chb.2022.107419

Delmerico, J. (2019). *Are we ready for autonomous drone racing? the uzh-fpv drone racing dataset.* IEEE. .DOI: 10.1109/ICRA.2019.8793887

Deo, N., & Anjankar, A. (2023, May 23). (n.d.). Artificial Intelligence With Robotics in Healthcare: A Narrative Review of Its Viability in India. *Cureus*, 15(5), e39416. DOI: 10.7759/cureus.39416 PMID: 37362504

Dhilip Kumar, V., Kanagachidambaresan, G. R., Chyne, P., & Kandar, D. (2022). Extended Communication Range for Autonomous Vehicles using Hybrid DSRC/WiMAX Technology. *Wireless Personal Communications*, 123(3), 2301–2316. DOI: 10.1007/s11277-021-09242-0

Diao, Y. (2022). *Drone authentication via acoustic fingerprint.* ACM. .DOI: 10.1145/3564625.3564653

Dogru, S., & Marques, L. (2020). Pursuing Drones with Drones Using Millimeter Wave Radar. *IEEE Robotics and Automation Letters*, 5(3), 4156–4163. DOI: 10.1109/LRA.2020.2990605

Domeyer, J., Lee, J., & Toyoda, H. (2020). Vehicle automation–other road user com- munication and coordination: Theory and mechanisms. *IEEE Access : Practical Innovations, Open Solutions*, 8, 19860–19872. DOI: 10.1109/ACCESS.2020.2969233

Doncieux, S., Chatila, R., Straube, S., & Kirchner, F. (2022). Human-centered ai and robotics. *Ai Perspectives*, 4(1), 1. DOI: 10.1186/s42467-021-00014-x

Doroudi, S. (2022). The intertwined histories of artificial intelligence and educa- tion. *International Journal of Artificial Intelligence in Education*, 33(4), 885–928. DOI: 10.1007/s40593-022-00313-2

Dufourd, D., & Dalgalarrondo, A. (2006, April). Integrating human/robot interaction into robot control architectures for defense applications. In *1th National Conference on Control Architecture of Robots, April* (pp. 6-7). IEEE.

Dumanska, I., Vasylkivskyi, D., Zhurba, I., Pukhalska, Y., Matviiets, O., & Goncharuk, A. (2021). Dronology and 3d printing as a catalyst for international trade in industry 4.0. *WSEAS Transactions on Environment and Development*, 17, 740–757. DOI: 10.37394/232015.2021.17.71

Duong, C. (2017). *Temporal non-volume preserving approach to facial age-progression and age-invariant face recognition.* IEEE. .DOI: 10.1109/ICCV.2017.403

Dutta, S. (2023). *Perspective chapter: digital inclusion of the farming sector using drone technology.* InTech Open. .DOI: 10.5772/intechopen.108740

Dwivedi, K., Govindarajan, P., Srinivasan, D., Keerthi Sanjana, A., Selvanambi, R., & Karuppiah, M. (2023). Intelligent Autonomous Drones in Industry 4.0. In Sarveshwaran, V., Chen, J. I.-Z., & Pelusi, D. (Eds.), *Artificial Intelligence and Cyber Security in Industry 4.0* (pp. 133–163). Springer Nature. DOI: 10.1007/978-981-99-2115-7_6

Ebrahimzadeh, A. (2019). *Tactile Internet over Fiber-Wireless Enhanced HetNets using Edge Intelligence* [Doctoral dissertation, Institut National de la Recherche Scientifique, Canada].

Ebrahimzadeh, A., & Maier, M. (2019). Delay-constrained teleoperation task scheduling and assignment for human+ machine hybrid activities over FiWi enhanced networks. *IEEE Transactions on Network and Service Management*, 16(4), 1840–1854. DOI: 10.1109/TNSM.2019.2937020

Edwards, D., Subramanian, N., Chaudhuri, A., Morlacchi, P., & Zeng, W. (2023). Use of delivery drones for humanitarian operations: Analysis of adoption barriers among logistics service providers from the technology acceptance model perspective. *Annals of Operations Research*, 335(3), 1645–1667. DOI: 10.1007/s10479-023-05307-4 PMID: 37361062

Egan, C., Blackwell, B., Fern'andez-Juricic, E., & Klug, P. (2020). Testing a key assumption of using drones as frightening devices: Do birds perceive drones as risky? *The Condor*, 122(3), duaa014. DOI: 10.1093/condor/duaa014

Eisenberg, D. A., Alderson, D. L., Kitsak, M., Ganin, A., & Linkov, I. (2018). Network foundation for command and control (C2) systems: Literature review. *IEEE Access : Practical Innovations, Open Solutions*, 6, 68782–68794. DOI: 10.1109/ACCESS.2018.2873328

Elbasheer, M., Longo, F., Mirabelli, G., Nicoletti, L., Padovano, A., & Solina, V. (2023). Shaping the role of the digital twins for human-robot dyad: Connotations, scenarios, and future perspectives. *IET Collaborative Intelligent Manufacturing*, 5(1), e12066. DOI: 10.1049/cim2.12066

Elendu, C., Amaechi, D. C., Elendu, T. C., Jingwa, K. A., Okoye, O. K., John Okah, M., Ladele, J. A., Farah, A. H., & Alimi, H. A. (2023). Ethical implications of ai and robotics in healthcare: A review. *Medicine*, 102(50), e36671. DOI: 10.1097/MD.0000000000036671 PMID: 38115340

Eli-Chukwu, N. (2019). Applications of artificial intelligence in agriculture: A re- view. *Engineering Technology & Applied Science Research*, 9(4), 4377–4383. DOI: 10.48084/etasr.2756

Elmenreich, W., & Pitzek, S. (2001). Using sensor fusion in a time-triggered network. *IECON'01. 27th Annual Conference of the IEEE Industrial Electronics Society (Cat. No.37243)*. IEEE. DOI: 10.1109/IECON.2001.976510

Emaminejad, N., & Akhavian, R. (2022). Trustworthy AI and robotics: Implications for the AEC industry. *Automation in Construction*, 139, 104298. DOI: 10.1016/j.autcon.2022.104298

Esposito, M., Crimaldi, M., Cirillo, V., Sarghini, F., & Maggio, A. (2021). Drone and sensor technology for sustainable weed management: A review. *Chemical and Biological Technologies in Agriculture*, 8(1), 18. DOI: 10.1186/s40538-021-00217-8

Eu, K. Q., & Phang, S. K. (2023). Automated Parcel Loading-Unloading Mechanism Design for Delivery UAV. *Journal of Physics: Conference Series*, 2523(1), 012016. DOI: 10.1088/1742-6596/2523/1/012016

Fan, Z., Yang, H., Liu, F., Liu, L., & Han, Y. (2022). Reinforcement learning method for target hunting control of multi-robot systems with obstacles. *International Journal of Intelligent Systems*, 37(12), 11275–11298. DOI: 10.1002/int.23042

Farooq, M., Sohail, O., Abid, A., & Rasheed, S. (2022). A survey on the role of iot in agri- culture for the implementation of smart livestock environment. *IEEE Access : Practical Innovations, Open Solutions*, 10, 9483–9505. DOI: 10.1109/ACCESS.2022.3142848

Fathy, G., Hassan, H., Sheta, W., Omara, F., & Nabil, E. (2021). A novel no-sensors 3d model reconstruction from monocular video frames for a dynamic environment. *PeerJ. Computer Science*, 7, e529. DOI: 10.7717/peerj-cs.529 PMID: 34084931

Feizi, N. (2021). Robotics and ai for teleoperation, tele-assessment, and tele-training for surgery in the era of covid-19: existing challenges, and future vision. *Frontiers in Robotics and AI,8*. DOI: 10.3389/frobt.2021.610677

Fernandes, S. V., & Ullah, M. S. (2022). A Comprehensive Review on Features Extraction and Features Matching Techniques for Deception Detection. *IEEE Access : Practical Innovations, Open Solutions*, 10, 28233–28246. DOI: 10.1109/ACCESS.2022.3157821

Fiestas Lopez Guido, J. C., Kim, J. W., Popkowski Leszczyc, P. T., Pontes, N., & Tuzovic, S. (2024). Retail robots as sales assistants: How speciesism moderates the effect of robot intelligence on customer perceptions and behaviour. *Journal of Service Theory and Practice*, 34(1), 127–154. DOI: 10.1108/JSTP-04-2023-0123

Filho, E., Gomes, F., Monteiro, S., Severino, R., Penna, S., Koubaa, A., & Tovar, E. (2023). A drone secure handover architecture validated in a software in the loop environ- ment. *Journal of Physics: Conference Series*, 2526(1), 012083. DOI: 10.1088/1742-6596/2526/1/012083

Filho, F., Heldens, W., Kong, Z., & Lange, E. (2019). Drones: Innovative technology for use in precision pest management. *Journal of Economic Entomology*, 113(1), 1–25. DOI: 10.1093/jee/toz268 PMID: 31811713

Floreano, D., & Wood, R. J. (2015). Science, technology and the future of small autonomous drones. *Nature*, 521(7553), 460–466. DOI: 10.1038/nature14542 PMID: 26017445

Foehn, P., Brescianini, D., Kaufmann, E., Cieslewski, T., Gehrig, M., Muglikar, M., & Scaramuzza, D. (2021). Alphapilot: Autonomous drone racing. *Autonomous Robots*, 46(1), 307–320. DOI: 10.1007/s10514-021-10011-y PMID: 35221535

Fox, S. (2018). Cyborgs, robots and society: Implications for the future of soci- ety from human enhancement with in-the-body technologies. *Technologies*, 6(2), 50. DOI: 10.3390/technologies6020050

Frese, U., Wagner, R., & Röfer, T. (2010). A SLAM Overview from a User's Perspective. *KI - Künstliche Intelligenz, 24*(3), 191–198. DOI: 10.1007/s13218-010-0040-4

Frogner, L., Hellfeldt, K., Ångström, A. K., Andershed, A. K., Källström, Å., Fanti, K. A., & Andershed, H. (2022). Stability and change in early social skills development in relation to early school performance: A longitudinal study of a Swedish cohort. *Early Education and Development*, 33(1), 17–37. DOI: 10.1080/10409289.2020.1857989

Gallardo-Camacho, J., & Breijo, V. (2020). Relationships between law enforcement authorities and drone journalists in spain. *Media and Communication*, 8(3), 112–122. DOI: 10.17645/mac.v8i3.3097

Gallud, J. A., Carreño, M., Tesoriero, R., Sandoval, A., Lozano, M. D., Durán, I., Penichet, V. M. R., & Cosio, R. (2023). Technology-enhanced and game based learning for children with special needs: A systematic mapping study. *Universal Access in the Information Society*, 22(1), 227–240. DOI: 10.1007/s10209-021-00824-0 PMID: 34248457

Gao, Y. (2021). *Space robotics and autonomous systems: technologies, advances and applications*. Institution of Engineering and Technology.

Gao, K., Xiao, H., Qu, L., & Wang, S. (2022). Optimal interception strategy of air defence missile system considering multiple targets and phases. *Proceedings of the Institution of Mechanical Engineers. Part O, Journal of Risk and Reliability*, 236(1), 138–147. DOI: 10.1177/1748006X211022111

Garcia, F., & Rachelson, E. (2013). Markov Decision Processes. In *Markov Decision Processes in Artificial Intelligence* (pp. 1–38). John Wiley & Sons, Ltd. DOI: 10.1002/9781118557426.ch1

Gasparetto, A., & Scalera, L. (2019). From the unimate to the delta robot: The early decades of industrial robotics. *History of Mechanism and Machine Science*, 37, 284–295. DOI: 10.1007/978-3-030-03538-9_23

Gauri, P., & Van Eerden, J. (2019). *What the Fifth Industrial Revolution is and why it matters*. Europeansting.com.

Gavrilovi'c, M., Muhović, A., & Pavlović, N. (2024). Analysis of the application of modern technologies in agriculture in three balkan countries and the impact on biodiversity. *Romanian Agricultural Research*, 41, 79–90. DOI: 10.59665/rar4108

Geetha, D. (2024). The future is now: ai powers next-generation robots. *Interantional Journal of Scientific Research in Engineering and Management*. .DOI: 10.55041/IJSREM28890

Gervasi, R., Barravecchia, F., Mastrogiacomo, L., & Franceschini, F. (2023). Applications of affective computing in human-robot interaction: State-of-art and challenges for manufacturing. *Proceedings of the Institution of Mechanical Engineers. Part B, Journal of Engineering Manufacture*, 237(6-7), 815–832. DOI: 10.1177/09544054221121888

Ghafghazi, S., Carnett, A., Neely, L., Das, A., & Rad, P. (2021, October). AI-Augmented Behavior Analysis for Children With Developmental Disabilities: Building Toward Precision Treatment. *IEEE Systems, Man, and Cybernetics Magazine*, 7(4), 4–12. DOI: 10.1109/MSMC.2021.3086989

Ghamari, M., Rangel, P., Mehrubeoglu, M., Tewolde, G. S., & Sherratt, R. S. (2022). Unmanned Aerial Vehicle Communications for Civil Applications: A Review. *IEEE Access : Practical Innovations, Open Solutions*, 10, 102492–102531. DOI: 10.1109/ACCESS.2022.3208571

Ghita, M., Neckebroek, M., Muresan, C., & Copot, D. (2020). Closed-loop control of anesthe- sia: Survey on actual trends, challenges and perspectives. *IEEE Access : Practical Innovations, Open Solutions*, 8, 206264–206279. DOI: 10.1109/ACCESS.2020.3037725

Gill, S. S., Wu, H., Patros, P., Ottaviani, C., Arora, P., Pujol, V. C., Haunschild, D., Parlikad, A. K., Cetinkaya, O., Lutfiyya, H., Stankovski, V., Li, R., Ding, Y., Qadir, J., Abraham, A., Ghosh, S. K., Song, H. H., Sakellariou, R., Rana, O., & Buyya, R. (2024). Modern computing: Vision and challenges. *Telematics and Informatics Reports*, 13, 100116. DOI: 10.1016/j.teler.2024.100116

Giordan, D., Adams, M. S., Aicardi, I., Alicandro, M., Allasia, P., Baldo, M., De Berardinis, P., Dominici, D., Godone, D., Hobbs, P., Lechner, V., Niedzielski, T., Piras, M., Rotilio, M., Salvini, R., Segor, V., Sotier, B., & Troilo, F. (2020). The use of unmanned aerial vehicles (UAVs) for engineering geology applications. *Bulletin of Engineering Geology and the Environment*, 79(7), 3437–3481. DOI: 10.1007/s10064-020-01766-2

Girshick, R. (2015). *Fast R-CNN*. 1440–1448. https://openaccess.thecvf.com/content_iccv_2015/html/Girshick_Fast_R-CNN_ICCV_2015_paper.html

Gong, X., Qian, L., Ge, W., & Yan, J. (2020). Research on electronic brake force distribution and anti-lock brake of vehicle based on direct drive electro hydraulic actuator. *International Journal of Automotive Engineering*, 11(2), 22–29. DOI: 10.20485/jsaeijae.11.2_22

Gong, X., Yue, X., & Liu, F. (2020). Performance analysis of cooperative NOMA networks with imperfect CSI over Nakagami-m fading channels. *Sensors (Basel)*, 20(2), 424. DOI: 10.3390/s20020424 PMID: 31940864

Gonzalez-Jimenez, H. (2018). Taking the fiction out of science fiction: (self-aware) robots and what they mean for society, retailers and marketers. *Futures*, 98, 49–56. DOI: 10.1016/j.futures.2018.01.004

Griffin, L. (2005). Optimality of the basic colour categories for classification. *Journal of the Royal Society, Interface*, 3(6), 71–85. DOI: 10.1098/rsif.2005.0076 PMID: 16849219

Guan, B. (2023). Global application of the atmospheric river scale. *Journal of Geophysical Research Atmospheres,128*. DOI: 10.1029/2022JD037180

Guo, J., Ahmad, I., & Chang, K. (2020). Classification, positioning, and tracking of drones by hmm using acoustic circular microphone array beamforming. *EURASIP Journal on Wireless Communications and Networking*, 2020(1), 9. DOI: 10.1186/s13638-019-1632-9

Guo, X., Zeng, T., Wang, Y., & Jie, Z. (2019). Fuzzy topsis approaches for assess- ing the intelligence level of iot-based tourist attractions. *IEEE Access : Practical Innovations, Open Solutions*, 7, 1195–1207. DOI: 10.1109/ACCESS.2018.2881339

Gupta, B. B., Gaurav, A., Chui, K. T., & Arya, V. (2023). Optimized Edge- cCCN Based Model for the Detection of DDoS Attack in IoT Environment. In *International Conference on Edge Computing*, (pp. 14–23). Springer.

Gupta, B. B., Gaurav, A., Chui, K. T., & Arya, V. (2023). Optimized Edge- cCCN Based Model for the Detection of DDoS Attack in IoT Environment. In *International Conference on Edge Computing*. Springer.

Gupta, A., Singh, S. K., Gupta, B. B., Chopra, M., & Gill, S. S. (2023). Evaluating the Sustainable COVID-19 Vaccination Framework of India Using Recurrent Neural Networks. *Wireless Personal Communications*, 133(1), 73–91. DOI: 10.1007/s11277-023-10751-3

Gupta, B. B., Gaurav, A., Chui, K. T., Arya, V., & Choi, C. (2024). Au- toencoders Based Optimized Deep Learning Model for the Detection of Cyber Attack in IoT Environment. In *2024 IEEE International Conference on Consumer Electronics (ICCE)*, (pp. 1–6). IEEE.

Gupta, L., Jain, R., & Vaszkun, G. (2016). Survey of important issues in uav communication networks. *IEEE Communications Surveys and Tutorials*, 18(2), 1123–1152. DOI: 10.1109/COMST.2015.2495297

Gupta, R., Kumari, A., & Tanwar, S. (2020). Fusion of blockchain and artificial intelligence for secure drone networking underlying 5g communications. *Transactions on Emerging Telecommunications Technologies*, 32(1), e4176. DOI: 10.1002/ett.4176

Gupta, S. (2021). *Internet of Things (IOT)*. Hands-On Sensing, Actuating, and Output Modules in Robotics.

Hadiwardoyo, S. A., Hernández-Orallo, E., Calafate, C. T., Cano, J. C., & Manzoni, P. (2018). Experimental characterization of UAV-to-car communications. *Computer Networks*, 136, 105–118. DOI: 10.1016/j.comnet.2018.03.002

Haenlein, M., & Kaplan, A. (2019). A brief history of artificial intelligence: On the past, present, and future of artificial intelligence. *California Management Review*, 61(4), 5–14. DOI: 10.1177/0008125619864925

Hafeez, A., Husain, M. A., Singh, S. P., Chauhan, A., Khan, M. T., Kumar, N., Chauhan, A., & Soni, S. K. (2023). Implementation of drone technology for farm monitoring & pesticide spraying: A review. *Information Processing in Agriculture*, 10(2), 192–203. DOI: 10.1016/j.inpa.2022.02.002

Hahn, N., Mwakatobe, A., Konuche, J., Souza, N., Keyyu, J., Goss, M., Chang'a, A., Palminteri, S., Dinerstein, E., & Olson, D. (2016). Unmanned aerial vehicles mitigate hu- man–elephant conflict on the borders of tanzanian parks: A case study. *Oryx*, 51(3), 513–516. DOI: 10.1017/S0030605316000946

Haidegger, T., Galambos, P., & Rudas, I. J. (2019, April). Robotics 4.0–Are we there yet? In *2019 IEEE 23rd International Conference on Intelligent Engineering Systems (INES)* (pp. 000117-000124). IEEE.

Halamek, L. P., Cady, R. A., & Sterling, M. R. (2019). Using briefing, simulation and debriefing to improve human and system performance. *Seminars in Perinatology*. Elsevier. DOI: 10.1053/j.semperi.2019.08.007

Hameed, B., Shah, M., Pietropaolo, A., Coninck, V., Naik, N., Skolarikos, A., & Somani, B. (2023). The technological future of percutaneous nephrolithotomy: A young academic urologists endourology and urolithiasis working group update. *Current Opinion in Urology*, 33(2), 90–94. DOI: 10.1097/MOU.0000000000001070 PMID: 36622261

Han, L. (2015). *A study on flexible vibratory feeding system based on halcon machine vision software*. IEEE. .DOI: 10.2991/isrme-15.2015.3

Han, D., Chen, W., & Liu, J. (2021). Energy-Efficient UAV Communications Under Stochastic Trajectory: A Markov Decision Process Approach. *IEEE Transactions on Green Communications and Networking*, 5(1), 106–118. DOI: 10.1109/TGCN.2020.3016266

Handa, S. (2015). *Human-Machine Interaction for Unmanned Surface Systems* [Doctoral dissertation, University of Illinois at Urbana-Champaign].

Han, T., Ribeiro, I., Magaia, N., Preto, J., Segundo, A., Macedo, A., & Muhammad, K. (2021). Emerging drone trends for blockchain-based 5g networks: Open issues and future perspectives. *IEEE Network*, 35(1), 38–43. DOI: 10.1109/MNET.011.2000151

Ha, Q. P., Yen, L., & Balaguer, C. (2019). Robotic autonomous systems for earthmoving in military applications. *Automation in Construction*, 107, 102934. DOI: 10.1016/j.autcon.2019.102934

Hartmann, W., Fishlock, V., & Leslie, A. (2021). First guidelines and suggested best pro-tocol for surveying african elephants (loxodonta africana) using a drone. *Koedoe*, 63(1). DOI: 10.4102/koedoe.v63i1.1687

Hart, P. E. (2022). An Artificial Intelligence Odyssey: From the Research Lab to the Real World. *IEEE Annals of the History of Computing*, 44(1), 57–72. DOI: 10.1109/MAHC.2021.3077417

Hassabis, D., Kumaran, D., Summerfield, C., & Botvinick, M. (2017). Neuroscience-inspired artificial intelligence. *Neuron*, 95(2), 245–258. DOI: 10.1016/j.neuron.2017.06.011 PMID: 28728020

Hassan, A., Biaggi, A., Asaad, M., Andejani, D., Li, J., Selber, J., & Butler, C. (2022a). Development and assessment of machine learning models for individualized risk as- sessment of mastectomy skin flap necrosis. *Annals of Surgery Open : Perspectives of Surgical History, Education, and Clinical Approaches*, 278, e123–e130. DOI: 10.1097/SLA.0000000000005386 PMID: 35129476

Hassan, A., Biaggi-Ondina, A., Rajesh, A., Asaad, M., Nelson, J., Coert, J., Mehrara, B., & Butler, C. (2022b). Predicting patient-reported outcomes following surgery using machine learning. *The American Surgeon*, 89(1), 31–35. DOI: 10.1177/00031348221109478 PMID: 35722685

He, H., Gray, J., Cangelosi, A., Meng, Q., McGinnity, T. M., & Mehnen, J. (2022). The Challenges and Opportunities of Human-Centered AI for Trustworthy Robots and Autonomous Systems. *IEEE Transactions on Cognitive and Developmental Systems*, 14(4), 1398–1412. DOI: 10.1109/TCDS.2021.3132282

Hemanth, A., Umamaheswari, K., Pogaku, A. C., Do, D.-T., & Lee, B. M. (2020). Outage performance analysis of reconfigurable intelligent surfaces-aided NOMA under presence of hardware impairment. *IEEE Access : Practical Innovations, Open Solutions*, 8, 212156–212165. DOI: 10.1109/ACCESS.2020.3039966

Herlambang, H., Purba, H., & Jaqin, C. (2021). Development of machine vision to increase the level of automation in indonesia electronic component industry. *Journal Européen des Systèmes Automatisés*, 54(2), 253–262. DOI: 10.18280/jesa.540207

He, S., Wang, Y., & Liu, H. (2022). Image information recognition and classification of ware- housed goods in intelligent logistics based on machine vision technology. *TS. Traitement du Signal*, 39(4), 1275–1282. DOI: 10.18280/ts.390420

He, T., Dong, C., Li, Y., & Yin, H. (2021). Motion state classification for micro-drones via modified mel frequency cepstral coefficient and hidden markov mode. *Electronics Letters*, 58(4), 164–166. DOI: 10.1049/ell2.12384

Hinkley, T., Brown, H., Carson, V., & Teychenne, M. (2018). Cross sectional associations of screen time and outdoor play with social skills in preschool children. *PLoS One*, 13(4), e0193700. DOI: 10.1371/journal.pone.0193700 PMID: 29617366

Hiraguri, T., Kimura, T., Endo, K., Ohya, T., Takanashi, T., & Shimizu, H. (2023). Shape classification technology of pollinated tomato flowers for robotic implementation. *Scientific Reports*, 13(1), 2159. DOI: 10.1038/s41598-023-27971-z PMID: 36750598

Hodge, V. J., Hawkins, R., & Alexander, R. (2021). Deep reinforcement learning for drone navigation using sensor data. *Neural Computing & Applications*, 33(6), 2015–2033. DOI: 10.1007/s00521-020-05097-x

Hodgson, J., Mott, R., Baylis, S., Pham, T., Wotherspoon, S., Kilpatrick, A., Segaran, R., Reid, I., Terauds, A., & Koh, L. (2018). Drones count wildlife more accurately and precisely than humans. *Methods in Ecology and Evolution*, 9(5), 1160–1167. DOI: 10.1111/2041-210X.12974

Hoffmann, M., & Pfeifer, R. (2018). Robots as powerful allies for the study of embodied cognition from the bottom up. *arXiv preprint arXiv:1801.04819*.

Höflinger, F., Müller, J., Zhang, R., Reindl, L. M., & Burgard, W. (2013). A Wireless Micro Inertial Measurement Unit (IMU). *IEEE Transactions on Instrumentation and Measurement*, 62(9), 2583–2595. DOI: 10.1109/TIM.2013.2255977

Hong, Y., Lian, J., Li, X., Min, J., Wang, Y., Freeman, L., & Deng, X. (2021). *Statistical perspectives on reliability of artificial intelligence systems*. https://doi.org//arxiv.2111.05391.DOI: 10.48550

Horeis, T. F., Kain, T., Müller, J.-S., Plinke, F., Heinrich, J., Wesche, M., & Decke, H. (2020). A Reliability Engineering Based Approach to Model Complex and Dynamic Autonomous Systems. *2020 International Conference on Connected and Autonomous Driving (MetroCAD)*, (pp. 76–84). IEEE. DOI: 10.1109/MetroCAD48866.2020.00020

Hosokawa, R., & Katsura, T. (2017). A longitudinal study of socio-economic status, family processes, and child adjustment from preschool until early elementary school: The role of social competence. *Child and Adolescent Psychiatry and Mental Health*, 11(1), 1–28. DOI: 10.1186/s13034-017-0206-z

Hu, H. (2024). *Position control of mobile robot based on deep reinforcement learning*. SPIE. .DOI: 10.1117/12.3024732

Huang, S., Zhang, X., Chen, N., Ma, H., Zeng, J., Fu, P., Nam, W.-H., & Niyogi, D. (2022). Generating high-accuracy and cloud-free surface soil moisture at 1 km resolution by point-surface data fusion over the Southwestern US. *Agricultural and Forest Meteorology*, 321, 108985. DOI: 10.1016/j.agrformet.2022.108985

Huang, X. (2020). Quality of service optimization in wireless transmission of industrial Internet of Things for intelligent manufacturing. *International Journal of Advanced Manufacturing Technology*, 107(3), 1007–1016. DOI: 10.1007/s00170-019-04288-8

Hu, B., Gaurav, A., Choi, C., & Almomani, A. (2022). Evaluation and com- parative analysis of semantic web-based strategies for enhancing educational system devel- opment. [IJSWIS]. *International Journal on Semantic Web and Information Systems*, 18(1), 1–14. DOI: 10.4018/IJSWIS.302895

Hulstaert, E., & Hulstaert, L. (2019). Artificial intelligence in dermato-oncology: A joint clinical and data science perspective. *International Journal of Dermatology*, 58(8), 989–990. DOI: 10.1111/ijd.14511 PMID: 31149729

Husak, E. (2021). *Actuators in Service Robots*.

Idrissi, M., Salami, M., & Annaz, F. (2022). A Review of Quadrotor Unmanned Aerial Vehicles: Applications, Architectural Design and Control Algorithms. *Journal of Intelligent & Robotic Systems*, 104(2), 22. DOI: 10.1007/s10846-021-01527-7

Iovino, M., Scukins, E., Styrud, J., Ögren, P., & Smith, C. (2022). A survey of Behavior Trees in robotics and AI. *Robotics and Autonomous Systems*, 154, 104096. DOI: 10.1016/j.robot.2022.104096

Ivanova, K., Gallasch, G. E., & Jordans, J. (2016). *Automated and autonomous systems for combat service support: scoping study and technology prioritisation*. Defence Science and Technology Group Edinburgh SA Australia.

Jaakkola, H. (2019). *Artificial intelligence yesterday, today and tomorrow*. IEEE.

Jacob, S. (2019). *Effective ground-truthing of supervised machine learning for drone classification*. IEEE. .DOI: 10.1109/RADAR41533.2019.171322

Jacobsen, R. H., Matlekovic, L., Shi, L., Malle, N., Ayoub, N., Hageman, K., Hansen, S., Nyboe, F. F., & Ebeid, E. (2023). Design of an Autonomous Cooperative Drone Swarm for Inspections of Safety Critical Infrastructure. *Applied Sciences (Basel, Switzerland)*, 13(3), 3. DOI: 10.3390/app13031256

Janeera, D. A., Gnanamalar, S. S. R., Ramya, K. C., & Kumar, A. G. A. (2021). Internet of Things and Artificial Intelligence-Enabled Secure Autonomous Vehicles for Smart Cities. In Kathiresh, M., & Neelaveni, R. (Eds.), *Automotive Embedded Systems: Key Technologies, Innovations, and Applications* (pp. 201–218). Springer International Publishing. DOI: 10.1007/978-3-030-59897-6_11

Jang, H., Hinton, H., Jung, W., Lee, M., Kim, C., Park, M., Lee, S., Park, S., & Ham, D. (2022). In- sensor optoelectronic computing using electrostatically doped silicon. *Nature Electronics*, 5(8), 519–525. DOI: 10.1038/s41928-022-00819-6

Javaid, A., Alduais, A., Shullar, M. H., Baroudi, U., & Alnaser, M. (2024). Monocular-based collision avoidance system for unmanned aerial vehicle. *IET Smart Cities*, 6(1), 1–9. DOI: 10.1049/smc2.12067

Javaid, M., Haleem, A., Pratap Singh, R., Suman, R., & Rab, S. (2022). Significance of machine learning in healthcare: Features, pillars and applications. *International Journal of Intelligent Networks*, 3, 58–73. DOI: 10.1016/j.ijin.2022.05.002

Javaid, M., Haleem, A., Singh, R. P., Rab, S., & Suman, R. (2022). Significant applications of Cobots in the field of manufacturing. *Cognitive Robotics*, 2, 222–233. DOI: 10.1016/j.cogr.2022.10.001

Javaid, M., Haleem, A., Singh, R. P., & Suman, R. (2021). Substantial capabilities of robotics in enhancing industry 4.0 implementation. *Cognitive Robotics*, 1, 58–75. DOI: 10.1016/j.cogr.2021.06.001

Javaid, M., Haleem, A., Singh, R. P., & Suman, R. (2022). Enabling flexible manufacturing system (FMS) through the applications of industry 4.0 technologies. *Internet of Things and Cyber-Physical Systems*, 2, 49–62. DOI: 10.1016/j.iotcps.2022.05.005

Jensen, S., Akhter, M., Azim, S., & Rasmussen, J. (2021). The predictive power of regression models to determine grass weed infestations in cereals based on drone imagery—Statistical and practical aspects. *Agronomy (Basel)*, 11(11), 2277. DOI: 10.3390/agronomy11112277

Jeon, J., Park, J., & Jeong, Y. (2020). Dynamic analysis for iot malware de- tection with convolution neural network model. *IEEE Access : Practical Innovations, Open Solutions*, 8, 96899–96911. DOI: 10.1109/ACCESS.2020.2995887

Jeyabalan, V., Nouvet, E., Meier, P., & Donelle, L. (2020). Context-specific chal- lenges, opportunities, and ethics of drones for healthcare delivery in the eyes of program managers and field staff: A multi-site qualitative study. *Drones (Basel)*, 4(3), 44. DOI: 10.3390/drones4030044

Jiang, H. (2021). Learning for a robot: deep reinforcement learning, imitation learning, transfer learning. *Sensors,21*. .DOI: 10.3390/s21041278

Jiao, J., Yuan, L., Tang, W., Deng, Z., & Wu, Q. (2017). A post-rectification approach of depth images of kinect v2 for 3d reconstruction of indoor scenes. *ISPRS International Journal of Geo-Information*, 6(11), 349. DOI: 10.3390/ijgi6110349

Johal, W., Gatos, D., Yanta¸c, A., & Obaid, M. (2022). Envisioning social drones in education. *Frontiers in Robotics and AI*, 9, 666736. DOI: 10.3389/frobt.2022.666736 PMID: 36093212

Joksimovic, S., Ifenthaler, D., Marrone, R., De Laat, M., & Siemens, G. (2023). Opportunities of artificial intelligence for supporting complex problem-solving: Findings from a scoping review. *Computers and Education: Artificial Intelligence*, 4, 100138. DOI: 10.1016/j.caeai.2023.100138

Jones, A. M., Rigling, B., & Rangaswamy, M. (2016). Signal-to-interference-plus-noise-ratio analysis for constrained radar waveforms. *IEEE Transactions on Aerospace and Electronic Systems*, 52(5), 2230–2241. DOI: 10.1109/TAES.2016.150511

Joo, H. J., & Jeong, H. Y. (2020). A study on eye-tracking-based Interface for VR/AR education plat- form. *Multimedia Tools and Applications*, 79(23-24), 16719–16730. DOI: 10.1007/s11042-019-08327-0

Kabir, H., Tham, M.-L., & Chang, Y. C. (2023). Internet of robotic things for mobile robots: Concepts, technologies, challenges, applications, and future directions. *Digital Communications and Networks*, 9(6), 1265–1290. DOI: 10.1016/j.dcan.2023.05.006

Kadek Dwi Wahyuadnyana, P. A. D. (2022)... . *Parallel Control System PD-SMCNN for Robust Auton- omous Mini-Quadcopter.*, (July), c1–c1. DOI: 10.1109/ISITIA56226.2022.9855344

Kamat, A. (2022). Uncovering in- terrelationships between barriers to unmanned aerial vehicles in human- itarian logistics. *Operations Management Research, 15,* 1134–1160. .DOI: 10.1007/s12063-021-00235-7

Kangunde, V., Jamisola, R.Jr, & Theophilus, E. (2021). A review on drones con- trolled in real-time. *International Journal of Dynamics and Control*, 9(4), 1832–1846. DOI: 10.1007/s40435-020-00737-5 PMID: 33425650

Karam, S. N., Bilal, K., Shuja, J., Rehman, F., Yasmin, T., & Jamil, A. (2022). RETRACTED: Inspection of unmanned aerial vehicles in oil and gas industry: critical analysis of platforms, sensors, networking architecture, and path planning. *Journal of Electronic Imaging*, 32(1), 011006. DOI: 10.1117/1. JEI.32.1.011006

Kardasz, P., & Doskocz, J. (2016). Drones and Possibilities of Their Using. *Journal of Civil & Environmental Engineering*, 6(3). DOI: 10.4172/2165-784X.1000233

Karnouskos, S. (2021). Symbiosis with artificial intelligence via the prism of law, robots, and society. *Artificial Intelligence and Law*, 30(1), 93–115. DOI: 10.1007/s10506-021-09289-1

Kataria, A., & Puri, V. (2022). AI- and IoT-based hybrid model for air quality prediction in a smart city with network assistance. *IET Networks*, 11(6), 221–233. DOI: 10.1049/ntw2.12053

Katsura, U., Matsumoto, K., Kawamura, A., Ishigami, T., Okada, T., & Kurazume, R. (2019). Spatial change detection using normal distributions transform. *Robomech Journal*, 6(1), 20. DOI: 10.1186/s40648-019-0148-8

Keating, R., Säily, M., Hulkkonen, J., & Karjalainen, J. (2019). Overview of positioning in 5G new radio. *Paper presented at the 2019 16th International Symposium on Wireless Communication Systems (ISWCS)*. IEEE. DOI: 10.1109/ISWCS.2019.8877160

Kennedy, K., Alexander, L., Alexander, L., Landis, R., Landis, R., Linne, D., & Sims, J. (2011, September). NASA Technology Area 07 Human Exploration Destination Systems Roadmap. In *AIAA SPACE 2011 Conference & Exposition* (p. 7255). IEEE.

Khade, G., Kumar, S., & Bhattacharya, S. (2012). Classification of web pages on attractiveness: A supervised learning approach. *2012 4th International Conference on Intelligent Human Computer Interaction (IHCI)*. IEEE. DOI: 10.1109/IHCI.2012.6481867

Khanam, S., Tanweer, S., & Khalid, S. S. (2022). Future of internet of things: Enhancing cloud-based iot using artificial intelligence. [IJCAC]. *International Journal of Cloud Applications and Computing*, 12(1), 1–23. DOI: 10.4018/IJCAC.297094

Khokthong, W., Zemp, D., Irawan, B., Sundawati, L., Kreft, H., & Hölscher, D. (2019). Drone- based assessment of canopy cover for analyzing tree mortality in an oil palm agroforest. *Frontiers in Forests and Global Change*, 2, 12. DOI: 10.3389/ffgc.2019.00012

Kiencke, U., Nielsen, L., Sutton, R., Schilling, K., Papageorgiou, M., & Asama, H. (2006). The impact of automatic control on recent developments in transportation and vehicle systems. *Annual Reviews in Control*, 30(1), 81–89. DOI: 10.1016/j.arcontrol.2006.02.001

Kim, B., Nam, S., Jin, Y., & Seo, K. (2020). Simulation framework for cyber-physical production system: Applying concept of lvc interoperation. *Complexity*, 2020, 1–11. DOI: 10.1155/2020/4321873

Kim, J. (2022). The Interconnectivity of Heutagogy and Education 4.0 in Higher Online Education. *Canadian Journal of Learning and Technology*, 48(4), 1–17. DOI: 10.21432/cjlt28257

Kiohara, P., de Souza, R., Ivo, F. S., Mippo, N. T., Coutinho, O. L., Pérennou, A., & Quintard, V. (2021). Microwave Photonic Approach to Antenna Remote on Airborne Radar Warning Receiver System. *Paper presented at the 2021 SBMO/IEEE MTT-S International Microwave and Optoelectronics Conference (IMOC)*. IEEE. DOI: 10.1109/IMOC53012.2021.9624933

Kipnis, E., McLeay, F., Grimes, A., Saille, S., & Potter, S. (2022). Service robots in long-term care: A consumer-centric view. *Journal of Service Research*, 25(4), 667–685. DOI: 10.1177/10946705221110849

Kiran, M. A., Pasupuleti, S. K., & Eswari, R. (2022). Efficient pairing-free identity-based signcryption scheme for cloud-assisted iot. [IJCAC]. *International Journal of Cloud Applications and Computing*, 12(1), 1–15. DOI: 10.4018/IJCAC.305216

Kleinberg, J., Lakkaraju, H., Leskovec, J., Ludwig, J., & Mullainathan, S. (2017). Human decisions and machine predictions*. *The Quarterly Journal of Economics*. DOI: 10.1093/qje/qjx032 PMID: 29755141

Knoblauch, A., Rosa, S., Sherman, J., Blauvelt, C., Matemba, C., & Maxim, L. (2019). Bi-directional drones to strengthen healthcare provision: Experiences and lessons from madagascar, malawi and senegal. *BMJ Global Health*, 4(4), e001541. DOI: 10.1136/bmjgh-2019-001541 PMID: 31413873

Knox, J. (2020). Artificial intelligence and education in china. *Learning, Media and Technology*, 45(3), 298–311. DOI: 10.1080/17439884.2020.1754236

Kogelis, M., Fuge, Z. J., Herron, C. W., Kalita, B., & Leonessa, A. (2022). *Design Of Low-Level Hardware For A Multi-Layered Control Architecture* (Vol. 5). ASME. DOI: 10.1115/IMECE2022-94614

Kolachalama, V. B. (2022). Machine learning and pre-medical education. *Artificial Intelligence in Medicine*, 129, 102313. DOI: 10.1016/j.artmed.2022.102313 PMID: 35659392

Komorkiewicz, M., Chin, A., Skruch, P., & Szelest, M. (2023). Intelligent data handling in current and next-generation automated vehicle development—A review. *IEEE Access : Practical Innovations, Open Solutions*, 11, 32061–32072. DOI: 10.1109/ACCESS.2023.3258623

Kono, H., Katayama, R., Takakuwa, Y., Wen, W., & Suzuki, T. (2019). Activation and spreading sequence for spreading activation policy selection method in transfer reinforce- ment learning. *International Journal of Advanced Computer Science and Applications*, 10(12). DOI: 10.14569/IJACSA.2019.0101202

Konstantoudakis, K., Christaki, K., Tsiakmakis, D., Sainidis, D., Albanis, G., Dimou, A., & Daras, P. (2022). Drone control in ar: An intuitive system for single-handed gesture control, drone tracking, and contextualized camera feed visualization in augmented reality. *Drones (Basel)*, 6(2), 43. DOI: 10.3390/drones6020043

Korkmaz, G., Ekici, E., Özgüner, F., & Özgüner, Ü. (2004). Urban multi-hop broadcast protocol for inter-vehicle communication systems. *Proceedings of the 1st ACM International Workshop on Vehicular Ad Hoc Networks*, (pp. 76–85). ACM. DOI: 10.1145/1023875.1023887

Kose, T., & Sakata, I. (2017). *Identifying technology advancements and their linkages in the field of robotics research*. IEEE. DOI: 10.23919/PICMET.2017.8125283

Koub^aa, A., Allouch, A., Alajlan, M., Javed, Y., Belghith, A., & Khalgui, M. (2019). Mi- cro air ve-hicle link (mavlink) in a nutshell: A survey. *IEEE Access : Practical Innovations, Open Solutions*, 7, 87658–87680. DOI: 10.1109/ACCESS.2019.2924410

Kouroupa, A., Laws, K. R., Irvine, K., Mengoni, S. E., Baird, A., & Sharma, S. (2022). The use of social robots with children and young people on the autism spectrum: A systematic review and meta-analysis. *PLoS One*, 17(6), e0269800. DOI: 10.1371/journal.pone.0269800 PMID: 35731805

Koutalakis, P. D., Tzoraki, O. A., Prazioutis, G. I., Gkiatas, G. T., & Zaimes, G. N. (2021). Can Drones Map Earth Cracks? Landslide Measurements in North Greece Using UAV Photogrammetry for Nature-Based Solutions. *Sustainability (Basel)*, 13(9), 9. DOI: 10.3390/su13094697

Kovari, B. & Ebeid, E. (2021). *Mpdrone: fpga-based platform for intelligent real-time au- tonomous drone operations*. IEEE. .DOI: 10.1109/SSRR53300.2021.9597857

Kozima, H., Michalowski, M. P., & Nakagawa, C. (2009). Keepon: A playful robot for research, therapy, and entertainment. *International Journal of Social Robotics*, 1(1), 3–18. DOI: 10.1007/s12369-008-0009-8

Krashen, S. (2009). *Principles and practice in second language acquisition* (Internet edition). Oxford, UK: Pergamon. https://sdkrashen.com/content/books/principles_and_practice.pdf

Kubota, T., & Kunii, Y. (2009, April). Intelligent guidance of mobile explorer for Planetary Robotic Exploration. In *2009 IEEE International Conference on Mechatronics* (pp. 1-6). IEEE. DOI: 10.1109/ICMECH.2009.4957151

Kumar, N. (2014). *High-temperature motors*. Research Gate.

Kumar, S. (2021). Impact of artificial intelligence and service robots in tourism and hospitality sector: current use & future trends. *Administrative Development a Journal of Hipa Shimla,8*, 59–83. DOI: 10.53338/ADHIPA2021.V08.Si01.04

Kumar, S., Karnani, G., Gaur, M. S., & Mishra, A. (2021). Cloud Security using Hybrid Cryptography Algorithms. *2021 2nd International Conference on Intelligent Engineering and Management (ICIEM)*, (pp. 599–604). IEEE. DOI: 10.1109/ICIEM51511.2021.9445377

Kumar, S., Singh, S. K., & Aggarwal, N. (2023, September). Sustainable Data Dependency Resolution Architectural Framework to Achieve Energy Efficiency Using Speculative Parallelization. In *2023 3rd International Conference on Innovative Sustainable Computational Technologies (CISCT)* (pp. 1-6). IEEE. DOI: 10.1109/CISCT57197.2023.10351343

Kumar, S., Singh, S., Aggarwal, N., & Aggarwal, K. (2021a, December 26). *Efficient speculative par-allelization architecture for overcoming speculation overheads*.

Kumar, T., Mahrishi, M., & Meena, G. (2022). A comprehensive review of recent automatic speech summarization and keyword identification techniques. *Artificial Intelligence in Industrial Applications: Approaches to Solve the Intrinsic Industrial Optimization Problems*, 111-126.

Kumar, A. (2020). Drone Proliferation and Security Threats: A Critical Analysis. *Indian Journal of Asian Affairs*, 33(1/2), 43–62.

Kumar, I., Rawat, J., Mohd, N., & Husain, S. (2021). Opportunities of artificial intelli- gence and machine learning in the food industry. *Journal of Food Quality*, 2021, 1–10. DOI: 10.1155/2021/4535567

Kumari, P., Shankar, A., Behl, A., Pereira, V., Yahiaoui, D., Laker, B., Gupta, B. B., & Arya, V. (2024). Investigating the barriers towards adoption and im- plementation of open innovation in healthcare. *Technological Forecasting and Social Change*, 200, 123100. DOI: 10.1016/j.techfore.2023.123100

Kumar, R., Ai, H., Beuth, J. L., & Rosé, C. P. (2010). Socially capable conversational tutors can be effective in collaborative learning situations. In *Intelligent Tutoring Systems: 10th International Conference,* (pp. 156–164). Springer.

Kumar, R., Singh, S. K., & Lobiyal, D. K. (2023). Communication structure for Vehicular Internet of Things (VIoTs) and review for vehicular networks. In *Automation and Computation*. CRC Press. DOI: 10.1201/9781003333500-34

Kumar, R., Singh, S. K., Lobiyal, D. K., Chui, K. T., Santaniello, D., & Rafsanjani, M. K. (2022). A novel decentralized group key management scheme for cloud-based vehicular IoT networks. [IJCAC]. *International Journal of Cloud Applications and Computing*, 12(1), 1–34. DOI: 10.4018/IJCAC.311037

Kumar, R., Singh, S. K., Lobiyal, D. K., Kumar, S., & Jawla, S. (2024). Security Metrics and Authentication-based RouTing (SMART) Protocol for Vehicular IoT Networks. *SN Computer Science*, 5(2), 236. DOI: 10.1007/s42979-023-02566-7

Kumar, S. S., Singh, S., Aggarwal, N., & Aggarwal, K. (2021). Efficient speculative parallelization architecture for overcoming speculation overheads. *International Conference on Smart Systems and Advanced Computing (Syscom-2021),* (pp. 132–138). IEEE.

Kumar, S., Dai, Y., & Li, H. (2021). Superpixel soup: Monocular dense 3d reconstruction of a complex dynamic scene. *IEEE Transactions on Pattern Analysis and Machine Intelligence*, 43(5), 1705–1717. DOI: 10.1109/TPAMI.2019.2955131 PMID: 31765303

Kumar, S., Singh, S. K., & Aggarwal, N. (2023). Speculative parallelism on multicore chip architecture strengthen green computing concept: A survey. In *Advanced computer science applications* (pp. 3–16). Apple Academic Press. DOI: 10.1201/9781003369066-2

Kumar, S., Singh, S. K., Aggarwal, N., Gupta, B. B., Alhalabi, W., & Band, S. S. (2022). An efficient hardware supported and parallelization architecture for intelligent systems to overcome speculative overheads. *International Journal of Intelligent Systems*, 37(12), 11764–11790. DOI: 10.1002/int.23062

Kumar, S., Singh, S., Aggarwal, N., & Aggarwal, K. (2021b). Evaluation of automatic parallelization algorithms to minimize speculative parallelism overheads: An experiment. *Journal of Discrete Mathematical Sciences and Cryptography*, 24(5), 1517–1528. DOI: 10.1080/09720529.2021.1951435

Kumar, S., Tiwari, P., & Zymbler, M. (2019). Internet of Things is a revolutionary approach for future technology enhancement: A review. *Journal of Big Data*, 6(1), 111. DOI: 10.1186/s40537-019-0268-2

Kumbhojkar, N. R., & Menon, A. B. (2022). Integrated predictive experience management framework (IPEMF) for improving customer experience: In the era of digital transformation. [IJCAC]. *International Journal of Cloud Applications and Computing*, 12(1), 1–13. DOI: 10.4018/IJCAC.2022010107

Kümmerle, R., Steder, B., Dornhege, C., Ruhnke, M., Grisetti, G., Stachniss, C., & Kleiner, A. (2009). On measuring the accuracy of SLAM algorithms. *Autonomous Robots*, 27(4), 387–407. DOI: 10.1007/s10514-009-9155-6

Kunze, L., Hawes, N., Duckett, T., Hanheide, M., & Krajn'ık, T. (2018). Artificial intelligence for long-term robot autonomy: A survey. *IEEE Robotics and Automation Letters*, 3(4), 4023–4030. DOI: 10.1109/LRA.2018.2860628

Kupervasser, O., Kutomanov, H., Levi, O., Pukshansky, V., & Yavich, R. (2020). Using deep learning for visual navigation of drone with respect to 3d ground objects. *Mathematics*, 8(12), 2140. DOI: 10.3390/math8122140

Kuru, K. (2021). Conceptualisation of human-on-the-loop haptic teleoperation with fully autonomous self-driving vehicles in the urban environment. *IEEE Open Journal of Intelligent Transportation Systems*, 2, 448–469. DOI: 10.1109/OJITS.2021.3132725

La Porte, T. R. (2019). The United States Air Traffic System: Increasing Reliability in the Midst of Raped Growth 1. *The development of large technical systems*, 215-244.

Lacava, G., Marotta, A., Martinelli, F., Saracino, A., La Marra, A., Gil-Uriarte, E., & Vilches, V. M. (2021). Cybsersecurity Issues in Robotics. *Journal of Wireless Mobile Networks, Ubiquitous Computing and Dependable Applications*, 12(3), 1–28.

Lagkas, T., Argyriou, V., Bibi, S., & Sarigiannidis, P. (2018). Uav iot framework views and challenges: Towards protecting drones as "things". *Sensors (Basel)*, 18(11), 4015. DOI: 10.3390/s18114015 PMID: 30453646

Lahmeri, M., Kishk, M., & Alouini, M. (2021). Artificial intelligence for uav-enabled wire- less networks: A survey. *IEEE Open Journal of the Communications Society*, 2, 1015–1040. DOI: 10.1109/OJCOMS.2021.3075201

Lahoz-Monfort, J., & Magrath, M. (2021). A comprehensive overview of technolo- gies for species and habitat monitoring and conservation. *Bioscience*, 71(10), 1038–1062. DOI: 10.1093/biosci/biab073 PMID: 34616236

Landowska, A., Karpus, A., Zawadzka, T., Robins, B., Erol Barkana, D., Kose, H., Zorcec, T., & Cummins, N. (2022). Automatic emotion recognition in children with autism: A systematic literature review. *Sensors (Basel)*, 22(4), 1649. DOI: 10.3390/s22041649 PMID: 35214551

Lappas, V., Shin, H. S., Tsourdos, A., Lindgren, D., Bertrand, S., Marzat, J., Piet-Lahanier, H., Daramouskas, Y., & Kostopoulos, V. (2022). Autonomous Unmanned Heterogeneous Vehicles for Persistent Monitoring. *Drones (Basel)*, 6(4), 1–27. DOI: 10.3390/drones6040094

Larsen, G. D., & Johnston, D. W. (2024). Growth and opportunities for drone surveillance in pinniped research. *Mammal Review*, 54(1), 1–12. DOI: 10.1111/mam.12325

Lawrence, I. D., Vijayakumar, R., & Agnishwar, J. (2023). Dynamic Application of Unmanned Aerial Vehicles for Analyzing the Growth of Crops and Weeds for Precision Agriculture. In *Artificial Intelligence Tools and Technologies for Smart Farming and Agriculture Practices* (pp. 115–132). IGI Global. DOI: 10.4018/978-1-6684-8516-3.ch007

Le Nhu Ngoc Thanh, H., & Hong, S. K. (2018). Quadcopter robust adaptive second order sliding mode control based on PID sliding surface. *IEEE Access : Practical Innovations, Open Solutions*, 6, 66850–66860. DOI: 10.1109/ACCESS.2018.2877795

Lee, J. S., Ham, Y., Park, H., & Kim, J. (2022). Challenges, tasks, and opportunities in teleoperation of excavator toward human-in-the-loop construction automation. *Automation in Construction*, 135, 104119. DOI: 10.1016/j.autcon.2021.104119

Lee, M., Siewiorek, D., Smailagic, A., Bernardino, A., & Badia, S. (2022). Enabling ai and robotic coaches for physical rehabilitation therapy: Iterative design and evaluation with therapists and post-stroke survivors. *International Journal of Social Robotics*, 16(1), 1–22. DOI: 10.1007/s12369-022-00883-0

Lee, M., & Yusuf, S. (2022). Mobile robot navigation using deep reinforcement learning. *Processes (Basel, Switzerland)*, 10(12), 2748. DOI: 10.3390/pr10122748

Liang, C. J., Wang, X., Kamat, V. R., & Menassa, C. C. (2021). Human–robot collaboration in construction: Classification and research trends. *Journal of Construction Engineering and Management*, 147(10), 03121006. DOI: 10.1061/(ASCE)CO.1943-7862.0002154

Liao, M., Tang, H., Li, X., Pandi, V., & Arya, V. (2024). A lightweight network for abdominal multi-organ segmentation based on multi-scale context fusion and dual self-attention. *Information Fusion*, 108, 102401. DOI: 10.1016/j.inffus.2024.102401

Liao, S., Wu, J., Li, J., Bashir, A., & Yang, W. (2021). Securing collaborative environment monitoring in smart cities using blockchain enabled software-defined internet of drones. *Ieee Internet of Things Magazine*, 4(1), 12–18. DOI: 10.1109/IOTM.0011.2000045

Li, B., Feng, Y., Xiong, Z., Yang, W., & Liu, G. (2021). Research on AI security enhanced encryption algorithm of autonomous IoT systems. *Information Sciences*, 575, 379–398. DOI: 10.1016/j.ins.2021.06.016

Li, C., Zhang, Y., Xie, R., Hao, X., & Huang, T. (2021). Integrating edge computing into low earth orbit satellite networks: Architecture and prototype. *IEEE Access : Practical Innovations, Open Solutions*, 9, 39126–39137. DOI: 10.1109/ACCESS.2021.3064397

Licardo, J. T., Domjan, M., & Orehovački, T. (2024). Intelligent Robotics—A Systematic Review of Emerging Technologies and Trends. *Electronics (Basel)*, 13(3), 3. DOI: 10.3390/electronics13030542

Lichtman, A., & Nair, M. (2015). Humanitarian uses of drones and satellite im- agery analysis: The promises and perils. *AMA Journal of Ethics*, 17(10), 931–937. DOI: 10.1001/journalofethics.2015.17.10. stas1-1510 PMID: 26496056

Li, D. (2020). Ergodic capacity of intelligent reflecting surface-assisted communication systems with phase errors. *IEEE Communications Letters*, 24(8), 1646–1650. DOI: 10.1109/LCOMM.2020.2997027

Li, L., Li, W., Wang, J., Chen, X., Peng, Q., & Huang, W. (2023). UAV Trajectory Optimization for Spectrum Cartography: A PPO Approach. *IEEE Communications Letters*, 27(6), 1575–1579. DOI: 10.1109/LCOMM.2023.3265214

Lindblom J, Ziemke T. (2002). Social Situatedness of Natural and Artificial Intelligence: Vygotsky and Beyond. *Adaptive Behavior, 11*(2), 79-96.

Lindsay, G. W. (2021). Convolutional Neural Networks as a Model of the Visual System: Past, Present, and Future. *Journal of Cognitive Neuroscience*, 33(10), 2017–2031. DOI: 10.1162/jocn_a_01544 PMID: 32027584

Lingenfelser, F., Wagner, J., & André, E. (2011). A systematic discussion of fusion techniques for multi-modal affect recognition tasks. *Proceedings of the 13th International Conference on Multimodal Interfaces*, (pp. 19–26). IEEE. DOI: 10.1145/2070481.2070487

Liu, X. (2022). Sustainable oil palm resource assessment based on an enhanced deep learning method. *Energies, 15*. DOI: 10.3390/en15124479

Liu, D., Wang, Z., Lu, B., Cong, M., Yu, H., & Zou, Q. (2020). A reinforcement learning- based framework for robot manipulation skill acquisition. *IEEE Access : Practical Innovations, Open Solutions*, 8, 108429–108437. DOI: 10.1109/ACCESS.2020.3001130

Liu, T., Bai, G., Tao, J., Zhang, Y.-A., & Fang, Y. (2024). A Multistate Network Approach for Resilience Analysis of UAV Swarm considering Information Exchange Capacity. *Reliability Engineering & System Safety*, 241, 109606. DOI: 10.1016/j.ress.2023.109606

Liu, X. (2008). Airborne LiDAR for DEM generation: Some critical issues. *Progress in Physical Geography*, 32(1), 31–49. DOI: 10.1177/0309133308089496

Liu, Y., Ding, Z., Elkashlan, M., & Poor, H. V. (2016). Cooperative non-orthogonal multiple access with simultaneous wireless information and power transfer. *IEEE Journal on Selected Areas in Communications*, 34(4), 938–953. DOI: 10.1109/JSAC.2016.2549378

Liu, Y., Hassan, K., Karlsson, M., Pang, Z., & Gong, S. (2019). A data-centric in- ternet of things framework based on azure cloud. *IEEE Access : Practical Innovations, Open Solutions*, 7, 53839–53858. DOI: 10.1109/ACCESS.2019.2913224

Li, Z., Liu, F., Yang, W., Peng, S., & Zhou, J. (2022). A Survey of Convolutional Neural Networks: Analysis, Applications, and Prospects. *IEEE Transactions on Neural Networks and Learning Systems*, 33(12), 6999–7019. DOI: 10.1109/TNNLS.2021.3084827 PMID: 34111009

Long, R., Liang, Y.-C., Pei, Y., & Larsson, E. G. (2021). Active reconfigurable intelligent surface-aided wireless communications. *IEEE Transactions on Wireless Communications*, 20(8), 4962–4975. DOI: 10.1109/TWC.2021.3064024

Louw, L., & Droomer, M. (2019). Development of a low cost machine vision based quality control system for a learning factory. *Procedia Manufacturing*, 31, 264–269. DOI: 10.1016/j.promfg.2019.03.042

Lubold, N., Walker, E., Pon-Barry, H., & Ogan, A. (2018). Automated pitch convergence improves learning in a social, teachable robot for middle school mathematics. In *Artificial Intelligence in Education: 19th International Conference*. Springer.

Lu, S., Tsakalis, K., & Chen, Y. (2022). Development and Application of a Novel High-order Fully Actuated System Approach: Part I. 3-DOF Quadrotor Control. *IEEE Control Systems Letters*, 7, 1177–1182. DOI: 10.1109/LCSYS.2022.3232305

Lutz, C., Sch¨ottler, M., & Hoffmann, C. (2019). The privacy implications of social robots: Scoping review and expert interviews. *Mobile Media & Communication*, 7(3), 412–434. DOI: 10.1177/2050157919843961

Lu, Y., Xue, Z., Xia, G., & Zhang, L. (2018). A survey on vision-based uav navigation. *Geo-Spatial Information Science*, 21(1), 21–32. DOI: 10.1080/10095020.2017.1420509

Lv, L., Ni, Q., Ding, Z., & Chen, J. (2016). Application of non-orthogonal multiple access in cooperative spectrum-sharing networks over Nakagami-$ m $ fading channels. *IEEE Transactions on Vehicular Technology*, 66(6), 5506–5511. DOI: 10.1109/TVT.2016.2627559

Lv, Y., Ai, Z., Chen, M., Gong, X., Wang, Y., & Lu, Z. (2022). High-resolution drone detection based on background difference and sag-yolov5s. *Sensors (Basel)*, 22(15), 5825. DOI: 10.3390/s22155825 PMID: 35957382

Lyons, M., Brandis, K., Callaghan, C., McCann, J., Mills, C., Ryall, S., & Kingsford, R. (2018). Bird interactions with drones, from individuals to large colonies. *Australian Field Ornithology*, 35, 51–56. DOI: 10.20938/afo35051056

Ma, B., Wu, J., Lai, E., & Hu, S. (2021, December). PPDTSA: Privacy-preserving deep transformation self-attention framework for object detection. In *2021 IEEE Global Communications Conference (GLOBECOM)* (pp. 1-5). IEEE. DOI: 10.1109/GLOBECOM46510.2021.9685855

Mackenzie, R. (2018). Sexbots: Customizing them to suit us versus an ethi- cal duty to created sentient beings to minimize suffering. *Robotics (Basel, Switzerland)*, 7(4), 70. DOI: 10.3390/robotics7040070

Madadi, Y., Delsoz, M., Khouri, A. S., Boland, M., Grzybowski, A., & Yousefi, S. (2024). Applications of artificial intelligence-enabled robots and chatbots in oph- thalmology: Recent advances and future trends. *Current Opinion in Ophthalmology*, 35(3), 238–243. DOI: 10.1097/ICU.0000000000001035 PMID: 38277274

Maghazei, O. (2022). *Drones in manufacturing: opportunities and challenges*. IEOM Society. .DOI: 10.46254/EU05.20220073

Maghazei, O., & Netland, T. (2019). Drones in manufacturing: Exploring opportunities for research and practice. *Journal of Manufacturing Technology Management*, 31(6), 1237–1259. DOI: 10.1108/JMTM-03-2019-0099

Mahdi, I. (2021). Evaluation of robot professor technology in teaching and business. *Information Technology in Industry*, 9, 1182–1194. .DOI: 10.17762/itii.v9i1.255

Mahmud, S., Lin, X., & Kim, J. H. (2020, January). Interface for human machine interaction for assistant devices: A review. In *2020 10th Annual computing and communication workshop and conference (CCWC)* (pp. 0768-0773). IEEE. DOI: 10.1109/CCWC47524.2020.9031244

Majeed, R., Abdullah, N., & Mushtaq, M. (2021). Iot-based cyber-security of drones using the naïve bayes algorithm. *International Journal of Advanced Computer Science and Applications*, 12(7). DOI: 10.14569/IJACSA.2021.0120748

Majeed, R., Abdullah, N., Mushtaq, M., Umer, M., & Nappi, M. (2021). Intelligent cyber-security system for iot-aided drones using voting classifier. *Electronics (Basel)*, 10(23), 2926. DOI: 10.3390/electronics10232926

Malche, T., Maheshwary, P., Tiwari, P. K., Alkhayyat, A. H., Bansal, A., & Kumar, R. (2023). Efficient solid waste inspection through drone-based aerial imagery and tinyml vision model. *Transactions on Emerging Telecommunications Technologies*, 35(4), e4878. DOI: 10.1002/ett.4878

Mangsatabam, R. (2018). *Control Development for Autonomous Landing of Quadcopter on Moving Platform*.

Manuylova, N. B., & Bulychev, S. N. (2023). Space monitoring in precision agriculture. *IOP Conference Series. Earth and Environmental Science*, 1154(1), 012043. DOI: 10.1088/1755-1315/1154/1/012043

Mao, S., He, S., & Wu, J. (2020). Joint UAV position optimization and resource scheduling in space-air-ground integrated networks with mixed cloud-edge computing. *IEEE Systems Journal*, 15(3), 3992–4002. DOI: 10.1109/JSYST.2020.3041706

Marin, L. (2015). *Deploying drones in policing southern european borders: constraints and challenges for data protection and human rights*. Springer. DOI: 10.1007/978-3-319-23760-2

Marshall, J. A., Sun, W., & L'Afflitto, A. (2021). A survey of guidance, navigation, and control systems for autonomous multi-rotor small unmanned aerial systems. *Annual Reviews in Control*, 52(July), 390–427. DOI: 10.1016/j.arcontrol.2021.10.013

Marturano, F., Martellucci, L., Chierici, A., Malizia, A., Giovanni, D., D'Errico, F., Gaudio, P., & Ciparisse, J. (2021). Numerical fluid dynamics simulation for drones' chemical detection. *Drones (Basel)*, 5(3), 69. DOI: 10.3390/drones5030069

Mathiassen, K., Schneider, F. E., Bounker, P., Tiderko, A., Cubber, G. D., Baksaas, M., Główka, J., Kozik, R., Nussbaumer, T., Röning, J., Pellenz, J., & Volk, A. (2021). Demonstrating interoperability between unmanned ground systems and command and control systems. *International Journal of Intelligent Defence Support Systems*, 6(2), 100–129. DOI: 10.1504/IJIDSS.2021.115236

Ma, Y., Li, M., Liu, Y., Wu, Q., & Liu, Q. (2022). Active Reconfigurable Intelligent Surface for Energy Efficiency in MU-MISO Systems. *IEEE Transactions on Vehicular Technology*, 72(3), 4103–4107. DOI: 10.1109/TVT.2022.3221720

McEnroe, P., Wang, S., & Liyanage, M. (2022). A Survey on the Convergence of Edge Computing and AI for UAVs: Opportunities and Challenges. *IEEE Internet of Things Journal*, 9(17), 15435–15459. DOI: 10.1109/JIOT.2022.3176400

McGinn, C. & Kelly, K. (2015). *Towards an embodied system-level architecture for mobile robots.* Research Gate.

Mellinger, D., Lindsey, Q., Shomin, M., & Kumar, V. (2011). Design, modeling, estimation and control for aerial grasping and manipulation. *IEEE International Conference on Intelligent Robots and Systems*, (pp. 2668–2673). IEEE. DOI: 10.1109/IROS.2011.6094871

Mengi, G., Singh, S. K., Kumar, S., Mahto, D., & Sharma, A. (2023). Automated Machine Learning (AutoML): The Future of Computational Intelligence. In Nedjah, N., Martínez Pérez, G., & Gupta, B. B. (Eds.), *International Conference on Cyber Security, Privacy and Networking (ICSPN 2022)* (pp. 309–317). Springer International Publishing. DOI: 10.1007/978-3-031-22018-0_28

Menouar, H., Gu¨ven¸c, I. ˙., Akkaya, K., Uluagac, A., Kadri, A., & Tuncer, A. (2017). Uav-enabled intelligent transportation systems for the smart city: Applications and challenges. *IEEE Communications Magazine*, 55(3), 22–28. DOI: 10.1109/MCOM.2017.1600238CM

Mercier-Laurent, E. (2021). *Can artificial intelligence effectively support sustainable development?* Springer. ¡0.DOI: 10.1007/978-3-030-80847-1

Merkert, R., & Bushell, J. (2020). Managing the drone revolution: A systematic lit- erature review into the current use of airborne drones and future strategic direc- tions for their effective control. *Journal of Air Transport Management*, 89, 101929. DOI: 10.1016/j.jairtraman.2020.101929 PMID: 32952321

Michael, K., Abbas, R., Roussos, G., Scornavacca, E., & Fosso-Wamba, S. (2020). Ethics in AI and Autonomous System Applications Design. *IEEE Transactions on Technology and Society*, 1(3), 114–127. DOI: 10.1109/TTS.2020.3019595

Michels, M., Hobe, C., Ahlefeld, P., & Mußhoff, O. (2021). The adoption of drones in german agriculture: A structural equation model. *Precision Agriculture*, 22(6), 1728–1748. DOI: 10.1007/s11119-021-09809-8

Millner, N., Newport, B., Sandbrook, C., & Simlai, T. (2024). Between monitoring and surveillance: Geographies of emerging drone tech- nologies in contemporary conservation. *Progress in Environmental Geography*, 3(1), 17–39. DOI: 10.1177/27539687241229739

Milne-Ives, M., Selby, E., Inkster, B., Lam, C., & Meinert, E. (2022). Artificial intelligence and machine learning in mobile apps for mental health: A scoping review. *PLOS Digital Health*, 1(8), e0000079. DOI: 10.1371/journal.pdig.0000079 PMID: 36812623

Minbaleev, A. (2022). The concept of "artificial intelligence" in law. *Bulletin of Udmurt University Series Economics and Law,32*. .DOI: 10.35634/2412-9593-2022-32-6-1094-1099

Ming, R., Jiang, R., Luo, H., Lai, T., Guo, E., & Zhou, Z. (2023). Comparative Analysis of Different UAV Swarm Control Methods on Unmanned Farms. *Agronomy (Basel)*, 13(10), 10. DOI: 10.3390/agronomy13102499

Minot, D. (2021). *Robot-Assisted Instruction for Children with Autism: How Can Robots Be Used in Special Education?* Autism Spectrum News.

Mirzaeinia, A., & Hassanalian, M. (2019). Minimum-cost drone–nest matching through the kuhn–munkres algorithm in smart cities: Energy management and efficiency enhancement. *Aerospace (Basel, Switzerland)*, 6(11), 125. DOI: 10.3390/aerospace6110125

Mishra, A., & Kong, K. T. C. H. (2024). Tempered Image Detection Using ELA and Convolutional Neural Networks. In *2024 IEEE International Conference on Consumer Electronics (ICCE)*, (pp. 1–3). IEEE. DOI: 10.1109/ICCE59016.2024.10444440

Mitrea, G., & Lecturer, S. (2020). Drones - ethical and legal issues in civil and military research as a future opportunity. *JESS*, 4(1), 83–98. DOI: 10.18662/jess/4.1/30

Mohd Daud, S. M. S., Mohd Yusof, M. Y. P., Heo, C. C., Khoo, L. S., Chainchel Singh, M. K., Mahmood, M. S., & Nawawi, H. (2022). Applications of drone in disaster management: A scoping review. *Science & Justice*, 62(1), 30–42. DOI: 10.1016/j.scijus.2021.11.002 PMID: 35033326

Mohsan, S. A. H., Othman, N. Q. H., Li, Y., Alsharif, M. H., & Khan, M. A. (2023). Unmanned aerial vehicles (UAVs): Practical aspects, applications, open challenges, security issues, and future trends. *Intelligent Service Robotics*, 16(1), 109–137. DOI: 10.1007/s11370-022-00452-4 PMID: 36687780

Mohsan, S. A. H., Zahra, Q., Khan, M. A., Alsharif, M. H., Elhaty, I. A., & Jahid, A. (1593). ul A., Khan, M. A., Alsharif, M. H., Elhaty, I. A., & Jahid, A. (2022). Role of Drone Technology Helping in Alleviating the COVID-19 Pandemic. *Micromachines*, 13(10), 1593. DOI: 10.3390/mi13101593 PMID: 36295946

Molina, N. (2018b). *A wireless method for drone identification and monitoring using ais technology.* IEEE. .DOI: 10.23919/URSI-AT-RASC.2018.8471616

Molina, M., Campos, V., Montagud, M., & Molina, B. (2018). Ethics for civil indoor drones: A qualitative analysis. *International Journal of Micro Air Vehicles*, 10(4), 340–351. DOI: 10.1177/1756829318794004

Moniz, A., Boavida, N., & Candeias, M. (2022). Changes in productivity and labour relations: Artificial intelligence in the automotive sector in portugal. *International Journal of Automotive Technology and Management*, 22(2), 1. DOI: 10.1504/IJATM.2022.10046022

Moon, H. (2019). Challenges and implemented technologies used in autonomous drone racing. *Intelligent Service Robotics, 12*. DOI: 10.1007/s11370-018-00271-6

Mourtzis, D., Angelopoulos, J., & Panopoulos, N. (2021). Smart manufacturing and tactile internet based on 5G in industry 4.0: Challenges, applications and new trends. *Electronics (Basel)*, 10(24), 3175. DOI: 10.3390/electronics10243175

Mozaffari, M., Kasgari, A., Saad, W., Bennis, M., & Debbah, M. (2019). Beyond 5g with uavs: Foundations of a 3d wireless cellular network. *IEEE Transactions on Wireless Communications*, 18(1), 357–372. DOI: 10.1109/TWC.2018.2879940

Mozaffari, M., Saad, W., Bennis, M., Nam, Y., & Debbah, M. (2019). A tutorial on uavs for wireless networks: Applications, challenges, and open problems. *IEEE Communications Surveys and Tutorials*, 21(3), 2334–2360. DOI: 10.1109/COMST.2019.2902862

Muhammad, A., Elhattab, M., Arfaoui, M. A., & Assi, C. (2023). Optimizing age of information in ris-empowered uplink cooperative noma networks. *IEEE Transactions on Network and Service Management*.

Munteanu, I. S., & Ungureanu, L. M. (2022). Analysis Of The Evolution And Prospects Of Introducing Robots And Artificial Intelligence In The Activities Of The Modern Competitive Society. *International Journal of Mechatronics and Applied Mechanics*https://api.semanticscholar.org/CorpusID:250544380

Muslimov, T. (2023). Curl-Free Vector Field for Collision Avoidance in a Swarm of Autonomous Drones. In Ronzhin, A., Sadigov, A., & Meshcheryakov, R. (Eds.), *Interactive Collaborative Robotics* (pp. 369–379). Springer Nature Switzerland. DOI: 10.1007/978-3-031-43111-1_33

Mustafa, A. (2015). *General dynamic scene reconstruc- tion from multiple view video*. IEEE. .DOI: 10.1109/ICCV.2015.109

Mustafa, A., Volino, M., Kim, H., Guillemaut, J., & Hilton, A. (2020). Temporally coherent general dynamic scene reconstruction. *International Journal of Computer Vision*, 129(1), 123–141. DOI: 10.1007/s11263-020-01367-2

Nabi, W., & Xu, B. (2021). Applications of artificial intelligence and ma- chine learning approaches in echocardiography. *Echocardiography (Mount Kisco, N.Y.)*, 38(6), 982–992. DOI: 10.1111/echo.15048 PMID: 33982820

Nair, A., & Thampi, S. (2023). A location-aware physical unclonable function and chebyshev map-based mutual authentication mechanism for internet of surveillance drones. *Concurrency and Computation*, 35(19), e7564. DOI: 10.1002/cpe.7564

Natarajan, K. (2018). *Hand gesture controlled drones: an open source library*. IEEE. .DOI: 10.1109/ICDIS.2018.00035

Neumann, M. M. (2023). Bringing Social Robots to Preschool: Transformation or Disruption? *Childhood Education*, 99(4), 62–65. DOI: 10.1080/00094056.2023.2232283

Nguyen, N. (2021). *Augmented reality and human factors applications for the neurosurgical operating room*. IEEE. .DOI: 10.32920/ryerson.14643753.v1

Nguyen, H. A., & Ha, Q. P. (2023). Robotic autonomous systems for earthmoving equipment operating in volatile conditions and teaming capacity: A survey. *Robotica*, 41(2), 486–510. DOI: 10.1017/S0263574722000339

Nguyen, T. L., & Do, D. T. (2018). Power allocation schemes for wireless powered NOMA systems with imperfect CSI: An application in multiple antenna–based relay. *International Journal of Communication Systems*, 31(15), e3789. DOI: 10.1002/dac.3789

Nhi, N. T. U., & Le, T. M. (2022). A model of semantic-based image retrieval using C-tree and neighbor graph. [IJSWIS]. *International Journal on Semantic Web and Information Systems*, 18(1), 1–23. DOI: 10.4018/IJSWIS.295551

Nichols, R. K., Mumm, H. C., Lonstein, W. D., Ryan, J. J., Carter, C., & Hood, J.-P. (2019). *Unmanned Aircraft Systems in the Cyber Domain*. New Prairie Press.

Nichols, R. K., Mumm, H. C., Lonstein, W. D., Ryan, J. J., Carter, C., & Hood, J.-P. (2020). *Counter unmanned aircraft systems technologies and operations*. New Prairie Press.

Ninh, N. (2022). Navigation for drones in gps-denied environ- ments based on vision processing. *ACSIS, 33,* 25-27. .DOI: 10.15439/2022R46

Noble, S. M., Mende, M., Grewal, D., & Parasuraman, A. (2022). The Fifth Industrial Revolution: How harmonious human–machine collaboration is triggering a retail and service [r] evolution. *Journal of Retailing,* 98(2), 199–208. DOI: 10.1016/j.jretai.2022.04.003

Nof, S. Y., Ceroni, J., Jeong, W., & Moghaddam, M. (2015). *Revolutionizing Collaboration through e-Work, e-Business, and e-Service* (Vol. 2). Springer. DOI: 10.1007/978-3-662-45777-1

Norré, M. (2020). Evaluation of a Word Prediction System in an Augmentative and Alternative Communication for Disabled People. *Journal,* 81(1-4), 49–54. http://iieta. org/journals/mmc_c. DOI: 10.18280/mmc_c.811-409

Novitzky, P., Kokkeler, B., & Verbeek, P. (2018). The dual-use of drones. *Tijdschrift Voor Veiligheid,* 17(1-2), 79–95. DOI: 10.5553/TvV/187279482018017102007

Ntalakas, A., Dimoulas, C., Kalliris, G., & Veglis, A. (2017). Drone journal- ism: Generating immersive experiences. *Journal of Media Critiques,* 3(11), 187–199. DOI: 10.17349/jmc117317

Nugraha, R., Jeyakodi, D., & Mahem, T. (2016). Urgency for legal framework on drones: Lessons for indonesia, india, and thailand. *Indonesia Law Review,* 6(2), 137. DOI: 10.15742/ilrev.v6n2.229

O'Sullivan, S., Nevejans, N., Allen, C., Blyth, A., L'eonard, S., Pagallo, U., Holzinger, K., Holzinger, A., Sajid, M., & Ashrafian, H. (2019). Legal, regulatory, and ethical frame- works for development of standards in artificial intelligence (ai) and autonomous robotic surgery. *International Journal of Medical Robotics and Computer Assisted Surgery,* 15(1), e1968. DOI: 10.1002/rcs.1968 PMID: 30397993

Okafor, N. (2022). *Business demand for a Cloud enterprise data warehouse in electronic Healthcare Computing.* IGI Global.

Okafor, N. (2022). *Business demand for a Cloud enterprise data warehouse in electronic Healthcare Computing: Issues and developments.*

Ollero, A., Boverie, S., Goodall, R., Sasiadek, J., Erbe, H., & Zuehlke, D. (2005). MECHATRONICS, ROBOTICS AND COMPONENTS FOR AUTOMATION AND CONTROL IFAC CC MILESTONE REPORT. *IFAC Proceedings Volumes, 38*(1), 1-13.

Ollero, A., Boverie, S., Goodall, R., Sasiadek, J., Erbe, H., & Zühlke, D. (2006). Mechatronics, robotics and components for automation and control: IFAC milestone report. *Annual Reviews in Control,* 30(1), 41–54. DOI: 10.1016/j.arcontrol.2006.02.002

Otto, A., Agatz, N., Campbell, J., Golden, B., & Pesch, E. (2018). Optimization approaches for civil applications of unmanned aerial vehicles (uavs) or aerial drones: A survey. *Networks,* 72(4), 411–458. DOI: 10.1002/net.21818

Ottun, A., Yin, Z., Liyanage, M., Boerger, M., Asadi, M., Hui, P., Tarkoma, S., Tcholtchev, N., Nurmi, P., & Flores, H. (2022). *Toward trustworthy and responsible autonomous drones in future smart cities.* TechRxiv. DOI: 10.36227/techrxiv.21444102

P'erez, L., Rodr'ıguez, ´. I., Rodr'ıguez, N., Usamentiaga, R., & Garc'ıa, D. (2016). Robot guidance using machine vision techniques in industrial environments: A comparative review. *Sensors (Basel)*, 16(3), 335. DOI: 10.3390/s16030335 PMID: 26959030

Pagter, J. (2021). Speculating about robot moral standing: On the constitution of social robots as objects of governance. *Frontiers in Robotics and AI*, 8, 769349. DOI: 10.3389/frobt.2021.769349 PMID: 34926591

Paik, P., Thudi, S., & Atashzar, S. F. (2022). Power-based velocity-domain variable structure passivity signature control for physical human-(tele) robot interaction. *IEEE Transactions on Robotics*, 39(1), 386–398. DOI: 10.1109/TRO.2022.3197932

Palinkas, I., Pekez, J., Desnica, E., Rajic, A., & Nedelcu, D. (2022). Analysis and optimization of uav frame design for manufacturing from thermoplastic materials on fdm 3d printer. *Materiale Plastice*, 58(4), 238–249. DOI: 10.37358/MP.21.4.5549

Panesar, A., & Panesar, A. (2021). Machine learning and AI ethics. *Machine Learning and AI for Healthcare: Big Data for Improved Health Outcomes*, 207-247.

Panja, S. K. S. (2021). Human Factors of Vehicle Automation. In *Autonomous Driving and Advanced Driver-Assistance Systems (ADAS)*. CRC Press.

Papakostas, G. A., Sidiropoulos, G. K., Papadopoulou, C. I., Vrochidou, E., Kaburlasos, V. G., Papadopoulou, M. T., Holeva, V., Nikopoulou, V.-A., & Dalivigkas, N. (2021). Social robots in special education: A systematic review. *Electronics (Basel)*, 10(12), 1398. DOI: 10.3390/electronics10121398

Pappachan, P. Sreerakuvandana, & Rahaman, M. (2024). Conceptualizing the Role of Intellectual Property and Ethical Behaviour in Artificial Intelligence. In B. Gupta & F. Colace (Eds.), *Handbook of Research on AI and ML for Intelligent Machines and Systems* (pp. 1-26). IGI Global. https://doi.org/ DOI: 10.4018/978-1-6684-9999-3.ch001

Paramythis, A., & Loidl-Reisinger, S. (2003). Adaptive learning environments and e-learning standards. In *Second European conference on e-learning* (Vol. 1, No. 2003, pp. 369–379).

Park, G., Jeon, G.-Y., Sohn, M., & Kim, J. (2022). A Study of Recommendation Systems for Supporting Command and Control (C2) Workflow. *Journal of Internet Computing and Services*, 23(1), 125–134.

Parnell, K. J., Fischer, J. E., Clark, J. R., Bodenmann, A., Galvez Trigo, M. J., Brito, M. P., Divband Soorati, M., Plant, K. L., & Ramchurn, S. D. (2023). Trustworthy UAV Relationships: Applying the Schema Action World Taxonomy to UAVs and UAV Swarm Operations. *International Journal of Human-Computer Interaction*, 39(20), 4042–4058. DOI: 10.1080/10447318.2022.2108961

Pasquinelli, M., & Joler, V. (2020). The nooscope manifested: Ai as instrument of knowledge extractivism. *AI & Society*, 36(4), 1263–1280. DOI: 10.1007/s00146-020-01097-6 PMID: 33250587

Patel, J., Fioranelli, F., & Anderson, D. (2018). Review of radar classification and rcs charac- terisation techniques for small uavs or drones. *IET Radar, Sonar & Navigation*, 12(9), 911–919. DOI: 10.1049/ iet-rsn.2018.0020

Paucar, C., Morales, L., Pinto, K., Sánchez, M., Rodríguez, R., Gutierrez, M., & Palacios, L. (2018). Use of drones for surveillance and reconnaissance of military areas. *Developments and Advances in Defense and Security: Proceedings of the Multidisciplinary International Conference of Research Applied to Defense and Security (MICRADS 2018)*. Springer. DOI: 10.1007/978-3-319-78605-6_10

Pearcey, S., Gordon, K., Chakrabarti, B., Dodd, H., Halldorsson, B., & Creswell, C. (2021). Research Review: The relationship between social anxiety and social cognition in children and adolescents: a systematic review and meta-analysis. *Journal of Child Psychology and Psychiatry, and Allied Disciplines*, 62(7), 805–821. DOI: 10.1111/jcpp.13310 PMID: 32783234

Pech, M., Vrchota, J., & Bednář, J. (2021). Predictive maintenance and intelligent sensors in smart factory: Review. *Sensors (Basel)*, 21(4), 1470. DOI: 10.3390/s21041470 PMID: 33672479

Pecora, F. (2014). "Is Model-Based Robot Programming a Mirage? A Brief Sur- vey of AI Reasoning in Robotics." *KI -. Kunstliche Intelligenz*, 28(4), 255–261. DOI: 10.1007/s13218-014-0325-0

Peeters, A., & Haselager, P. (2019). Designing virtuous sex robots. *International Journal of Social Robotics*, 13(1), 55–66. DOI: 10.1007/s12369-019-00592-1

Peñalvo, F. J. G., Maan, T., Singh, S. K., Kumar, S., Arya, V., Chui, K. T., & Singh, G. P. (2022). Sustainable Stock Market Prediction Framework Using Machine Learning Models. [IJSSCI]. *International Journal of Software Science and Computational Intelligence*, 14(1), 1–15. DOI: 10.4018/IJSSCI.313593

Peñalvo, F. J. G., Sharma, A., Chhabra, A., Singh, S. K., Kumar, S., Arya, V., & Gaurav, A. (2022). Mobile Cloud Computing and Sustainable Development: Opportunities, Challenges, and Future Directions. [IJCAC]. *International Journal of Cloud Applications and Computing*, 12(1), 1–20. DOI: 10.4018/IJCAC.312583

Peres, R., Jia, X., Lee, J., Sun, K., Colombo, A., & Barata, J. (2020). Industrial artificial intelligence in industry 4.0 - systematic review, challenges and outlook. *IEEE Access : Practical Innovations, Open Solutions*, 8, 220121–220139. DOI: 10.1109/ACCESS.2020.3042874

Piquero, J., Sybingco, E., Chua, A., Say, M., Crespo, C., Rivera, R., & Roque, M. (2021). A novel implementation of an autonomous human following drone using lo- cal context. *International Journal of Automation and Smart Technology*, 11(1), 2147–2147. DOI: 10.5875/ausmt.v11i1.2147

Plioutsias, A., Karanikas, N., & Chatzimihailidou, M. (2017). Hazard analysis and safety requirements for small drone operations: To what extent do popular drones embed safety? *Risk Analysis*, 38(3), 562–584. DOI: 10.1111/risa.12867 PMID: 28768049

Pluchino, P., Pernice, G. F. A., Nenna, F., Mingardi, M., Bettelli, A., Bacchin, D., Spagnolli, A., Jacucci, G., Ragazzon, A., Miglioranzi, L., Pettenon, C., & Gamberini, L. (2023). Advanced workstations and collaborative robots: Exploiting eye-tracking and cardiac activity indices to unveil senior workers' mental workload in assembly tasks. *Frontiers in Robotics and AI*, 10, 1275572. DOI: 10.3389/frobt.2023.1275572 PMID: 38149058

Podlubne, A., & Gohringer, D. (2023). A Survey on Adaptive Computing in Robotics: Modelling, Methods and Applications. *IEEE Access : Practical Innovations, Open Solutions*, 11, 53830–53849. DOI: 10.1109/ACCESS.2023.3281190

Pogaku, A. C., Do, D.-T., Lee, B. M., & Nguyen, N. D. (2022). UAV-assisted RIS for future wireless communications: A survey on optimization and performance analysis. *IEEE Access : Practical Innovations, Open Solutions*, 10, 16320–16336. DOI: 10.1109/ACCESS.2022.3149054

Poikonen, S., & Campbell, J. (2020). Future directions in drone routing research. *Networks*, 77(1), 116–126. DOI: 10.1002/net.21982

Poláková, M., Suleimanová, J. H., Madzík, P., Copuš, L., Molnárová, I., & Polednová, J. (2023). Soft skills and their importance in the labour market under the conditions of Industry 5.0. *Heliyon*, 9(8), e18670. DOI: 10.1016/j.heliyon.2023.e18670 PMID: 37593611

Polo, J., Hornero, G., Duijneveld, C., García, A., & Casas, O. (2015). Design of a low-cost Wireless Sensor Network with UAV mobile node for agricultural applications. *Computers and Electronics in Agriculture*, 119, 19–32. DOI: 10.1016/j.compag.2015.09.024

Popescu, S. M., Mansoor, S., Wani, O. A., Kumar, S. S., Sharma, V., Sharma, A., Arya, V. M., Kirkham, M. B., Hou, D., Bolan, N., & Chung, Y. S. (2024). Artificial intelligence and IoT driven technologies for environmental pollution monitoring and management. *Frontiers in Environmental Science*, 12, 1336088. DOI: 10.3389/fenvs.2024.1336088

Prasad, B., Kanojia, R., Mishra, P., Singh, P., & Rathi, V. (2024). *Ad- vancement of actuators in today's world*. AcSIR.

Premebida, C. (2019). *Intelligent robotic perception systems*. InTech Open. .DOI: 10.5772/intechopen.79742

Pugliese, L. (2016). Adaptive learning systems: Surviving the storm. *EDUCAUSE Review*, 10(7).

Puglisi, A., Caprì, T., Pignolo, L., Gismondo, S., Chilà, P., Minutoli, R., Marino, F., Failla, C., Arnao, A. A., Tartarisco, G., Cerasa, A., & Pioggia, G. (2022, June 25). Social Humanoid Robots for Children with Autism Spectrum Disorders: A Review of Modalities, Indications, and Pitfalls. *Children (Basel, Switzerland)*, 9(7), 953. DOI: 10.3390/children9070953 PMID: 35883937

Qian, R., Sengan, S., & Juneja, S. (2022). English language teaching based on big data analytics in augmentative and alternative communication system. *International Journal of Speech Technology*, 25(2), 409–420. DOI: 10.1007/s10772-022-09960-1

Qin, R., & Gruen, A. (2020). The role of machine intelligence in photogrammetric 3d mod- eling – an overview and perspectives. *International Journal of Digital Earth*, 14(1), 15–31. DOI: 10.1080/17538947.2020.1805037

Qiu, R., Li, D., Ibez, A., Xu, Z., & Tarazn, R. (2023). Intent-based deployment for robot applica- tions in 5g-enabled non-public networks. *ITU Journal : ICT Discoveries*, 4(1), 209–220. DOI: 10.52953/AYMI1991

Qu, C., Sorbelli, F. B., Singh, R., Calyam, P., & Das, S. K. (2023). Environmentally-Aware and Energy-Efficient Multi-Drone Coordination and Networking for Disaster Response. *IEEE Transactions on Network and Service Management*, 20(2), 1093–1109. DOI: 10.1109/TNSM.2023.3243543

Ragab, M. (2022). A drones optimal path planning based on swarm intelligence algorithms. *Computers, Materials & Continua*, 72, 365–380. DOI: 10.32604/cmc.2022.024932

Rahman, S. (2019). *Millimeter-wave radar micro-doppler fea- ture extraction of consumer drones and birds for target discrimination.* ACM. .DOI: 10.1117/12.2518846

Rahman, D., Sitorus, A., & Condro, A. (2021). From coastal to montane forest ecosystems, using drones for multi-species research in the tropics. *Drones (Basel)*, 6(1), 6. DOI: 10.3390/drones6010006

Ramachandran, A. (2016). *Robot's Sensors and Instrumentation.*

Rane, N., Choudhary, S., & Rane, J. (2023). Education 4.0 and 5.0: Integrating Artificial Intelligence (AI) for personalized and adaptive learning. *SSRN* 4638365. DOI: 10.2139/ssrn.4638365

Raponi, F., Moscetti, R., Monarca, D., Colantoni, A., & Massantini, R. (2017). Monitoring and optimization of the process of drying fruits and vegetables using computer vision: A review. *Sustainability (Basel)*, 9(11), 2009. DOI: 10.3390/su9112009

Raptis, T., Passarella, A., & Conti, M. (2019). Data management in indus-try 4.0: State of the art and open challenges. *IEEE Access : Practical Innovations, Open Solutions*, 7, 97052–97093. DOI: 10.1109/ACCESS.2019.2929296

Rastogi, A., Singh, S., Sharma, A., & Kumar, S. (2017). *Capacity and Inclination of High Performance Computing in Next-Generation Computing.*

Rathee, G., Garg, S., Kaddoum, G., Choi, B., & Hossain, M. (2020). Trusted orchestra- tion for smart decision-making in internet of vehicles. *IEEE Access : Practical Innovations, Open Solutions*, 8, 157427–157436. DOI: 10.1109/ACCESS.2020.3019795

Ravankar, A. (2018). Autonomous map- ping and exploration with unmanned aerial vehicles using low cost sensors. IEEE. .DOI: 10.3390/ecsa-5-05753

Rawat, B., Bist, A., Apriani, D., Permadi, N., & Nabila, E. (2022). Ai based drones for se- curity concerns in smart cities. *Aptisi Transactions on Management (Atm)*, 7(2), 125–130. DOI: 10.33050/atm.v7i2.1834

Razmi, H., & Afshinfar, S. (2019). Neural network-based adaptive sliding mode control design for position and attitude control of a quadrotor UAV. *Aerospace Science and Technology*, 91, 12–27. DOI: 10.1016/j.ast.2019.04.055

Rebolo-Ifr'an, N., Grilli, M., & Lambertucci, S. (2019). Drones as a threat to wildlife: Youtube comple- ments science in providing evidence about their effect. *Environmental Conservation*, 46(3), 205–210. DOI: 10.1017/S0376892919000080

Reddy, K., Gharde, P., Tayade, H., Patil, M., Reddy, L. S., & Surya, D. (2023, December 12). (n.d.). Advancements in Robotic Surgery: A Comprehensive Overview of Current Utilizations and Upcoming Frontiers. *Cureus*, 15(12), e50415. DOI: 10.7759/cureus.50415 PMID: 38222213

Rejeb, A., Abdollahi, A., Rejeb, K., & Treiblmaier, H. (2022). Drones in agriculture: A review and bibliometric analysis. *Computers and Electronics in Agriculture*, 198, 107017. DOI: 10.1016/j.com-pag.2022.107017

Resnik, D., & Elliott, K. (2018). Using drones to study human beings: Ethical and regulatory issues. *Science and Engineering Ethics*, 25(3), 707–718. DOI: 10.1007/s11948-018-0032-6 PMID: 29488061

Rezwan, S., & Choi, W. (2022). Artificial intelligence approaches for uav nav- igation: Recent advances and future challenges. *IEEE Access : Practical Innovations, Open Solutions*, 10, 26320–26339. DOI: 10.1109/ACCESS.2022.3157626

Ribeiro, R., Ramos, J., Safadinho, D., Reis, A., Rabad˜ao, C., Barroso, J., & Pereira, A. (2021). Web ar solution for uav pilot training and usability testing. *Sensors (Basel)*, 21(4), 1456. DOI: 10.3390/s21041456 PMID: 33669733

Riley, I. (2022). *Improving the expected performance of self- organization in a collective adaptive system of drones using stochastic multiplayer games*. IEEE. .DOI: 10.24251/HICSS.2022.918

Robertson, L., Alici, G., Mun˜oz, A., & Michael, K. (2019). Engineering-based design method- ology for embedding ethics in autonomous robots. *Proceedings of the IEEE*, 107(3), 582–599. DOI: 10.1109/JPROC.2018.2889678

Robinson, N., Tidd, B., Campbell, D., Kulić, D., & Corke, P. (2023). Robotic vision for human-robot interaction and collaboration: A survey and systematic review. *ACM Transactions on Human-Robot Interaction*, 12(1), 1–66. DOI: 10.1145/3570731

Rodriguez-Amat, J. & Duller, N. (2019). *Responsibility and resistance*. Springer. .DOI: 10.1007/978-3-658-26212-9

Rojas, L. C. V. (2015). *Temporarily Distributed Hierarchy in Unmanned Vehicles Swarms*.

Rosete, A. (2020). *Service robots in the hospitality industry: an exploratory literature review*. Springer. ₁3.DOI: 10.1007/978-3-030-38724-2

Rudd, I. (2022). Leveraging artificial intelligence and robotics to improve mental health. *IAJ*. .DOI: 10.32370/IAJ.2710

Ruwaimana, M. (2018). The advantages of using drones over space-borne imagery in the mapping of mangrove forests. *PLoS One*, 13, e0200288. DOI: 10.1371/journal.pone.0200288 PMID: 30020959

Ryu, J., Clements, J., & Neufeld, J. (2022). Low-cost live insect scouting drone: Idrone bee. *Journal of Insect Science*, 22(4), 5. DOI: 10.1093/jisesa/ieac036 PMID: 35793373

S˜onmez, M. (2022). *Unmanned aerial vehicles – classification, types of composite materials used in their structure and applications*. ICAMS. .DOI: 10.24264/icams-2022.I.11

Sætra, H., & Danaher, J. (2022). To each technology its own ethics: The problem of ethical proliferation. *Philosophy & Technology*, 35(4), 93. DOI: 10.1007/s13347-022-00591-7

Sahoo, L., Panda, S. K., & Das, K. K. (2022). A Review on Integration of Vehicular Ad-Hoc Networks and Cloud Computing. [IJCAC]. *International Journal of Cloud Applications and Computing*, 12(1), 1–23. DOI: 10.4018/IJCAC.300771

Saini, T., Kumar, S., Vats, T., & Singh, M. (2020). Edge computing in cloud computing environment: opportunities and challenges. In *International Conference on Smart Systems and Advanced Computing (Syscom-2021)*. Research Gate.

Samantaray, R. (2023). *AI and blockchain fundamentals*. IGI Global. .DOI: 10.4018/979-8-3693-0659-8.ch001

Şandru, V. (2016). *Performances of Air Defence Systems measured with AHP-SWOT analysis*. Paper presented at the Forum Scientiae Oeconomia.

Sanguino, T., & Webber, P. (2018). Making image and vision effortless: Learning method- ology through the quick and easy design of short case studies. *Computer Applications in Engineering Education*, 26(6), 2102–2115. DOI: 10.1002/cae.22003

Santangeli, A., Chen, Y., Kluen, E., Chirumamilla, R., Tiainen, J., & Loehr, J. (2020). In- tegrating drone-borne thermal imaging with artificial intelligence to locate bird nests on agricultural land. *Scientific Reports*, 10(1), 10993. DOI: 10.1038/s41598-020-67898-3 PMID: 32665596

Santoso, F., & Finn, A. (2023). An In-Depth Examination of Artificial Intelligence-Enhanced Cybersecurity in Robotics, Autonomous Systems, and Critical Infrastructures. *IEEE Transactions on Services Computing*, 1–18. DOI: 10.1109/TSC.2023.3331083

Sassis, L., Kefala-Karli, P., Sassi, M., & Zervides, C. (2021). Exploring medical students' and faculty's perception on artificial intelligence and robotics. a questionnaire survey. *Journal of Artificial Intelligence for Medical Sciences*, 2(1-2), 76–84. DOI: 10.2991/jaims.d.210617.002

Sauer, M. (2010). *Mixed-reality for enhanced robot teleoperation*. Universität Würzburg.

Scassellati, B., Boccanfuso, L., Huang, C. M., Mademtzi, M., Qin, M., Salomons, N., Ventola, P., & Shic, F. (2018, August 22). Improving social skills in children with ASD using a long-term, in-home social robot. *Science Robotics*, 3(21), eaat7544. DOI: 10.1126/scirobotics.aat7544 PMID: 33141724

Schranz, M., Di Caro, G. A., Schmickl, T., Elmenreich, W., Arvin, F., Şekercioğlu, A., & Sende, M. (2021). Swarm Intelligence and cyber-physical systems: Concepts, challenges and future trends. *Swarm and Evolutionary Computation*, 60, 100762. DOI: 10.1016/j.swevo.2020.100762

Sebbane, Y. B. (2018). *Intelligent autonomy of UAVs: advanced missions and future use*. CRC Press. DOI: 10.1201/b22485

Seizovic, A., Thorpe, D., & Goh, S. (2022). Emergent behavior in the battle management system. *Applied Artificial Intelligence*, 36(1), 2151183. DOI: 10.1080/08839514.2022.2151183

Sekhar, B. V. D. S., Udayaraju, P., Kumar, N. U., Sinduri, K. B., Ramakrishna, B., Babu, B. S. S. V. R., & Srinivas, M. S. S. S. (2023). Artificial neural network-based secured communication strategy for vehicular ad hoc network. *Soft Computing*, 27(1), 297–309. DOI: 10.1007/s00500-022-07633-4

Setia, H., Chhabra, A., Singh, S. K., Kumar, S., Sharma, S., Arya, V., Gupta, B. B., & Wu, J. (2024). Securing the road ahead: Machine learning-driven DDoS attack detection in VANET cloud environments. *Cyber Security and Applications*, 2, 100037. DOI: 10.1016/j.csa.2024.100037

Shafii, N., Saudi, A., Chyang, P., Abu, I., Kamarudin, M., & Saudi, H. (2019). Application of chemo-metrics techniques to solve environmental issues in malaysia. *Heliyon*, 5(10), e02534. DOI: 10.1016/j.heliyon.2019.e02534 PMID: 31667387

Shahroom, A., & Hussin, N. (2018). Industrial revolution 4.0 and education. *International Journal of Academic Research in Business & Social Sciences*, 8(9). DOI: 10.6007/IJARBSS/v8-i9/4593

Shah, S. A., Lakho, G. M., Keerio, H. A., Sattar, M. N., Hussain, G., Mehdi, M., Vistro, R. B., Mahmoud, E. A., & Elansary, H. O. (2023). Application of Drone Surveillance for Advance Agriculture Monitoring by Android Application Using Convolution Neural Network. *Agronomy (Basel)*, 13(7), 7. DOI: 10.3390/agronomy13071764

Shakhatreh, H., Sawalmeh, A. H., Al-Fuqaha, A., Dou, Z., Almaita, E., Khalil, I., Othman, N. S., Khreishah, A., & Guizani, M. (2019). Unmanned Aerial Vehicles (UAVs): A Survey on Civil Applications and Key Research Challenges. *IEEE Access : Practical Innovations, Open Solutions*, 7, 48572–48634. DOI: 10.1109/ACCESS.2019.2909530

Shan, L., Kagawa, T., Ono, F., Li, H., & Kojima, F. (2019). Machine learning-based field data analysis and modeling for drone communications. *IEEE Access : Practical Innovations, Open Solutions*, 7, 79127–79135. DOI: 10.1109/ACCESS.2019.2922544

Sharma, A., Singh, S. K., Badwal, E., Kumar, S., Gupta, B. B., Arya, V., Chui, K. T., & Santaniello, D. (2023). Fuzzy Based Clustering of Consumers' Big Data in Industrial Applications. *2023 IEEE International Conference on Consumer Electronics (ICCE)*, (pp. 01–03). IEEE. DOI: 10.1109/ICCE56470.2023.10043451

Sharma, A. (2024). Revolutionizing Healthcare Systems: Synergistic Multimodal Ensemble Learning & Knowledge Transfer for Lung Cancer Delineation & Taxonomy. In *2024 IEEE International Conference on Consumer Electronics (ICCE)*, (pp. 1–6). IEEE. DOI: 10.1109/ICCE59016.2024.10444476

Sharma, A., Jain, A., Gupta, P., & Chowdary, V. (2021). Machine learning applica- tions for precision agriculture: A comprehensive review. *IEEE Access : Practical Innovations, Open Solutions*, 9, 4843–4873. DOI: 10.1109/ACCESS.2020.3048415

Sharma, A., Singh, S. K., Chhabra, A., Kumar, S., Arya, V., & Moslehpour, M. (2023). A Novel Deep Federated Learning-Based Model to Enhance Privacy in Critical Infrastructure Systems. [IJSSCI]. *International Journal of Software Science and Computational Intelligence*, 15(1), 1–23. DOI: 10.4018/IJSSCI.334711

Sharma, A., Singh, S. K., Kumar, S., Chhabra, A., & Gupta, S. (2023). Security of Android Banking Mobile Apps: Challenges and Opportunities. In Nedjah, N., Martínez Pérez, G., & Gupta, B. B. (Eds.), *International Conference on Cyber Security, Privacy and Networking (ICSPN 2022)* (pp. 406–416). Springer International Publishing. DOI: 10.1007/978-3-031-22018-0_39

Sharma, S. K., Singh, S. K., & Panja, S. C. (2021). Human Factors of Vehicle Automation. In *Autonomous Driving and Advanced Driver-Assistance Systems (ADAS)*. CRC Press. DOI: 10.1201/9781003048381-17

Shazly, S., Trabuco, E., Ngufor, C., & Famuyide, A. (2022). Introduction to ma- chine learning in obstetrics and gynecology. *Obstetrics and Gynecology*, 139(4), 669–679. DOI: 10.1097/AOG.0000000000004706 PMID: 35272300

Shen, J., Yang, B., Dudley, J. J., & Kristensson, P. O. (2022, March). Kwickchat: A multi-turn dialogue system for aac using context-aware sentence generation by bag-of-keywords. In *27th International Conference on Intelligent User Interfaces* (pp. 853-867). DOI: 10.1145/3490099.3511145

Shi, W., Zheng, J., Sheng, Q., Wang, Q., Wang, L., & Li, Q. (2023). Illumination modelling for reconstructing the machined surface topography. *International Journal of Advanced Manufacturing Technology*, 125(11-12), 4975–4987. DOI: 10.1007/s00170-023-10925-0

Shuford, J. (2024). Deep Reinforcement Learning Unleashing the Power of AI in Decision-Making. *Journal of Artificial Intelligence General Science (JAIGS), 1*(1). DOI: 10.60087/jaigs.v1i1.36

Shukla, J. (2017). *Effectiveness of so-cially assistive robotics during cognitive stimulation interventions: impact on caregivers.* IEEE. .DOI: 10.1109/ROMAN.2017.8172281

Shukla, A., & Karki, H. (2016). Application of robotics in offshore oil and gas industry—A review Part II. *Robotics and Autonomous Systems*, 75, 508–524. DOI: 10.1016/j.robot.2015.09.013

Shukla, A., & Karki, H. (2016). Application of robotics in onshore oil and gas industry—A review Part I. *Robotics and Autonomous Systems*, 75, 490–507. DOI: 10.1016/j.robot.2015.09.012

Sidiq, M. S. (2023, January 23). *The Ethics of Machine Learning: Understanding the Role of Developers and Designers.* HackerNoon. https://hackernoon.com/the-ethics-of-machine-learning-understanding-the -role-of-developers-and-designers

Sifakis, J., & Harel, D. (2023). Trustworthy Autonomous System Development. *ACM Transactions on Embedded Computing Systems, 22*(3). DOI: 10.1145/3545178

Siju, N. (2022). Applications of drone technology in construction projects: a systematic literature review. *International Journal of Research -Granthaalayah,10*, 1–14. .DOI: 10.29121/granthaalayah.v10. i10.2022.4810

Silva, R. (2018). *Machine vision systems for industrial quality control inspections.* Springer. ₅8.DOI: 10.1007/978-3-030-01614-2

Silver, D., Lever, G., Heess, N., Degris, T., Wierstra, D., & Riedmiller, M. (2014). Deterministic Policy Gradient Algorithms. *Proceedings of the 31st International Conference on Machine Learning*, (pp. 387–395). MLR. https://proceedings.mlr.press/v32/silver14.html

Simlai, T. (2021). *Digital surveillance technologies in conservation and their social implications.* Springer. .DOI: 10.1093/oso/9780198850243.003.0012

Sindiramutty, S. (2024). *Eyes in the sky.* IGI Global. .DOI: 10.4018/979-8-3693-0774-8.ch017

Sindiramutty, S. (2024). *Future trends and emerging threats in drone cybersecurity.* Springer. .DOI: 10.4018/979-8-3693-0774-8.ch007

Singh, A., & Singh, S. Kr., & Mittal, A. (2022). A Review on Dataset Acquisition Techniques in Gesture Recognition from Indian Sign Language. In P. Verma, C. Charan, X. Fernando, & S. Ganesan (Eds.), *Advances in Data Computing, Communication and Security* (pp. 305–313). Springer Nature. DOI: 10.1007/978-981-16-8403-6_27

Singh, I., & Singh, S. Kr., Kumar, S., & Aggarwal, K. (2022). Dropout-VGG Based Convolutional Neural Network for Traffic Sign Categorization. In M. Saraswat, H. Sharma, K. Balachandran, J. H. Kim, & J. C. Bansal (Eds.), *Congress on Intelligent Systems* (pp. 247–261). Springer Nature. DOI: 10.1007/978-981-16-9416-5_18

Singh, I., & Singh, S. Kr., Kumar, S., & Aggarwal, K. (2022). Dropout-VGG Based Convolutional Neural Network for Traffic Sign Categorization. In M. Saraswat, H. Sharma, K. Balachandran, J. H. Kim, & J. C. Bansal (Eds.), *Congress on Intelligent Systems* (pp. 247–261). Springer Nature. https://doi.org/DOI: 10.1007/978-981-16-9416-5_18\

Singh, I., Singh, S. K., Singh, R., & Kumar, S. (2022). Efficient Loop Unrolling Factor Prediction Algorithm using Machine Learning Models. *2022 3rd International Conference for Emerging Technology (INCET)*, (pp. 1–8). IEEE. DOI: 10.1109/INCET54531.2022.9825092

Singh, S. Kr., Sharma, S. K., Singla, D., & Gill, S. S. (2022). Evolving Requirements and Application of SDN and IoT in the Context of Industry 4.0, Blockchain and Artificial Intelligence. In *Software Defined Networks* (pp. 427–496). John Wiley & Sons, Ltd. https://doi.org/DOI: 10.1002/9781119857921.ch13

Singh, M., Singh, S. K., Kumar, S., Madan, U., & Maan, T. (2023). Sustainable Framework for Metaverse Security and Privacy: Opportunities and Challenges. In Nedjah, N., Martínez Pérez, G., & Gupta, B. B. (Eds.), *International Conference on Cyber Security, Privacy and Networking (ICSPN 2022)* (pp. 329–340). Springer International Publishing. DOI: 10.1007/978-3-031-22018-0_30

Singh, R., Singh, S. K., Kumar, S., & Gill, S. S. (2022). SDN-Aided Edge Computing-Enabled AI for IoT and Smart Cities. In *SDN-Supported Edge-Cloud Interplay for Next Generation Internet of Things*. Chapman and Hall/CRC. DOI: 10.1201/9781003213871-3

Singh, S. K. (2021). *Linux Yourself: Concept and Programming*. Chapman and Hall/CRC. DOI: 10.1201/9780429446047

Singh, S. K., Kumar, A., Gupta, S., & Madan, R. (2011). Architectural performance of WiMAX over WiFi with reliable QoS over wireless communication. *International Journal of Advanced Networking and Applications*, 3(1), 1017.

Singh, S., Sharma, S., Singla, D., & Gill, S. S. (2022). *Evolving Requirements and Application of SDN and IoT in the Context of Industry 4.0*. Blockchain and Artificial Intelligence.

Siwach, G., & Li, C. (2024). Unveiling the Potential of Natural Language Processing in Collaborative Robots (Cobots): A Comprehensive Survey. *2024 IEEE International Conference on Consumer Electronics (ICCE)*, (pp. 1–6). IEEE. DOI: 10.1109/ICCE59016.2024.10444393

Smagh, N. S. (2020). Intelligence, surveillance, and reconnaissance design for great power competition. *Congressional Research Service, 46389*.

Sole, J., Centelles, R., Freitag, F., & Meseguer, R. (2022). Implementation of a lora mesh library. *IEEE Access: Practical Innovations, Open Solutions, 10*, 113158–113171. DOI: 10.1109/ACCESS.2022.3217215

Song, D., Goldberg, K., & Chong, N. Y. (2016). *Springer Handbook of Robotics: Networked Robots*. Springer.

Song, H., Bai, J., Yi, Y., Wu, J., & Liu, L. (2020). Artificial intelligence enabled Internet of Things: Network architecture and spectrum access. *IEEE Computational Intelligence Magazine*, 15(1), 44–51. DOI: 10.1109/MCI.2019.2954643

Song, J. G., Lee, J. H., & Park, I. S. (2021). Enhancement of cooling performance of naval combat management system using heat pipe. *Applied Thermal Engineering*, 188, 116657. DOI: 10.1016/j.applthermaleng.2021.116657

Soori, M., Arezoo, B., & Dastres, R. (2023). Artificial intelligence, machine learning and deep learning in advanced robotics, a review. *Cognitive Robotics*, 3, 54–70. DOI: 10.1016/j.cogr.2023.04.001

Sorantin, E. (2021). The augmented radiologist: artificial intelligence in the practice of radiology. *Pediatric Radiology,52.* .DOI: 10.1007/s00247-021-05177-7

Spitale, M., Axelsson, M., Kara, N., & Gunes, H. (2023, August). Longitudinal evolution of coachees' behavioural responses to interaction ruptures in robotic positive psychology coaching. In *2023 32nd IEEE International Conference on Robot and Human Interactive Communication (RO-MAN)* (pp. 315-322). IEEE.

Sprock, T. (2018). *Self-similar architectures for smart manufacturing and logistics systems.*

Sravya, P. (n.d.). Influence of Artificial Intelligence in Robotics. In *Artificial Intelligence and Knowledge Processing.* CRC Press.

St"ocker, C., Bennett, R., Nex, F., Gerke, M., & Zevenbergen, J. (2017). Review of the current state of uav regulations. *Remote Sensing (Basel)*, 9(5), 459. DOI: 10.3390/rs9050459

Stark, D., Vaughan, I., Evans, L., Kler, H., & Goossens, B. (2017). Combining drones and satellite tracking as an effective tool for informing policy change in riparian habitats: A proboscis monkey case study. *Remote Sensing in Ecology and Conservation*, 4(1), 44–52. DOI: 10.1002/rse2.51

Stefanova, V., & Komitov, G. (2021). Overview of the machine vision methods at agrorobots. *Agrarni Nauki*, 13(30), 13–19. DOI: 10.22620/agrisci.2021.30.002

Stoddard, B. (2022). *Designing and Evaluating a User Interface for Multi-Robot Furniture.*

Strawbridge, D. (2022). Civil drone ethics and sustainability. *Proceedings of the Wellington Faculty of Engineering Ethics and Sustainability Symposium.* Victoria University. DOI: 10.26686/wfeess.vi.7660

Suarez-Ibarrola, R., & Miernik, A. (2020). Prospects and challenges of artificial intelligence and computer science for the future of urology. *World Journal of Urology*, 38(10), 2325–2327. DOI: 10.1007/s00345-020-03428-0 PMID: 32910230

Suh, J. (2018). Drones: How They Work, Applications, and Legal Issues. *Georgetown Law Technology Review*, 3, 502.

Suwardhi, D., Trisyanti, S., Virtriana, R., Syamsu, A., Jannati, S., & Halim, R. (2022). Heritage smart city mapping, planning and land administration (hestya). *ISPRS International Journal of Geo-Information*, 11(2), 107. DOI: 10.3390/ijgi11020107

Svaigen, A. R., Boukerche, A., Ruiz, L. B., & Loureiro, A. A. F. (2023). Security in the Industrial Internet of Drones. *IEEE Internet of Things Magazine*, 6(3), 110–116. DOI: 10.1109/IOTM.001.2200260

Swamy, S., & Raju, K. (2020). An empirical study on system level aspects of internet of things (iot). *IEEE Access: Practical Innovations, Open Solutions*, 8, 188082–188134. DOI: 10.1109/ACCESS.2020.3029847

Syriopoulou-Delli, C. K., & Eleni, G. (2022). Effectiveness of different types of Augmentative and Alternative Communication (AAC) in improving communication skills and in enhancing the vocabulary of children with ASD: A review. *Review Journal of Autism and Developmental Disorders*, 9(4), 493–506. DOI: 10.1007/s40489-021-00269-4

T'oth, Z., Caruana, R., Gruber, T., & Loebbecke, C. (2022). The dawn of the ai robots: Towards a new framework of ai robot accountability. *Journal of Business Ethics*, 178(4), 895–916. DOI: 10.1007/s10551-022-05050-z

Taha, B., & Shoufan, A. (2019). Machine learning-based drone detection and classification: State-of-the-art in research. *IEEE Access: Practical Innovations, Open Solutions*, 7, 138669–138682. DOI: 10.1109/ACCESS.2019.2942944

Tahir, N., & Parasuraman, R. (2023). Mobile Robot Control and Autonomy Through Collaborative Simulation Twin. *arXiv preprint arXiv:2303.06172*.

Takahashi, Y., Okada, K., Hoshino, T., & Anme, T. (2015, August 12). Developmental Trajectories of Social Skills during Early Childhood and Links to Parenting Practices in a Japanese Sample. *PLoS One*, 10(8), e0135357. DOI: 10.1371/journal.pone.0135357 PMID: 26267439

Talaat, F. M., Ali, Z. H., Mostafa, R. R., & El-Rashidy, N. (2024, January 4). Real-time facial emotion recognition model based on kernel autoencoder and convolutional neural network for autism children. *Soft Computing*, 28(9-10), 6695–6708. DOI: 10.1007/s00500-023-09477-y

Taleb, S. M., Meraihi, Y., Mirjalili, S., Acheli, D., Ramdane-Cherif, A., & Gabis, A. B. (2023). Mesh Router Nodes Placement for Wireless Mesh Networks Based on an Enhanced Moth–Flame Optimization Algorithm. *Mobile Networks and Applications*, 28(2), 518–541. DOI: 10.1007/s11036-022-02059-6

Tanaka, F., & Matsuzoe, S. (2012). Children teach a care-receiving robot to promote their learning: Field experiments in a classroom for vocabulary learning. *Journal of Human-Robot Interaction*, 1(1), 78–95. DOI: 10.5898/JHRI.1.1.Tanaka

Tang, B., Chen, L., Sun, W., & Lin, Z. (2022). Review of surface defect detec- tion of steel products based on machine vision. *IET Image Processing*, 17(2), 303–322. DOI: 10.1049/ipr2.12647

Tang, J., Duan, H., & Lao, S. (2023). Swarm intelligence algorithms for multiple unmanned aerial vehicles collaboration: A comprehensive review. *Artificial Intelligence Review*, 56(5), 4295–4327. DOI: 10.1007/s10462-022-10281-7

Tang, W., Chen, M. Z., Dai, J. Y., Zeng, Y., Zhao, X., Jin, S., Cheng, Q., & Cui, T. J. (2020). Wireless communications with programmable metasurface: New paradigms, opportunities, and challenges on transceiver design. *IEEE Wireless Communications*, 27(2), 180–187. DOI: 10.1109/MWC.001.1900308

Tan, H., & Dijken, S. (2023). Dynamic machine vision with retinomorphic photomemristor- reservoir computing. *Nature Communications*, 14(1), 2169. DOI: 10.1038/s41467-023-37886-y PMID: 37061543

Tanveer, M., Zahid, A., Ahmad, M., Baz, A., & Alhakami, H. (2020). Lake-iod: Lightweight authenticated key exchange protocol for the internet of drone environment. *IEEE Access : Practical Innovations, Open Solutions*, 8, 155645–155659. DOI: 10.1109/ACCESS.2020.3019367

Tatsidou, E., Tsiamis, C., Karamagioli, E., Boudouris, G., Pikoulis, A., Kakalou, E., & Pikoulis, E. (2019). Reflecting upon the humanitarian use of unmanned aerial vehicles (drones). *Swiss Medical Weekly*. DOI: 10.4414/smw.2019.20065 PMID: 30950503

Taye, M. M. (2023). Understanding of Machine Learning with Deep Learning: Architectures, Workflow, Applications and Future Directions. *Computers*, 12(5), 5. DOI: 10.3390/computers12050091

Taylor, D. L., Yeung, M., & Bashet, A. Z. (2021). *Personalized and adaptive learning. Innovative Learning Environments in STEM Higher Education: Opportunities*. Challenges, and Looking Forward.

Tedre, M., Toivonen, T., Kahila, J., Vartiainen, H., Valtonen, T., Jormanainen, I., & Pears, A. (2021). Teaching machine learning in K–12 classroom: Pedagogical and technological trajectories for artificial intelligence education. *IEEE Access : Practical Innovations, Open Solutions*, 9, 110558–110572. DOI: 10.1109/ACCESS.2021.3097962

Tella, A., & Ajani, Y. (2022). Robots and public libraries. *Library Hi Tech News*, 39(7), 15–18. DOI: 10.1108/LHTN-05-2022-0072

Telli, K., Kraa, O., Himeur, Y., Ouamane, A., Boumehraz, M., Atalla, S., & Mansoor, W. (2023). A Comprehensive Review of Recent Research Trends on Unmanned Aerial Vehicles (UAVs). *Systems*, 11(8), 8. DOI: 10.3390/systems11080400

Tesei, A., Luise, M., Pagano, P., & Ferreira, J. (2021). Secure Multi-access Edge Computing Assisted Maneuver Control for Autonomous Vehicles. *2021 IEEE 93rd Vehicular Technology Conference (VTC2021-Spring)*, (pp. 1–6). IEEE. DOI: 10.1109/VTC2021-Spring51267.2021.9449087

Tezza, D., & Andujar, M. (2019). The state-of-the-art of human–drone interaction: A survey. *IEEE Access : Practical Innovations, Open Solutions*, 7, 167438–167454. DOI: 10.1109/ACCESS.2019.2953900

Thandavarayan, G., Sepulcre, M., & Gozalvez, J. (2020). Cooperative perception for connected and automated vehicles: Evaluation and impact of congestion control. *IEEE Access : Practical Innovations, Open Solutions*, 8, 197665–197683. DOI: 10.1109/ACCESS.2020.3035119

Thangavel, K. (2023). *Trusted autonomous operations of distributed satellite systems for earth observation missions* [Doctoral dissertation, RMIT University].

Theodorou, A., Wortham, R. H., & Bryson, J. J. (2017). Designing and implementing transparency for real time inspection of autonomous robots. *Connection Science*, 29(3), 230–241. DOI: 10.1080/09540091.2017.1310182

Tiddi, I., Bastianelli, E., Daga, E., d'Aquin, M., & Motta, E. (2020). Robot–city interaction: Mapping the research landscape—a survey of the interactions between robots and modern cities. *International Journal of Social Robotics*, 12(2), 299–324. DOI: 10.1007/s12369-019-00534-x

Tiderko, A., Bachran, T., Hoeller, F., & Schulz, D. (2008). RoSe—A framework for multicast communication via unreliable networks in multi-robot systems. *Robotics and Autonomous Systems*, 56(12), 1017–1026. DOI: 10.1016/j.robot.2008.09.004

Tolcha, Y., Montanaro, T., Conzon, D., Schwering, G., Maselyne, J., & Kim, D. (2021). Towards interoperability of entity-based and event-based iot platforms: The case of ngsi and epcis standards. *IEEE Access : Practical Innovations, Open Solutions*, 9, 49868–49880. DOI: 10.1109/ACCESS.2021.3069194

Toncic, J. (2021). Advancing a critical artificial intelligence theory for schooling. *Teknokultura Revista De Cultura Digital Y Movimientos Sociales,19*, 13–24. .DOI: 10.5209/tekn.71136

Tran, T. (2022). Management and regulation of drone operation in urban en- vironment: a case study. *Social Sciences, 11.* . https://doi.org/.DOI: 10.3390/socsci11100474

Tran, C. N. N., Tat, T. T. H., Tam, V. W. Y., & Tran, D. H. (2023). Factors affecting intelligent transport systems towards a smart city: A critical review. *International Journal of Construction Management*, 23(12), 1982–1998. DOI: 10.1080/15623599.2022.2029680

Trevelyan, J. P., Kang, S.-C., & Hamel, W. R. (2008). Robotics in Hazardous Applications. In Siciliano, B., & Khatib, O. (Eds.), *Springer Handbook of Robotics* (pp. 1101–1126). Springer. DOI: 10.1007/978-3-540-30301-5_49

Tuncer, O., & Cirpan, H. A. (2022). Target priority based optimisation of radar resources for networked air defence systems. *IET Radar, Sonar & Navigation*, 16(7), 1212–1224. DOI: 10.1049/rsn2.12255

Tuo, Y. (2021). *How artificial intelligence will change the future of tourism industry: the practice in China*. Springer. .DOI: 10.1007/978-3-030-65785-7

Tzafestas, S. (2018). Roboethics: Fundamental concepts and future prospects. *Information (Basel)*, 9(6), 148. DOI: 10.3390/info9060148

Umar, T. (2020). Applications of drones for safety inspection in the gulf cooperation coun- cil construction. *Engineering, Construction, and Architectural Management*, 28(9), 2337–2360. DOI: 10.1108/ECAM-05-2020-0369

Ünal, H. T., & Başçiftçi, F. (2020). Using Evolutionary Algorithms for the Scheduling of Aircrew on Airborne Early Warning and Control System. *Defence Science Journal*, 70(3), 240–248. DOI: 10.14429/dsj.70.15055

Upadhyay, J., Rawat, A., & Deb, D. (2021). Multiple drone navigation and for- mation using selective target tracking-based computer vision. *Electronics (Basel)*, 10(17), 2125. DOI: 10.3390/electronics10172125

Uphoff, M. K. (2023). *Social emotional learning and its needs and benefits for students with down syndrome, autism spectrum disorder, and emotional behavioral disorder* [Master's thesis, Bethel University]. Spark Repository.

USAE, R. N. (2020). Table Stakes of the Advanced Battle Management System. *Air & Space Power Journal*, 34(3), 81–86.

Vajgel, B. (2021). Development of intelligent robotic process automation: a utility case study in brazil. *IEEE Access, 9*.DOI: 10.1109/ACCESS.2021.3075693

Van Chien, T., Tu, L. T., Chatzinotas, S., & Ottersten, B. (2020). Coverage probability and ergodic capacity of intelligent reflecting surface-enhanced communication systems. *IEEE Communications Letters*, 25(1), 69–73. DOI: 10.1109/LCOMM.2020.3023759

Van den Berghe, R., Verhagen, J., Oudgenoeg-Paz, O., Van der Ven, S., & Leseman, P. (2019). Social robots for language learning: A review. *Review of Educational Research*, 89(2), 259–295. DOI: 10.3102/0034654318821286

Vanak, J. (2022). Artificial intelligence and medicine. *Science Insights*, 41(1), 567–575. DOI: 10.15354/si.22.re068

Vanhée, L., Jeanpierre, L., & Mouaddib, A. I. (2021, September). Optimizing Requests for Support in Context-Restricted Autonomy. In *2021 IEEE/RSJ International Conference on Intelligent Robots and Systems (IROS)* (pp. 6434-6440). IEEE. DOI: 10.1109/IROS51168.2021.9636240

Vas, E., Lescro̎el, A., Duriez, O., Boguszewski, G., & Gr'emillet, D. (2015). Approaching birds with drones: First experiments and ethical guidelines. *Biology Letters*, 11(2), 20140754. DOI: 10.1098/rsbl.2014.0754 PMID: 25652220

Vats, T., Singh, S. K., Kumar, S., Gupta, B. B., Gill, S. S., Arya, V., & Alhalabi, W. (2023). Explainable context-aware IoT framework using human digital twin for healthcare. *Multimedia Tools and Applications*, 83(22), 62489–62490. DOI: 10.1007/s11042-023-16922-5

Vogt, P., van den Berghe, R., De Haas, M., Hoffman, L., Kanero, J., Mamus, E., . . . Pandey, A. K. (2019, March). Second language tutoring using social robots: a large-scale study. In *2019 14th ACM/IEEE International Conference on Human-Robot Interaction (HRI)* (pp. 497-505). Ieee. DOI: 10.1109/HRI.2019.8673077

Vrochidou, E., Oustadakis, D., Kefalas, A., & Papakostas, G. (2022). Computer vision in self-steering tractors. *Machines (Basel)*, 10(2), 129. DOI: 10.3390/machines10020129

Wang, J. (2015). *Research on medical image processing*. IEEE. .DOI: 10.2991/icmii-15.2015.119

Wang, C. (2022). The application of machine vision in intelligent manufacturing. *Highlights in Science Engineering and Technology*, 9, 47–50. DOI: 10.54097/hset.v9i.1714

Wang, H., Chen, S., Yuan, B., Liu, J., & Sun, X. (2021). Liquid metal transformable machines. *Accounts of Materials Research*, 2(12), 1227–1238. DOI: 10.1021/accountsmr.1c00182

Wang, L., Zhang, Y., Zhang, H., An, K., & Hu, K. (2023a). A deep learning-based experi- ment on forest wildfire detection in machine vision course. *IEEE Access : Practical Innovations, Open Solutions*, 11, 32671–32681. DOI: 10.1109/ACCESS.2023.3262701

Wang, N. (2020). "we live on hope...": Ethical considerations of humanitarian use of drones in post-disaster nepal. *IEEE Technology and Society Magazine*, 39(3), 76–85. DOI: 10.1109/MTS.2020.3012332

Wang, N. (2021). "as it is africa, it is ok"? ethical considerations of development use of drones for delivery in malawi. *IEEE Transactions on Technology and Society*, 2(1), 20–30. DOI: 10.1109/TTS.2021.3058669

Wang, N., Christen, M., & Hunt, M. (2021). Ethical considerations associated with "hu- manitarian drones": A scoping literature review. *Science and Engineering Ethics*, 27(4), 51. DOI: 10.1007/s11948-021-00327-4 PMID: 34342721

Wang, N., Christen, M., Hunt, M., & Biller-Andorno, N. (2022). Supporting value sensitivity in the humanitarian use of drones through an ethics assessment framework. *International Review of the Red Cross*, 104(919), 1397–1428. DOI: 10.1017/S1816383121000989

Wang, P. (2019). On defining artificial intelligence. *Journal of Artificial General Intelligence*, 10(2), 1–37. DOI: 10.2478/jagi-2019-0002

Wang, S., Sung, R., Reinholz, D., & Bussey, T. (2023b). Equity analysis of an augmented reality-mediated group activity in a college biochemistry classroom. *Journal of Research in Science Teaching*, 60(9), 1942–1966. DOI: 10.1002/tea.21847

Wang, X., Yang, Z., Chen, G., & Liu, Y. (2023). A Reinforcement Learning Method of Solving Markov Decision Processes: An Adaptive Exploration Model Based on Temporal Difference Error. *Electronics (Basel)*, 12(19), 19. DOI: 10.3390/electronics12194176

Wang, Y. (2021). When artificial intelligence meets educational leaders' data-informed decision-making: A cautionary tale. *Studies in Educational Evaluation*, 69, 100872. DOI: 10.1016/j.stueduc.2020.100872

Wang, Y., & Ye, T. (2022). Applications of artificial intelligence enhanced drones in distress pavement, pothole detection, and healthcare monitoring with service delivery. *Journal of Engineering*, 2022, 1–16. DOI: 10.1155/2022/7733196

Waqas, M. (2023). *Unmanned aerial vehicles (uavs) in modern agriculture*. Springer. .DOI: 10.4018/978-1-6684-9231-4.ch006

Watkins, C. J. C. H., & Dayan, P. (1992). Q-learning. *Machine Learning*, 8(3), 279–292. DOI: 10.1007/BF00992698

Wazid, M. (2020). *Private blockchain- envisioned security framework for ai-enabled iot-based drone-aided healthcare services*. ACM. .DOI: 10.1145/3414045.3415941

Weber, S., & Knaus, F. (2017). Using drones as a monitoring tool to detect evidence of winter sports activities in a protected mountain area. *Eco.Mont*, 9(1), 30–34. DOI: 10.1553/eco.mont-9-1s30

Wehmeyer, M. L. (2022). From segregation to strengths: A personal history of special education. *Phi Delta Kappan*, 103(6), 8–13. DOI: 10.1177/00317217221082792

Weichlein, T., Zhang, S., Li, P., & Zhang, X. (2023). Data flow control for network load balancing in ieee time sensitive networks for automation. *IEEE Access : Practical Innovations, Open Solutions*, 11, 14044–14060. DOI: 10.1109/ACCESS.2023.3243286

Wen, H., Xie, Z., Wu, Z., Lin, Y., & Feng, W. (2024). Exploring the future application of UAVs: Face image privacy protection scheme based on chaos and DNA cryptography. *Journal of King Saud University. Computer and Information Sciences*, 36(1), 101871. DOI: 10.1016/j.jksuci.2023.101871

West, J., & Bowman, J. (2016). The domestic use of drones: An ethical analysis of surveillance issues. *Public Administration Review*, 76(4), 649–659. DOI: 10.1111/puar.12506

Widodo, A. M., Wijayanto, H., Wijaya, I. G. P. S., Wisnujati, A., & Musnansyah, A. (2023). Analyzing Coverage Probability of Reconfigurable Intelligence Surface-aided NOMA. *JOIV: International Journal on Informatics Visualization*, 7(3), 839–846. DOI: 10.30630/joiv.7.3.2054

Wilcock, G. (2022). *Conversational ai and knowledge graphs for social robot interaction*. IEEE. .DOI: 10.1109/HRI53351.2022.9889583

Williams, J. (2015). Distant intimacy: Space, drones, and just war. *Ethics & International Affairs*, 29(1), 93–110. DOI: 10.1017/S0892679414000793

Wilson, J., Tickle-Degnen, L., & Scheutz, M. (2020). Challenges in designing a fully au- tonomous socially assistive robot for people with parkinson's disease. *ACM Transactions on Human-Robot Interaction*, 9(3), 1–31. DOI: 10.1145/3379179

Wolff, F., Wickord, L. C., Rahe, M., & Quaiser-Pohl, C. M. (2023). Effects of an intercultural seminar using telepresence robots on students' cultural intelligence. *Computers & Education: X Reality, 2*, 100007.

Wood, L. J., Robins, B., Lakatos, G., Syrdal, D. S., Zaraki, A., & Dautenhahn, K. (2019, March 1). Developing a protocol and experimental setup for using a humanoid robot to assist children with autism to develop visual perspective taking skills. *Paladyn : Journal of Behavioral Robotics*, 10(1), 167–179. DOI: 10.1515/pjbr-2019-0013

Wortham, R. H. (2020). *Transparency for Robots and Autonomous Systems: Fundamentals, technologies and applications*. Institution of Engineering and Technology. DOI: 10.1049/PBCE130E

Wu, D., Rosen, D., Wang, L., & Schaefer, D. (2015). Cloud-based design and manufacturing: A new paradigm in digital manufacturing and design innovation. *Computer Aided Design*, 59, 1–14. DOI: 10.1016/j.cad.2014.07.006

Wu, J., Guo, S., Huang, H., Liu, W., & Xiang, Y. (2018). Information and communications technologies for sustainable development goals: State-of-the-art, needs and perspectives. *IEEE Communications Surveys and Tutorials*, 20(3), 2389–2406. DOI: 10.1109/COMST.2018.2812301

Wynsberghe, A. (2019). Drones in humanitarian contexts, robot ethics, and the human–robot interaction. *Ethics and Information Technology*, 22(1), 43–53. DOI: 10.1007/s10676-019-09514-1

Xiao, G., Zhong, X., Weixian, S., Xia, Q., & Chen, W. (2018). Research on the dimensional accuracy measurement method of cylindrical spun parts based on machine vision. *Matec Web of Conferences, 167*. IEEE. DOI: 10.1051/matecconf/201816703010

Xie, Y., Li, P., Nedjah, N., Gupta, B. B., Taniar, D., & Zhang, J. (2023). Pri- vacy protection framework for face recognition in edge-based Internet of Things. *Cluster Computing*, 26(5), 3017–3035. DOI: 10.1007/s10586-022-03808-8

Xing, B., & Marwala, T. (2018). Smart maintenance for human–robot interaction. *Studies in Systems, Decision and Control.* Springer.

Xu, X., Yang, J., & Tang, Z. (2015). *Data virtualization for coupling command and control (C2) and combat simulation systems.* Paper presented at the Advances in Image and Graphics Technologies: 10th Chinese Conference, IGTA 2015, Beijing, China. DOI: 10.1007/978-3-662-47791-5_22

Xuan-Mung, N., Nguyen, N. P., Nguyen, T., Pham, D. B., Vu, M. T., Thanh, H. L. N. N., & Hong, S. K. (2022). Quadcopter Precision Landing on Moving Targets via Disturbance Observer-Based Controller and Autonomous Landing Planner. *IEEE Access : Practical Innovations, Open Solutions*, 10(July), 83580–83590. DOI: 10.1109/ACCESS.2022.3197181

Xu, P., Chen, G., Pan, G., & Di Renzo, M. (2020). Ergodic secrecy rate of RIS-assisted communication systems in the presence of discrete phase shifts and multiple eavesdroppers. *IEEE Wireless Communications Letters*, 10(3), 629–633. DOI: 10.1109/LWC.2020.3044178

Yadav, N., Singh, S. K., & Sharma, D. (2023). Forecasting Air Pollution for Environment and Good Health Using Artificial Intelligence. *2023 3rd International Conference on Innovative Sustainable Computational Technologies (CISCT)*, (pp. 1–5). IEEE. DOI: 10.1109/CISCT57197.2023.10351334

Yan, C., Xiang, X., & Wang, C. (2020). Towards Real-Time Path Planning through Deep Reinforcement Learning for a UAV in Dynamic Environments. *Journal of Intelligent & Robotic Systems*, 98(2), 297–309. DOI: 10.1007/s10846-019-01073-3

Yang, B., & Kristensson, P. O. (2023, September). Designing, Developing, and Evaluating AI-driven Text Entry Systems for Augmentative and Alternative Communication Users and Researchers. In *Proceedings of the 25th International Conference on Mobile Human-Computer Interaction* (pp. 1-4). ACM. DOI: 10.1145/3565066.3609738

Yang, L., Meng, F., Zhang, J., Hasna, M. O., & Di Renzo, M. (2020). On the performance of RIS-assisted dual-hop UAV communication systems. *IEEE Transactions on Vehicular Technology*, 69(9), 10385–10390. DOI: 10.1109/TVT.2020.3004598

Yang, R., Zhang, W., Tiwari, N., Yan, H., Li, T., & Cheng, H. (2022). Multimodal Sensors with Decoupled Sensing Mechanisms. *Advancement of Science*, 9(26), 2202470. DOI: 10.1002/advs.202202470 PMID: 35835946

Yao, C., Kao, C., & Lin, J. (2023). Drone for dynamic monitoring and tracking with intelligent image analysis. *Intelligent Automation & Soft Computing*, 36(2), 2233–2252. DOI: 10.32604/iasc.2023.034488

Yaramala, D., Khan, D., Vasanthakumar, N., Koteswararao, K., Sridhar, D., & Ansari, M. (2022). Application of internet of things (iot) and artificial intelligence in unmanned aerial vehicles. *International Journal of Electrical and Electronics Research*, 10(2), 276–281. DOI: 10.37391/ijeer.100237

Yazid, Y., Ez-Zazi, I., Guerrero-González, A., El Oualkadi, A., & Arioua, M. (2021). UAV-Enabled Mobile Edge-Computing for IoT Based on AI: A Comprehensive Review. *Drones (Basel)*, 5(4), 4. DOI: 10.3390/drones5040148

Ye, X., Song, W., Hong, S., Kim, Y., & Yoo, N. (2022). Toward data interoperability of enter- prise and control applications via the industry 4.0 asset administration shell. *IEEE Access : Practical Innovations, Open Solutions*, 10, 35795–35803. DOI: 10.1109/ACCESS.2022.3163738

Yi, H., Liu, T., & Lan, G. (2024). The key artificial intelligence technologies in early childhood education: A review. *Artificial Intelligence Review*, 57(1), 12. DOI: 10.1007/s10462-023-10637-7

Ying, H. (2021). Big data techniques for clinical image analysis. *International Journal of Advanced Information and Communication Technology*, 213–221. .DOI: 10.46532/ijaict-202108029

Yin, Y., Zheng, P., Li, C., & Wang, L. (2023). A state-of-the-art survey on Augmented Reality-assisted Digital Twin for futuristic human-centric industry transformation. *Robotics and Computer-integrated Manufacturing*, 81, 102515. DOI: 10.1016/j.rcim.2022.102515

Yulianto, A., Yuniar, D., & Prasetyo, Y. (2022). Navigation and guidance for autonomous quadcopter drones using deep learning on indoor corridors. *Jurnal Jartel Jurnal Jaringan Telekomunikasi*, 12(4), 258–264. DOI: 10.33795/jartel.v12i4.422

Yun, S., Shin, J., Kim, D., Kim, C. G., Kim, M., & Choi, M. T. (2011). Engkey: Tele-education robot. In *Social Robotics: Third International Conference*. Springer.

Zamora-Antun˜ano, M., Luque-Vega, L., Carlos-Mancilla, M., Hern'andez-Quesada, R., V'azquez, N., Carrasco-Navarro, R., Gonz'alez-Guti'errez, C., & Aguilar-Molina, Y. (2022). Methodology for the development of augmented reality applications: Medara. drone flight case study. *Sensors (Basel)*, 22(15), 5664. DOI: 10.3390/s22155664 PMID: 35957223

Zatsiorsky, V. M., & Duarte, M. (1999). Instant equilibrium point and its migration in standing tasks: Rambling and trembling components of the stabilogram. *Motor Control*, 3(1), 28–38. DOI: 10.1123/mcj.3.1.28 PMID: 9924099

Zeng, Y., Lyu, J., & Zhang, R. (2019). Cellular-connected uav: Potential, chal- lenges, and promising technologies. *IEEE Wireless Communications*, 26(1), 120–127. DOI: 10.1109/MWC.2018.1800023

Zeng, Y., & Zhang, R. (2017). Energy-efficient UAV communication with trajectory optimization. *IEEE Transactions on Wireless Communications*, 16(6), 3747–3760. DOI: 10.1109/TWC.2017.2688328

Zhai, Z., Dai, X., Duo, B., Wang, X., & Yuan, X. (2022). Energy-efficient UAV-mounted RIS assisted mobile edge computing. *IEEE Wireless Communications Letters*, 11(12), 2507–2511. DOI: 10.1109/LWC.2022.3206587

Zhang, T., Li, Q., Zhang, C. S., Liang, H. W., Li, P., Wang, T. M., & Wu, C. (2017). Current trends in the development of intelligent unmanned autonomous systems. *Frontiers of information technology & electronic engineering*, 18, 68-85.

Zhang, D., & Chang, W. (2023). A Novel Semantic Segmentation Approach Using Improved SegNet and DSC in Remote Sensing Images. *International Journal on Semantic Web and Information Systems*, 19(1), 1–17. DOI: 10.4018/IJSWIS.332769

Zhang, D., Shen, J., Li, S., Gao, K., & Gu, R. (2021). I, robot: Depression plays differ- ent roles in human–human and human–robot interactions. *Translational Psychiatry*, 11(1), 438. DOI: 10.1038/s41398-021-01567-5 PMID: 34420040

Zhang, K., & Zhang, H. (2022). Machine learning modeling of environmentally rel- evant chemical reactions for organic compounds. *ACS ES&T Water*, 4(3), 773–783. DOI: 10.1021/acsestwater.2c00193

Zhang, L., Zhao, H., Hou, S., Zhao, Z., Xu, H., Wu, X., Wu, Q., & Zhang, R. (2019). A survey on 5g millimeter wave communications for uav-assisted wireless networks. *IEEE Access : Practical Innova- tions, Open Solutions*, 7, 117460–117504. DOI: 10.1109/ACCESS.2019.2929241

Zhang, T., Li, Q., Zhang, C., Liang, H., Li, P., Wang, T., Li, S., Zhu, Y., & Wu, C. (2017). Current trends in the development of intelligent unmanned autonomous sys- tems. *Frontiers of Information Technology & Electronic Engineering*, 18(1), 68–85. DOI: 10.1631/FITEE.1601650

Zhang, W., & Li, G. (2018). Detection of multiple micro-drones via cadence velocity diagram analysis. *Electronics Letters*, 54(7), 441–443. DOI: 10.1049/el.2017.4317

Zhang, W., & Wang, Z. (2021). Theory and practice of VR/AR in K-12 science education—A systematic review. *Sustainability (Basel)*, 13(22), 12646. DOI: 10.3390/su132212646

Zhang, X., Fan, K., Hou, H., & Liu, C. (2022). Real-time detection of drones using channel and layer pruning, based on the yolov3-spp3 deep learning algorithm. *Micromachines*, 13(12), 2199. DOI: 10.3390/mi13122199 PMID: 36557498

Zhang, X., Tlili, A., Nascimbeni, F., Burgos, D., Huang, R., Chang, T. W., Jemni, M., & Khribi, M. K. (2020). Accessibility within open educational resources and practices for disabled learners: A systematic literature review. *Smart Learning Environments*, 7(1), 1–19. DOI: 10.1186/s40561-019-0113-2

Zhang, Y., Al-Fuqaha, A., Humar, I., & Wan, J. (2022). Special Issue on Advanced Sensors and Sensing Technologies in Robotics. *IEEE Sensors Journal*, 22(18), 17334. DOI: 10.1109/JSEN.2022.3198783 PMID: 36346095

Zhang, Y., Liu, M., Guo, J., Wang, Z., Wang, Y., Liang, T., & Singh, S. K. (2023). Optimal Revenue Analysis of the Stubborn Mining Based on Markov Decision Process. In Xu, Y., Yan, H., Teng, H., Cai, J., & Li, J. (Eds.), *Machine Learning for Cyber Security* (pp. 299–308). Springer Nature Switzerland. DOI: 10.1007/978-3-031-20099-1_25

Zhang, Z., Wu, J., Dai, J., & He, C. (2020). A novel real-time penetration path planning algorithm for stealth UAV in 3D complex dynamic environment. *IEEE Access : Practical Innovations, Open Solutions*, 8, 122757–122771. DOI: 10.1109/ACCESS.2020.3007496

Zhang, Z., Zeng, K., & Yi, Y. (2024). Blockchain-Empowered Secure Aerial Edge Computing for AIoT Devices. *IEEE Internet of Things Journal*, 11(1), 84–94. DOI: 10.1109/JIOT.2023.3294222

Zheng, Y. (2023). *Research on target localization method based on binocular vision technology*. SPIE. .DOI: 10.1117/12.3007964

Zheng, Z., Xie, S., Dai, H.-N., Chen, X., & Wang, H. (n.d.). Blockchain Challenges and Opportunities. *Survey (London, England)*.

Zhou, J., & Chen, F. (2023). AI ethics: From principles to practice. *AI & Society*, 38(6), 2693–2703. DOI: 10.1007/s00146-022-01602-z

Zhou, Y., Tang, Y., & Zhao, X. (2019). A novel uncertainty management approach for air combat situation assessment based on improved belief entropy. *Entropy (Basel, Switzerland)*, 21(5), 495. DOI: 10.3390/e21050495 PMID: 33267209

Zhu, D., Bu, Q., Zhu, Z., Zhang, Y., & Wang, Z. (2024). Advancing autonomy through lifelong learning: A survey of autonomous intelligent systems. *Frontiers in Neurorobotics*, 18, 1385778. DOI: 10.3389/fnbot.2024.1385778 PMID: 38644905

Zirar, A., Ali, S. I., & Islam, N. (2023). Worker and workplace Artificial Intelligence (AI) coexistence: Emerging themes and research agenda. *Technovation*, 124, 102747. DOI: 10.1016/j.technovation.2023.102747

Zitar, R., Kassab, M., Fallah, A., & Barbaresco, F. (2023). *Bird/drone de- tection and classification using classical and deep learning methods.* Authorea. DOI: 10.22541/au.168075364.45332093/v1

Ziv, Y. (2013, February). Social information processing patterns, social skills, and school readiness in preschool children. *Journal of Experimental Child Psychology*, 114(2), 306–320. DOI: 10.1016/j.jecp.2012.08.009 PMID: 23046690

Zohdy, B. (2019). *Machine vision application on science and industry.* IGI Global. .DOI: 10.4018/978-1-5225-5751-7.ch008

Zwickle, A., Farber, H., & Hamm, J. (2018). Comparing public concern and support for drone regulation to the current legal framework. *Behavioral Sciences & the Law*, 37(1), 109–124. DOI: 10.1002/bsl.2357 PMID: 30004141

About the Contributors

Brij B. Gupta received the PhD degree from Indian Institute of Technology (IIT) Roorkee, India. In more than 19 years of his professional experience, he published over 500 papers in journals/conferences including 35 books and 10 Patents with over 30,000 citations. He has received numerous national and international awards including Canadian Commonwealth Scholarship (2009), Best Faculty Award (2018 & 2019), etc. He is also selected in the 2021 and 2020 Stanford University's ranking of the world's top 2% scientists. He is also a visiting/adjunct professor with several universities worldwide. He is also an IEEE Senior Member (2017) and also selected as 2021 Distinguished Lecturer in IEEE CTSoc. Dr Gupta is also serving as Member-in-Large, Board of Governors, IEEE Consumer Technology Society (2022-2024). At present, Prof. Gupta is working as Director, International Center for AI and Cyber Security Research and Innovations, and Professor with the Department of Computer Science and Information Engineering (CSIE), Asia University, Taiwan. His research interests include cyber security, cloud computing, artificial intelligence, blockchain technologies, social media and networking.

***

Kwok Tai Chui received the B.Eng. degree in electronic and communication engineering - Business Intelligence Minor and Ph.D. degree from City University of Hong Kong. He had industry experience as Senior Data Scientist in Internet of Things (IoT) company. He is with the Department of Electronic Engineering and Computer Science, School of Science and Technology, at Hong Kong Metropolitan University as an Assistant Professor. He was the recipient of 2nd Prize Award (Postgraduate Category) of 2014 IEEE Region 10 Student Paper Contest. Also, he received Best Paper Award in IEEE The International Conference on Consumer Electronics-China, in both 2014 and 2015. He has more than 100 research publications including edited books, book chapters, journal papers, and conference papers. His research interests include computational intelligence, data science, energy monitoring and management, intelligent transportation, smart metering, healthcare, machine learning algorithms and optimization.

Purwadi Agus Darwito earned an Engineer (Ir.) degree in 1987 from ITS, majoring in Engineering Physics. Master of Science (M.Sc.) degree from the University of Indonesia in 1992, Computer Science Study Program. Doctorate degree (Dr.) in 2013 from the Electrical Engineering Study Program (ITS). Currently, he is a permanent lecturer in the Department of Engineering Physics (ITS) in the Instrumentation, Control, and Optimization Laboratory (IKO) and teaches several courses in the field of instrumentation and control. Actively teaches and researches in the field of artificial intelligence-based modeling and control, especially motor control used in quadcopters and quadplanes. Currently, research is developing an autonomous quadcopter that can reach targets based on given coordinates while avoiding various disturbances. Several scientific works were presented at international seminars and published in indexed international journals, such as the Mathematical Modeling of Engineering Problems (MMEP) journal published by the International Information and Engineering Technology Association (IIETA).

Shaurya Katna is persuing computer science and engineering at Punjab University. Love to play sports. State champion and National player in roller skating from Chandigarh.

Christian Kaunert is Professor of International Security at Dublin City University, Ireland. He is also Professor of Policing and Security, as well as Director of the International Centre for Policing and Security at the University of South Wales. In addition, he is Jean Monnet Chair, Director of the Jean Monnet Centre of Excellence and Director of the Jean Monnet Network on EU Counter-Terrorism (www .eucter.net).

Gunawan Nugroho is a lecturer and a researcher in the Department of Engineering Physics at Institut Teknologi Sepuluh Nopember (ITS) and has been a member of the staff since 2002. He received both B.Eng. and M.Eng. degree in the area of Aerodynamics from the same university and a Ph.D degree in the area of Fluid Dynamics from Universiti Teknologi Petronas (UTP). He is the principal investigator on research programs in the Department of Engineering Physics, ITS. His current research interest in fluid dynamics are in the area of Navier-Stokes and energy equations related to the parametrical aspects of engineering applications.

Princy Pappachan is a postdoctoral researcher at the College of Education, National Chengchi University, Taiwan. She was employed as an Assistant Professor at the Department of Foreign Languages and Literature at Asia University, Taiwan. Additionally, she has served on the International Advisory Committee for ICRAMLET22 and was an International Advisory Member of MIMSE2023. She holds a PhD in Applied Linguistics from the Centre of Applied Linguistics and Translation Studies (CALTS) at the University of Hyderabad (UoH), an M.Phil. in Cognitive Science from the Centre for Cognitive Science and Neuroscience (CNCS) at UoH. Her primary research focus areas are Generative Artificial Intelligence (AI), cognitive science, Multilingual Education (MLE) in Sustainable Development Research and English Language Teaching (ELT). She has also published several papers in peer-reviewed journals and is on the Editorial Review Board of the International Journal of Teacher Education and Professional Development (IJTEPD).

Siwada Piyakanjana earned a bachelor's degree in Science and Communication Disorders from the Faculty of Medicine at Mahidol University in Thailand. After graduation, Siwada worked as a speech and language pathologist at HRH Princess MahaChakri Sirindhorn Medical Center for three years, assisting people with communication challenges and empowering them to live fulfilling lives. Siwada also pursued a Master's degree in Psychology at Asia University in Taiwan, motivated by her experiences in speech therapy and a desire to learn more about the complexities of the human mind and behavior. She is currently conducting research on the relationship between parenting styles and emotional and behavioral problems in children with learning disabilities. Her research aims to benefit the academic community and influence interventions and support systems for vulnerable children and families.

Eko Prasetyo was born in Magelang, Central Java, Indonesia in 1967. He received the B.S. degree in electrical engineering from Universitas Gadjah Mada, Yogyakarta, Indonesia, in 1991 and the M.S.

degree in the same department and university in 2011. He received the Ph.D. degree in computer science and information engineering at Asia University, Taichung, Taiwan. After completing his bachelor's degree, he embarked on a career in the oil and gas sector, where he served as an Exploitation Engineer and Field Automation Supervisor. In 1997, he transitioned to the Construction and Software industries. In 2001, he joined a Software Development company to build a Geographic Information System for the Distribution Network of the State Electricity Company in Central Java and Yogyakarta. In 2007, he joined the Cisco Networking Academy in Indonesia and the Asia Pacific region, where he currently serves as an Instructor Trainer and Academy Advisor. In 2011, He joined the Information Technology Department at Universitas Muhammadiyah Yogyakarta, where he continues to work. His current research focuses on Network Security, Data Mining, and Artificial Intelligence.

Mosiur Rahaman, a Postdoctoral Researcher at the International Center for AI and Cyber Security Research and Innovations (CCRI) at Asia University, Taiwan, holds a PhD in Computer Science and Information Engineering and master's degrees in Business Management and Computer Science and Engineering. With prior roles as a Research Assistant in AI and an Assistant Professor in Computer Science, he has extensive experience in academia. Dr. Rahman has been actively involved in international conferences, serving in significant roles, and is a part-time lecturer in Humanities and Social Science. His research interests include AI, Neural Networks, and Security, and he is a dedicated peer reviewer for over 50 journals, showcasing his commitment to research and academic excellence.

Bhupinder Singh working as Professor at Sharda University, India. Also, Honorary Professor in University of South Wales UK and Santo Tomas University Tunja, Colombia. His areas of publications as Smart Healthcare, Medicines, fuzzy logics, artificial intelligence, robotics, machine learning, deep learning, federated learning, IoT, PV Glasses, metaverse and many more. He has 3 books, 139 paper publications, 163 paper presentations in international/national conferences and seminars, participated in more than 40 workshops/FDP's/QIP's, 25 courses from international universities of repute, organized more than 59 events with international and national academicians and industry people's, editor-in-chief and co-editor in journals, developed new courses. He has given talks at international universities, resource person in international conferences such as in Nanyang Technological University Singapore, Tashkent State University of Law Uzbekistan; KIMEP University Kazakhstan, All'ah meh Tabatabi University Iran, the Iranian Association of International Criminal law, Iran and Hague Center for International Law and Investment, The Netherlands, Northumbria University Newcastle UK, Taylor's University Malaysia, AFM Krakow University Poland, European Institute for Research and Development Georgia, Business and Technology University Georgia, Texas A & M University US name a few. His leadership, teaching, research and industry experience is of 16 years and 3 Months. His research interests are health law, criminal law, research methodology and emerging multidisciplinary areas as Blockchain Technology, IoT, Machine Learning, Artificial Intelligence, Genome-editing, Photovoltaic PV Glass, SDG's and many more.

Sunil Singh has areas of expertise in high-performance computing, Linux/Unix, Data Mining, Internet of Things, Machine Learning, Computer Architecture & Organization, Embedded System and Computer Network. He has published more than 160 research papers in reputed International/National journals, conferences, and workshops. He has also received 07 patents granted and 02 patents published, and some are in the pipeline too. His textbook, titled "Linux Yourself: Concept & Programming", was published by Taylor and Francis (CRC Press) in August 2021. He is very active as an ACM professional member and also contributed to the Eminent Speaker Program (ESP) of ACM India. He is a reviewer of several renowned national and international research journals and reviewed more than 640+ papers, and a member of professional bodies such as ACM, IEEE, IE, IDES, LMISTE, ACEEE, IACSIT, and IAENG.

Sreerakuvandana is an Assistant Professor of English, specializing in Cognitive Linguistics at the Department of AI-ML and Cybersecurity at Jain deemed-to-be University. She received her Bachelor's, Master's and Doctorate degree from the University of Hyderabad, specializing in Linguistics and Cognitive Science. Her research areas include Artificial Intelligence, English Language Teaching, Sociolinguistics, and Psycholinguistics. She also has a range of academic publications in the areas of her interest to her credit. She currently serves as a guest editor for the Athens Journal of Education and is an International Advisory Member of MIMSE23 and is a guest editor for the same.

Karisma Trinanda Putra graduated with a bachelor's degree and a master's degree from Electrical Engineering Polytechnic Institute of Surabaya (2012) and Sepuluh Nopember Institute of Technology (2015), respectively. He received his Doctoral degree from Asia University, Taiwan, in January (2022). He is currently an Assistant Professor and Head Department of Electrical Engineering, Faculty of Engineering, Universitas Muhammadiyah Yogyakarta, Indonesia. His current research interests include the development of deep learning techniques, image processing, sensor networks, compressed sensing, internet of medical things, in particular in computational biology and medical data analysis. Topics he is working on include deep network inference on PM2.5 propagation, explainable analysis of medical data using XAI, and deep learning on microcontrollers for clinical decision support system.

Kadek Dwi Wahyuadnyana, a PhD candidate in the Engineering Physics Department-ITS, specializes in developing intelligent control algorithms for robotic systems. Beyond his academic pursuits, he maintains a full-time role as a robot researcher at Beehive Drones, a startup dedicated to advancing drone technology in Indonesia. Beyond robotics, Kadek applies his control expertise to contribute to energy sustainability. In 2021, he secured a full scholarship from GSEP Canada (through the ESED program), a company committed to sustainability and renewable energy, for the Master's program he is enrolled in. Next, in 2023, Kadek and his team earned a gold medal in a paper competition focusing on electricity, EVs, solar and grid systems, and renewable energy integration. With a deep understanding of mathematics, physics, and engineering, Kadek seamlessly integrates these three domains to generate innovative ideas for academia and the broader community interested in control systems.

Agung Mulyo Widodo obtained a PhD in Computer Science and Information Engineering, Asia University, Taiwan, in 2022, and currently works as a lecturer and researcher at the Faculty of Computer Science, Esa Unggul University, Jakarta, Indonesia. He has worked as a research assistant to his advising professor while a PostDoc Research Fellow at the Artificial Intelligence and Information Security Laboratory, Department of Computer Science and Information Engineering (CSIE), Asia University, Taiwan. Previous projects were with Caltex Pacific Indonesia (CPI), Siemens-Telecommunications, Nokia-Siemens-Networks (NSN). Currently, his research focuses on wireless communications technologies including NOMA, OMA, NB-IoT, backscatter systems, cognitive radio, smart antennas, artificial intelligence, and information security.

Andika Wisnujati received the B.S. and M.S degree in Mechanical Engineering from Universitas Gadjah Mada, Yogyakarta, Indonesia in 2008 and 2014. He is also received the Ph.D degree in Computer Science and Information Engineering, Asia University, Taiwan in 2023. He had been the Assistant Professor in Department of Automotive Engineering Technology, Universitas Muhammadiyah Yogyakarta, Indonesia since 2012. His research interests include Artificial Intelligence and Material Science. His previous research includes Neural Network and Automatic Control in Antilock Braking System and Grey Prediction Model in Welding Distortion.

Index

A

Active-Reconfigurable Intelligent Surfaces 187, 195, 196, 197, 198, 199, 208, 216, 217, 218, 219, 220, 221, 222, 223

Adaptive learning environments 236, 252

AI 1, 2, 3, 4, 5, 6, 7, 8, 9, 10, 11, 12, 13, 14, 15, 17, 18, 19, 20, 23, 24, 25, 26, 27, 28, 29, 31, 32, 35, 36, 37, 38, 39, 40, 43, 47, 48, 50, 52, 55, 56, 72, 74, 77, 78, 81, 82, 85, 86, 87, 89, 92, 93, 94, 95, 96, 98, 99, 100, 101, 102, 103, 104, 105, 106, 107, 108, 109, 111, 112, 113, 114, 115, 125, 126, 127, 128, 129, 132, 133, 134, 153, 159, 160, 163, 164, 165, 167, 172, 173, 174, 175, 176, 177, 178, 179, 180, 182, 183, 184, 187, 231, 233, 235, 240, 241, 242, 245, 246, 247, 249, 250, 251, 252, 253, 254, 255, 258, 261, 268, 271, 273, 274, 275, 278, 279, 280, 282, 283, 284, 300, 303, 307, 317, 319, 323, 324, 331, 342

AI-based drones 8, 43

AI-powered Decision-Making 115

Algorithmic Transparency 159, 160, 162, 168, 175, 180

Artificial Intelligence 1, 2, 3, 7, 9, 10, 11, 12, 13, 14, 15, 17, 18, 19, 24, 28, 29, 30, 31, 32, 34, 36, 37, 49, 50, 53, 55, 56, 72, 73, 82, 85, 86, 87, 92, 94, 95, 96, 97, 98, 100, 101, 102, 106, 107, 111, 112, 113, 114, 117, 125, 127, 132, 134, 136, 137, 138, 139, 140, 143, 146, 150, 152, 160, 163, 164, 172, 174, 176, 182, 183, 184, 199, 228, 232, 233, 234, 235, 236, 240, 241, 242, 243, 244, 245, 246, 247, 248, 249, 250, 251, 252, 253, 254, 258, 271, 272, 274, 278, 279, 280, 282, 283, 284, 300, 303, 307, 316, 317, 323, 324, 325, 331, 332, 344, 345

C

Communication Systems 4, 19, 37, 44, 102, 115, 116, 117, 121, 125, 126, 127, 128, 129, 131, 132, 133, 134, 195, 196, 197, 198, 199, 201, 202, 203, 204, 205, 206, 207, 208, 216, 221, 223, 226, 227, 228, 258, 267, 314

D

Drones 1, 2, 3, 4, 5, 6, 7, 8, 9, 10, 13, 14, 15, 33, 34, 35, 36, 37, 38, 39, 40, 41, 42, 43, 44, 45, 46, 47, 48, 49, 50, 51, 52, 53, 55, 72, 83, 85, 86, 87, 96, 97, 98, 99, 100, 101, 102, 103, 104, 106, 107, 109, 111, 112, 114, 115, 116, 118, 119, 120, 121, 122, 123, 124, 125, 126, 127, 128, 129, 130, 131, 132, 133, 134, 136, 137, 138, 139, 141, 142, 143, 144, 147, 148, 149, 150, 152, 153, 154, 155, 156, 157, 159, 160, 161, 162, 163, 170, 175, 179, 180, 182, 184, 188, 189, 191, 195, 209, 210, 227, 257, 258, 259, 261, 265, 266, 268, 269, 270, 271, 273, 274, 275, 277, 278, 280, 281, 282, 284, 285, 286, 287, 288, 289, 290, 291, 294, 296, 297, 299, 300, 331, 332, 333, 334, 335, 336, 337, 338, 339, 340, 341, 342, 343, 344, 345, 346

Drone Swarms 115, 116, 117, 118, 119, 120, 121, 122, 124, 125, 126, 127, 128, 130, 131, 132, 133

E

Edge Computing 29, 34, 49, 50, 92, 115, 117, 125, 126, 127, 128, 129, 135, 136, 163, 182, 184, 185, 229, 258, 261, 272, 274, 275, 279, 282, 283, 302, 303, 342

Energy efficiency 44, 55, 133, 135, 183, 200, 204, 220, 221, 223, 227, 281, 301, 319

Ergodic Capacity 187, 201, 202, 203, 204, 205, 206, 207, 208, 227, 228

Extended flight times 55, 223

H

Human-Robot Collaboration 174, 308, 310, 319, 320, 325

I

Image Processing 74, 91, 94, 137, 139, 140, 141, 143, 144, 148, 150, 151, 156

Industrial Drones 86, 258, 268, 270, 275, 285, 286, 287, 288, 289, 290, 291, 294, 299

Industrial Robots 4, 18, 19, 26, 86, 88, 89, 90, 93, 111, 138, 159, 160, 161, 162, 163, 170, 174, 180, 182

Industry 2, 3, 6, 8, 9, 11, 12, 13, 14, 15, 18, 19, 28, 34, 45, 54, 86, 87, 88, 96, 112, 113, 131, 134, 138, 140, 152, 157, 159, 171, 178, 181, 182, 184, 224, 232, 259, 270, 275, 278, 280, 284, 289, 300, 303, 328, 329, 334, 335, 337, 342

Intelligent Drones 86, 96, 97, 137, 159, 160, 161, 162, 163, 170, 175, 180, 331, 332, 339

IoT 2, 4, 7, 8, 9, 15, 19, 29, 30, 33, 34, 35, 36, 38, 39, 43, 44, 45, 46, 47, 49, 50, 51, 53, 88, 92, 94, 96, 104, 114, 132, 134, 136, 148, 150, 153, 163, 182, 183, 184, 196, 199, 225, 248, 271, 272, 273,

Printed in the United States
by Baker & Taylor Publisher Services